Christian Freksa Markus Knauff
Bernd Krieg-Brückner Bernhard Nebel
Thomas Barkowsky (Eds.)

Spatial Cognition IV

Reasoning, Action, Interaction

International Conference Spatial Cognition 2004
Frauenchiemsee, Germany, October 11-13, 2004
Revised Selected Papers

AF214267

Springer

Series Editors

Jaime G. Carbonell, Carnegie Mellon University, Pittsburgh, PA, USA
Jörg Siekmann, University of Saarland, Saarbrücken, Germany

Volume Editors

Christian Freksa
Bernd Krieg-Brückner
Thomas Barkowsky
Universität Bremen, FB 3, Mathematik und Informatik
SFB/TR 8 Spatial Cognition, Bibliothekstr. 1, 28359 Bremen, Germany
E-mail: {freksa, bkb, barkowsky}@sfbtr8.uni-bremen.de

Markus Knauff
Albert-Ludwigs-Universität Freiburg, Abt. Kognitionswissenschaft
SFB/TR 8 Spatial Cognition, Friedrichstr. 50, 79098 Freiburg, Germany
E-mail: knauff@cognition.iig.uni-freiburg.de

Bernhard Nebel
Albert-Ludwigs-Universität Freiburg, Institut für Informatik
SFB/TR 8 Spatial Cognition, Georges-Köhler-Allee, Geb. 52
79110 Freiburg, Germany
E-mail: nebel@informatik.uni-freiburg.de

Library of Congress Control Number: 2005921106

CR Subject Classification (1998): I.2.4, I.2, J.2, J.4, E.1, I.3, I.7, I.6

ISSN 0302-9743
ISBN 3-540-25048-4 Springer Berlin Heidelberg New York

This work is subject to copyright. All rights are reserved, whether the whole or part of the material is concerned, specifically the rights of translation, reprinting, re-use of illustrations, recitation, broadcasting, reproduction on microfilms or in any other way, and storage in data banks. Duplication of this publication or parts thereof is permitted only under the provisions of the German Copyright Law of September 9, 1965, in its current version, and permission for use must always be obtained from Springer. Violations are liable to prosecution under the German Copyright Law.

Springer is a part of Springer Science+Business Media

springeronline.com

© Springer-Verlag Berlin Heidelberg 2005
Printed in Germany

Typesetting: Camera-ready by author, data conversion by Scientific Publishing Services, Chennai, India
Printed on acid-free paper SPIN: 11397977 06/3142 5 4 3 2 1 0

Preface

This is the fourth volume in a series of books dedicated to basic research in spatial cognition. Spatial cognition is a field that investigates the connection between the physical spatial world and the mental world. Philosophers and researchers have proposed various views concerning the relation between the physical and the mental worlds: Plato considered pure concepts of thought as separate from their physical manifestations while Aristotle considered the physical and the mental realms as two aspects of the same substance. Descartes, a dualist, discussed the interaction between body and soul through an interface organ and thus introduced a functional view that presented a challenge for the natural sciences and the humanities. In modern psychology, the relation between the physical and the cognitive space has been investigated using thorough experiments, and in artificial intelligence we have seen views as diverse as 'problems can be solved on a representation of the world' and 'a representation of the world is not necessary.'

Today's spatial cognition work establishes a correspondence between the mental and the physical worlds by studying and exploiting their interaction; it investigates how mental space and spatial "reality" join together in understanding the world and in interacting with it. The physical and representational aspects are equally important in this work. Almost all topics of cognitive science manifest themselves in spatial cognition. A special feature of spatial cognition is that the spatial dimensions in the physical world are accessible to most of the human sensory systems and to a great variety of technical sensors and measuring approaches that provide information about the spatial environment. Thus, on one hand, mental phenomena can be investigated using methods of the natural sciences and using experimental methods from psychology. On the other hand, they can be explored through the behavior of artificial systems in space, through formal methods for dealing with spatial knowledge, and through computational investigations. Ideally, these different approaches are strongly interconnected.

After almost 20 years of research dedicated to spatial language, conceptualization of spatial relations, representation of spatial knowledge, spatial and spatio-temporal reasoning, spatial reference systems, cultural differences in conceptualizing space, spatial memory, neural mechanisms of spatial cognition, localization of spatial functions in the brain, spatial attention, and robot navigation, spatial cognition has become a well-established interdisciplinary research field within the disciplines of cognitive science. Structured cross-disciplinary research initiatives in Germany, Europe, and in the USA were instrumental in bringing different research communities together through workshops and conferences in this area.

In 2002, the German Academic Exchange Service (DAAD) provided funds in the framework of the Future Investment Program to establish an *International Quality Network on Spatial Cognition* (IQN) that connects major research teams in the field worldwide and to provide an infrastructure for scientific exchange and training. In 2003, the Deutsche Forschungsgemeinschaft (DFG) established the *Transregional Collaborative Research Center on Spatial Cognition* (SFB/TR 8) at the Universities of Bremen and Freiburg to carry out basic research on the integration and specialization of approaches to spatial reasoning, spatial action, and spatial interaction.

The SFB/TR 8 organized the international conference *Spatial Cognition 2004* held in October 2004 at the abbey Frauenwörth on the island of Frauenchiemsee in Bavaria, Germany. Fifty contributions were submitted in response to the conference call. After a thorough peer-review process carried out by the international program committee of the conference, 27 contributions were selected for oral presentation and for publication in this proceedings volume; 14 contributions on work in progress were selected for poster presentation.

This volume presents contributions by 67 authors from 10 countries on 4 continents on a large spectrum of interdisciplinary work on descriptions of space, on spatial mental models and maps, on spatio-temporal representation and reasoning, on route directions, wayfinding in natural and virtual environments, and spatial behavior, and on robot mapping and piloting.

Many people contributed to the success of the *Spatial Cognition 2004* conference. First of all, we thank the members of the review committee of the SFB/TR 8 Armin Cremers, Rüdiger Dillmann, Max Egenhofer, Ulrich Furbach, Werner Kuhn, Elke van der Meer, Michael Richter, Helge Ritter, Ipke Wachsmuth, Wolfgang Wahlster, Jürgen Wehland, and Martin Wirsing, as well as the program officers Gerit Sonntag and Bettina Zirpel of the Deutsche Forschungsgemeinschaft for their excellent guidance and support. We thank all authors for their careful work and for observing our tight deadlines in an exemplary fashion. We thank the reviewers for their careful work, their excellent suggestions, and their speedy reviews. We also thank the members of our support staff Eva Räthe, Dagmar Sonntag, Marion Stubbemann, and Sandra Budde for their competent and smooth organization of the conference and for editorial support; we thank Frank Dylla and Dominik Engel for maintaining the conference management system. Special thanks are due to Ms. Scholastica McQueen for her friendly reception at the abbey Frauenwörth and for her dedicated assistance. Finally, we thank Alfred Hofmann and his staff at Springer for their continuing support of our book series.

October 2004

Christian Freksa
Markus Knauff
Bernd Krieg-Brückner
Bernhard Nebel
Thomas Barkowsky

Review Committee

Marco Aiello
Marios Avraamides
Philippe Balbiani
John Bateman
Brandon Bennett
Sven Bertel
Thomas Bittner
Wilfried Brauer
Wolfram Burgard
Didac Busquets
Markus Büttner
Anthony G. Cohn
Kenny Coventry
Matt Duckham
Ivo Duentsch
Max Egenhofer
Dominik Engel
Susan L. Epstein
Carola Eschenbach
Sara Fabrikant
Scott Farrar
Aidan Feeney
Kerstin Fischer
Dieter Fox
Andrew Frank
Udo Frese
Ulrich Furbach
Bärbel Garsoffky
Merideth Gattis
Sabine Gillner
Janice Glasgow
Gabriela Goldschmidt
Reg Golledge
Hans Werner Güsgen
Steffen Gutmann
Christopher Habel
Udo Hahn
Dirk Hähnel
Mary Hegarty
Robin Hirsch
Stephen Hirtle
Robin Hörnig
Kathleen Hornsby
Ludger Hovestadt
Georg Jahn
Margaret Jefferies
Barbara Kaup
Kristian Kersting
Frank Kirchner
Roberta Klatzky
Alexander Kleiner
Alexander Klippel
Markus Knauff
Boicho Kokinov
Werner Kuhn
Bart Kuijpers
Ben Kuipers
Lars Kulik
Peter B. Ladkin
Longin Latecki
Gerard Ligozat
Jack Loomis
Hanspeter Mallot
David Mark
Timothy P. McNamara
Tobias Meilinger
Bernd Meyer
Joseph Modayil
Daniel R. Montello
Reinhard Moratz
Till Mossakowski
Ian Pratt-Hartmann
Marco Ragni
Martin Raubal
Jochen Renz
Kai-Florian Richter
Thomas Röfer
Kerstin Schill
Christoph Schlieder
Inessa Seifert
Gunter Senft
Barry Smith

Hartwig Hochmair
Bernhard Hommel
Gerhard Strube
Karim Tahboub
Thora Tenbrink
Sabine Timpf
Rudolph Triebel
Barbara Tversky
Andre Vandierendonck
Laure Vieu
Steffen Werner
Jan Wiener
Cyrill Stachniss
Klaus Stein
Stephan Winter
Stefan Wölfl
Diedrich Wolter
Frank Wolter
Mike Worboys
Michael Zakharyashchev
Emile van der Zee
Hubert Zimmer
Alexander Zipf

Related Book Publications

Meng, L., Zipf, A. & Reichenbacher, T., eds., *Map-based mobile services – Theories, methods and implementations,* Springer, 2005.

Gero, J. S., Tversky, B. & Knight, T., eds., *Visual and spatial reasoning in design III*, Key Centre of Design Computing and Cognition, University of Sydney, 2004.

Blackwell, A., Marriott, K. & Shimojima, A., eds., *Diagrammatic representation and inference,* LNAI 2980, Springer, 2004.

Egenhofer, M. J., Freksa, C. & Miller, H. J., eds., *Geographic information science, Third International Conference GIScience 2004,* LNCS 3234, Springer, 2004.

Freksa, C., Brauer, W., Habel, C. & Wender, K. F., eds., *Spatial cognition III – Routes and navigation, human memory and learning, spatial representation and spatial learning,* LNAI 2685, Springer, 2003.

Kuhn, W., Worboys, M. & Timpf, S., eds., *Spatial information theory: Foundations of geographic information science,* LNCS 2825, Springer, 2003.

Barkowsky, T., *Mental representation and processing of geographic knowledge – A computational approach,* LNAI 2541, Springer, Berlin 2002.

Egenhofer, M. J. & Mark, D. M., eds., *Geographic information science*, LNCS 2478, Springer, 2002.

Hegarty, M., Meyer, B. & Narayanan, N. H., eds., *Diagrammatic representation and inference*, LNCS 2317, Springer, 2002.

Coventry, K. & Olivier, P., eds., *Spatial language: Cognitive and computational perspectives*, Kluwer, Dordrecht, 2002.

Renz, J., *Qualitative spatial reasoning with topological information*, LNAI 2293, Springer, 2002.

Montello, D. R., ed., *Spatial information theory: Foundations of geographic information science*, LNCS 2205, Springer, 2001.

Gero, J. S., Tversky, B. & Purcell, T., eds., *Visual and spatial reasoning in design II,* Key Centre of Design Computing and Cognition, University of Sydney, 2001.

Freksa, C., Brauer, W., Habel, C. & Wender, K. F., eds., *Spatial cognition II – Integrating abstract theories, empirical studies, formal methods, and practical applications*, LNAI 1849, Springer, 2000.

Habel, C. & von Stutterheim, C., Hrsg., *Räumliche Konzepte und sprachliche Strukturen,* Niemeyer, Tübingen, 2000.

Habel, C. & Werner, S., eds., Special issue on spatial reference systems, *Spatial Cognition and Computation,* Vol. 1, No. 4, 1999.

Gero, J. S. & Tversky, B., eds., *Visual and spatial reasoning in design,* Key Centre of Design Computing and Cognition, University of Sydney, 1999.

Freksa, C. & Mark, D. M., eds., *Spatial information theory – Cognitive and computational foundations of geographic information science,* LNCS 1661, Springer, 1999.

Freksa, C., Habel, C. & Wender, K. F., eds., *Spatial cognition,* LNAI 1404, Springer, 1998.

Egenhofer, M. J. & Golledge, R. G., eds., *Spatial and temporal reasoning in geographic information systems,* Oxford University Press, Oxford, 1997.

Hirtle, S. C. & Frank, A. U., eds., *Spatial information theory: A theoretical basis for GIS*, LNCS 1329, Springer, 1997.

Burrough, P. & Frank, A., eds., *Geographic objects with indeterminate boundaries,* Taylor & Francis, London, 1996.

Frank, A. U. & Kuhn, W., eds., *Spatial information theory: A theoretical basis for GIS*, LNCS 988, Springer, 1995.

Frank, A. U. & Campari, I., eds., *Spatial information theory: A theoretical basis for GIS*, LNCS 716, Springer, 1993.

Frank, A. U., Campari, I. & Formentini, U., eds., *Theories and methods of spatio-temporal reasoning in geographic space*, LNCS 639, Springer, 1992.

Mark, D. M. & Frank, A. U., eds., *Cognitive and linguistic aspects of geographic space*, Kluwer, Dordrecht, 1991.

Freksa, C. & Habel, C., Hrsg., *Repräsentation und Verarbeitung räumlichen Wissens,* Informatik-Fachberichte 245, Springer, 1990.

Table of Contents

Route Directions, Wayfinding, and Spatial Behavior

Descriptions of Space – Prepositions and Reference

Mental Models, Diagrams, and Maps

Finding the Way Inside: Linking Architectural Design Analysis and Cognitive Processes

Christoph Hölscher[1], Tobias Meilinger[1,3], Georg Vrachliotis[1,2], Martin Brösamle[1], and Markus Knauff[1,3]

[1] University of Freiburg, Centre for Cognitive Science,
Friedrichstr. 50, 79098 Freiburg, Germany
{hoelsch, martinb}@cognition.uni-freiburg.de
[2] ETH Zurich, Department of Architecture,
ETH Hoenggerberg, CH-8093 Zurich, Switzerland
vrachliotis@arch.ethz.ch
[3] Max-Planck-Institute for Biological Cybernetics,
Spemannstr. 38, 72076 Tuebingen, Germany
{tobias.meilinger, markus.knauff}@tuebingen.mpg.de

Abstract. The paper is concerned with human wayfinding in public buildings. Two main aspects of wayfinding difficulties are considered: architectural features of the building and cognitive processes of the agent. We conducted an empirical study in a complex multi-level building, comparing performance measures of experienced and inexperienced participants in different wayfinding tasks. Thinking aloud protocols provide insights into navigation strategies, planning phases, use of landmarks and signage, and measures of survey knowledge. Specific strategies for navigation in multi-level buildings, like the floor strategy, are identified and evaluated. An architectural analysis of the building is provided and possible causes for navigation problems are discussed. Different architectural features of the building are investigated with respect to human spatial cognition and usability issues. Finally we address potential benefits for the architectural design process and discuss options for further research.

1 Introduction

Many people have problems finding their way around public buildings such as airports, hospitals, offices or university buildings. The problem may partially lie in their spatio-cognitive abilities, but also in an architecture that only rudimentarily accounts for human spatial cognition. We aim to make progress towards linking architectural design and human spatial cognition research. The paper begins with an overview of previous work on wayfinding cognition and describes empirical methods from cognitive psychology available to investigate human indoor navigation behavior. In the main part of the paper we report on an empirical investigation in which twelve participants solved way-finding problems in a complex multi-level building. Half of the participants were very familiar with the building, the other half visited the site the first time. We provide a detailed architectural analysis of the building and relate the results to architectural design, human spatial cognition research, and indoor-wayfinding.

C. Freksa et al. (Eds.): Spatial Cognition IV, LNAI 3343, pp. 1–23, 2005.
© Springer-Verlag Berlin Heidelberg 2005

2 Elements of Indoor Wayfinding

A pioneering study on indoor navigation was conducted by Best (1970), who first identified fundamental aspects of a building's route network, like choice points, directional changes and distances as relevant predictors of wayfinding difficulties in complex buildings. Numerous studies, especially in the *Environmental Psychology* community, have since investigated wayfinding difficulties in settings such as airports (e.g., Raubal, 2002), shopping malls (Dogu & Erkip, 2000) or hospitals (Haq & Zimring, 2003).

Weisman's (1981) pivotal paper identifies four general classes of environmental variables that shape wayfinding situations: visual access, the degree of architectural differentiation, the use of signs and room numbers, and floorplan configuration. Further studies pointed to the impact of layout complexity on both wayfinding performance and cognitive mapping (Gärling et al., 1986; O'Neill 1991a/b).

Gärling et al. (1983) point out that familiarity with a building has substantial impact on wayfinding performance, as does visual access within the building: If large parts of the building are immediately visible and vistas connect the parts of the building, people have to rely less on stored spatial knowledge and can rely on information directly available in their field of vision, a notion inspired by Gibson (1979).

The role of familiarity with and knowledge about a building is also stressed by Hunt (1984) and Moeser (1988), showing how specific training of sequential routes or survey knowledge can boost navigation performance in complex buildings like nursing homes or hospitals.

Only very few researchers have explicitly discussed usability issues of buildings. Werner and Long (2003) are among these few and point to opportunities for improving the mental representation of a building's structure. Butler et al. (1993) explicitly present a usability study into the effects of graphical information, showing not only positive impacts of signage and floor maps but also providing guidelines for improving signage design.

While floorplan complexity and visual access have been defined rather informally in the literature discussed above (e.g., by subjective ratings), the Space Syntax movement has introduced formalized, graph-based accounts of complexity and visibility (Peponis et al., 1990; Hillier & Hanson, 1984). Calculations based on these representations reveal that the connective structure of rooms and circulation areas in a building largely determines wayfinding behavior. Haq and Zimring (2003) recently reported strong correlations between topological connectedness of locations in a hospital building with route choices of visitors both in unguided exploration and in directed search tasks. Note that research along this methodology is generally based on correlations of building layout and aggregate movement patterns, thus providing no immediate understanding of individual cognitive processes (Penn, 2003).

2.1 Wayfinding in Three-Dimensional Structures

Almost all controlled studies into wayfinding performance and building complexity have limited themselves to investigating movement and orientation in the horizontal plane of isolated floor levels (with notable exceptions like Hunt, 1984; Moeser, 1988). Soeda et al. (1997) observed wayfinding performance in tasks involving vertical level changes. They found people losing their orientation due to vertical travel, supporting more informal results of Passini (1992). Soeda et al. identified another challenge of multi-level buildings: Wayfinders assume that the topology of the floorplans of different levels is identical, an assumption that can lead to severe wayfinding difficulties.

In Section 3.1 of the paper we provide a building analysis revealing that our setting could be similarly prone to challenges based on multi-level properties. Therefore, our investigations into both the navigation performance of test participants as well as their mental processes explicitly focus on the above-mentioned aspects. Montello and Pick (1993), although not investigating wayfinding behavior directly, present evidence that humans have trouble correctly aligning vertical spaces in pointing tasks. We also expect wayfinders to have trouble integrating survey knowledge of different floors. Properly connecting mental floorplans at transition points like staircases or elevators may also be further impaired by difficulties of maintaining one's heading due to the rapid direction changes involved in stair climbing.

2.2 Investigating Cognitive Processes with Verbal Reports

According to Passini (1992), wayfinding tasks can be described as sequences of wayfinding decisions. The decisions are anticipated and organized in planning processes. Depending on the available knowledge (in the current vista and in memory), plans are formed in a hierarchical fashion, i.e., organized around major decisions like general route choices and, especially, level changes. Planning varies with respect to the degree of completeness: If the wayfinder possesses sufficient knowledge about the building, a complete plan can be formed. Otherwise, only partial planning can commence ("sub-tasks" in Passini's terminology), postponing local decisions until further information comes into sight. Planning involves mental simulation of the route and forming expected images to be met along the way. Later, while travelling a route, expected images are matched with actual vistas, be it landmarks or position checks with the outside.

The majority of experimental studies on human wayfinding behavior and related cognitive competencies are based on direct observation of user behavior. We agree with Passini (1992) that the collection of behavioural data can successfully be complemented with verbal reports of task-concurrent thoughts to get a comprehensive picture, especially in exploratory studies. Hence, we introduce verbal reports of wayfinders as an additional data source. The *thinking aloud method* of collecting verbalisations concurrent with task performance is an established method for tapping into those cognitive processes that can be verbally accessed (Ericsson & Simon, 1993). Verbal reports are especially useful for investigating the reasoning steps involved in complex real-world domains, e.g., navigation in information spaces (Hölscher & Strube, 2000). Thinking aloud has only occasionally been employed in spatial cognition research. For example, Cornell et al. (2003) were able to show that verbal reports provide insights about the reasoning processes involved in self-estimations of spatial competence as well as during wayfinding tasks.

Passini (1992) based his seminal qualitative investigations into wayfinding processes on the extensive analysis of individual wayfinding episodes and the verbal comments of his test participants. Our study aims at a somewhat more formalized approach to qualitative verbal data by quantifying occurrences of verbal reports and comparing these with behavioural measures like time, distance, pointing accuracy and objective route choice since verbal reports of, for example, strategic decisions alone may not be sufficiently reliable. In multi-level buildings with complex floorplans involving inconsistencies and dead-end routes, planning processes and adequate route choice strategies should be very important for wayfinding success. Therefore, our thinking aloud analysis of cognitive processes focuses on the degree of planning, the type of environmental information perused (signs, visual access, etc.) and strategic reasoning.

2.3 Wayfinding Strategies for Complex Buildings

Authors like Weisman (1981) or Lawton (1996) have analyzed wayfinding strategies as to what degree they rely on different types of knowledge. Spatial knowledge is commonly distinguished into three levels (Siegel & White, 1975). In the context of this study it can be assumed that finding destinations inside the building requires all three types of spatial knowledge: landmarks identify one's own position and relevant navigational choice points, route knowledge connects distinguishable landmarks, while survey knowledge integrates routes and guides high-level decisions for route selection and general direction.

In a building with a complex network like in Figure 2, the general notion of survey knowledge – in the sense of correct positional information about the metric spatial position of destinations – representing the most advanced and valuable information may not hold. In fact, knowing the routes through the maze of levels and vertical and horizontal corridors may be even more important, especially since seemingly direct routes may be blocked by dead-ends in the building.

Wayfinding strategies like the *least-angle strategy* (Hochmair & Frank, 2002) or *region-based strategies* (Wiener, Schnee & Mallot, 2004) have been described for two-dimensional outdoor settings. But how do people incorporate their available knowledge in wayfinding strategies in multi-level buildings? We propose a distinction of three strategies for finding one's way in cases with incomplete information:

- The *central point strategy* of finding one's way by sticking as much as possible to well-known parts of the building, like the main entry hall and main connecting corridors, even if this requires considerable detours.
- The *direction strategy* of choosing routes that head towards and lead to the *horizontal* position of the goal as directly as possible, irrespective of level-changes.
- The *floor strategy* of first finding one's way to the floor of the destination, irrespective of the horizontal position of the goal.

Mapping these strategies to other accounts, the *least-angle strategy* can be directly related to the *direction strategy* in our classification. In a more abstract sense, the *region-based „fine-to-coarse" strategy* of – ceteris paribus – preferring paths that quickly bring one into the region of a destination, is compatible with the *floor strategy*, if you assume floor levels as organizing principles in the mental representation of multi-level buildings (cf. Montello & Pick, 1993).

2.4 Knowledge About the Environment

The application of the strategies defined above clearly requires access to information about the building. With the complexity of the environment the relevant types of knowledge can become quite intertwined. To address this, we look into the knowledge requirements from three perspectives:

First, the overall familiarity of the wayfinders with the building is controlled for by comparing a group of novices to the building to a group of repeat visitors. Second, for each task and participant we take into account their degree of familiarity with the specific task destination. And third, survey knowledge about the building is identified for each participant in a pointing task.

This design, combined with verbal reports and task performance measures, will allow us to address a set of research questions related to building complexity, strategic choices and spatial knowledge as well as methodological concerns:

- What cognitive processes can be identified in verbal reports of wayfinding tasks and how do they relate to performance? How are verbal reports of strategic choices related to objective measures of route choice?
- What is the role of planning and navigation strategies in multi-level settings?
- Do cognitive processes vary with task characteristics such as difficulty?
- How does expertise about the building, knowledge about goal locations and survey knowledge affect a) verbalized cognitive processes, b) navigation strategies and c) task performance?

3 Methods

3.1 Participants

In this part of the paper we report on an empirical investigation conducted with participants of an annual summer school for human and machine intelligence which takes place at the Heinrich-Lübke Haus, a conference centre in Günne, Germany. Seven women and five men were asked if they would volunteer in a wayfinding experiment. Six of them were familiar with the building. These experts[1] had previously visited the one-week conference at least two times and therefore knew the building well. The six novices were unfamiliar with the building when this years' conference started. Their sessions took place within the first three days after their arrival. The participants were in their mid-twenties to mid-thirties and were all native German speakers.

3.2 Building Analysis

The conference centre was built in 1970. We explore the ground floor (level 0) of the multi-functional building to exemplify the general characteristics and spatial organi-

[1] Note that we do not use the terms "experts" and "novices" in the usual psychological sense, but rather as a distinction between participants who visited the site before and those who were there the first time.

zation of the layout (see Figure 1). The common layout consists of various simple geometrical elements that are arranged in a complex and multi-faceted architectural setting. In the theory of architectural design, building structures can be formally understood from diverse points of views, as a group of voids or solids (Mitchell, 1990). Consequently, this building is subdivided into a well-designed group of solids with void space between them. Additionally, each group of solids implies various functions, e.g., the living quarters (C) have a quadratic design style and the communication area (D) a hexagonal design style. With this in mind the building can be architecturally categorized as an "indoor city" (Uzzell, 1995) as it is composed of a small ensemble of units and a large public circulation area. The main path of walking through the building is an axial one rather than a cyclical one, which means one has to pass the central point (B) frequently when traveling between areas.

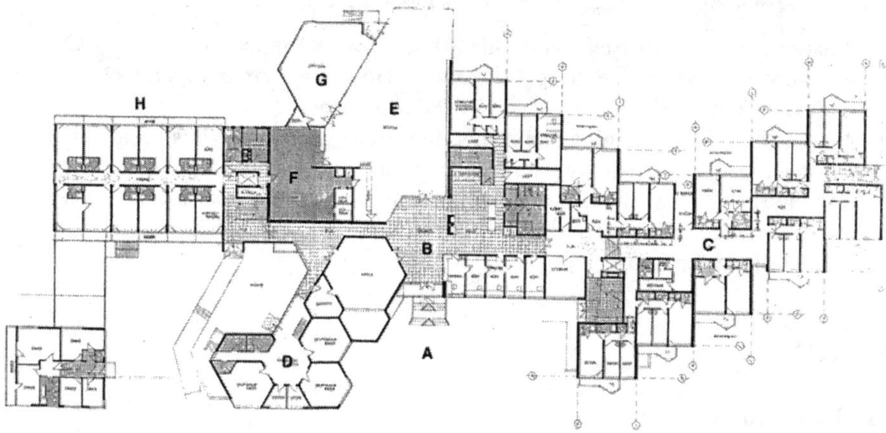

Fig. 1. ground plan: (A) public entrance (B) entrance hall (C) living quarters (D) Commons – communication and conversation area (E) dining-room (F) kitchen (G) coffee bar (H) lecture rooms

Changing floors in the building exemplifies its spatial complexity and vertical impenetrability. As one can see in Figure 2 the layout of the hallways on every floor seems to be one and the same, but is actually different for each floor. For example, the configuration of the ground floor (level 0) and the basement (level -1) differs significantly. As a result of this counter-intuitive layout, the user has to repeatedly look for a new and unknown route on every level.

3.3 Procedure

In this building, the participants' task was to find six locations. The participants were filmed with a camera and had to verbalise their thoughts. Between wayfinding tasks they had to point to four locations they had previously visited in order to assess their survey knowledge. The whole experiment lasted about 45 minutes including the instruction, as well as an interview and debriefing after the experiment.

Fig. 2. The floors of the building with circulation areas. Stairways are illustrated as vertical connections. Starting points and goals of the navigation tasks are marked by numbers (example: "1" marks the starting point for task 2 and "2" marks its goal)

First, the participants were instructed to think aloud while performing the tasks of the experiment and not to pay attention to the camera. During the whole experiment, they were not allowed to use floor maps or ask other people for advice, but they were allowed to use signs or to look out of the window for orientation as long as they stayed inside. For most task instructions the experimenter just mentioned the goal such as "Find room number 308".

All participants received the tasks in the same order, as each destination point is the start location for the following task, making randomization unfeasible. Throughout this paper, navigation tasks are identified by numbers, pointing tasks by capitals:

1 From outside the building, the participants were shown a wooden *anchor* sculpture inside the living quarters. They had to find it from the main entrance without leaving the building again.
2 The goal was to find *room 308.*
3 Participants had to navigate to the *bowling alley.* It was located in the cellar of the building, where the locations for all leisure activities were to be found.
4 The *swimming pool* could also be found there.
A From the swimming pool the participants had to point to the *anchor*, the destination of the first task.
B After moving a few meters away from the swimming pool the participants were asked to point to the *forecourt* in front of the main entrance.
5 The participants had to navigate their way to the *lecture room number four.*
C From a point close to (or near) the lecture rooms, the participants had to point to the *bowling alley.*
6 The final navigation task's destination was the *billiard table.*
D From the billiard table they had to point back to the *lecture rooms.*

3.4 Measures

Performance: For each task, the shortest route as well as a list of reasonable route alternatives was determined beforehand. Reasonable routes are defined as neither containing cycles nor dead ends or obvious detours. Navigation performance was measured with six variables:

- time to complete the task, taken from the video. Extra time, e.g., stops with explanations because of experimental issues was subtracted
- stops
- getting lost, i.e., number of leaving a *reasonable route alternative,* detour behavior
- distance covered
- distance covered divided by length of the shortest possible route. This parameter expresses the proportion of superfluous way independent of task length. For example, a value of 1.35 can be interpreted as walking 35% farther than necessary.
- speed is distance covered divided by the time to reach the goal

Subjective Measures: The second group of measures classified the participants' verbal comments. To do so, the walked route of every task was first drawn into the plans of the building. This was also used to determine distances of routes and superfluous way after getting lost (see above). The verbal codes and stops were written beside this drawn route at the location they were mentioned. The coding scheme for classifying the verbal comments was developed according to Krippendorff (1980). An initial coding scheme was developed based on a pilot session to determine what types of verbalisations can be related to categories of theoretical interest. Step-by-step the coding scheme was adjusted where necessary so that categories could be reliably recognized by independent raters, based on the video sequences of four participants. The process was repeated until a sufficient inter-rater reliability with a kappa value of .7 ("substantial" reliability according to Landis & Koch, 1977) was reached. To reduce coding error, every participant was coded twice and in case of disagreement one consensual rating was achieved. In addition to the verbalization categories, the participants' remarks about their strategies were collected for every task. Out of the mentioned strategies for each task, the preferred one was identified by the raters where possible. Four subjectively preferred strategies could be identified: The already described direction, floor and central point strategy (see 2.3) and, in addition to that, the "route is well-known" strategy when participants mentioned walking a route completely familiar to them.

Survey Knowledge: From their current position the participant had to point his/her arm in the direction of a location previously visited during this experiment. From the video the position and pointing direction were transferred to a map in which the angular deviation to the correct direction was determined. Taking into account that the pointing error is to the right (negative angle) or to the left (positive angle), the mean is a measure of the systematic error, which is specific to each pointing task. The unsystematic error can be measured by the standard deviation.

4 Results

First, aspects of the process of navigation as expressed in the verbalisation and their interrelations to performance are presented. Second, the tasks are compared according to these measures, the familiarity with the goal locations and to the strategies used to solve them. Then differing navigation processes according to the strategies are shown. Finally, the influence of expertise and survey knowledge on verbalized cognitive processes, navigation strategies and task performance is presented.

Table 1. Shows the performance in each task and the average performance and standard deviation across all tasks

	task 1	task 2	task 3	task 4	Task 5	task 6	mean	Sd
time [s]	226	78	159	34	103	81	112	78
stops [n]	2.8	0.4	1.7	0.3	0.5	0.9	1.1	1.80
getting lost [n]	0.7	0.1	0.5	0.0	0.3	0.2	0.3	0.57
distance [m]	168	84	127	40	113	87	102	58
way/shortest way	1.68	1.24	1.71	1.00	1.08	1.50	1.36	0.59
speed [m/s]	0.74	1.08	0.81	1.28	1.12	1.10	1.03	0.29

In the two rightmost columns of Table 1 the average performance and standard deviation per task are shown. At a speed of one meter per second the participants needed almost two minutes to cover the average 100 meter distance which is 36% more than the shortest possible way. They stopped about once per task and lost their way 0.3 times.

The verbalisations mentioned during these tasks are shown in Table 2. 40% of all verbalisations were reflections mainly about the building. 22% refer to partial planning, 12% to landmark checks during plan execution (like "here is the fire place") and 9% to usage of signs. Remaining categories each make up for 5% or less of the utterances.

But how can the verbalisations, as indicators of the navigation process, be connected to performance? Correlations of variables were computed by aggregating the variables to average values per participant. Here a cluster of highly correlated variables could be identified (19 of 20 pairwise correlations of verbalisations and performance measures, $r>.25$, $p<.04$): participants who rarely planned a complete route but often a partial one, and those who reflected a lot on the building and often experienced a failure of their plan also showed a bad performance: They stopped often and often got lost and therefore covered a longer distance absolute and relative to the shortest possible. So they needed more time to reach the goal. None of the other verbalisations showed a substantial connection to performance measures.

4.1 Tasks

Do the processes differ according to the tasks' affordances? And do the tasks – as intended - cover a broad range of difficulty? To answer the last question performance was compared between tasks in an ANOVA for each dependent measure.

Table 2. The verbalisation categories are described, their frequency and proportion shown. An asterisk* marks a significant difference in average frequency between tasks ($p<.05$), a cross† marks a statistical trend ($p<.10$)

verbalisation category	Description	Frequ-ency [n]	pro-potion [%]
complete plan†	A complete plan covers a path from the current location to the destination of the current task	13	3
partial plan*	A non-complete plan contains uncertainty and/or covers only parts of a complete path	87	22
search	Systematic number-based search, e.g., to find a room	17	4
correct reflection	Reflections about the building that are correct.	18	5
false reflection	Reflections about the building that are *incorrect*.	7	2
reflection*	General reflections and assumptions, not only about the building	130	33
Alternatives*	Consideration of more than one possible route to the goal	16	4
failed plan†	Failure of a pursued plan	11	3
identify landmark	Recognition of a known landmark in sight	48	12
outside orientation	Use of the outside space for orientation	14	4
sign	Participants mention a sign in sight	34	9
sum		395	100

The tasks differed in all performance measures (see Table 1, all $F(5, 65)>3.0$, $p<.016$). The most difficult task was task 1, finding an anchor shown from outside of the building. The participants stopped and got lost most often and they covered the longest distance at the lowest average speed.[2] Both in task 1 and task 3, the second most difficult task, the covered distance was 70% longer than in the shortest possible route. In task 3 the participants had to go to the bowling alley where many alternative routes were available. Here stopping and getting lost happened second most often, and speed was the second lowest. By the same variables, task 6 can be considered third in its degree of difficulty. The easiest task was task 4 (pool). No one got lost, there was no superfluous distance covered, stops were least frequent and therefore the speed was highest. The performance on tasks number 2 and 5 fell in between the rest. So there was a variation in task difficulty as intended.

Table 3. Shows the average verbalisation frequency per 1000 seconds

frequency/1000s	task 1	task 2	task 3	task 4	task 5	task 6	mean
complete plan	2.0	0	1.0	9.0	2.7	3.5	3.1
partial plan	10	14.5	11.9	6.3	8.3	9.9	10.2
search*	0	11.1	0	0	4.9	0	2.7
correct reflection	1.8	1.0	2.8	3.0	2.8	5.5	2.8
false reflection	1.0	0	0.6	0	1.4	0	0.5
reflection	16.1	19.7	16.3	21.1	15.4	14.7	17.2
alternatives*	1.0	0	3.5	0	1.5	5.3	1.9
failed plan	2.1	0	0.4	0	0.5	1.4	0.7
Identify landmark*	6.1	19.1	5.3	2.0	2.1	3.3	6.3
outside orientation*	4.7	0	1.1	0	0.8	0	1.0
Sign	0.4	1.2	5.3	14.7	3.7	4.1	4.9

To answer the question for different processes according to these tasks, the verbalizations were compared (see Table 3). To be able to compare verbalizations between tasks of different length, not the absolute frequency but the frequency per time was taken. More or less identical results can be expected with frequency per distance covered, as this is correlated with time per task at $r=.90$ ($p<.001$). As task difficulty was controlled, the verbalisation categories previously identified to indicate bad performance showed no *further* differences. Nonetheless verbalised processes did differ according to the affordances of the tasks: systematic search only occurred in task 2 and 5 where rooms had to be identified by number (ANOVA between tasks, $F(5, 65)=12.6$, $p<.001$). In task 2 and 4 no consideration of alternatives was mentioned: in task 4 all participants took the same route, while in task 2 the participants seemed not to notice available alternatives at all ($F(5, 65)=3.34$, $p=.010$). The identification of landmarks primarily occurred during task 2, where participants checked the levels while climbing the staircase ($F(5, 65)=8.68$, $p<.001$). Orientation on the outside of the building was most prominent during task 1 where the goal was shown from the out-

[2] Stops and getting lost can be considered dependent of length of the task, but normalising them on navigation time or the shortest possible way did not produce a different pattern of results and so the average per task, which is easier to interpret, was taken. From a theoretical point of view, this parameter is also favourable, as the number of intersections, number of turns, etc., are more important for difficulty than length of the route.

side (F(5, 65)=7.86, p<.001). So the character of different tasks can be found in the verbalised processes. Can familiarity with the goal explain this?

The familiarity with the goal differed between the tasks (see Table 4, Chi-Square test, $\chi^2(10)=32.2$, p<.001). In the difficult tasks 1 & 3 the goals were quite unknown whereas the relatively easy tasks 4 and 5 were quite well-known. Surprisingly, in the third most difficult task, task 6, the goal was as well-known as in the easiest task 4. And in the relatively easy task 2 the goal was quite unknown to most of the participants. Why this? To find an explanation, strategies have to be taken into account.

Table 4. Shows the familiarity with the goal in the different tasks

Frequency	task 1	Task 2	task 3	task 4	task 5	task 6	sum
goal unknown	9	5	4	2	0	2	22
goal imprecisely known	0	5	1	1	5	1	13
goal familiar	3	2	7	9	7	9	37

4.2 Strategies

Most of the participants voiced remarks concerning the strategy they use to find their goal. Sometimes they switched their strategy during a task, but in 61 cases a preferred strategy could be identified by the raters.

As shown in Table 5 the choice of the strategy was dependent on the familiarity with the goal ($\chi^2(6)=16.6$, p=.011). To run a well-known route was only possible when the goal was familiar. If the goal was either unknown or known imprecisely, participants had to choose a different strategy: The direction strategy of walking to the assumed horizontal position of a goal as directly as possible; the floor strategy of first walking to the assumed floor of the goal; or the central point strategy of going back to a well-known central point or a route element and trying to find the goal from there. Each of these strategies was chosen roughly with the same frequency.

Table 5. Shows the frequencies of preferred strategies depending on familiarity with the goal

frequency	unknown goal	Imprecisely known goal	familiar goal	sum
direction strategy	5	3	6	14
floor strategy	6	4	9	19
central point strategy	6	4	4	14
route is well-known	0	0	14	14
sum	17	11	33	61

Going back to the tasks, strategy selection in the individual tasks can be considered. Different strategies were chosen in different tasks (not shown here, χ^2 (15)=56.9, p<.001). In the easiest task, number 4, all identified strategies walked the well-known route. In the two most difficult and often unknown tasks 1 and 3, many participants chose a direction strategy. Contrarily, in the also often unknown task 2, the floor strategy was chosen most frequently. Assuming that the floor strategy is efficient, its application might explain the good results in this task.

To test this, performance according to the preferred strategy has to be considered. As strategy choice was dependant on the tasks and the tasks differ in difficulty, the influence of the tasks had to be partialed out, i.e., controlled statistically as a covariate in an ANOVA. So the benefit of the strategies could be compared independently of the tasks. As shown in Table 6 best performance was achieved when walking a well-known route (except stops all $F(3, 56)>3.1$, $p<.035$). Here, the absolute and relative distance as well as time was shortest, speed highest and getting lost occurred least often. When using the direction strategy or the central point strategy, the absolute and relative distance as well as time measures indicated the worst performance. With a central point strategy you try to walk known routes and therefore can walk quite fast without getting lost. But as the route is longer than in other strategies, it takes longer to reach the goal. With the direction strategy it's easier to get lost and re-orienting oneself might take time, so the average speed leads to the same amount of time needed to reach the goal as in the central point strategy, even though the distance is shorter. Floor strategy was better than these two with respect to distance and time, but clearly less so than walking a known route.

Table 6. Shows average performance and verbalisation per task solved with the preferred strategy. The influence of task difficulty is partialed out, so slightly negative values can occur

partialed out means	direction strategy	floor strategy	central point strategy	route is well-known
performance				
time [s] *	145	113	140	67
stops [n]	1.50	1.62	1.05	0.18
getting lost [n] *	0.69	0.35	0.23	0.03
distance [m] *	119	97	142	68
way/shortest way*	1.38	1.33	1.86	1.06
speed [m/s] *	0.86	0.96	1.04	1.29
verbalisations				
complete plan*	0.13	0.05	0.08	0.58
partial plan†	1.37	1.57	1.60	0.40
search*	0.11	0.29	0.60	0.05
correct reflection*	0.67	0.18	0.05	0.19
false reflection	0.27	0.14	0.02	0.02
reflection*	2.57	1.99	2.19	0.68
alternatives	0.31	0.33	0.34	-0.03
failed plan	0.33	0.19	0.09	0.03
identify landmark†	0.65	0.78	0.98	0.25
outside orientation*	0.63	0.03	0.15	0.11
sign*	0.30	0.01	1.42	-0.01

The differences between the strategies can also be identified in the navigation process itself, manifested in the verbalisations (all described differences $F(3,56)>2.9$, $p<.044$). Again, walking a known route was quite different from the other strategies: participants most often planned their route completely and all other processes were verbalised less often. Presumably these participants just relied on their readily stored (route) knowledge and did not need further reasoning. Participants using a central point strategy most often searched systematically, used signs most often and tended to identify landmarks most often ($F(3,56)=2.58$, $p=.062$) as well as planning their route only partially ($F(3,56)=2.56$, $p=.059$). Participants using a direction strategy mentioned the highest number of correct reflections and general reflections.

Similar results according to performance and verbalisations could be found when the selected route alternative was considered instead of the subjective mentioning of a strategy. Even if a well-known route can not be assigned to a specific route, subjective direction, floor and central point strategy are highly correlated with the objective choice of route.

To summarize, strategies could be identified on a subjective and an objective level. The shortest and fastest way to reach a goal was to walk a well-known route. If that was not possible – e.g., because the goal was unknown – the floor strategy was the best alternative in our scenario. Walking via a central point or going directly in the assumed direction of the goal led to worse performance.

4.3 The Role of Experience

The straightforward way to assess the importance of knowledge for navigating a building is to compare novices with experts who know a building well. Experts are assumed to show better performance – is this true? Indeed, experts performed better (see Table 7). They got lost less often, covered shorter distance (absolute & relative), with greater speed, and therefore reached the goal quicker (all $t(10)>2.23$, $p<.05$).

Experts performed better in reaching a goal. But can this difference be traced back to different processes during navigation? As shown in Table 7 experts more often completely planned their route (unless stated otherwise, all $t(10)>2.26$, $p<.048$), whereas novices tended towards more partial planning ($t(10)=1.91$, $p=.085$). There was a trend for novices to utter more reflections ($t(10)=1.92$, $p=.084$) and to identify more landmarks ($t(10)=2.13$, $p=.059$). Novices also needed to search more as well as to orient themselves more towards signs and the outside of the building.

So experts were able to rely on their (route-related) knowledge for execution whereas novices needed to process more local information from the building and from outside. Can this difference also be found in the choice of strategies? Indeed, novices and experts differed in their preferred strategies (see Table 8, $\chi^2(3)=19.0$, $p<.001$). Novices most often chose the central point strategy and almost never walked a well-known route, whereas experts almost never chose a central point strategy and most often either walked a well-known route or used a floor strategy. The direction strategy was equally used by both groups.

Table 7. Shows means and standard deviations of the performance and verbalisations of novices and experts

	novices		experts	
Performance	m	sd	m	sd
time [s] *	128	22	95	21
stops [n]	1.36	0.69	0.78	0.80
getting lost [n] *	0.42	0.17	0.17	0.21
distance [m] *	115	16	89	17
way/shortest way*	1.55	0.22	1.17	0.16
speed [m/s] *	0.96	0.06	1.10	0.09
Verbalisations				
complete plan*	0.03	0.07	0.35	0.33
partial plan†	1.56	0.70	0.88	0.52
search*	0.36	0.22	0.11	0.14
correct reflection	0.08	0.14	0.45	0.49
false reflection	0.06	0.09	0.14	0.22
reflection†	2.36	1.12	1.26	0.87
alternatives	0.28	0.23	0.17	0.15
failed plan	0.25	0.27	0.06	0.09
identify landmark†	0.86	0.19	0.48	0.40
outside orientation*	0.31	0.16	0.08	0.09
sign*	0.92	0.96	0.03	0.07

Taken together, experts more often relied on their knowledge and they walked a well-known route that they had completely planned in advance. If that was not possible, they chose another efficient strategy, the floor strategy. With their knowledge experts did not have to collect as much information from their surroundings as novices, who had to search and look at signs as well as looking outside. This led to a clearly better performance.[3]

Table 8. Shows the frequencies of strategy selection in novices and experts

	novices	experts	sum
direction strategy	8	6	14
floor strategy	7	12	19
central point strategy	13	1	14
route is well-known	2	12	14
Sum	30	31	61

[3] A similar comparison between women and men did not reveal any gender differences.

4.4 Survey Knowledge

If survey knowledge is the crucial factor for the good navigation performance, novices and experts should differ in their pointing performance.

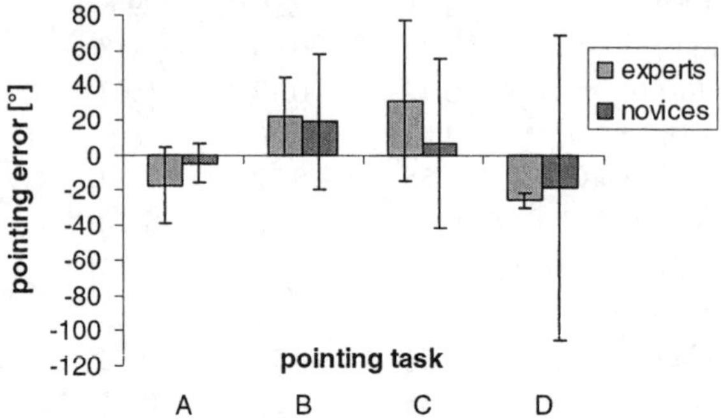

Fig. 3. shows the pointing errors in experts and novices in the four pointing tasks. Pointing to the left of the right direction resulted in a positive error, pointing to the right in a negative one. The systematic pointing error is displayed in the mean deviation from the right pointing direction, the unsystematic error in the standard deviation

But in the four pointing tasks no difference could be found in the systematic error expressed in the mean pointing error (although these tests are not orthogonal, see Figure 3, all t(10)>1.21, p>.252). For the unsystematic error expressed in the standard deviation, there was a trend in pointing task A for a smaller pointing error in novices (F(5,5)=3.90, p<.10) and there was a smaller pointing error in experts in task D (F(5,5)=388, p<.001). So, except for task D, no indication of better survey knowledge in experts was found[4].

Still, even if experts might not show a better performance because of survey knowledge, survey knowledge might be important for navigation. To test the direct influence of survey knowledge on navigation, the sample was bisected into good vs. bad pointers according to their average absolute pointing error across the four tasks. But no differences could be revealed for navigation performance measures (all t(10)>1.30, p>.221). Even among the eleven verbalisation categories only a single difference was found: good pointers uttered more correct reflections (t(10)=2.60, p=.026). Survey knowledge did not explain differences in performance and verbalisation.

Looking at the four pointing tasks, yet another interesting result was revealed. In task B the systematic error differed from zero (t(10)=2.38, p=.036) and there was a trend in task A to do so (t(10)=2.17, p=.053). Due to the large standard deviation in novices, there was no reliable systematic error in task D. But looking at experts separately (t(4)=14.1, p<.001) or applying a binomial sign test ignoring variance for both

[4] An additional analysis of absolute pointing error as a combined measure of systematic and unsystematic error revealed the same pattern of results.

groups together (10 out of 12 < 0, p=.039) a difference was revealed. In task B the error can be explained by the huge size of the target location, a place. In task A it can be explained with the greater distance from the pointing place. But the errors in task D remain surprising as this was the only pointing task which could be solved by path integration: the participants just had to remember the direction of the starting point of their last navigation task. As this was not possible in the other tasks one would expect the best results in task D, but not the worst ones. But taking into account that this was the only task where the parts of the building the participants pointed from and to did not lie at a right angle to each other but at 60° (see Figure 4, left), the systematic error can be explained. A person remembering a 90° angle instead of the right 60° one would locate him/herself standing on the start of the (dotted) arrow to the right and not at the start of the arrow to the left. From this position the mean pointing direction would be quite accurate. Similar results are found in pointing (e.g., Thorndyke & Hayes-Roth, 1982) and in map drawing (e.g., Gillner & Mallot, 1998).

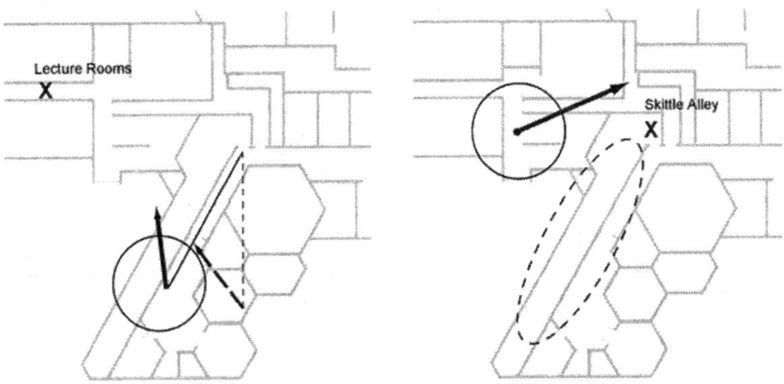

Fig. 4. shows starting (circle) and goal point (cross) in pointing task D (left). The mean pointing direction is marked with the arrow in the circle. If you assume that the participants remembered a right angle between the parts of the building and not the correct 60°, pointing from the assumed place (dotted line and arrow) is quite accurate. Pointing performance for task C is shown on the right side

5 Discussion

The present study was conducted to explore wayfinding strategies in a complex indoor environment and their relations to the knowledge and experience available to the participants. Throughout the study an architect, author G.V., has monitored our process. The experiment has provided both quantitative analyses of behavioral and verbal data, presented above, as well as the opportunity to observe deficits of the building with respect to wayfinding usability. In the next sub-section we first discuss the main quantitative results. Then we link the experimental data collection with architectural design expertise and insights about building usability issues collected from the test participants in extensive debriefing interviews. These *lessons learned* from an architectural perspective are completed with a glance at further research issues.

5.1 Discussion of Quantitative Results

Complete planning is found to be associated with good performance, while reflecting, partial planning and re-planning can be tied to poor performance. Verbal reports alone must be interpreted with caution as they are restricted to consciously accessible aspects of cognitive processes (Ericsson & Simon, 1993). Thus it is important to note that in our study we have identified wayfinding strategies on a *subjective* and an *objective* level with converging results: The shortest and fastest way to reach a goal is by using one's knowledge to walk a well-known route, as most experts do. If that is not possible, for example, because the goal is unknown, experts also choose the floor strategy which is the best alternative in our scenario. Walking via a central point like most novices do or going directly in the assumed goal direction leads to worse performance. With their knowledge experts do not have to collect as much information from their surroundings like novices who have to search and look at signs as well as outside. This clearly leads to a better performance. Survey knowledge as measured here can not account for these differences. Even with experts systematic errors in survey knowledge prevail, which could be explained by them interpreting oblique turns as being orthogonal. No gender differences could be found.

Overall, novices verbalise more. Assuming that this requires more (cognitive) resources and therefore makes novices slower, could explain their poor performance. But referring to the strategies, one reason for poor performance is them taking long and winding routes like in the central point strategy or getting lost as in the direction strategy. Slowness alone can not account for that.

Why could this difference not be explained by survey knowledge? Is it the small number of participants? For other variables reliable effects can be found, and even the direction of the differences often is not in favour of experts. Maybe measuring pointing *after* the navigation task is the reason. Previously existing differences in survey knowledge could account for the better navigation performance in experts. But by walking the route novices were able to acquire this survey knowledge, reduce the difference and perform equally well in the pointing task afterwards. To test that, pointing performance must be measured before navigating a route. Another approach would ask for the more global sense of direction, known to account for individual differences in strategies, pointing and navigational performance (e.g., Hegarty et al., 2002). Finally, it is possible that survey knowledge is not as much of a key issue in reaching a goal as route knowledge is. Meilinger and Knauff (submitted) were able to show that survey knowledge (in the form of maps) can be less helpful for wayfinding than concrete route knowledge (in the form of verbal descriptions) even in outdoor settings. Indoors, this may be even more pronounced, since dead-ends and limited connectedness of floors and path make survey and direction-related knowledge even less useful here. So the strategy exclusively dependent on survey knowledge – the direction strategy – is accompanied with getting lost and relatively bad performance. Also, searching systematically is not associated with bad performance and the two tasks including systematic search are solved quite well.

As a design consequence, the floor strategy, which is most efficient for unknown goals, should be supported by easy transitions between the floors. Also the systematic search is to be taken into account with systematic room numbers or informative signs.

5.2 Exploring Architectural Space

Architecture as the science and art of building generally deals with the design, construction and conceptualization of built space. It greatly influences the comprehension and knowledge of orientation and navigation systems. Akin (2002) clarifies that the architect aims to construct buildings as complex systems of numerous architectural dimensions. To develop an adequate and satisfactory compromise is an essentially spatial task. Architectural space is not generated on a blank sheet, but constantly in respect to the present environment and consequently in a high-dimensional decision space (Bertel, Freksa & Vrachliotis, 2004).

Emphasizing the idea of movement as a central theme in the theory of architectural design, Le Corbusier declared in 1962 (p. 30): "To experience architectural space truthfully it is necessary to perambulate and stride the building." So perception of a built environment is described as a dynamic process of movement caused by the fact that we do not experience the spatial layout of a building as a static structure. We discover architectural shapes and layouts literally step by step. Thus, from a user's perspective several points of environmental ability, legibility (Lynch, 1960) and imageability (Passini, 1992) are essential to understand and interpret building layouts, e.g., landmarks, routes, paths and walkways, and to differentiate shapes and forms, configured space and building topology, and the close relation between inside and outside space. On the one hand, David Stea might be right when he assumes"...the idea or image of a building is as important as the building itself' (Stea, 1974, p. 157). On the other hand, our data suggest that more abstract representations also help to navigate through complex buildings. Such representations might have nothing to do with visual mental images. But they work efficiently. This interpretation is in agreement with the findings by Meilinger and Knauff (submitted) and by results from another field of spatial cognition research showing that the role of visual mental images for human spatial cognition is often overestimated (Knauff, Fangmeier, Ruff, & Johnson-Lairds, 2003; Knauff & Johnson-Laid, 2002).

Understanding a building from its inside structure and spatial organization requires making ones way through the building. Thus, in theories of building design, the idea of architectural experience and the meanings of walkways have a very close relationship. From Space Syntax's point of view walkways seems to be the most fundamental aspect of architectural space, not only for investigating pedestrian movement in designed environments but for general exploring, discovering and learning about architectural settings. In order to provide useful spatial points of reference, the differentiation and discrimination of shapes is the most central property in planning an architectural setting. Although symmetry and similarity are very well-known features in the history of architecture, they contrast with the indispensable need of distinguishing multi-faceted environments. Symmetrical architectural settings are principally one of the foremost difficulties in spatial problem solving processes (Remolina & Kuipers, 2004). Yet, they can be helpful in interpreting vertical information of space, e.g., for spatial reasoning within multi-level buildings (Montello & Pick, 1993).

5.3 Analysis of Usability Hotspots in the Conference Facility

The functional dilemma of the building for wayfinding is prominently caused by the problematic arrangement of complex decision points, their linking paths, the position and design of stairways, vertical incongruence of floors, incomprehensible signage, and too few possibilities for monitoring interior and exterior landmarks. Consequently, the building as a whole gives the impression of a three-dimensional maze. We discuss several design aspects in detail:

Indiscernible entrance hall: For public buildings the entrance hall is one of the most important points in the layout. The public entrance (see Fig. 1, A) as well as the large entrance hall (B), the two central points of the conference center, are comparatively indiscernible, although they are centrally positioned in the general configuration of the building. It is essential for the entrance hall to be readable as such in order to be able to cognitively structure the route network. However, this function is not properly met, which imposes a usability deficit on the building as a whole.

Incongruent floors: In the planning of complex buildings architects have to pay attention to the uncomplicated and insightful organization of floors. The floors of the conference center give the impression of matching one another, but in fact the hallways are considerably different (see Fig. 2). From wayfinding research and a building usability point of view, this a) prompts improper assumptions in the users about the route networks and b) hampers the mental alignment of levels. Pointing task C (bowling alley, see Fig. 4, right) illustrates the problem: Although the bowling alley is directly ahead and extends to the right, participants systematically point left, presumably because they misalign their current position with respect to the floor below, due to inconsistent hallways (ground floor vs. basement) in this area.

Poorly located stairways: Normally, stairways of a building represent its functional framework. The five small stairways in the conference center are not evenly dispersed and not perceptively placed. The frequently used stairway near the entrance hall is particularly counter-intuitively located (Fig. 1). Consequently, not only the impractical location of the entrance hall but also the stairway has a negative effect on the building's usability. Users do not readily perceive a main stairway to the upper floors.

Disorientating design of the stairways: The design of the stairways has a significant effect on orientation and navigation processes in buildings. Using the foremost stairway (near the entrance hall), there are a lot of spatial twists and turns without an opportunity for controlling one's location. This deficit is due to the lack of visual access to the outside, which would help to improve spatial updating. Frequently, users reported being very disoriented after using this stairway.

Dead ends: It is very important in architecture and particularly for public buildings such as universities, hospitals or conference centers to pay attention to always provide an alternative route to any navigational decision. But there are many locations that can be characterized as "dead space", "dead ends" or "blind alleys" (Fig. 1 & 2). For example, the public area surrounded by the living quarters leads to a dark and uncomfortable corridor. Users will not expect the stairways at the end of the corridor (far right in Fig. 2) and thus miss relevant route choices and feel lost in dead ends.

No distinguishable interior building structure: To understand a building layout both the exterior and the interior structure of a public building has to be effortlessly understood. Looking at the floorplan (see Fig.1), the dissimilarity of geometrical shapes and architectural forms would appear to be helpful for the users to orientate themselves.

But in fact, when actually crossing the building, the different subsections are no longer readily recognizable, leading to a lack of visual differentiation.

No survey places: Especially within complex spatial settings architects and designers have to create places of survey and overview to allow users to build well-integrated spatial knowledge. Even on the ground floor of this conference center there are not enough areas of open space to familiarize oneself with the environment, neither with the interior space (e.g., visual axis) nor with the exterior surroundings (e.g., inside-outside relationship).

No differentiation of public and private space: When planning multi-functional public buildings architects have to bear in mind to separate private, or personal from public space. This rule serves the purpose of integrating two diverse spatial systems within one building. There are a lot of mistaken public and private areas within the conference center which results in disorientating the user and the production of unnecessary dead ends. Therefore public spaces have to be clearly indicated both by architectural layout and signage.

5.4 Future Research

First of all, our study demonstrated the general usefulness of verbal data for systematic statistical analyses of cognitive processes in wayfinding. We see great opportunity to further investigate the very rich data provided by the verbal reports, especially to drill down into individual usability problems and local route choices on a task-by-task basis, allowing for additional support of the usability analysis provided in the previous paragraphs. Also, we wish to extend the analysis by collecting comparative data in other complex settings, particularly ensembles of inter-connected buildings (e.g., on a university campus or hospital site) as we suspect that the usability challenges of complex indoor route networks are magnified by multi-level *inter-building* connections.

Based on the present study we hope to intensify the cooperation of cognitive scientists and architectural designers. Providing guidelines for improving wayfinding friendliness is clearly a practical goal of our research. It will be worthwhile to develop more detailed methods to support usability from the early planning stages on, in order to avoid costly design mistakes. Besides using virtual reality techniques for testing layout prototypes, we can imagine augmenting Space-Syntax-type layout analyses with the techniques presented here to identify usability deficits.

It is clear that all these attempts will have to be embedded in a sound understanding of the architectural practitioners' working methods if cognitive science wants to be taken seriously as a partner in architectural planning.

Acknowledgements

The research was supported by grants from the DFG in the Transregional Collaborative Research Center SFB/TR 8. Markus Knauff is supported by a Heisenberg Award from the Deutsche Forschungsgemeinschaft (DFG).The authors are very grateful to the participants of the wayfinding experiment for their cooperation, to Anna Widiger for support with data analysis and to Kristen Drake for proofreading the manuscript.

References

Akin, Ö., (2002). Variants in Design Cognition, in C. Eastman, M McCracken, and W, Newstetter (eds), *Knowing and Learning Cognition in Design Education*, Elsevier, Amsterdam, pp. 105-124.

Bertel, S., Freksa, C., & Vrachliotis, G., (2004). Aspectualize and conquer in architectural design, *Proceedings of the 3rd Conference at Visual and Spatial Reasoning in Design 04 at MIT, Boston.*

Best, G. (1970). Direction finding in large buildings. In D. V. Canter (Ed.), *Architectural Psychology - Proceedings of the conference at Dalandhui* (pp. 72-75). London: RIBA.

Butler, D. L., Acquino, A. L., Hissong, A. A., & Scott, P. A. (1993). Wayfinding by newcomers in a complex building. *Human Factors*, *35*(1), 159-173.

Cornell, E. H., Sorenson, A., & Mio, T. (2003). Human sense of direction and wayfinding. *Annals of the Association of American Geographers*, *93*(2), 399-425.

Dogu, U., & Erkip, F. (2000). Spatial factors affecting wayfinding and orientation: A case study in a shopping mall. *Environment and Behavior*, *32*(6), 731-755.

Ericsson, K. A., & Simon, H. A. (1993). *Protocol analysis: verbal reports as data.* Cambridge, MA: MIT Press.

Gärling, T., Böök, A., & Lindberg, E. (1986). Spatial orientation and wayfinding in the designed environment. *Journal of Architectural Planning Research*, *3*, 55-64.

Gärling, T., Lindberg, E., & Mäntylä, T. (1983). Orientation in Buildings: Effects of familiarity, Visual Access, and Orientation Aids. *Journal of Applied Psychology*, *68*(1), 177-186.

Gillner, S. & Mallot, H.A. (1998). Navigation and Acquisition of Spatial Knowledge in a virtual Maze. *Journal of Cognitive Neuroscience, 10,* 445-463.

Gibson, J. J. (1979). *The ecological approach to visual perception.* Boston: Houghton-Mifflin.

Haq, S., & Zimring, C. (2003). Just down the road a piece: The development of topological knowledge of building layouts. *Environment and Behavior*, *35*(1), 132-160.

Hegarty, M.; Richardson, A.E.; Montello, D.R.; Lovelace, K. & Subbiah, I. (2002). Development of a self-report measure of environmental spatial ability. *Intelligence, 30,* 425-447.

Hillier, B., & Hanson, J. (eds.). (1984). *The social logic of space.* New York: Cambridge University Press.

Hochmair, H., & Frank, A.U. (2002). Influence of estimation errors on wayfinding decisions in unknown street networks - analyzing the least-angle strategy. *Spatial Cognition and Computation* 2(4): 283-313.

Hölscher, C., & Strube, G. (2000). Web search behavior of Internet experts and newbies. *Computer Networks*, 33 (1-6), 337-346.

Hunt, M. E. (1984). Environmental learning without being there. *Environment and Behavior*, *16*(3) 307-334.

Knauff, M., Fangmeier, T., Ruff, C. C. & Johnson-Laird, P. N. (2003). Reasoning, models, and images: Behavioral measures and cortical activity. *Journal of Cognitive Neuroscience*, 4, 559-573.

Knauff, M. & Johnson-Laird, P. N. (2002). Visual imagery can impede reasoning. *Memory & Cognition*, 30, 363-371.

Krippendorf, K. (1980). *Content Analysis: An introduction to its methodology.* Beverly Hills, CA: Sage Publications.

Landis, J.R., & Koch, G.G. (1977). The measurement of observer agreement for categorical data. *Biometrics*, 33, 159-174.

Lawton, C. A. (1996). Strategies for indoor wayfinding: The role of orientation. *Journal of Environmental Psychology*, *16*(2), 137-145.

Le Corbusier (1962). *An die Studenten, Die Charta d'Athénes*, Reinbek bei Hamburg.

Lynch, K. (1960). *The Image of the city,* MIT Press, Cambridge

Meilinger, T., & Knauff, M., (submitted). *Asking for the way of using a map – What is better? A field experiment on human wayfinding in urban environments.*

Mitchell, W., J. (1990). The logic of architecture – Design, computation and cognition, MIT Press, Cambridge, MA

Moeser, S. D. (1988). Cognitive Mapping in a Complex Building. *Environment and Behavior, 20*(1), 21-49.

Montello, D. R., & Pick, H. L. (1993). Integrating knowledge of vertically aligned large-scale spaces. *Environment and Behavior, 25*(4), 457-484.

O'Neill, M. J. (1991a). Effects of signage and floorplan configuration on wayfinding accuracy. *Environment and Behavior, 23*(5), 553-574.

O'Neill, M. J. (1991b). Evaluation of a conceptual model of architectural legibility. *Environment and Behavior, 23*(3), 259-284.

Passini, R. (1992). *Wayfinding in architecture* (2nd ed.). New York: Van Nostrand Reinhold Company

Penn, A. (2003). Space syntax and spatial cognition: Or, why the axial line? *Environment and Behavior, 35*(1), 30-65.

Peponis, J., Zimring, C., & Choi, Y. K. (1990). Finding the building in wayfinding. *Environment and behavior, 22*(5), 555-590.

Raubal, M. (2002). Wayfinding in Built Environments: The Case of Airports. In *IfGIprints* (Vol. 14). Münster: Institut für Geoinformatik, WWU Münster.

Remolina, E., & Kuipers, B. (2004). Towards a general theory of topological maps, *Artificial Intelligence* 152, 47-104,

Siegel, A.W. & White, S. (1975). The development of spatial representations of large-scale environments. In Reese, H.W. (ed) *Advances in Child Development and Behaviour.* New York: Academic Press.

Soeda, M., Kushiyama, N., Ohno, R. (1997). Wayfinding in Cases with Vertical Motion. *Proceedings of MERA 97: Intern. Conference on Environment-Behavior Studies*, 559-564.

Stea, D., (1974). Architecture in the Head: Cognitive mapping, in Lang, J., Burnette, C., Moleski, W., & D. Vachon, (Eds.) *Designing for human behavior: Architecture and the behavioural sciences.* Stroudsburg, 1974

Thorndyke, P.W. & Hayes-Roth, B. (1982). Differences in spatial knowledge acquired from maps and navigation. *Cognitive Psychology, 14,* 560-589

Uzzel, D.L. (1995). The Myth of the Indoor City, *Journal of Environmental* Psychology, 15, 4, 299-310

Weisman, J. (1981). Evaluating architectural legibility: Way-finding in the built environment. *Environment and Behavior, 13*(2), 189-204.

Werner, S., & Long, P. (2003). Cognition meets Le Corbusier - Cognitive principles of architectural design. In C. Freksa, W. Brauer, C. Habel & K. F. Wender (Eds.), *Spatial Cognition III* (pp. 112-126). Heidelberg: Springer Verlag.

Wiener, J. M., Schnee, A., & Mallot, H. A. (2004). *Use and interaction of navigation strategies in regionalized environments* (Tech Report). Max-Planck-Inst. für biol. Kybernetik.

Modelling Wayfinding in Public Transport: Network Space and Scene Space

Urs-Jakob Rüetschi and Sabine Timpf

Department of Geography, University of Zürich, Switzerland
{uruetsch, timpf}@geo.unizh.ch

Abstract. Wayfinding in the public transportation infrastructure takes place on traffic networks. These consist of lines that are interconnected at nodes. The network is the basis for routing decisions; it is usually presented in maps and through digital interfaces. But to the traveller, the stops and stations that make up the nodes are at least as important as the network, for it is there that the complexity of the system is experienced. These observations suggest that there are two cognitively different environments involved, which we will refer to as network space and scene space. *Network space* consists of the public transport network. *Scene space* consists of the environment at the nodes of the public transport system, through which travellers enter and leave the system and in which they change means of transport. We explore properties of the two types of spaces and how they interact to assist wayfinding. We also show how they can be modelled: for network space, graphs can be used; for scene space we propose a novel model based on cognitive schemata and partial orders.

1 Introduction

Human wayfinding is a daily activity and its importance for public transport is widely acknowledged, not least in light of the continually growing demand for transport. There is separate research on both these topics, human wayfinding and public transport, yet the combination of the two has remained an open niche; a notable exception is the work by Fontaine and Denis (1999) about route descriptions in subway stations.

Wayfinding is navigation with a focus on its cognitive component (Darken et al. 1999). We all accomplish many wayfinding tasks every day. In the most general of terms, wayfinding is purposeful (Golledge 1999) interaction (Allen 1999; Darken et al. 1999) with an environment, the purpose being to reach a place goal. Wayfinding is successful if the place goal can be reached within spatial and temporal constraints and facing the—usually unavoidable (Arthur and Passini 1992)—uncertainty (Allen 1999).

The wayfinding literature focuses on the process of wayfinding more than on the environment in which wayfinding takes place.[1] But if we understand wayfinding as an interaction between the wayfinder and an environment (Allen 1999;

[1] Most exceptions to this general rule seem to come from people involved in architecture, such as Lynch, his disciple Appleyard, as well as Arthur and Passini.

C. Freksa et al. (Eds.): Spatial Cognition IV, LNAI 3343, pp. 24–41, 2005.
© Springer-Verlag Berlin Heidelberg 2005

Darken et al. 1999), then this means that we have to focus on this environment. A first level of focus is our restriction to wayfinding in public transport, thus limiting the environment to the public transport system. The second level is an attempt to cluster this complex environment into categories based on what they mean for spatial cognition, that is, how humans acquire, organise, update, and use spatial knowledge (Knauff et al. 2002) about public transport.

Wayfinding is frequently assumed to take place on networks and this has implications for models about wayfinding, where graphs (as a formal representation of networks) are prevalent (e.g., Leiser and Zilbershatz 1989, Timpf et al. 1992, Car 1996, Raubal 2001, Winter 2002, Timpf and Kuhn 2003). This can be justified by the great importance of routes to human wayfinding and human life in general; indeed, humans exhibit what Kuipers (1979) referred to as "sequential behaviour." But there are environments in which there is no obvious network structure: on campuses, there are often open spaces between buildings and some of these open spaces and buildings belong to the natural sciences and some belong to the technical sciences, etc. In such cases it is not at all clear what the relation between routes and the underlying environment is, and hence the adequecy of network models has to be carefully studied.

In this paper we explore the distinction between network-structured spaces and spaces in which there is no evident network of paths. These two types of spaces are examined within the context of a rich and complex environment for wayfinding, i.e., the public transport system. The remainder of the paper is structured as follows: section 2 gives an overview of the public transport system as an environment for wayfindng, section 3 presents a wayfinding scenario in a large station, which is the starting point for an examination of the types of spaces described in section 4. Section 5 addresses the issue of models of the public transport environment. Finally, conclusions are drawn and areas for further research are identified.

2 Public Transport: An Environment for Wayfinding

Public transport constitutes a *designed environment*, hence it can be looked at from at least these two perspectives (Timpf 2002):

the designer's perspective, and	⟶ how is it conceived?
the traveller's perspective.	⟶ how is it perceived?

The designer's perspective is the view of the architects and engineers involved in the creation of public transport interchanges[2] and stops[3] as well as the design of the network and the timetable. Apart from a quick discussion of some basic

[2] We use the term "interchange" for large nodes in the public transport system, where several lines come together and various transfer possibilities within the system exist, that is, we follow the meaning put forward by Alexander et al. (1977, pattern 34).

[3] The term "stop" is used for small nodes in the public transport system, typically serviced by only one line.

facts about the public transport system, we will mostly neglect this perspective throughout the paper. From the traveller's perspective, this environment is unalterable. It is up to the traveller to negotiate his way: He takes the role of a wayfinder who is given a task (reach some destination) in a given environment (the public transport system).

Transportation, in general, refers to the movement of people, goods, information, and energy. In the context of this paper, we will focus on the transportation of people using public transport. Amazingly, there does not seem to be a widely accepted definition of public transport. But public transport can be characterised by some specific properties (Brändli 1984): public transport is the production of the service "transport" for masses of people, not just individuals; this service is completely fixed in space and time by means of the timetable; there is always a chauffeur, thus eliminating the need to drive oneself; and trips involve more than one means of transport (including walking), that is, passengers have to change at designated interchanges.

Based on these properties *public transport* denotes the set of services for the transportation of *people* according to a predefined *schedule* (fixing place and time) and subject to *published conditions of use*, employing *multiple modes* of transport.

For our purposes, however, this definition is still too vague. Public transport as it exists today is mostly organised according to the *line operation* principle, and this form of organisation is expected to remain prevalent in the future (Maier and Atzkern 1992). Brändli (2001) defines line operation as the servicing of a fixed sequence of stops with predefined departure times. This service is publicly accessible subject to some transportation regulations. Furthermore, it is subject to four bindings, namely to

1. the network of roads, contact wires, or tracks;
2. the lines, which use the road, track, or contact wire network;
3. the stops; and
4. the timetable.

These bindings influence how the traveller interacts with the system.

From a traveller's point of view, a journey using public transport consists of several elements, as shown in fig. 1. The access, transfer, and egress nodes (together with interconnections, lines, tariffs, a schedule, and services) are part of the *public transport system* (or "the system" for short). Because of the binding to stops, people have to access (and leave) the system using some other mode of transport, like walking or cycling. Because of the binding to lines, transfers are inevitable for most trips. And because of the binding to a timetable, passengers are unlikely to access the system at arbitrary times;[4] rather, they have to plan in advance, look for services and connecting services, and try to optimise travel time, travel cost, route complexity (Timpf and Heye 2002), and other criteria (Golledge 1999).

[4] Brändli (1984, 2001) notes that the arrival of passengers at stops is independent of the schedule if this stop is being serviced in intervals of no more than 7 minutes.

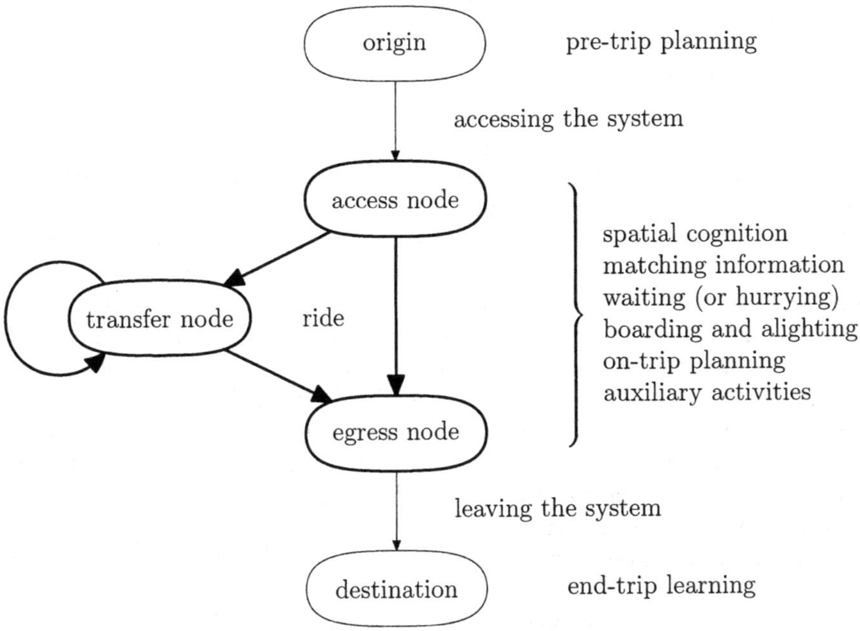

Fig. 1. Elements of a journey using public transport (after a figure in Brändli (2001)), with relevant actions and reasoning processes added

Journey elements can be assigned to the three contexts of the travel process (Infopolis 2 Consortium 1999):

- ▷ *pre-trip*: the planning context, in which people have to plan the trip while still at the origin
- ▷ *on-trip*: the tracking context, in which people verify that the journey runs according to plan and in which adjustments may be made; this happens during the trip and especially at the access, transfer, and egress nodes
- ▷ *end-trip*: after leaving the system, travellers assess the trip and have gained additional knowledge for future trips

These contexts constitute the logistical framework of any journey.

3 Wayfinding in Public Transport: A Scenario

The following scenario illustrates a traveller's wayfinding activity in an environment created by a public transport system, including his many and varying interactions with that system. The scenario is situated in Zürich but prototypical for transfers in European train stations.

Fred, a business man, has to go from Bern to Oerlikon, which is a part of the city of Zürich. This trip requires a transfer at Zürich main station. Fred knows this, because he used the Internet to look up the best connection:[5]

[5] `http://www.sbb.ch/`

Bern	dep 07:47	track 6		IC 911
Zürich main station	arr 08:56	track 13		
Zürich main station	dep 09:06	track 21/22		S5 18530
Zürich Oerlikon	arr 09:12	track 6		

His train arrives at 08:56 (on time) in Zürich main station. The second leg of his trip consists of a short ride using the "S5" line of the local rapid-transit railway, departing at 09:06. Unfortunately, Fred forgot to take note where this train departs. He does not mind, since such information is subject to change anyway. Fred starts looking for a departure board, which is easy to find: it is in the same eye-catching yellow color as everywhere in Switzerland (and probably elsewhere). He discovers that his train departs on track "21/22". He had expected a simple number like "7" or "22", not something that looks like a fraction. Fred walks along the platform towards the darker side, because that is where he expects the station building and more information to be. Moreover, this is the direction where most people are walking. Happily, this strategy works fine and Fred can indeed find a sign with "21/22" on it, pointing down an escalator. After some other twists and escalators, Fred has finally reached the connecting train, somewhere in the underground.

Allen (1999) proposed three wayfinding tasks: travel to familiar destinations, exploratory travel, and travel to novel destinations. Fred is facing the third task, travel to a novel destination. He tries to solve his wayfinding problem using a mixture of piloting between signs containing "21/22" and oriented search (if there is no such sign in sight). Fred's search is not merely sign-following, it is also guided by the architectural layout of the station: he walks along the platform, approaches and enters the building, and goes down some escalators.

4 Networks and Scenes

The scenario with Fred's transfer in the Zürich main station shows, that he successfully accomplished his wayfinding task using information about two aspects of the public transport system:

1. Information about the network and the timetable.
   - Internet query about the route, departure and arrival times, and service names (like "S5"), given an origin and a destination, as well as a desired time of arrival
   - Local system information: the departure board contains a locally relevant subset of the overall timetable
2. Information about the layout of the interchange.
   - Local system information: the departure board also links timetable information to information about the interchange node: given a departure

time and a place goal, Fred can find the precise location of the departing train
- Signage: the collection of signs for guiding travellers and indicating where things are
- Architectural layout: the combination of elements like platforms, halls, departure boards and screens, escalators, as well as their accidental properties (e.g., "the darker side")

This division into two aspects is not arbitrary, but linked to the public transport system and how it presents itself to the traveller. Information of the first type is based on a representation of the public transport network and the timetable. Fred obtains a mediated view on this information through a digital interface (more traditionally, he would probably have consulted a printed network map and associated timetable). Information of the second type is not mediated—Fred obtained it on–trip and in direct interaction with the various "scenes" of which the interchange is composed.

4.1 Definitions

We may consider these types of information representative of two types of *spaces*. In the literature, there are many such classifications of spaces, including Montello's (1993) classification based on the scale level, Zubin's classification (reported in Mark et al. 1989) based on the possible (and impossible) interactions with the space, and Freundschuh and Egenhofer's (1997) classification that is based on a combination of different criteria, namely manipulability, the need for locomotion to experience a space, and the size of the space.

If we try to find single words that are mnemonic for the two types of spaces characterised by the information Fred called in, then we may settle on "network" and "scene" and define these spaces as follows:

Network Space is the space created by network and schedule engineers, but also by historical, social, and economical processes. Network space is a mediated space, presenting itself by means of maps and schedules, but also by audible announcements and tardiness. It exhibits a network structure.

Scene Space is a place of information and integration, of coordination and compensation. Scene space is directly experienced but documented only implicitly and within itself. In particular, information about scene spaces is (usually) not available via the Internet. It exhibits a hierarchical structure.

With respect to the domain of public passenger transport, network space comprises the network of lines and the timetable according to which these lines are operated. Scene space comprises the stops and interchange nodes, including their relation to the transport lines and their embedding into the surrounding environment. The bindings of public transport (in line operation) and the resulting elements of a journey (fig. 1) make an exposure of the traveller to both spaces unavoidable.

The dichotomous classification of the public transport environment is deeper than the scenario might suggest. An analysis of the properties of the two types of spaces in the following sections shows that there is a profound and cognitively significant distinction in many respects.

4.2 Properties with Respect to Lynch's "Environmental Image"

Perception of information about either type of space creates what Lynch (1960) called an "environmental image," but the process of the creation and the resulting elements of this image differ for the two space types. Lynch writes (p. 8):

> "An environmental image may be analyzed into three parts: identity, structure, and meaning. (...) A workable image requires first the identification of an object, which implies its distinction from other things, its recognition as a separable entity. (...) Second, the image must include the spatial or pattern relation of the object to the observer and to other objects. Finally, this object must have some meaning for the observer, whether practical or emotional."

An "environmental image" can also result from a mediated and abstract presentation, like a network map; indeed, public transport networks can hardly be apprehended without an abstract presentation since they are too large, in the extreme they are spanning the entire globe (airlines). Whether or not both direct and mediated experience results in equivalent environmental images is not clear. An experiment conducted by Lloyd et al. (1996) supports that information is encoded into the cognitive map in a perspective-free manner; other researchers give evidence that information is encoded in a perspective-specific manner (e.g. Presson et al. 1989).

In either case, a working environmental image must contain identifiable *entities*. In network space, these are the nodes and the links between the nodes, but also lines and (departure) times. When travelling on a network, at each node the traveller has to decide which link to take next. Therefore, these nodes are also referred to as *decision points* in the literature (see, e.g., Klippel 2003). Identities in scene space are not so obvious. It is easy to enumerate some such entities (platforms, buildings, signs, etc.), but hardly possible to define them.

Structure is about how these entities relate to each other (and to the wayfinder). Again, for networks this is simple and obvious, because the network in itself is a well defined structure, relating nodes to other nodes. Because of the binding to lines and to the timetable, this structure is indeed more complicated and dynamic than the networks in private transport. In scene space we propose the term *scene* as a *local spatial configuration* of smaller entities contained within the scene together with some qualitative spatial relations among these entities. Scenes are similar to Kuipers' (1979) views in that they are a description of a part of an environment, but they are more general in that they are not confined to one specific direction: scenes are direction-free.

As to the *meaning* of the environmental image, we follow the assumption that structure captures a lot of what an image means. For network space, this is especially evident: The network structure expresses connectivity, and allows for decisions about route choice. In scene space, meaning has to be tied to scenes or parts of scenes. They express a local overview and can communicate *affordances* (Gibson 1979) like "here you may enter" or "there you may turn around." An analysis of affordances for wayfinding can be found in works by Raubal (Raubal and Worboys 1999, Raubal 2001).

4.3 Level of Scale and Type of Space

There are many classifications of space in the literature. A review can be found in Freundschuh and Egenhofer (1997), where they also present a novel classification based on more than one criterion.

Most of the classifications presented are by scale. Kuipers, for example, defines large-scale space to be space whose structure cannot be observed from a single viewpoint (Kuipers 1978, p. 129). A more compact variation of this definition is: "the spatial structure that is larger than the sensory horizon of the agent" (Kuipers 2002, p. 40). This is true of both network space and scene space—in the latter case with the exception of a few very small bus and tramway stops.

A more detailed classification of types of spaces was presented by Zubin (reported in Mark et al. 1989, pp. 13–17). Based on the perception of (abstract) spatial objects, he defines four space types, named A through D. Types C and D are relevant in this context:

Type C: scenes that can be perceived from a single vantage point, but only by scanning; examples: a room, a small valley, the horizon.

Type D: territories such as forests, cities, countries, or the inside of a house that cannot be perceived as a unit, but small portions are perceived as type A or type C objects.

Both network space and scene space in public transport are instances of type D space, for they are beyond direct experience. Scene space is composed of "scenes" and each scene classifies as a Zubin type C space. Typical scenes in a station environment are a platform, an underpass, a station hall, and a ticket office. Though they vary in size, they are all regions that cannot be apprehended at a single glance, but that can be *scanned* from a single vantage point, hence they are type C spaces. Zubin's type C spaces were our motivation for the name "scene space." Actually, we should call it "scenes space," because it generally consists of more than one scene, but the inconvenient pronunciation made us stick with "scene space."

In comparison to Montello's (1993) well-known psychological spaces, public transport networks are on a *geographical scale*, that is, they are so large that they can hardly be apprehended without the help of symbolic representations, network maps in our case. Scene space is smaller, but still is much larger than the human body and surrounds it. Individual scenes can be apprehended from a single place without significant locomotion, and hence qualify as *vista spaces*. An entire station, being composed of individual scenes, can no longer be apprehended without locomotion and thus is an instance of *environmental space*.

But other than network space, scene space on the level of single stations can be apprehended without symbolic representations, given enough time to explore.

The classification by Freundschuh and Egenhofer (1997) makes the distinction between directly experienced and mediated spaces even more explicit by introducing the category "map space" that includes all symbolic representations of (large and small) spaces. These representations are subject to cartographic generalisation or, more generally, abstraction processes. This is clearly true of network maps, that usually are purely topological, and hence very abstract, representations.

What none of the classifications reviewed addresses is the time-variability of network space as discussed in sections 4.4 and 4.5.

4.4 Properties Based on Activities

From the designer's perspective (Sect. 2), both network space and scene space are explicitly documented in terms of plans, sketches, models, etc. These documents are indispensable tools for the process of the creation of both networks and interchange nodes, a means of communication between builders, architects, engineers, i.e., experts.

From the traveller's perspective, however, availability of information about network space and scene space is rather unbalanced: documentation about network space is abundant: we have network maps and timetables, traditionally in printed from, more recently even on the Internet. Spatially relevant subsets are posted at stops and in interchange nodes, and temporally relevant subsets are even available by audible announcements. Information from these sources taken together describe the topology and the dynamics of a public transport service.

Information about the stops and interchange nodes, i.e., about scene space, is sparse. The only information normally available is a list of facilities, like catering and left luggage services, but such a list lacks spatial information. For example, there is usually no way to say where the luggage lockers are relative to some platform, without actually being (or having been) at the relevant station. Even though sometimes station maps are put up, they are not available for pre-trip planning. Therefore, we say that scene space is implicitly documented.

This disproportionate availability of information about network space and scene space has an important consequence for travellers: *pre-trip* planning is only possible for the movement within network space; for scene space, *on-trip* planning is a necessity (unless, of course, the relevant access, transfer, and egress nodes are already known from prior trips—but even then, if there is a problem in network space, like a disrupted service or a delay, scene space is where and on-trip is when these problems have to be compensated for). On-trip planning also includes considerations about auxiliary activities besides travelling, such as shopping while waiting for a connecting service.

The two space types influence human wayfinding means. In the mediated network space, wayfinding can be reduced to the process of selecting one or several connecting lines from the public transport network. This is mainly subject to the route choice behaviour theory elaborated in Bovy and Stern (1990). It is different from travelling with a private car, where a sequence of route segments is chosen, largely independent of the four bindings mentioned in Sect. 2.

In scene space wayfinding is best described by terms like searching, following, exploring, and matching. There cannot be a choosing of paths from a network, since the environment constituted by interchange nodes is usually devoid of an evident network of paths. Rather, there are large spaces like halls and platforms, together with underpasses and corridors. In Allen's (1999) terms, people have to use piloting (between landmarks; in the scenario these were mostly signs) and oriented search (if an expected landmark in the succession cannot be seen).

Finally, time plays a different role in the two types of spaces. In network space, time has an absolute meaning in that it defines when trains and busses depart or arrive on an absolute time scale. In scene space, however, time is measured on a relative scale. Fred's connecting train (cf. scenario in Sect. 2) departed at 09:06, which is a feature of the particular network space created by the Swiss railway system. On the other hand, when Fred started looking for the place of departure, he had scarcely ten minutes to find his (intermediate) goal.

In summary, there are many properties that can be used to describe network space and scene space as two cognitively different environments. Table 1 lists these properties and how they differ for the two spaces. The sequence of elements of a public transport journey (Fig. 1) has indicated that both spaces are necessarily involved in any trip, but there are other links between network space and scene space to be explored in the next section.

Table 1. Comparison of network space and scene space

Property	**Network Space**	**Scene Space**
Scale (Zubin)	type D	type D, composed of type C
Scale (Montello)	geographical	vista/environmental
Space type	map space	environmental space
Planning	pre-trip	on-trip
Process	selection	searching, following, exploring
Experience	mediated	direct
Time	absolute	relative
Structure	lines/nodes	scenes
Documentation	explicit	implicit

4.5 Interactions Between Network Space and Scene Space

Network space and scene space can be considered individually, but our application domain, public transport, causes them to be linked in some interesting ways.

1. Scene space ties the public transport lines. The lines of public transport are tied together at the interchange nodes of the system and thus become what is described as a network. Without this "glue," the lines would be much like spaghetti in a plate. The distinction between linear elements (the lines) and linking elements (the nodes) is already present in Lynch's (1960) classification of spatial elements in the environmental image: paths and edges come together at nodes and regions. In public transport there is a third component required to build a network: the timetable. If there is no connecting service, then the corresponding link in the network is essentially missing.

2. Scene space penetrates network space. The traveller in public transport is always surrounded by scenes, even while sitting in a train coach, a tramway, or a bus. When there is a network, then it is an abstraction of what the traveller experiences. This experience can be (and is) reinforced by presenting the network in readily prepared maps and diagrams, which in turn makes scene space more navigable.

3. Network space controls behaviour in scene space. People navigate in stations mainly to reach some train or other transport vehicle. Since these vehicles arrive and depart at specific places at specific times, network space directly controls important behavioural parameters of people in scene space. In case of temporary problems such as delayed or cancelled services, this control is particularly evident. Travellers may suddenly find themselves with much more (or much less) time than they had anticipated. In this way, network space with its absolute notion of time as defined by the timetable induces the relative role time plays in scene space.

Therefore, as far as public transport is concerned, network space and scene space are not two independent spaces, but two *aspects* (Berendt et al. 1998) of the same environment for wayfinding and spatial reasoning.

5 Modelling

It is tempting to use graphs to model environments for wayfinding. A graph is an abstract structure consisting of *nodes* and *edges* connecting some of the nodes. There are many different types of graphs and a discussion is beyond the scope of this paper.

We maintain that, whenever graphs are used to model spatial configurations, three questions should be asked and answered:

1. What are the nodes? What is their cognitive significance?
2. What are the edges? What is their cognitive significance?
3. What nodes are connected and why?

For network space, answering these questions is straightforward because both, nodes and of edges, are present in our everyday notion of public transport networks: nodes correspond to the interchanges and stops of the system, and edges correspond to the links between the stops and interchanges. Whenever we look at a network map of some public transport operator, we see nodes and edges.

Answering these questions for scene space is hard. We may try fitting a graph into a station, but it is not at all clear where to place the nodes, nor how many nodes there shall be placed. Does a platform correspond to one node or to several nodes? How about a station hall or a station square?

Since there is an established modelling formalism for network space, namely graphs, the remainder of this section focuses on modelling scene space.

5.1 Modelling Scene Space

It seems the primary problem with modelling scene space is that there is *no network*. This does not mean, that humans do not build networks in their minds while their experience with a particular scene space (a station for example) increases. Kuipers' (1977, 1978) TOUR model and its further development, the Spatial Semantic Hierarchy (Kuipers 2000) are interesting in the way they model the creation of a topological network from input that does not have a network structure at all: views and actions (turn and move) that result in new views.

We propose a model of scene space called *schematic geometry* (see also Rüetschi and Timpf 2004). This model is based on cognitive schemata and partial orders. The cognitive schemata build on Johnson's (1987) image schemata and thus provide a grounding in human cognition, while partial orders are in accordance with the common believe that human spatial knowledge is hierarchically structured (see, e.g., Hirtle 1995) and mirror the hierarchical wayfinding decision plans (Arthur and Passini 1992).

The cognitive schemata were defined with typical train stations in mind. They are (with references to Johnson's image schemata typeset in small caps):

ROOM: a CONTAINER which is "enterable," affords support (it is "stand-on-able"), and is bounded; examples include a pedestrian underpass or a big station hall

REGION: a (soft-) bounded area or SURFACE affording support, like a "shopping area" or a "station square"

COLLECTION: a set of things belonging spatially or functionally together like the whole station or a "platform area"

GATEWAY: a LINK affording "walk-through-ability" and which is consciously experienced when travelled through, like a door (similar to gateways in Chown et al. 1995)

ULINK (unconscious link): a LINK affording "walk-through-ability" in such an immediate and intuitive way that taking the link is unconscious to the wayfinder; this occurs, for instance, if a station hall and the platform area are not separated from each other; nevertheless the wayfinder can easily tell where he/she is

ITEM: a fixture, an object, a catchall for whatever might be relevant but none of the above, like a newspaper kiosk or a timetable

To approximate some actual or fictional station (i.e, scene space), we take instances of these cognitive schemata and arrange them in a partial order of membership, that is, we construct a part-of structure in the sense described by Alexander (1965). The result is what we call the schematic geometry; it represents the functional and spatial structure of the station it represents. To each element in the partial order we may attach arbitrary attributes such as names, surface properties, or lightning conditions.

The image schemata that underly this model provide a means to establish a spatial structure among the vague concept of scenes, because image schemata are rich in internal structure. This also means that not every partial order constitutes a valid schematic geometry: some consistency rules, derived from the image schemata, have to be obeyed (for details see Rüetschi and Timpf 2004).

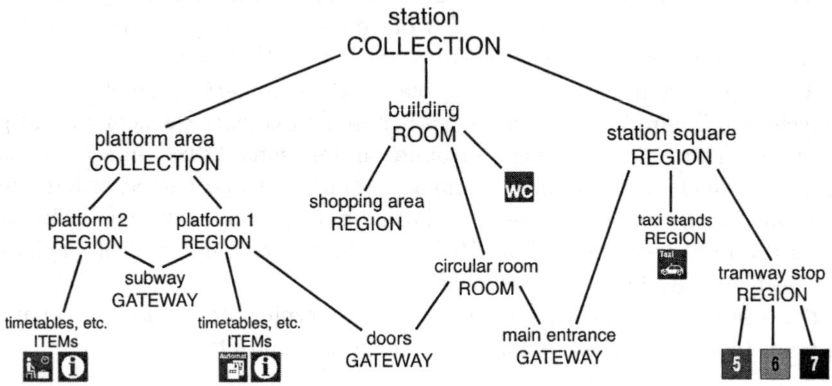

Fig. 2. Schematic geometry as a model of scene space, applied to a small station

Figure 2 shows the schematic geometry that has been created for a small station in Zürich. Labels in upper case denote the cognitive schema that is instantiated. The whole station is considered to be a COLLECTION, which consists of the platform area, the station building, and the station square. This allows the representation of multiple locations. For example, when someone is alighting from a tramway, the person finds him- or herself standing at the tramway stop and at the same time on the station square. We call this *inheritance of location*. The station square is linked with the station building by means of the main entrance (an instance of a GATEWAY). To the wayfinder this means that he or she can enter the building, but probably only after first finding the entrance by looking around.

The movement of the wayfinder in the schematic geometry traces what we call a *route*, which is a sequence of individual scenes and links between scenes. These links can be specific or abstract. Specific links are close to the bottom elements in a schematic geometry and instances of GATEWAYS (or ULINKS, but that does not occur in the example in fig. 2). Abstract links occur above the scenes they link and typically are instances of abstract ROOMS, REGIONS, or COLLECTIONS.

If a wayfinder has specific knowledge, then he or she can use a specific route, otherwise he or she has to resort to an abstract route, which will become specific at the time it is being followed. This sequence of scenes and gateways is consistent with the high importance that Chown (2000) attributed to gateways for the understanding of spaces. An interesting observation is that the intrinsically linear concept of routes arises from scene space, where there is, in general and in the example, no network of paths.

Besides inheritance of location, other cognitively relevant concepts can be defined in a schematic geometry. For example, a scene can be defined as the largest region (without crossing a room) or the smallest room in which the wayfinder may be. If the wayfinder is at the tramway stop (fig. 2), then the scene is the station square and what it contains, that is, the taxi stands, the main entrance, and the tramway stop. If, however, the wayfinder is inside the circular room, then the scene consists of this room and the two gateways.

As fig. 2 indicates, the schematic geometry takes the form of a graph and thus the same form as is traditionally used for networks. From a formal point of view, this coincidence is manifest, but when we consider the semantics of both formal structures, differences become apparent: the network graph expresses connectivity, the schematic geometry expresses membership among scenes and their elements. The membership relation naturally leads to partial orders, and because every partial order is equivalent to a graph (but not vice versa), the schematic geometry has the structure of a graph.

Other models of space that do not assume a network structure include Kuipers' (2000) Spatial Semantic Hierarchy and the PLAN model (Chown et al. 1995), as well as some models based on partial orders, for example the one used by Schlieder and Werner (2003) for their investigation of intentional behaviour and the Pattern Language by Alexander et al. (1977), which is closest to our model, but far more informal.

6 Conclusions and Outlook

In public transport, there are network-structured and scene-structured spaces. They are defined and distinguished from each other by a set of specific properties, including differences in the level of scale, the wayfinding processes, the possibility to plan ahead, the role of time, and—most important—the intrinsic spatial structure. Network space has a network structure and scene space is built around the deliberately vague concept of scenes as local spatial configurations.

These properties influence any modelling attempt in that they determine what is essential and what is accidential in models of wayfinding in public transport. For network space, graphs are an appropriate formalism, and graph-based models of networks are numerous, within and beyond the spatial cognition community. Models for scene space, however, must not assume a network when there is none, instead, the concept of scenes and their hierarchical arrangement can be exploited. We did so using cognitive schemata and partial orders in a model called schematic geometry.

The main contribution of this paper then is to provide a classification of the environment constituted by the public transport system into two *aspects*, each with a set of properties that can be assumed to hold. A second contribution is to illustrate how image schemata can be used to provide structure where at first there seems to be no obvious structural principle at all, such as in stations.

An interesting direction for future work is to check whether the concept of network space and scene space generalises to domains other than public transport, probably wayfinding in general. For example, when driving a car, this takes place on the street network, hence in a space that has a network structure. At the same time, however, the car driver is in a scene-structured space, for he can perceive only such small parts of the street networks, that they hardly constitute a network: road segments, lanes, junctions, signs, besides many on- and off-road landmarks that are not part of the network proper but of great help to the driver. To test if such a generalisation is feasible, we have to at least check the various properties of the two space types compiled in table 1. Time (absolute/relative) and documentation (explicit/implicit) are probably hard to maintain, that is, they characterise the two space types in the domain of public transport but not in general.

Another direction for future work is the schematic geometry. As it currently stands, it can only represent the coarse spatial structure of scene space. What is missing in particular are concepts to represent the dynamics of wayfinding, that is, the matching of information from the immediate surroundings with previous knowledge of the environment (Arthur and Passini 1992). We intend to define extensions to the model using the concept of homomorphisms, or structure-preserving maps, between schematic geometries. Another deficiency of the schematic geometry is, that it cannot epxress spatial relations within scenes: for example, figure 2 shows that both the "circular room" and the "shopping area" are inside the building, but gives no information about their relative position *within* the building. Possible applications are in architectural planning, where the schematic geometry approach could be used as a tool to reveal and analyse structural properties of future and existing buildings, and in traffic simulations, where there is a tendency to do micro simulations (looking at the individuals) but changing means of transport is still subject to great simplifications.

Finally, and with respect to the original problem of wayfinding in public transport, we should note that any model of scene space must depend on network space and vice versa (cf. sec. 4.5). According to fig. 1, the wayfinder in public transport is always exposed to both network space and scene space. In scene space, the primary wayfinding goal is to reach places of departure in time, such as a specific platform. This means, that all movement along routes in scene space is governed by spatial and temporal constraints established by network space and its schedule. Our schematic geometry is not yet prepared to handle these interactions between the two space types and neither is any other model we are aware of.

Acknowledgements

The authors are grateful for the thorough comments by an anonymous reviewer.

References

Alexander, Ch. (1965): A city is not a tree. *Architectural Forum* **122**(1), pp. 58–62 and **122**(2), pp. 58–62.

Alexander, Ch., Ishikawa, S. and Silverstein, M. (1977): *A Pattern Language: towns, buildings, construction.* New York: Oxford University Press.

Allen, G. (1999): Spatial Abilities, Cognitive Maps, and Wayfinding. Pages 46–80 in *Wayfinding behavior: cognitive mapping and other spatial processes*, edited by R. G. Golledge. Baltimore, Maryland: John Hopkins University Press.

Arthur, P. and Passini, R. (1992): *Wayfinding. People, Signs, and Architecture.* New York: McGraw-Hill.

Berendt, B., Barkowsky, Th., Freksa, Ch., Kelter, S. (1998): Spatial Representation with Aspect Maps. Pages 313–336 in *Spatial Cognition*, edited by Ch. Freksa, Ch. Habel, and K. F. Wender. Berlin: Springer-Verlag, Lecture Notes in Artificial Intelligence 1404.

Bovy, P. H. L. and Stern, E. (1990): *Route Choice: Wayfinding in Transport Networks.* Dordrecht, The Netherlands: Kluwer Academic Publishers.

Brändli, H. (1984): *Angebote des öffentlichen Verkehrs.* Skriptum, Institut für Verkehrsplanung und Transporttechnik, Swiss Federal Institute of Technology, Zürich.

Brändli, H. (2001): *Konfliktstellen Individualverkehr ⟷ Öffentlicher Linienverkehr*, draft technical report, Institut für Verkehrsplanung und Transporttechnik, Swiss Federal Institute of Technology, Zürich.

Car, A. (1996): *Hierarchical Spatial Reasoning. Theoretical Consideration and its Application to Modeling Wayfinding.* Ph. D. thesis, Technical University Vienna.

Chown, E. (2000): Gateways: An approach to parsing spatial domains. ICML 2000 Workshop on Machine Learning of Spatial Knowledge.

Chown, E., Kaplan, S., and Kortenkamp, D. (1995): Prototypes, Location, and Associative Networks (PLAN): Towards a Unified Theory of Cognitive Mapping. *Cognitive Science* **19**, pp. 1–51.

Darken, R. P., Allard, T., and Achille, L. B. (1999): Spatial Orientation and Wayfinding in Large-Scale Virtual Spaces II. *Presence* **8**(6), pp. iii–vi.

Fontaine, S., Denis, M. (1999): The Production of Route Instructions in Undderground and Urban Environments. Pages 83–94 in *Spatial Information Theory: Cognitive and Computational Foundations of Geographic Information Science. Proceedings COSIT'99*, edited by Ch. Freksa and D. M. Mark. Berlin: Springer-Verlag, Lecture Notes in Computer Science 1661.

Freundschuh, S. M. and Sharma, M. (1996): Spatial Image Schemata, Locative Terms, and Geographic Spaces in Children's Narrative: Fostering Spatial Skills in Children. *Cartographica* **32**(2), pp. 38–49.

Gibson, J. J. (1979): *The Ecological Approach to Visual Perception.* London: Lawrence Erlbaum Associates.

Golledge, R. G. (1999): Human Wayfinding and Cognitive Maps. Pages 5–45 in *Wayfinding behavior: cognitive mapping and other spatial processes*, edited by R. G. Golledge. Baltimore, Maryland: John Hopkins University Press.

Hirtle, St. (1995): Representational Structures for Cognitive Space: Trees, Ordered Trees and Semi-Lattices. Pages 327–340 in *Spatial Information Theory: A Theoretical Basis for GIS. Proceedings COSIT'95*, edited by A. U. Frank and W. Kuhn. Berlin: Springer-Verlag, Lecture Notes in Computer Science 988.

Infopolis 2 Consortium (1999): *Needs of travellers: An Analysis Based on the Study of Their Tasks and Activities.* Brussels: Commission of the European Communities, DG XIII.

Johnson, M. (1987): *The Body in the Mind: The Bodily Basis of Meaning, Imagination, and Reason.* University of Chicago Press.

Klippel, A. (2003): *Wayfinding Choremes: Conceptualizing Wayfinding and Route Direction Elements.* Ph.D. thesis, University of Bremen.

Knauff, M., Schlieder, Ch., Freksa, Ch. (2002): Spatial Cognition: From Rat-Research to Multifunctional Spatial Assistance Systems. *Künstliche Intelligenz*, no. 4, pp. 5–9.

Kuipers, B. J. (1977): *Representing Knowledge of Large-Scale Space.* Ph.D. thesis, M.I.T. Artificial Intelligence Laboratory.

Kuipers, B. J. (1978): Modeling Spatial Knowledge. *Cognitive Science* **2**, pp. 129–153.

Kuipers, B. J. (1979): On Representing Commonsense Knowledge. Pages 393–408 in *Associative Networks: The Representation and Use of Knowledge by Computers*, edited by N. V. Findler, New York: Academic Press.

Kuipers, B. J. (2000): The Spatial Semantic Hierarchy. *Artificial Intelligence* **119**, pp. 191–233.

Kuipers, B. J. (2002): Interview in *Künstliche Intelligenz* **2002**(4), p. 40

Leiser, D. and Zilbershatz, A. (1989): The traveller: A computational model of spatial network learning. *Environment and Behavior* **21**(4), pp. 435–463.

Lloyd, R., Cammack, R., and Holliday, W. (1996): Learning Environments and Switching Perspectives. *Cartographica* **32**(2), pp. 5–17.

Lynch, K. (1960): *The Image of the City.* Cambridge, Massachusetts: M.I.T. Press.

Maier, J. and Atzkern, H.-D. (1992): *Verkehrsgeographie.* Stuttgart: Teubner.

Mark, D. M., Frank, A. U., Egenhofer, M. J., Freundschuh, S. M., McGranaghan, M., and White, R. M. (1989): Languages of Spatial Relations: Initative 2 Specialist Meeting Report. NCGIA Technical Paper 89-2.

Montello, D. R. (1993): Scale and Multiple Psychologies of Space. Pages 312–321 in *Spatial Information Theory: A theoretical basis for GIS*, edited by A. U. Frank and I. Campari. Berlin: Springer-Verlag, Lecture Notes in Computer Science 716.

Presson, C. C., DeLange, N., and Hazelrigg, M. (1989): Orientation specificity in spatial memory: What makes a path different from a map of the path? Pages 887-897 in *Journal of Experimental Psychology: Learning, Memory, and Cognition.*

Raubal, M. (2001): *Agent-based simulation of human wayfinding: A perceptual model for unfamiliar buildings.* Ph.D. thesis, Technical University Vienna.

Raubal, M. and Worboys, M. (1999): A Formal Model of the Process of Wayfinding in Built Environments. Pages 381–399 in *Spatial Information Theory. Proceedings COSIT'99*, edited by Ch. Freksa and D. M. Mark. Berlin: Springer-Verlag, Lecture Notes in Computer Science 1661.

Rüetschi, U.-J. and Timpf, S. (2004): Schematic Geometry of Public Transport Spaces for Wayfinding. Pages 191–203 in *Geoinformation und Mobilität.* Tagungsband der Münsteraner GI-Tage, edited by M. Raubal, A. Sliwinski, and W. Kuhn.

Schlieder, Ch. and Werner, A. (2003): Interpretation of Intentional Behavior in Spatial Partonomies. Pages 401–414 in *Spatial Cognition III*, edites by Ch. Freksa, W. Brauer, Ch. Habel, and K. F. Wender. Berlin: Springer-Verlag, Lecture Notes in Artificial Intelligence 2685.

Timpf, S., Volta, G. S., Pollock, D. W., and Egenhofer, M. J. (1992): A Conceptual Model of Wayfinding Using Multiple Levels of Abstractions. Pages 348–367 in *Theories and Methods of Spatio-Temporal Reasoning in Geographic Space*, edited by A. U. Frank, I. Campari, and U. Formentini. Heidelberg-Berlin: Springer Verlag, LNCS 639.

Timpf, S. (2002): Ontologies of Wayfinding: a traveler's perspective. *Networks and Economics* **2**(1), pp. 9–22.

Timpf, S. and Heye, C. (2002): Complexity of Routes in multi-modal wayfinding (extended abstract). Presented at *GIScience 2002*, Boulder, CO, USA.

Timpf, S. and Kuhn, W. (2003): Granularity Transformations in Wayfinding. Pages 77–88 in *Spatial Cognition III*, edited by Ch. Freksa, W. Brauer, Ch. Habel, and K. F. Wender. Berlin: Springer-Verlag, Lecture Notes in Artificial Intelligence 2685.

Winter, St. (2002): Route Specifications with a Linear Dual Graph. Pages 329–338 in *Advances in Spatial Data Handling*, edited by D. Richardson and P. Oosterom. Berlin: Springer-Verlag.

Isovists as a Means to Predict Spatial Experience and Behavior

Jan M. Wiener[1,2] and Gerald Franz[2]

[1] Cognitive Neuroscience, University of Tübingen,
Auf der Morgenstelle 28, 72076 Tübingen, Germany
jan.wiener@tuebingen.mpg.de
http://www.kyb.mpg.de/~malte

[2] Max Planck Institute for Biological Cybernetics,
Spemannstr. 38, 72076 Tübingen, Germany
gerald.franz@tuebingen.mpg.de
http://www.kyb.mpg.de/~gf

Abstract. Two experiments are presented studying interrelations between spatial properties of environments and both experience and spatial behavior. In order to systematically study such interrelations, a generic description of space is required that provides comparability between arbitrarily shaped environments and captures behaviorally relevant properties of space. In this study the suitability of isovist derived measurands for this purpose was explored. Isovist-based descriptions of 16 virtual indoor scenes were correlated with behavioral data from two experimental tasks. For both tasks, an active navigation task and a rating of experiential qualities, strong correlations between subjects' behavior and measurands derived from isovist analysis were found. The general outcomes suggest that isovist measurands are indeed a promising means to predict the experience of space and spatial behavior for the chosen experimental tasks.

1 Introduction

Spatial properties of architecture influence subjective experience as well as spatial behavior. Several theories, mainly from environmental psychology, explain human behavior and experience by their interdependency with the environment. For example, evolutionary based theories of environmental preferences such as "prospect and refuge" [1] or "defensible space" [17] suggest that certain spatial settings were advantageous for the survival of a species and therefore corresponding preferences enhanced its fitness. Also the influence of selected features of space on human navigation behavior has been tested in several studies. For example, O'Neill [18] demonstrated that wayfinding performance decreased with increasing plan complexity. Wener and Long [23] have shown that the misalignment of local reference systems impaired users' ability to integrate spatial information across multiple places. Janzen, Herrmann, Katz, and Schweitzer [12] investigated the influence of oblique angled intersections within an environment on wayfinding performance. When navigating arrow-fork intersections, subjects' error rate depended on which branch they entered the intersection (see also [13]). Wiener and Mallot [24] have revealed an influence of environmental regions on human navigation and route planning behavior.

C. Freksa et al. (Eds.): Spatial Cognition IV, LNAI 3343, pp. 42–57, 2005.
© Springer-Verlag Berlin Heidelberg 2005

While the truth of the initial statement is therefore beyond any doubt, few theories and empirical studies have aimed at analyzing the corresponding interrelations comprehensively, but have rather made use of qualitative descriptions of certain selected spatial situations. Therefore they are often difficult to compare and do not provide a basis for systematic spatial analysis. In order to study the relations between physical properties of spaces and both spatial experience and behavior systematically, generic formal descriptions of space are required that provide comparability between arbitrarily shaped environments and capture biologically and psychologically relevant properties of the environment.

In the following section, several description sytems are briefly reviewed. Afterwards, two experiments are presented that test an isovist based description system for its ability to capture behaviorally relevant properties of space.

2 Background

Several disciplines already offer description systems and models for aspects of spatial environments. For example, in architectural construction, buildings are specified by a combination of lists of constructive elements (walls, windows, columns, etc.) and scale plans. While this description of the *array of architectural elements* is quite elaborated and standardized, the formal structure is graphically and thus not quantitatively represented, and therefore cannot be directly compared. In architectural theory *compositional approaches* [15, 6, 16] developed more or less formal languages based on basic geometric primitives. By combining and/or transforming these primitives, more complex forms and structures can be generated. While these approaches may allow to retrace steps of the genesis of form from the top-down perspective of the designer, they are not ideal for an analytical description of the mere shape, since a decomposition of given forms is often difficult and ambiguous.

Another approach to describe environments rises from phenomenology. In everyday language, non-trivial forms are often compared using intermediate concepts such as complexity and regularity. In empirical aesthetics these properties are termed *collative variables* that have been defined as introspective assessment criteria of the structural properties of a stimulus array [3, 4, 25]. However, while collative properties offer a basis for comparing a wide range of objects and environments introspectively, they cannot be directly derived from the stimuli and therefore lack the objectivity of "real physical" properties.

In response to the reported shortcomings, the technique of *space syntax* has been developed [10, 8, 9]. Space syntax is a set of technologies for the analysis of spatial configurations using simple graphs solely consisting of paths and nodes. This analytical reduction of space to mere structural mathematical information facilitates a calculation of characteristic values such as connectivity, centrality, control level that can be directly compared. One aim of space syntax has always been the identification of variables that determine the social meaning and behavioral relevance of spaces. Original space syntax has been developed to analyze large-scale spatial configurations from the room layout of building complexes to whole cities. Hence, spatial properties of environments smaller than rooms were not adequately represented. Additionally, the initial analytical operations depended on an often ambiguous and therefore somehow arbitrary manual decomposition of space into convex subspaces and axes.

For analyzing spatial characteristics of smaller environments, Benedict [2] has proposed *isovists* as objectively determinable basic elements. Isovists are viewshed polygons that capture spatial properties by describing the visible area from a given observation point. In order to better describe the spatial characteristics of an environment as a whole, Turner, Doxa, O'Sullivan, and Penn [21] have proposed the technique of *visibility graph analysis,* that integratively considers multiple positions within an environment. This technique offers further second-order measurands like for example on visual stability that may be relevant for locomotion and navigation. A more detailed description of isovist analysis and visibility graph techniques as considered in this study is given in Section 4.3.

Originally derived from abstract spatial analysis, the relevance of isovists and visibility graphs was not initially backed by psychophysical empirical findings. However, isovists describe spatial properties from an inside beholder-centered perspective, and there is first empirical evidence that they capture environmental properties of space that are relevant for spatial behavior and experience. For example, case studies on spatial behavior in the Tate Gallery [11, 22] have revealed high correlations between visibility graph measurands and the statistical dispersal of visitors. Furthermore, in a recent study Franz, von der Heyde,and Bülthoff [7] compared experiential qualities of arbitrarily shaped architectural spaces with isovist measurands. They found that already a few isovist derivatives describing visual characteristics of the observation points were widely sufficient to explain the variance in the affective appraisals of the environments. Nevertheless, to the authors' knowledge, elementary studies, for example on the perceptibility of isovists, that shed some light on their biological foundations are still missing.

3 Objective

The overall aim of this exploratory study was to investigate interrelations between spatial properties of indoor environments and both spatial experience and behavior. In particular, likely predictor variables for experience and behavior in space should be identified. As stated above, isovist analysis provides a generic description of the form of architectural spaces from an inside beholder-centered perspective and may therefore offer suitable measurands that capture behaviorally relevant aspects of space. This hypothesis was tested by two experimental tasks in a set of 16 virtual indoor scenes: in an active navigation task subjects were asked to navigate to positions that either maximized or minimized the visible area. A subsequent semantic differential rating quantified the experiential qualities of the scenes. The analysis then tested for interrelations between characteristic values derived from the isovists and the behavioral data.

4 General Material and Methods

4.1 Experimental Setup

The empirical study was conducted using a virtual reality experimental setup. Virtual reality simulations combine flexibility, controlled laboratory conditions, and a good degree of perceptual realism [5] and therefore allow the systematic variation of spatial properties of the experimental environments. The experiments were conducted at the virtual reality facilities of the Max Planck Institute for Biological Cybernetics, Tübingen. The virtual environment models used in both experiments were created using AutoCAD and 3ds max (discreet). A detailed description of the virtual environments is given below. The visual scenery was rendered in realtime on standard PCs (1.0 GigaHz Pentium Pro, nVidia GForce 4 graphics card), running a C++ simulation software that was designed and programmed especially for psychophysical virtual reality experiments[1]. Subjects experienced the virtual environments in the egocentric perspective. The visual scenery was displayed with a simulated field of view of 90°x73°. Subjects were seated in front of a 21" standard CRT screen at a distance of approximately 50 cm; they interacted with the simulation using a joypad (Logitech Wing Man Rumble Pad).

4.2 The Virtual Environments

The study was based on a set of sixteen virtual indoor scenes that was derived from stimuli used by Franz et al. [7]. The scenes represented spatial situations within a fictive art gallery, they were designed by varying the number of alcoves and connections to adjacent spaces of simple rectangular rooms. The maximally visible floor area was kept roughly constant. The floor plans of these indoor scenes are displayed in Figure 1. The walls of the indoor scenes were draped with unobtrusive paintings (46 portraits of Picasso's blue and pink period) to strengthen the art gallery character. Other surface properties as well as the lighting and illumination level were constant over all scenes. Figure 2 displays example screenshots of subjects' perspective during the experiments. Note that in contrast to Franz et al. [7], a different lighting model (ambient occlusion derived smooth diffuse illumination) was used in order to make the stimuli realtime-capable.

4.3 Formal Description of the Environments

As already stated in the introduction, a generic formal description of the virtual indoor scenes was required in order to relate subjects' spatial experience and behavior in both of the experiments to the form and structure of the corresponding spaces. For this purpose, isovist and visibility graph analysis appeared to offer a promising level of abstraction, since they translate perceptual and spatial properties of architectural space into simple polygons (see Figure 3). From the isovist polygons several quantitative descriptors can be derived that reflect physical properties of the corresponding space such as area, perimeter length, number of vertices, length of open or closed edges, etc. These basic measurands can be combined to generate further integrated values.

Isovists basically describe local physical properties of spaces with respect to certain standpoints. In order to overcome this limitation, Turner et al. [21] developed the technique of visibility graph analysis that integratively considers regional or global properties of a whole environment by computing the intervisibility of positions regularly distributed

[1] For details please refer to the *Virtual Environments Library* homepage: `http://www.kyb.mpg.de/prjs/facilities/velib`.

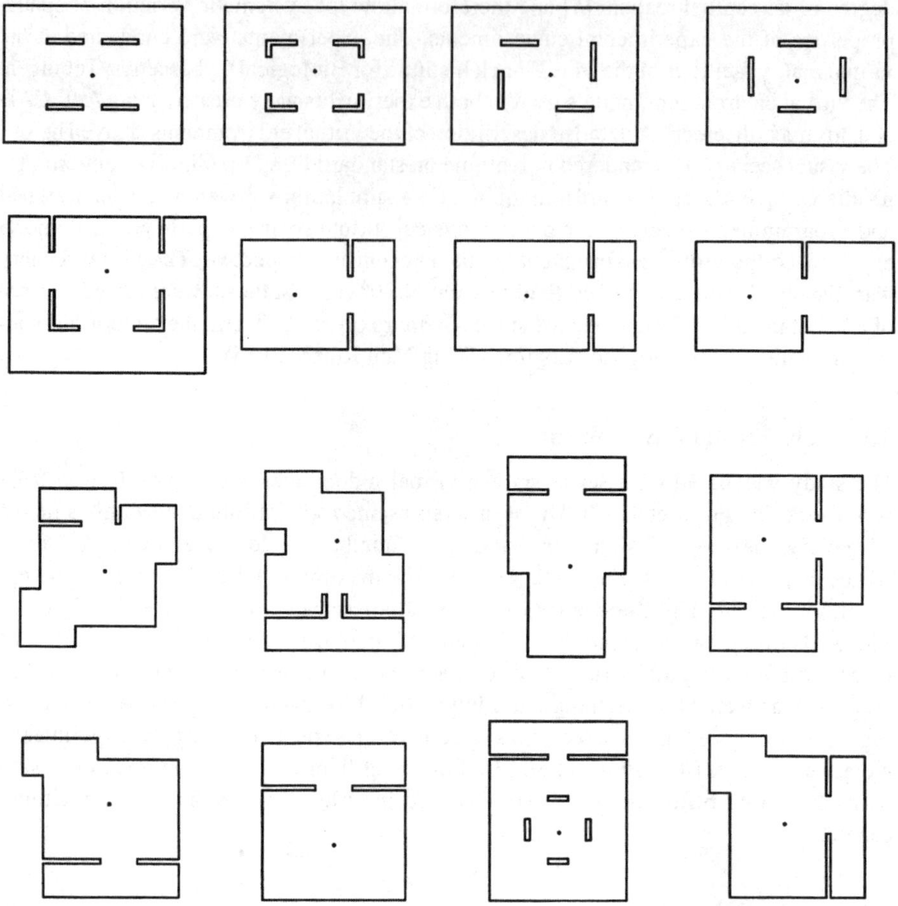

Fig. 1. Floor plans of the 16 virtual indoor scenes used in the experiments. The dot in the center of each room marks the starting position for the experimental tasks

Fig. 2. Three screenshots of the virtual indoor scenes as experienced by the subjects

over the whole environment. In order to make the computation more efficient, visibility graphs are not directly based on isovist polygons, but approximate their characteristics by a connectivity graph describing the intervisibility of multiple observation points. Typical visibility graph measurands are for instance neighborhood size (i.e. the number of directly connected graph vertices, corresponding to isovist area) and the clustering coefficient (i.e. the relative intervisibility within a neighborhood). Additionally, Psarra and Grajewski [19] have proposed further measurands that concentrate on the visibility graph boundaries (cf. openness, the relation between open and closed isovist boundaries, see below), because the boundaries are assumed to be the visually most important region of a viewshed.

For the analysis in this study, the techniques of isovist and visibility graph analysis were combined: the sixteen virtual indoor scenes were analyzed by calculating isovist measurands and visibility graphs on a 50 cm grid covering each environment. Since the correlation analysis required global characteristic values for each scene and measurand, the resulting values were averaged over each environment. A list of the isovist and visibility graph measurands that were calculated for the 16 indoor environments is given below. For the analysis a special isovist analysis tool was used, the tool is free software and available at `http://www.kyb.mpg.de/~gf/anavis`.

Isovist Derived Measurands Used in This Study. The following measurands were used to describe spatial characteristics of the simulated environments. They represented the best predictor variables according to the results of Franz et al. [7]. For a more detailed description of the measurands' mathematical and analytical background, refer to Turner et al. [21].

Neighborhood Size. The number of directly connected nodes of a visibility graph node, corresponding to the area of the isovist polygon.

Number of Vertices. The number of vertices making up the outline of an isovist polygon.

Openness. The ratio of open and closed edges of the isovist. Closed edges are defined by visible walls, open edges are generated by occlusions.

Jaggedness. The jaggedness of an isovist as an integrative measurand that is calculated mathematically as the squared isovist perimeter divided by the isovist area. It describes the convexity of an isovist polygon.

Clustering Coefficient. The clustering coefficient is a visibility graph measurand describing the relative intervisibility within an neighborhood. The clustering coefficient is calculated approximatively by dividing the sum of graph edges within a neighborhood by the squared neighborhood size.

Revelation Coefficient. The revelation coefficient describes the relative difference between a neighborhood and its adjacent neighborhoods. A low revelation coefficient indicates an area of high visual stability.

4.4 Statistical Analysis

All data were analyzed using the open source software mathematics packages 'Octave' (`http://www.octave.org`) and 'R' (`http://www.r-project.org`). For all

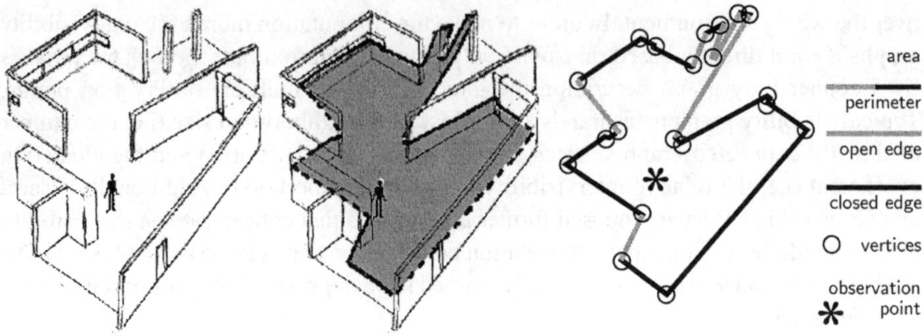

Fig. 3. Generating isovists: Left: a hypothetical indoor environment; middle: the gray area is visible from the person's observation point within the environment; right: the resulting isovist and its basic measurands

statistical analyses, the rating data was treated as even interval scaled. Correlation coefficients were calculated using linear Pearson's product moment correlation.

5 Experiment 1

5.1 Objective

In accordance with the overall objective of investigating interrelations between spatial properties and spatial behavior, the purpose of the experiment was twofold: First to test whether basic isovist properties can be perceived at all, and second, to explore correlations between global isovist measurands (see Section 4.3) and behavioral data. The behavioral data were gained both from a navigation task and a rating of experiential qualities in different virtual environments. It was hypothesized that the differently shaped environments used in this experiment systematically influenced subjects' behavior in both tasks. If the isovist measurands captured behaviorally relevant properties, significant correlations with the behavioral data were expected.

5.2 Method

Experimental Procedure. In each of the 16 indoor scenes (see Section 4.2 and Figure 1), subjects had to do a navigation task and a semantic differential rating task. Only after completing both experimental tasks, they proceeded to the next indoor scene. The order in which the 16 indoor scenes were presented was randomized for each subject. A complete experimental session had a duration of about 40 minutes.

The first experimental task was an **active navigation task**. At the beginning of this task, subjects were placed at the fixed starting position of the corresponding indoor scene (see Figure 1) facing a random direction. Subjects were then asked to navigate to the position within the scene that maximized the visible area (corresponding to maximal isovist area) as well as to the position within the scene that minimized the visible area (corresponding to minimal isovist area). Before the experiment, subjects were carefully instructed that their task was not to maximize or minimize the visible area with respect to their current heading direction, but the area revealed by a complete 360° rotation. During the experiment, the position that maximized the isovist area was referred to with the catchphrase *best overview place* and to the position that minimized the isovist area was referred to with the catchphrase *best hiding place*. The order in which subjects had to locate these two positions was randomized for each room. Subjects were instructed to solve the task quickly and as accurate as possible and to confirm a chosen position by pressing a button on the joypad. For each of the navigation tasks, subjects' final positions were recorded.

Table 1. English translations and original terms of the rating categories used in the semantic differential. The experiments were conducted in German language

Category	English low extreme	English high extreme	German low extreme	German high extreme
interestingness	boring	interesting	langweilig	interessant
pleasingness	unpleasant	pleasant	unangenehm	angenehm
beauty	ugly	beautiful	hässlich	schön
spaciousness	narrow	spacious	eng	weit
complexity	simple	complex	einfach	komplex
clarity	unclear	clear	unübersichtlich	übersichtlich

The second experimental task was a **rating of the experiential qualities** of the 16 scenes. At the beginning of each trial, subjects were automatically moved back to the initial starting position (roughly the center of the room: see Figure 1), again facing a random direction. After pressing a button on the joypad, subjects were confronted with the six ratings in a random sequence. The ratings were performed by manipulating an analog slider on the input device. In order to provide visual feedback, the scale and the currently selected value were displayed near the lower border of the screen. During the rating task, subjects were allowed to freely move through the environments.

Variables of Interest. During the **navigation task**, subjects were asked to move to the position that maximized the isovist area (*best overview place*) and to the position that minimized the isovist area (*best hiding place*). For each indoor scene, subjects' performance was evaluated by comparing the isovist area of the chosen positions with the isovist areas of the positions with the actual highest and lowest values.

The virtual indoor scenes differed with respect to the size of the isovists at the positions with the largest and smallest isovist area. In order to compare performance between different environments, subjects' navigation data were normalized according to the range of isovist sizes occurring in the particular scene (see Formula 1 and Formula 2). This performance measure ranges from 0 to 1. If subjects showed perfect behavior with respect to finding the positions that maximized and minimized the isovist area, performance was 1.

$$P_{max(r)} = \frac{Iso_{sub(r)} - Iso_{min(r)}}{Iso_{max(r)} - Iso_{min(r)}} \tag{1}$$

$$P_{min(r)} = 1 - \frac{Iso_{sub(r)} - Iso_{min(r)}}{Iso_{max(r)} - Iso_{min(r)}} \tag{2}$$

with:
r = idendity of virtual indoor scene
$P_{max(r)}$= performance for finding the position with the highest control value for room r
$P_{min(r)}$= performance for finding the position with the lowest control value for room r
$Iso_{sub(r)}$= size of isovist corresponding to subject's chosen position
$Iso_{min(r)}$= size of isovist corresponding to position with lowest control value for room r

The **rating task** was performed using the semantic differential scaling technique. Six dimensions of environmental experience were represented by pairs of oppositional adjectives (cf. Table 1). Subjects could differentiate their appraisals using a seven step Likert-like scale. The rating categories were selected to represent major dimensions of affective experience (pleasure, beauty, and interestingness), as well as denotative and collative properties that were expected to be potentially relevant for the navigation task (experienced spaciousness, clarity, and complexity). For the correlation analysis, the rating results of each scene were averaged by category over all subjects.

Participants. 16 subjects (8 female, 8 male) voluntarily participated in the experiment, they were paid 8 Euro per hour. Subjects were mostly university students at an age of 20-25 years.

5.3 Results

Navigation Performance. Overall, subjects showed comparable good performance (P) in finding the position with the smallest isovist area and the position with the largest isovist area (smallest isovist: P=.92 ± .02; largest isovist P=.90 ± .02, t-test: t=.96, df = 29.97, p=.3). In some of the virtual indoor scenes subjects reached performance measures over .97, in indoor scene 10 subjects actually reached 1 for finding the *best hiding place*, which means that all subjects found the position that minimized the visible area.

While performance of female and male subjects did not differ with respect to finding the *best overview place* (female: P=.88 ± .02, male: P=.91 ± .02, t-test: t=1.00, df=29.52, p=.3), male subjects showed better performance in finding the *best hiding place* as compared to female subjects (female: P=.88 ± .03, male: P=.96 ± .02, t-test: t=2.44, df=25.21, p=.02).

Figure 4 displays subjects' performance of finding the *best hiding place* and subjects' performance of finding the *best overview place* for each of the 16 indoor scenes separately.

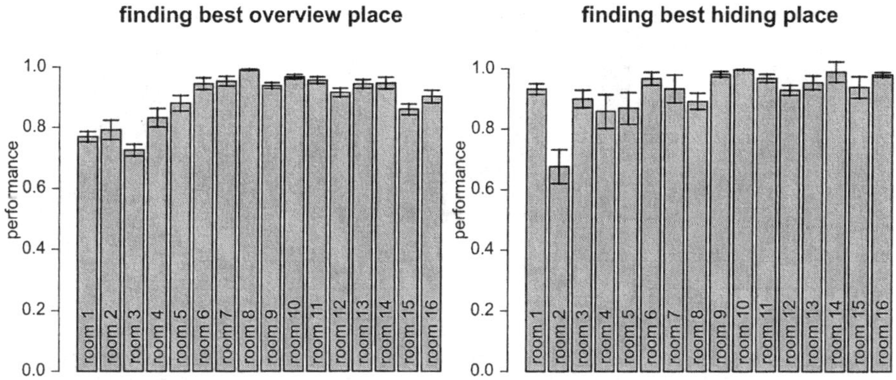

Fig. 4. Subjects' average performance per scene; left: finding the position that minimizes the isovist area (*best hiding place*), right: finding the position that maximizes the isovist area (*best overview place*). The error-bars display the standard error of the means

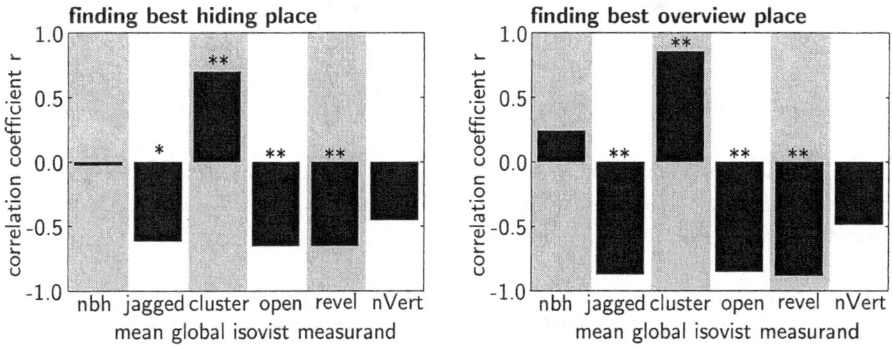

Fig. 5. Correlation between subjects' navigation performance and the isovist measurands neighborhood size (***nbh***), jaggedness (***jagged***), clustering coefficient (***cluster***), openness (***open***), revelation coefficient (***revel***) , and number of polygon vertices (***nVert***)

Correlations Between Navigation Performance and Isovist Measurands. Several strong correlations between subjects' performance and visibility graph measurands were found (see Figure 5). Subjects' performance in finding the *best hiding place* for the 16 indoor scenes significantly correlated with the global measures for jaggedness ($r=-.62$, $p=.01$), clustering coefficient ($r=.70$, $p<.01$), revelation ($r=-.65$, $p<.01$), and openness ($r=-.65$, $p<.01$), while performance did not significantly correlate with the global measures for neighborhood size ($r= -.03$, $p=.93$) and the number of isovist vertices ($r=-.45$, $p=.08$).

Subjects' performance in finding the *best overview place* for the 16 indoor scenes significantly correlated with the global measures for jaggedness ($r=-.87$, $p<.001$), clustering coefficient ($r=.86$, $p<.001$), revelation ($r=-.88$, $p<.001$), and openness ($r=-.85$, $p<.01$), while performance did not significantly correlate with the global measures for neighborhood size ($r=.24$, $p=.37$), and the number of isovist vertices ($r=-.49$, $p=.06$).

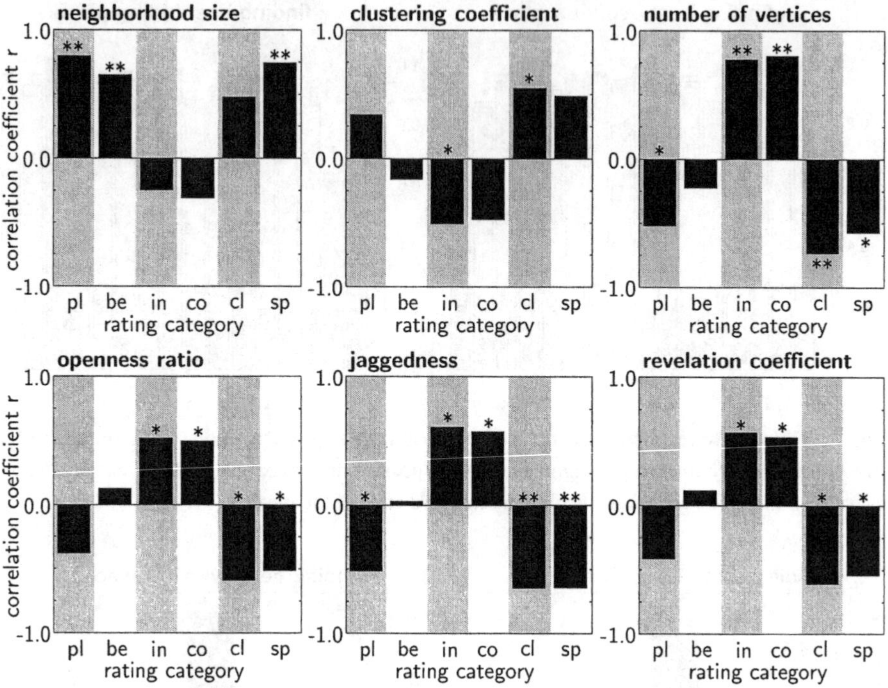

Fig. 6. Linear correlations between the selected global isovist measurands and averaged rated experiential qualities of the scenes in Experiment 1. The rating categories were ***pl***easure, ***be***auty, ***in***terestingness, ***co***mplexity, ***cl***arity, and ***sp***aciousness

It has to be noted that the global measurands jaggedness, openness, revelation, and clustering coefficient were highly intercorrelated in the 16 scenes ($r>.81$).

Rating Task. Also several strong correlations between the global isovist measurands describing the scenes and the corresponding averaged ratings were found (see Figure 6). Most prominently, average neighborhood size (corresponding to isovist area) was highly correlated with rated pleasure (correlation coefficient $r=.80$, $p<.01$), beauty ($r=.65$, $p<.01$), and spaciousness ($r=.74$, $p<.01$). Likewise, the average number of isovist polygon vertices turned out to be strongly interrelated with experienced complexity ($r=.81$, $p<.01$), interestingness ($r=.78$, $p<.01$), and clarity ($r=-.73$, $p<.01$). Additionally, several significant yet slightly lower correlations to the further characteristic values were found. However, due to the high level of intercorrelations within the measurands, they indicate the same statistical relations and are therefore not further discussed.

Correlations Between Navigation Performance and Ratings. A comparison between rated experience and subjects' performance in the navigation tasks rendered an uneven result for the two navigation tasks. For finding the best hiding place, no significant correlation with any rating dimension was found (explained variance $r<.11$). However, the environments in which subjects performed well in finding the best overview place were rated less interesting ($r=-.63$, $p<.01$), less complex ($r=-.54$, $p=.03$), but more clear ($r=.57$, $p=.02$) and spacious ($r=.58$, $p=.02$). Additionally, a moderate statistical relation between navigation performance and experienced pleasantness of the rooms was probable ($r=.45$, $p=.08$) , although this result was not significant.

5.4 Discussion

Overall, subjects showed remarkably good performance in both of the navigation tasks (finding the *best overview place* and finding the *best hiding place*), demonstrating that subjects were able to perceive the sizes of isovists very well. The basic initial hypothesis that isovists capture behaviorally relevant environmental properties was further supported by the result that the isovist measurands jaggedness, openness, revelation, and clustering coefficient were strongly correlated with the navigation performance. The question of how to qualitatively interpret these statistical relations is however not obvious. The high level of intercorrelations between the isovist derivatives jaggedness, openness, revelation, and clustering coefficient suggests that these measurands captured similar aspects of the environments and basically describe the same property. One possible interpretation could be based on jaggedness: Studies on polygon outlines [4] and building silhouettes [20] have found that the jaggedness measurand corresponds well to introspectively rated shape complexity. Pointing in the same direction, the results of the rating tasks showed positive correlations between jaggedness and rated complexity, and negative correlations between jaggedness and clarity. It is hard to explain, however, why number of vertices was the best predictor measurand for rated complexity, while the correlations with navigation performance were lower in comparison to jaggedness. Taken together, jaggedness, clustering coefficient, revelation, and openness ratio can be seen as measures describing similar aspects of environmental complexity. It is assumed that navigation in complex environments requires an increased mental or computational effort resulting in a negative influence on navigation performance.

The apparent statistical relations between the navigation task and the rating results may however be also interpreted in a different way: Since the navigation task preceded the ratings, the latter might have been influenced by the subjective experience of the former task. For example, the rated complexity of an indoor scene may basically mirror the effort or the subjectively perceived difficulty of the navigation tasks within that scene. This interpretation gains some support by the positive correlation between experienced pleasure and navigation performance, although this relation did not reach significance level. In order to test this alternative explanation, Experiment 2 was designed.

6 Experiment 2

6.1 Objective

This experiment was designed to discriminate between the alternative explanations of Experiment 1 (see Section 5.4). For this purpose, solely the rating task of Experiment 1 was repeated, the navigation task was skipped, and the ratings were done from a fixed central observation point. Comparing the rating results of the two experiments allowed to determine the impact of navigation on the experiential qualities in Experiment 1.

6.2 Method

Experimental Procedure and Variables of Interest. The procedure of this experiment was identical to the rating task of Experiment 1 (see Section 5.2), except for the fact that subjects' movements were restricted to rotational movements only. That is to say, subjects were stationary at the starting position marked in Figure 1. Subjects had to complete all six ratings for each room before they proceeded to the next scene. Again, the scenes were presented in random order. A complete experimental session had a duration of about 20 minutes.

Participants. 13 naive subjects (7 female, 6 male) voluntarily participated in the experiment, they were paid 8 Euro per hour. Subjects were mostly university students.

Analysis. The analysis compared the means and the variance of the samples between the experiments using a two sided t-test and tested for correlations. For the correlation analysis, the rating results of each scene were averaged by category over all subjects.

6.3 Results

No significant differences were found between the mean ratings of the two experiments (see Figure 7 left). If anything, a moderate tendency (p=0.22) was found that scenes were perceived as more interesting in Experiment 2. The ratings of the both sessions were all positively correlated (see Figure 7 right), the correlation coefficient r varied from .49 (beauty) to .88 (spaciousness and complexity). The overall variance between the scenes was almost identical in both experiments (cf. Figure 8 right). The variance within the scenes was very similar between the two conditions except of spaciousness (Figure 8 left): In Experiment 2 spaciousness ratings differed more between subjects than in Experiment 1 (p=.01, not corrected for multiple comparisons).

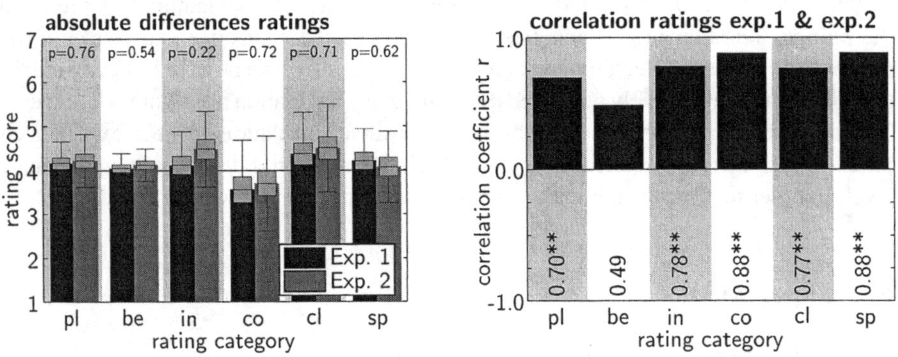

Fig. 7. Mean scores over all ratings and correlations between the ratings of Experiment 1 and Experiment 2. The rating categories were ***pl***easure, ***be***auty, ***in***terestingness, ***co***mplexity, ***cl***arity, and ***sp***aciousness

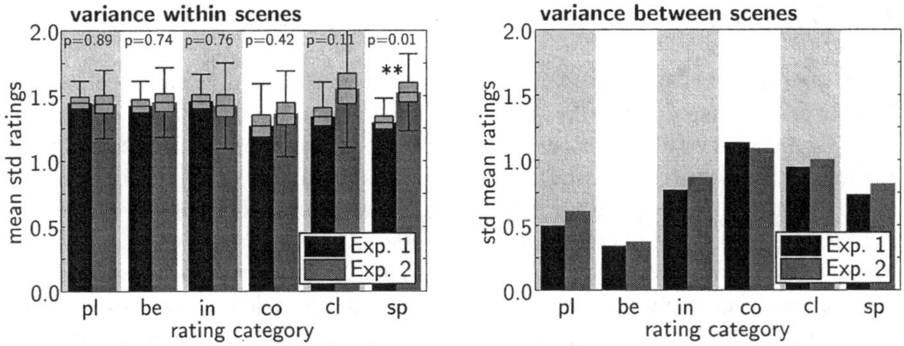

Fig. 8. Bar plots illustrating the variances of the ratings within the scenes (left) and between the scenes (right) of Experiment 1 (black) and Experiment 2 (green). The rating categories were ***pl***easure, ***be***auty, ***in***terestingness, ***co***mplexity, ***cl***arity, and ***sp***aciousness

6.4 Discussion

The high correlations between ratings of Experiment 1 and Experiment 2 together with the lack of significant absolute sample differences demonstrated that the average appraisals were very similar in both experiments. These results suggest that the navigation task including free exploration in Experiment 1 had little influence on the rating task. The potential negative influence of ego-motion on room interestingness could be interpreted in terms of the *mystery theory* [14] suggesting that spatial situations that only promise the gain of information when moving (as in Experiment 2) are more interesting than the same spatial situations after actual exploration (as in Experiment 1). Additionally, the rather low correlation in the beauty rating category (r=.49, p=.06) between experiments could indicate a moderate yet inconsistent influence of the navigation task on the experienced beauty. However, an analysis of the rating variance within and between the scenes offers a plausible alternative explanation for this potential influence on beauty: In both experiments the rating variance within the scenes was remarkably similar over all categories (Figure 8 left), while the variance between the scenes varied depending on the rating category (Figure 8 right). The differences of the mean ratings between the scenes were lowest in the beauty rating category, in other words, all scenes were perceived as being similarly beautiful. Hence, in the beauty category individual differences between the subjects had a much stronger influence on the correlation between the experiments than in the other ratings, and the apparent effect could therefore be explained by the small sample sizes.

The comparative analysis of Experiment 1 and Experiment 2 demonstrated that differences within the ratings were mainly caused by differences between the scenes, and were not an artefact caused by the navigation task. Altogether, remarkable similarities between the experiential qualities rated from a fixed position (Experiment 2) or after free navigation (Experiment 1) were found.

7 Conclusions

The experiments presented in this study investigated interrelations between spatial properties of environments on the one hand and spatial experience and behavior on the other hand. Taken together, the two experiments could demonstrate dominant influences of the environment on both experimental tasks. Beyond this qualitative statement, the technique of isovist analysis allowed to identify factors that were systematically related to both experimental tasks. For experiential qualities and navigation behavior, already single isovist measurands were sufficient to widely explain the variance in the behavioral data. The method of averaging isovist measurands over the complete indoor environments rendered meaningful and discriminatory global characteristic values. An additional indication for the behavioral relevance of isovists can be derived from subjects' remarkably good performance in the navigation task, demonstrating that the area of isovists was well perceivable.

These findings suggest that for further experiments it is worthwhile to translate qualitative descriptions and explanatory theories for spatial preferences and behavior such as "prospect and refuge" into empirically testable hypotheses that make use of isovist measurands. Of course, due to the limited number of tested scenes and the specific character of the navigation task, future work has to test the general validity of the specific findings both for a broader range of spatial situations and for different kinds of spatial behavior. Yet altogether the outcomes of this study suggest that isovist and visibility graph analysis, analyzing space from an inside beholder-centered perspective, provide generic descriptions of architectural spaces that have predictive power for subjects' spatial experience and behavior.

Acknowledgments

This work was supported by the Deutsche Forschungsgemeinschaft (MA 1038/9-1) and the Max Planck Institute for Biological Cybernetics, Tübingen. The authors express their thankfulness to Professor Dr. Heinrich H. Bülthoff and Professor Dr. Hanspeter A. Mallot for their support.

References

1. Jay Appleton. Prospects and refuges revisited. In *Environmental aesthetics: Theory, research, and application*, pages 27–44. Cambridge University Press, New York, 1988.
2. M. L. Benedict. To take hold of space: Isovists and isovists fields. *Environment and Planning B*, 6:47–65, 1979.
3. Daniel E. Berlyne. *Conflict, Arousal, and Curiosity*. McGraw-Hill, New York, 1960.
4. Daniel E. Berlyne. *Aesthetics and Psychobiology*. Appleton, New York, 1972.
5. Heinrich H. Bülthoff and Hendrik .A. H. C. van Veen. Vision and action in virtual environments: Modern psychophysics in spatial cognition research. In M. Jenkin and L. Harris, editors, *Vision and Attention*, pages 233–252. Springer, New York, 2001.
6. Francis D. K. Ching. *Architecture Form Space and Order*. Van Nostrand Reinhold, New York, 2nd edition, 1996.

7. Gerald Franz, Markus von der Heyde, and Heinrich H. Bülthoff. Predicting experiential qualities of architecture by its spatial properties. In Bob Martens and Alexander G. Keul, editors, *Evaluation in progress - Strategies for environmental research and implementation (IAPS 18 conference proceeedings on CD-ROM)*, 2004. http://www.kyb.mpg.de/publication.html?publ=2432.
8. B Hillier. *Space is the Machine*. Cambridge University Press, Cambridge, MA, 1996.
9. Bill Hillier. The common language of space: a way of looking at the social, economic and environmental functioning of cities on a common basis. http://www.spacesyntax.org /publications/commonlang.html, September 1998.
10. Bill Hillier and Julienne Hanson. *The Social Logic of Space*. Cambridge University Press, Cambridge, UK, 1984.
11. Bill Hillier, M. D. Major, J. Desyllas, K. Karimi, B. Campos, and T. Stonor. Tate gallery, millbank: a study of the existing layout and new masterplan proposal. Technical Report 5, Bartlett School of Graduate Studies, University College London, London, UK, 1996.
12. G. Janzen, T. Herrmann, S. Katz, and K. Schweizer. Oblique angled intersections and barriers: Navigating through a virtual maze. In C. Freksa, W. Brauer, C. Habel, and K. F. Wender, editors, *Spatial Cognition II. Integrating Abstract Theories, Empirical Studies, Formal Methods and Practical Applications*, number 1849 in Lecture Notes in Artificial Intelligence. Springer Verlag, Berlin, Germany, 2000.
13. G. Janzen, M. Schade, S. Katz, and T. Herrmann. Strategies for detour finding in a virtual maze: The role of the visual perspective. *Journal of Environmental Psychology*, 21:149–163, 2001.
14. Stephen Kaplan. Perception and landscape: conceptions and misconceptions. In *Environmental aesthetics: Theory, research, and application*, pages 45–55. Cambridge University Press, New York, 1988.
15. Rob Krier. *Über architektonische Komposition*. Klett-Cotta, Stuttgart, Germany, 1989.
16. Michael Leyton. *A generative theory of shape*. Springer, Berlin, Germany, 2001.
17. Oscar Newman. *Creating Defensible Space*. US Department of Housing and Urban Development, Washington, 1996.
18. M.J. O'Neill. Effects of familiarity and plan complexity on wayfinding in simulated buildings. *Journal of Environmental Psychology*, 12:319–327, 1992.
19. Sophia Psarra and Tadeusz Grajewski. Describing shape and shape complexity using local properties. In John Peponis, Jean Wineman, and Sonit Bafna, editors, *Proceedings 3rd International Space Syntax Symposium*, pages 28.1–28.16. A. Alfred Taubman College of Architecture and Urban Planning, University of Michigan, 2001.
20. Arthur E. Stamps. *Psychology and the aesthetics of the built environment*. Kluwer, Boston, MA, 2000.
21. Alasdair Turner, Maria Doxa, David O'Sullivan, and Alan Penn. From isovists to visibility graphs: a methodology for the analysis of architectural space. *Environment and Planning B*, 28(1):103–121, 2001.
22. Alasdair Turner and Alan Penn. Making isovists syntactic: isovist integration analysis. Paper presented at the 2nd International Symposium on Space Syntax, Brasilia., April 1999.
23. S. Werner and P. Long. Cognition meets le corbusier - cognitive principles of architectural design. *Lecture Notes in Artificial Intelligence*, 2685:112 – 126, 2003.
24. J.M. Wiener and H.A. Mallot. Fine-to-coarse route planning and navigation in regionalized environments. *Spatial Cognition and Computation*, 3(4):331–358, Dez 2003.
25. Joachim F. Wohlwill. Environmental aesthetics: The environment as a source of affect. In *Human Behavior and Environment*, volume 1, pages 37–85. Plenum Press, New York, 1976.

A Model for Context-Specific Route Directions

Kai-Florian Richter[1] and Alexander Klippel[2]

[1] Transregional Collaborative Research Center SFB/TR 8 Spatial Cognition,
Universität Bremen, Germany
richter@sfbtr8.uni-bremen.de
[2] Cooperative Research Centre for Spatial Information,
Department of Geomatics, University of Melbourne, Australia
aklippel@unimelb.edu.au

Abstract. Wayfinding, i.e. getting from some origin to a destination, is one of the prime everyday problems humans encounter. It has received a lot of attention in research and many (commercial) systems propose assistance in this task. We present an approach to route directions based on the idea to adapt route directions to route and environment's characteristics. The lack of such an adaptation is a major drawback of existing systems. Our approach is based on an information- and representation-theoretic analysis of routes and takes into account findings of behavioral research. The resulting systematics is the framework for the optimization process. We discuss the consequences of using an optimization process for generating route directions and outline its algorithmic realization.

1 Introduction

Getting from an origin A to a destination B is a prime problem in people's life. Efficiently solving this problem, i.e. determining a route between A and B and then purposively moving along that route, is called wayfinding (Golledge, 1999; Montello, in press). It has become a major research direction in many areas.

Wayfinding research can be organized in two broad areas: first, research that aims at shedding light on the question of how humans and other agents actually find their ways (e.g., Blades, 1991; Allen 1999); second, research that aims at supporting humans in the activity of finding a way (e.g., Wahlster et al., 2001; Heye & Timpf, 2003; Duckham & Kulik, 2003). Additionally, wayfinding can be differentiated in planning a route and actually following a route. In this contribution, we focus on supporting wayfinders in following a route.

The setting we are dealing with is wayfinding in outdoor environments where the movement occurs on a system of paths, like in city street networks or on footpaths in a park. Route directions are a primary means to guide someone in finding one's way. Here, route directions refer to instructions on how to follow a route; they are task-oriented specifications of the actions to be carried out to reach the destination (e.g., Klein, 1979; Tversky & Lee, 1998; Denis et al., 1999; Schweizer et al., 2000). Our approach complements research on incremental route directions (cf. Maaß, 1993;

C. Freksa et al. (Eds.): Spatial Cognition IV, LNAI 3343, pp. 58–78, 2005.
© Springer-Verlag Berlin Heidelberg 2005

Habel, 2003). We use the term route directions generically in this paper to refer to any form of instructions—verbal, graphical, gestures—for route following. In contrast to approaches designed to generate modality specific route directions, we present a computational model that generates *abstract route directions*, i.e. an abstract representation of the actions necessary to follow a route. The abstract representation may be *externalized* in different modalities, for example as verbal or graphical route directions or as gestures (see also Chomsky, 1986; Jackendoff, 1997; Tversky and Lee, 1999; Allen, 2003; Habel, 2003; Klippel, 2003; Klippel et al. submitted).

The article is structured as follows: we start with introducing a distinction between structure and function in wayfinding, which reflects the difference between features present in an environment and the role they take in the process of wayfinding. We argue why route directions benefit from taking into account their conceptual basis and introduce the concept of *context-specific route directions* (Section 3). A systematics of elements that can be exploited in generating abstract route directions is presented in Section 4. Section 5 motivates how this generation can be realized as an optimization process; Section 6 outlines a computational model for the optimization process and presents an example.

2 Structure and Function in Wayfinding

For the following argumentation, it is important to distinguish between the features physically present in an environment independent of any wayfinding actions and their role in the process of wayfinding. Klippel (2003) introduced the concepts of *structure* and *function* in wayfinding. With structure, he refers to an environment's physically present features; the structural level describes a static configuration of these features. Function denotes the relation of these structural elements to actions performed in the environment; the functional level demarcates those features relevant for a wayfinding action, i.e. it describes a dynamic situation and those parts of the structure that are demarcated by an action.

Accordingly, Klippel (2003, following Montello, in press) distinguishes between *path* and *route*. A path is a linear, unbounded feature in the environment upon which travel occurs. A route is a behavioral pattern; it has an origin and a destination and is directed and bounded. A route demarcates a path, i.e. it determines those parts of the paths—called *path-segments*—that are traversed while route following. We term the points where path-segments meet *branching points*. Paths, i.e. branching points and path-segments, form a *path-network*, a graph-like structure, which reflects the geometric layout of the paths in an environment, with branching points as nodes and path-segments as edges. On a functional level, we deal with *route-segments*, which correspond to those path-segments demarcated by a route. The point where two route-segments meet is termed *decision point*. At a decision point, a wayfinder needs to decide on the further direction to take; it corresponds to a branching point on the structural level. This does not imply that every decision point has to be mentioned explicitly in a route direction, as not every decision point requires the same attention by the wayfinder (see Section 4).

We consider decision points to be most pertinent for route directions. Following a route comprises two basic processes: getting to a decision point and, there, determining the further direction to take (e.g. Daniel & Denis, 1998). Route directions' main purpose is to support decision making in route following, i.e. providing information on how to proceed at a decision point. Hence, we concentrate on decision points in generating route directions; the basic representation underlying our model is a sequence of decision points.

The distinction between structure and function is also reflected in the generation of route directions (see Section 4.3). The features exploited in giving route directions are part of the structural level, i.e. they are features physically present in the environment. However, whether they are applicable for a specific route direction is determined by the functional level, i.e. in route following as well as in giving route directions the route itself demarcates those parts of an environment that are functionally relevant for the given task.

3 Conceptualizing Routes

Route directions provide instructions on how to get from A to B. Someone or some system generates them for somebody else as means of assistance; they are messages at this point. When a receiver of a message interprets or uses the instructions, they become information. Viewing route directions from this perspective allows for a representation theoretic analysis, i.e. the distinction of their syntactic, semantic, and pragmatic level (Richter et al., 2004; see also, e.g., MacEachren, 1995). The syntactic level comprises an analysis of the size of a message, for example, how many words are used in a verbal route direction, which relates to messages in an information-theoretic sense (Shannon & Weaver, 1949). On the semantic level, the processing of a message into information, i.e. the effort needed to interpret route directions, is analyzed.

Frank (2003) considers the pragmatic information content of route directions. He claims that two route directions for the same route that differ on the syntactic level, i.e. in the size of the messages, may be considered equal from a pragmatic point of view if they both lead agents to take identical routes. In this case both route directions lead to the same result, the wayfinder being at her goal, and to the same actions, the wayfinder took the same route using either of them. On the other hand, the same route direction may be different to different users, as users may differ in their knowledge of the environment or the task they try to perform with the directions given.

We are interested in the conceptual level of route directions. Two different route directions for the same route that are equal from a pragmatic perspective may well differ on the conceptual level. In order to use route directions successfully, a wayfinder needs to conceptualize the route she is about to encounter, or parts of it. As there can be differences with respect to the ease with which a route direction is understood and the extent to which it supports cognitive processes, the conceptualization of two route directions for the same route may differ. This difference may reside in the conceptualization itself. For example, the instruction 'go straight, straight, and then turn left' results in a different conceptualization than 'follow the signs to the train station'. Or the resulting conceptualizations are similar, but the differences reside in the processing of the route directions that leads to the conceptualization. For example, 'go straight, straight, and then turn left' may lead to the same conceptualization as 'turn left at the third intersection'; but the former requires more processing than the latter (cf. Klippel, 2003; cf. also Dale, et al., 2002, 2003). This is because the latter directions are already chunked, while in the former directions this chunking has to be done by the wayfinder herself (see Section 4.2; see also Miller, 1956; Cowan 2001).

In our approach, *conceptualization* of route directions is the (process of forming a) mental representation of a route. A route is represented as a sequence of decision point / action pairs. Hence, more precisely, conceptualization is the (process of forming a) mental representation of an (expected) decision point sequence with their accompanying actions. We aim at creating route directions that support this conceptualization. These route directions should be easy to process, i.e. they should support forming and processing a representation of the corresponding route. Consequently, route following also becomes easier as understanding a route direction is a prerequisite for using it (cf. Dale et al., 2003).

In order to generate such route directions, we need to account for the structure of the environment in which route following takes place. The structure of an environment influences the kind of instruction that can be given. Route directions depend on the embedding of the path—instantiated by the route—in the spatial structure surrounding the path, on the structure of that path itself, on path annotations, and on landmarks that are visible along the path. Additionally, the reference system used provides alternatives to describe actions needed to follow the route. These dependencies are reflected in the systematics of elements in route directions developed in this paper. Taking into account this systematics results in abstract route directions specifically adapted to a route's properties and the environmental characteristics. We coin the resulting route directions *context-specific route directions*[1].

This approach differs from other approaches. Duckham and Kulik (2003), for example, present an algorithm that modifies the classic AI search algorithm A*—used to find the shortest path—to calculate the route easiest to describe. They use a single way to describe a route and look for the optimal route given that description mechanism. The approach taken here starts with a given route and elicits abstract

[1] We introduce the term *context-specific route direction* to emphasize that our model explicitly adapts the generated directions to the situation at hand, i.e. to the current action to take along the route in the current surrounding environment. This reflects Dey and Abowd's (2000) definition of context, i.e. "[…] any information that can be used to characterize the situation of an entity" (p. 3). Our model provides several alternatives to describe the same action dependent on a route's properties and environmental characteristics. It differs, hence, from existing (internet) wayfinding assistance systems that employ strict, inflexible rules to any context. Such strict rules lead to effects like leaving a city when entering an inner-city highway and getting back to the same city when exiting that highway again, or the generation of new events just because the name of a street changes, not because the wayfinder needs to change her current action.

route directions for that route, which in turn are the basis for modality-adapted externalizations. Thus, our approach complements the approach of Duckham and Kulik.

It also differs from the CORAL system by Dale et al. (2002, 2003). Contrary to their approach we are not restricted to natural language output; the 'best' route direction also could mean a graphic representation or some mixed modality like map gestures (Hirtle, 2000) of a route or a part thereof. Common aspects are discussed, for example, in Section 4.2. The conceptualization approach taken here aims at an abstract representation formalism that forms the basis for various output formats or different realizations of the same output format.

Guhe et al. (2003) present an approach to generate abstract representations of motion events, which can be extended to the generation of route directions (Habel, 2003). They aim for a system that is able to describe events in a dynamically changing world. There are two main differences to our approach: first, as with the approach by Dale et al., their model is intended for natural language output; second, they focus on processing dynamic situations where information is acquired incrementally, and accordingly, their interest is in incremental route instructions.

4 A Systematics for Context-Specific Route Directions

The structure of an environment and its elements need to be considered in creating abstract route directions (ARDs) since it influences how instructions for route following can be given. This influence can be local, i.e. an environment's element is usable for one or a few ARDs, or global, i.e. an environment's structure is exploitable for several or most of the ARDs. Furthermore, the reference system used determines alternatives to give instructions. An action needed for route following can be described from the perspective of the wayfinder (egocentric references). Elements of the environment can be referred to in instructions (allocentric references), or some fixed references outside the environment may be used (absolute references). In these references, different elements of an environment and of a route may be employed for giving route directions. The elements are cataloged according to three levels in our systematics.

4.1 Levels of the Systematics

The three levels of the systematics reflect three categories of elements that can be used in giving route directions: global references, i.e. elements that are not part of the immediate surrounding, environmental structure, i.e. elements of an environment that impose a structure on that environment, and elements that belong to the path and the route.

Global References

This level comprises elements referred to in abstract route directions that are not part of the immediate surrounding environment in which the action takes place. These are references that rely on an absolute reference system, i.e. the direction referred to is the same everywhere in the environment and does not depend on a wayfinder's position. Most typical are cardinal directions like 'north', 'east', etc. Additionally, references to global landmarks, i.e. landmarks outside the surrounding environment, belong to this category if references to them are the same everywhere in the given environment. These landmarks are visible from many places of the environment or their location is everywhere unequivocally known, which makes them usable as reference objects (Sorrows & Hirtle, 1999). An example for a resulting instruction is 'towards the sea' with 'the sea' being an example of such a landmark.

Environmental Structure

There can be elements that are part of an environment, which have an influence on an environment as a whole. Such elements impose a structure on that environment, which leads to distinctive parts. Hence, these elements or a reference to the emerging distinctive parts can be exploited in abstract route directions. Since the parts are distinctive, they provide unambiguous direction information usable in giving route directions. An example for such an element is a slant; examples for such instructions are 'uphill' or 'downhill'.

Paths, Routes, and Landmarks

The third level comprises elements that relate to paths and routes. Landmarks and decision points are part of this level, as well as *annotations* along a path, which are elements that are set up to unequivocally identify that path, like street names, street signs, or markers. We also catalog the combination of several instructions into a single one to be on this level (Dale et al., 2003; Klippel et al., 2003).

Typically, instructions on this level use an egocentric reference frame, like 'turn right' or 'go straight'. That is, route directions refer to the locations of an intersection's branches relative to the wayfinder. We need to consider the configuration of a branching point when creating ARDs on this level. Since our aim is to generate ARDs that are unambiguous, the branch to take when following the route needs to be unequivocally identifiable. From a functional perspective, several branches at a decision point may broadly lead to the same direction and, therefore, the functionally relevant branch needs to be further specified. For example, if there are two branches leading to the left, they may be distinguished into 'half left'[2] and 'sharp left' and the resulting instruction may be 'turn half left'.

Landmarks are pertinent to route directions (cf., e.g., Denis et al., 1999). They influence the way instructions are given. For example, people refer more often to landmarks than to street names in generating route directions and they are more effective than street names in route guidance (cf. Tom & Denis, 2003). Structurally, landmarks can be point-like, linear, or areal. Point-like landmarks are located in small, restricted areas of an environment. Such a landmark is, for example, a salient building like a church. The other two, linear and areal landmarks, extend across an environment, like a river or a forest. We consider landmarks that influence route directions to be part of the route. We call them *routemarks* (cf. Krieg-Brückner et al, 1998). A routemark

[2] The German term *halb* [*links* / *rechts*] is not translatable directly to English. More appropriate natural language expressions would be *veer*, *bear*, *up to the* [*left* / *right*] etc.

can be either at a decision point, at a route-segment between two decision points (Hermann et al., 1998), or in some distant to, but visible from the route. We call the latter kind of routemark *distant routemarks* (cf. Lovelace et al., 1999).

Functionally, routemarks at decision points can be used to identify a decision point, for example, 'turn left *at* the church'. Routemarks between decision points may be employed to further describe the route and to function as confirmation that one is still on the right track ('you *pass* a church'). Distant routemarks, finally, are like beacons. Assuming they are visible while passing several decision points, they can be used as pointers to a certain direction. An example of such an instruction is '*towards* the church'. For the conceptualization of a turning action, the location of a landmark at a decision point is important. A routemark may be passed before the turn ("turn *after* the church"), after the turn ("turn *before* the church"), or the landmark may not be located at a functionally relevant branch of the decision point ("turn where the church is") (cf. Klippel, 2003). Routemarks before a turn are easily conceptualized as the turning action occurs immediately after them. They are, thus, a good identifier for a decision point. It is an open issue, though, what influence the latter two cases have on conceptualization and which additional parameters play a role here. We introduce and distinguish them for reasons of completeness in the systematics.

On a functional level, linear and areal landmarks can function either point-like or linear. In a point-like fashion, such landmarks identify a decision point and may indicate an action to be taken, for example 'turn right when coming to the river'. However, linear and areal landmarks not only identify a decision point, but may also allow combining several decisions into one decision. Examples for route directions that usually involve a linear pattern are 'follow the river' or 'walk along the forest'. They may determine the actions for several decision points. Therefore, such route directions require an additional qualifier that establishes the point until the instruction holds. An example for such a qualifier is 'until you reach the gas station'.

4.2 Chunking Instructions

Route directions provide instructions on how to proceed for every decision point. Yet, not every decision point and the accompanying action need to be mentioned explicitly. Often, it is possible to combine actions for several decision points into one route direction; this combination is an important mechanism in route directions and the conceptualization of routes. We call it *spatial chunking* (Klippel et al., 2003). Dale et al. (2003) refer to it as *segmentation*.

Spatial chunking groups several decision point / action pairs into a single segment; we call these segments *higher order route direction elements* (HORDE) (cf. Klippel, 2003). Dale et al. (2003) identify two segmentation principles: *landmark-based* and *path-based* segmentation. In landmark-based segmentation, landmarks at decision points delimit a part of the route to be followed; the route is decomposed into segments, each leading to such a landmark. Path-based segmentation is based on three features of paths—road status (highways, main roads, etc.), path length, and turn saliency (e.g. T-intersections). By employing any of these features or a combination thereof, routes can be segmented.

Klippel et al. (2003) differentiate three kinds of spatial chunking:

- Numerical chunking: Here, a sequence of several decision points that involve no direction change (termed *DP-*) and one decision point with a direction change (*DP+*) are combined into a single decision. This is done by counting the decision points until a direction change occurs, for example 'turn left at the third intersection'. Also, a sequence of decision points with equal direction changes can be grouped, for example 'turn twice right'.
- Landmark chunking: This kind of chunking is similar to numerical chunking. However, an unambiguous landmark identifying the DP+ is utilized to mark the point where a direction change occurs, instead of counting the DP-. An example for such a HORDE is 'turn right at the gas station'. The number of intermediate decision points is not specified in this kind of chunking.
- Structure chunking: In structure chunking, spatial structures that are unique in a given local environment are exploited. For example, the dead end of a T-intersection unequivocally marks the need for a direction change—one either needs to turn left or right as straight on is impossible. Hence, it is possible to chunk several DP- and the relevant DP+ located at such a structure into HORDE like 'turn right at the T-intersection'. An instruction like 'follow the river' also rests upon structure chunking as it combines actions for several decision points that are located along the river into a single one.

Klippel et al. (2003) present spatial chunking based on an egocentric reference frame and on instructions employing elements from the third level of the systematics used here. Abstract route directions on the other levels of the systematics can also be chunked, i.e. it is possible to combine sequences of several decision points with ARDs that employ other elements of the systematics. To pick up the example from the previous subsection again, the instruction 'go uphill' may combine actions for several decision points that happen to be in a line uphill into a single decision, which can also considered to be a HORDE. Usually, either landmark or structure chunking are used to combine ARD for several decision point / action pairs on the higher levels of the systematics. Just like with linear landmarks, there needs to be a qualifier that marks the end of such a HORDE, i.e., that denotes the point until an instruction based on such a HORDE holds.

4.3 Structure and Function in the Systematics

The distinction made between structure and function in wayfinding is also reflected in our systematics. The elements presented above are all part of the structural level, i.e. they are all part of either the path itself or the environment the path is embedded in—with an exception of global landmarks, which are nonetheless also clearly part of the structural level as references to them do not depend on a wayfinder's location in the environment. However, whether these elements are applicable in creating route directions for a route depends on the context set by the functional level.

The route, which is on the functional level, demarcates the functionally relevant parts of an environment, i.e. the path segments traversed while route following. It also determines the actions needed to follow that route; the corresponding sequence of decision point / action pairs represents the dynamic aspects of route following, especially the direction in which a wayfinder moves through the environment. Abstract route directions need to reflect this sequence and provide information on the directions to take. Consequently, only elements of the systematics that unequivocally denote these directions, i.e. allow a wayfinder to correctly orient herself, are applicable in generating ARDs.

4.4 Granularity in the Systematics

Granularity is one of the fundamental aspects of knowledge representation (cf. Hobbs, 1985). The elements of the systematics, which offer access to knowledge of an environment, are on different levels of granularity, i.e. ARDs generated using these elements provide information on how to follow a route on different levels of granularity. These changes in granularity reside between the levels of the systematics as well as within the levels. In our systematics, granularity refers to how closely abstract route directions are linked to individual decision point / action pairs and the corresponding branching point's configuration, i.e. to what extent ARDs abstract from a detailed description of a decision point / action pair itself.

The level of global references, which is the first level of the systematics, relates to the coarsest granularity. Referring to elements that are not part of the surrounding environment results in coarse direction information, which is not explicitly based on the structure of the environment itself. Such instructions just exploit that they lead to unequivocal choices at the decision points they hold for. ARDs on the level of environmental structure—the second level of the systematics—provide still coarse information on the further direction to take, but which explicitly takes into account an environment's structure and is, therefore, more closely related to the embedded path itself. Consequently, this kind of instruction is on a finer granularity level than those using elements of the systematics' first level.

The third level, which is the level of path, route, and landmarks, contains elements of the route itself. ARDs generated with these elements usually refer explicitly to decision points. The structure of these ARDs is therefore close to the decision point / action pairs themselves. Accordingly, we consider elements on the third level to be on the finest level of granularity.

A change of granularity occurs also within the different levels of the systematics. Chunking instructions obviously increases the degree of abstraction from individual decision point / action pairs. This holds for all three levels. The different kinds of chunking result in abstract route directions on different levels of granularity. While landmark and structure chunking combine a number of decision points that are not specified in the resulting instruction, i.e. they abstract from the exact number of decision points involved and may therefore provide a single instruction for a large part of the route, in numerical chunking the number of decision points is explicitly mentioned. Such instructions are only sensibly applicable for a small number of decision points (cf. Klippel, 2003). Finally, taking into account the configuration of a branching point, for example, by further qualifying a turning instruction is on the finest level of granularity as this is directly based on an individual decision point.

4.5 Implicit Verses Explicit Representation

We use a sequence that contains every decision point of a route as an underlying representation in our model. The application of chunking, however, combines several of these decision points into a single representation. Thus, the resulting representation does not necessarily contain every single decision point anymore. However, when following a route, a wayfinder needs to make a decision at every decision point encountered along the route. Therefore, she must be able to infer the decisions only implicitly represented in the route directions in order to know the further direction to take. Consequently, the route directions need to be correct, i.e. provide instructions on a route that leads from origin to destination, and complete, i.e. provide the instructions such that every decision necessary can be derived from them.

5 Generating Context-Specific Route Directions: An Optimization Problem

Our aim is to generate abstract route directions, which form the basis for real route directions that ease the conceptualization of a route. To this end, we need to choose from all possible ways to create abstract route directions for a route the one that best fits this aim. More precisely, for each decision point along the route we need to choose an (abstract) instruction on which action to perform that is most likely to ease the conceptualization. However, the kind of instruction to choose at a decision point may depend on the kind of instruction chosen for previous or following decision points, as, for example, decisions may be chunked. Thus, all different kinds of abstract route directions that are possible for a decision point need to be judged according to their consequences regarding conceptualization, taking into account the possible abstract route directions for other decision points of that route. The dependence of a local choice (an instruction for a single decision point) on the choices made elsewhere (instructions for other decision points) clearly shows that we are dealing with an optimization problem; we are looking for optimized abstract route directions for a given route. Accordingly, the sequence of decision point / action pairs needs to be processed and translated into optimized abstract route directions for each pair.

The kinds of abstract route directions that can be created for a decision point / action pair are based on the systematics presented in the last section. For the generation of abstract route directions, we explicitly exploit an environment and route's characteristics; this results in ARDs specifically adapted to these characteristics.

If we want to decide on which ARD for a decision point / action pair best fits our aim of an easy conceptualization, we need a measure to compare possible ARDs with respect to that aim. We need rules that define which kind of ARD to choose in which situation, i.e. an optimization criterion. Potentially, there is a huge number of such rules. There can be many kinds of abstract route directions applicable at the same time for a decision point, which need to be judged. Moreover, in combination with the potential dependency on ARDs chosen for other decision points, the number of rules needed increases even further. Thus, computationally, it is not sensible to have a specific rule for every situation that might occur in creating context-specific route directions. Instead, we need a heuristic, i.e. general rules that provide guidelines and sensible choices that can be applied when a specific situation occurs. As in most optimization problems, these general rules may in some cases result in abstract route directions that are not the best possible—though still good ones; but applying these heuristics makes the problem computationally feasible.

Several optimization criteria may be applicable. A first simple heuristic would be to use always the highest granularity level possible, i.e. to choose for each decision point the ARD that corresponds to an element of the systematics, which is—compared to all other possible elements—on the highest granularity level in the systematics. Other possible criteria include: minimal number of distinct parts, i.e. smallest number of chunks; abstract route directions based on no more than n elements of the systematics, i.e. the optimization results in abstract route directions that do not employ more than n different elements of the systematics; no more than n changes in the kind of instructions, i.e. in the resulting abstract route directions there is at most n times a switch from one element of the systematics to another; no more than n changes in the reference system used.

In order to create context-specific route directions we propose to aim at a minimal number of distinct parts with, everything else being equal, abstract route directions on the highest granularity levels possible. In the following, we will argue for this criterion.

First, from an information-theoretic perspective a small number of chunks reduces the amount of information that needs to be communicated, i.e. the size of the message decreases. A decrease of information leads to a decrease of memory load, i.e. the wayfinder needs to remember less information. To put it another way, a reduction in number of chunks results in a decreased amount of information explicitly represented and an increase of information that needs to be inferred. This relates to Grice's (1975) principles of communication, especially the ones he termed *quality* and *quantity*: the information provided needs to be correct and should not contain any details that are unnecessary for the message's purpose.

Second, the application of chunking and HORDE also reduces the processing involved in conceptualizing a route. As argued before, a principle of cognitive ergonomics is to combine several instructions into a higher-order instruction if possible. This clearly requires additional processing of the route directions, i.e. increases the cognitive load of a wayfinder. Since in the generation of context-specific route directions this chunking of single ARDs to higher-order route directions is already done, the wayfinder does not need to perform this herself anymore, which, accordingly, eases the cognitive processing of the route directions. That is, with context-specific route directions we provide instructions for route following that are easy to process and theoretically easy to memorize.

Furthermore, route directions on a high level of granularity reduce the problem of matching an expected decision point / action pair with the real environment. This kind of instruction is less prone to errors if the conceptualized decision point / action pair does not (exactly) match the actual situation in the environment. For example, an instruction 'turn left' might get a wayfinder into trouble if the actual configuration of branches met at an intersection does not seem to include a branch she considers leading to the left. While an instruction 'follow the signs to the train station' does not depend at all on the configuration of the intersections; all that is required is that there is actually a sign pointing in direction to the train station. Thus, with route directions on higher levels of granularity a wayfinder is not that strongly dependent on the environment meeting her conceptualization anymore.

It can also be argued that applying HORDE and providing route directions on a high level of granularity moves the task of wayfinding in an environment nearer to the task of planning a trip through an environment. Such route directions include fewer "real" decision points, i.e. fewer decision points where a wayfinder actively needs to remember a direction change (DP+). As HORDE combine several decision points into one decision, a wayfinder only needs to remember the point until the HORDE holds; all decisions in between can be inferred. 'Turn right at the third intersection', for example, indicates a direction change at the third intersection a wayfinder encounters; the information implicitly represented is that she has to keep the current direction at the first and second intersections. The advantages of HORDE are even more obvious when looking at instructions like 'follow the markers' for a hiking trail. Here, a single instruction suffices to lead a wayfinder to her goal; but following the markers may involve many direction changes while walking along the hiking trail. That is, such a HORDE on a high level of granularity may render decision points that actually involve a direction change, i.e. DP+, into decision points that do not require a change of action, i.e. practically turn them into DP-.

6 A Computational Model for Context-Specific Route Directions

The generation of context-specific route directions (CSRD) can be realized as an optimization problem. We need to find globally optimal abstract route directions for each decision point / action pair or chunks thereof, respectively, i.e. ARDs that are optimal with respect to the complete route, not just for a single decision point / action pair. In the last section, we presented the optimization criterion employed in our approach. In this section, we provide an overview on the computational part of our approach, which includes the algorithm used for finding the optimal CSRD. Additionally, we give an example of how the optimization process works and discuss how to deal with missing data.

6.1 A Computational Approach to Context-Specific Route Directions

For the automatic generation of CSRD we need information on the route in question, i.e. we need a representation of the environment that contains all information needed and allows us to compute a route from some origin to a destination. To this end, we employ a graph-like representation of the environment's path-network. The graph's edges represent the path-segments; nodes denote the branching points. The graph reflects the layout of the environment's paths, i.e. it preserves information on angles between branches and distances. In such a graph, we can calculate a route with any path-search method, like A* or Dijkstra's (1959) shortest path algorithm. The calculation results in a sequence of nodes that need to be traversed to get from an origin to a destination. This sequence corresponds to the sequence of decision points that is the underlying representation of our model. For the generation of context-specific route directions we need additional information on the elements of the systematics, for example on position, structure, and visibility of landmarks or on path annotations. Hence, we annotate the graph with this information (see Section 6.3 for a discussion on automatically extracting such information).

For the optimization process, we start with generating for each decision point all abstract route directions (ARDs) that are possible according to the systematics defined. Such ARDs represent a decision point and its accompanying action description based on the element used. The action description consists of a direction relation and, if one applies, that feature of the environment the relation refers to. Examples of ARDs are $(\mathtt{DP}_1, \mathtt{left})$, denoting a left turn at the first decision point of a route, or $(\mathtt{DP}_4, \mathtt{follow/marker})$, representing an instruction to follow the marker at the fourth decision point. In case of a routemark at a decision point, we also employ a relation to denote the position of that landmark: `after` is used as a relation to state that a turn occurs after a landmark is passed; `before` to state that a turn occurs before a landmark is passed, and the relation `at` is a generic term representing the presence of a landmark anywhere at a decision point (see Section 4.1). Thus, $(\mathtt{DP}_2, \mathtt{right/after\ church})$ denotes a right turn that can be further qualified using a landmark, here a church, which is passed before the turn occurs.

Each element of the systematics has a corresponding set of direction relations; these differ across the elements. For ARDs based on egocentric references, for example, we use the relations defined in the sector model presented in Klippel (2003), which has been further refined in behavioral experiments (e.g., Klippel et al., 2004). The model comprises three basic directions—`straight`, `left`, `right`—and two additional qualifiers for left and right—`half` and `sharp`—leading to seven different directions. As another example, global references are either represented with a cardinal direction—`north`, `east`, `south`, `west`—or with the relation `towards` combined with a referenced global landmark, like `towards/sea`.

The relations used in this approach, like `left` or `towards`, represent information on the direction to take at a decision point. This resembles the symbolic operators describing directional phrases as, for example, in Jackendoff (1990) or Eschenbach et al. (2000). However, it is important to note that all relations used and the abstract route directions, like $(\mathtt{DP}_2, \mathtt{right/after\ church})$, are by no means meant to be the actual (verbal) output of an assistance system. They are an abstract representation of the systematics' elements applicable for a given decision point, i.e. they represent possibilities of how a decision point / action pair can be described according to the systematics. We choose relation terms like `left` or `towards` because they are more readable than terms like `a`, `b`, `c`, and so on, but these terms need not be the terms used in an actual verbal output. The step to generate verbal or graphical route directions presented to a user, i.e. the transformation of the abstract route directions into concrete ones, is not covered in this research.

We check for each decision point which elements of the systematics are applicable and generate an abstract route direction based on this element. The annotations in the street-network's graph provide information on which elements can be used, for example, whether a landmark is located at a decision point or whether a global landmark is visible. This way, possible ARDs are generated resulting in a set of instructions for every decision point of the route.

Our aim is to find a minimal number of distinct parts in the abstract route directions on the highest granularity levels, i.e. in our route directions, we try to cover the complete route with as few chunks as possible while using elements of the systematics on the highest possible granularity levels. This resembles the approach of Dale et al. (2002): "... the general idea is to view messages as data objects corresponding to the largest distinct linguistic fragments we need in order to generate the variety of texts we are interested in" (p. 4). Different to Dale et al., our chunks are not necessarily "linguistic fragments" found in natural language route directions, but are derived from spatial data according to principles of HORDE.

We are looking for sub-sequences in the decision point sequence that share abstract route directions based on the same elements of the systematics and are chunkable. We apply the chunking rules as described in Section 4.2; these can be further refined to exclude results of the chunking process that are not sensible. Klippel (2003), for example, derived a set of rules in his wayfinding choreme route grammar for generating valid HORDE based on the direction model explained above. Other rules, like those by Dale et al. (2003) or the route direction principles by Denis (1997), can also be incorporated in our optimization process to prevent insensible chunks like 'right at the 21st intersection'.

For the optimization process, we choose the first ARD of the first decision point and calculate the union with the following decision points' sets of ARD until we encounter a decision point, which cannot be chunked with the previous ones according to the chunking rules employed. We then choose the next ARD of the first decision point and again try to chunk it with as many of the following decision points as possible. We repeat this until all abstract route directions of the first decision point have been processed. We continue with the second decision point, again building chunks with every possible ARD for that decision point. The process runs until we generated all chunks for every decision point of the route. Along this process, we keep track on which combination of chunks is minimal, i.e. covers the most decision points with the least number of chunks. The process can be implemented using dynamic programming; Table 1 summarizes it.

6.2 An Example

To clarify the idea of optimization, we present an example of our approach. We chose the Bürgerpark in Bremen—a big park in the center of the city. Route directions are generated from one of its entrances to one of the park's cafés. Fig. 1 shows a schematic map of the area; the chosen route is shown as a black line.

Table 1. The optimization process in an algorithmic description

```
For each DP in route,
  determine every ARD possible according to the
  systematics resulting in a set of ARD.

Start with first ARD of first DP,
  try to chunk it with as many following DPs as
  possible,
  store generated chunk.

Repeat with following ARDs of first DP,
  until all ARDs have been processed.

Store biggest chunk as current CSRD.

Repeat with following DPs.
  If,
    the biggest newly generated chunk does not overlap
      with current CSRD, add to CSRD.
  else if,
    the newly generated chunk covers more DPs than the
      one covering the current DP currently stored in
        CSRD, rebuild CSRD using new chunk.
  until all DPs have been processed.
```

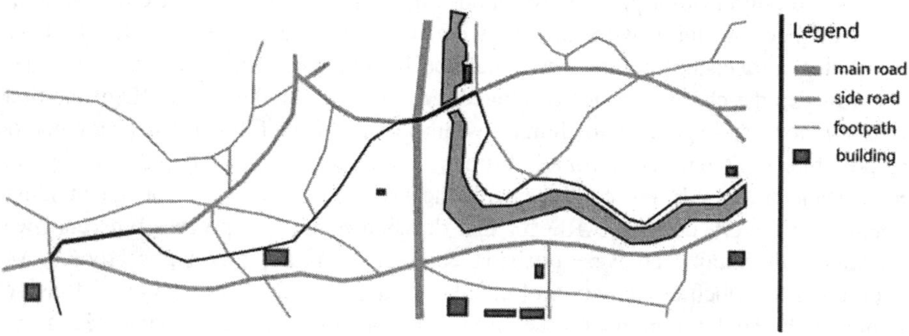

Fig. 1. A schematic map showing a part of the Bürgerpark in Bremen

The network of path segments is represented as a graph (see Fig. 2). The edges that correspond to the path-segments demarcated by the chosen route are shown as a bold line. The graph is also annotated with information on landmarks, like the buildings shown in Fig. 1; the annotations are not shown in Fig. 2. The route consists of ten decision points, i.e. ten branching points are passed when following this route.

As a first step in the optimization process, we determine for each decision point all ARDs that are possible according to the systematics (see Table 2). According to the sector model used (cf. Section 6.1; Klippel et. al, 2004), there is a half-right turn at the

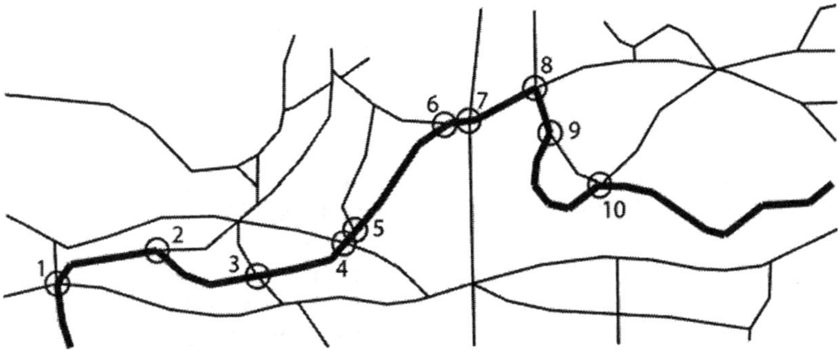

Fig. 2. The graph corresponding to the network of ways (see Fig. 1). The dots mark the decision points, which are numbered in the order of passage

first decision point. There is no further information available on this decision point in the graph, hence (DP_1, `half right`) is the only abstract route direction that can be generated for this decision point. The same holds for the second decision point. At decision point three, there is no change in direction; the direction relation applicable is `straight`. Here, additionally a landmark—a building in the park—visible from this decision point is in the direction of movement. It functions like a beacon as described in Section 4. This can be exploited resulting in two possible abstract route directions for this decision point: (DP_3, `straight`), (DP_3, `towards/building`).

We continue this process until all possible ARDs for all decision points have been generated; these are shown in Table 2. Decision points six, nine, and ten are worth a closer examination: the direction change at decision point six is slightly to the right (`half right`). However, since this turn is at a T-intersection, we can exploit this structural element rather than relying on the direction concept alone. The direction

Table 2. The route's decision points and their set of possible abstract route directions[3]

Decision point	Set of abstract route directions
1	{half right}
2	{half right}
3	{straight, towards/building}
4	{straight}
5	{straight, towards/building}
6	{half right, right/T-intersection}
7	{straight, straight/at bridge}
8	{right, right/after bridge, right/at building}
9	{half right, follow/river}
10	{half right, follow/river}

[3] We omitted cardinal directions in this example to keep it simple. No chunks based on cardinal directions would contribute to the resulting route directions; thus, this omission does not change the presented optimization process.

relation is coarsened to `right`—since at a T-intersection only a left or right turn is possible—and the structural element is added resulting in (DP_6,`right/T-intersection`). For decision points nine and ten, again `half right` is the relation to use according to the employed sector model; but here also a linear landmark—the river—is exploitable, as the route segment traveled resides along the river. There are, thus, again two possible abstract route directions: (DP_9,`half right`) and (DP_9,`follow/river`) (the same for decision point ten).

For the optimization process, we choose the first ARD in the set of the first decision point and try to chunk as many decision points as possible applying chunking rules for the kind of ARD chosen. Here, we can apply numerical chunking for the first two decision points, i.e. chunk both abstract route directions `half right`. This chunk is also the current optimal CSRD—since it is the only chunk so far. As there are no more ARDs for the first decision point, we continue with the second decision point. Its ARD cannot be chunked with any of the third decision point's ARDs. Therefore, we store this single abstract route direction for the second decision point; the CSRD still consists of the chunk generated for the first decision point.

Decision point three to six can be chunked using structure chunking (three times `straight`, followed by `right/T-intersection`). This is the biggest chunk that can be generated for these four decision points. Our CSRD now consists of two chunks; the first grouping decision points one and two, the second grouping decision points three to six. Finally, for the last four decision points, the best chunks we can generate are for seventh and eighth decision point (`straight/at bridge`, `right/after bridge`) and for ninth and tenth (`follow/river`, `follow/river`). Thus, the abstract route directions for the resulting CSRD consist of four chunks; they are summarized in Table 3. They contain the chunked decision points in the first part, and the direction relations that get chunked in the second part.

Table 3. The resulting chunks of decision points for the chosen route

{DP_1,DP_2;half right,half right}
{DP_3,DP_4,DP_5,DP_6;straight,straight,straight,right/T-intersection}
{DP_7,DP_8;straight/at bridge,right/after bridge}
{DP_9,DP_{10};follow/river,follow/river}

6.3 Availability of Data

The success of our approach relies on the availability of data on the environment route following takes place in. The underlying representation of our model is a sequence of decision points passed along a route. The graph used to calculate this sequence is derived, most typically, from GIS data sets like the ones provided by federal authorities. For the purpose of creating context-specific route directions, this graph needs to be further annotated with additional information, like street signs or landmarks.

While the availability of land-use data is fairly good these days and, thus, such a graph is readily available, additional data is not systematically available. Still, we argue that our approach calculates reasonable results even with missing data. In recent years, spatial, especially geographic data has increased tremendously in its importance for business; accordingly, more and more data gets collected. The automatic extraction of such data has become an important research issue. For example, there is work on extracting landmarks from land-use data and on automatically determining the saliency of landmarks (cf. Elias & Sester, 2003; Winter, 2003; Raubal & Winter, 2002, respectively).

Most importantly, even if such additional data is not completely available our approach still achieves good results. The optimization process as described above makes use of whatever data is available. It optimizes route directions according to this data. If only the underlying graph of the path network should be available, it is still possible to create route directions using an egocentric reference system, i.e. using directions like 'turn left' or 'go straight', and to apply spatial chunking to combine these into HORDE. Consequently, this still results in route directions, which are as good or even better than those generated by assistance systems available today.

7 Conclusions and Outlook

We present an approach to generate abstract route directions that explicitly takes into account a route's properties and environmental characteristics. This is a first step to *context-specific route directions*. They support the conceptualization of a route as they reflect cognitive principles of organizing spatial knowledge. To generate such route directions automatically, we employ an optimization process. This process aims at minimizing the number of distinct parts of route directions. Our model is based on a systematics of elements that can be employed in creating abstract route directions; this systematics reflects different levels of granularity and respects the distinction between structure and function in wayfinding.

Our claim is that context-specific route directions are easier to conceptualize, i.e. they allow forming a mental representation of a route that is easier to process and that better matches the actual route encountered. Hence, route following becomes easier. Our approach adapts abstract route directions to actual situations in the environment. Compared to existing approaches that use the same references and set of actions irrespective of the route, which may lead to inadequate, hard to use route directions, this is, thus, a step towards the goal of providing context-specific route directions that support cognitive processes (cf., e.g., Dale et al., 2003, for a critique on existing internet route-planners and Habel, 2003, for a discussion of benefits of multimodal route instructions). The approach differs from those that aim to specify natural language processes in that the scope of *conceptualization* is extended to information available in spatial data and conceptualization processes that are beyond those required for natural language generation.

Future work comprises an extension of the presented systematics. Furthermore, we need to evaluate different optimization criteria, applying both behavioral research and computational specification, to refine our approach. Finally, with some adaptation, our approach may also be usable to calculate routes through an environment that are optimized with respect to their ease of conceptualization, i.e. to already account for the proposed optimization in the path-search algorithm. This is in line with approaches like Duckham and Kulik's (2003), which try to overcome today's wayfinding assistance systems' limitations of just calculating shortest or fastest routes.

Acknowledgements

This research is carried out as a part of the Transregional Collaborative Research Center SFB/TR 8 Spatial Cognition. Funding by the Deutsche Forschungsgemeinschaft (DFG) is gratefully acknowledged. We like to thank Christian Freksa and Heike Tappe for helpful comments and discussions. We also thank three anonymous reviewers.

References

Allen, G.L. (1999). Spatial abilities, cognitive maps, and wayfinding. Bases for individual differences in spatial cognition and behavior. In R.G. Golledge (ed.), *Wayfinding behavior. Cognitive mapping and other spatial processes* (pp. 46-80). Baltimore: John Hopkins University Press.

Allen, G. L. (2003). Gestures accompanying verbal route directions: Do they point to a new avenue for examining spatial representations? *Spatial Cognition and Computation* 3:259-268.

Blades, M. (1991). Wayfinding theory and research: The need for a new approach. In D.M. Mark & A.U. Frank (eds.), *Cognitive and linguistic aspects of geographic space* (pp. 137-165). Dordrecht: Kluwer.

Chomsky, N. (1986). *Knowledge of language: Its nature, origin and use.* New York: Praeger.

Cowan, N. (2001). The magical number 4 in short term memory. A reconsideration of storage capacity. *Behavioral and Brain Sciences* 24:87-186.

Dale, R., Geldof, S., and Prost, J.-P. (2002). Generating more natural route descriptions. In *Proceedings of the 2002 Australasian natural language processing workshop.* Canberra, Australia.

Dale, R., Geldof, S., and Prost, J.-P. (2003). CORAL: Using natural language generation for navigational assistance. In *Proceedings of the 26th Australasian computer science conference (ACSC2003).* Adelaide, Australia.

Daniel, M-P. & Denis, M. (1998). Spatial descriptions as navigational aids: A cognitive analysis of route directions. *Kognitionswissenschaft* 7(1):45-52.

Denis, M. (1997). The description of routes: A cognitive approach to the production of spatial discourse. In *Cahiers de Psychologie Cognitive* 16(4):409-458.

Denis, M., Pazzaglia, F., Cornoldi, C., and Bertolo, L. (1999). Spatial discourse and navigation: An analysis of route directions in the city of Venice. *Applied Cognitive Psychology* 13(2):145-174.

Dey, A.K. & Abowd, G.D. (2000). Towards a better understanding of context and context-awareness. In *Proceedings of the workshop on the what, who, where, when, and how of context-awareness*. New York.

Dijkstra, E.W. (1959). A note on two problems in connexion with graphs. *Numerische Mathematik* 1:269-271.

Duckham, M. & Kulik, L. (2003). Simplest paths: Automated route selection for navigation. In W. Kuhn, M. Worboys, and S. Timpf (eds.), *COSIT 2003. Spatial information theory* (pp. 169-185). Berlin: Springer.

Elias, B. & Sester, M. (2003). Landmarken für Wegbeschreibungen - Identifikation, Extraktion und Visualisierung. *Kartographische Nachrichten* 53(2):51-57.

Eschenbach, C., Tschander, L., Habel, C., and Kulik, L. (2000). Lexical specifications of paths. In C. Freksa, W. Brauer, C. Habel, and K.F. Wender (eds.), *Spatial Cognition II. Integrating abstract theories, empirical studies, formal methods, and practical applications* (pp. 127-144). Berlin: Springer.

Frank, A.U. (2003). Pragmatic information content. How to measure the information in a route description. In M. Goodchild, M. Duckham, and M.Worboys (eds.), *Perspectives on geographic information science* (pp. 47-68). London: Taylor and Francis.

Golledge, R.G. (1999). *Wayfinding behavior. Cognitive mapping and other spatial processes.* Baltimore: John Hopkins University Press.

Grice, H.P. (1975). Logic and conversation. In P. Cole & J.L Morgan (eds.), *Speech acts* (pp. 41-58). Volume 3 of *Syntax and semantics*. New York: Academic Press.

Guhe, M., Habel, C., and Tschander, L. (2003). Describing motion events: Incremental representations for incremental processing. In M. Guhe, C. Habel, and L. Tschander (eds.), *Proceedings of the 5th international workshop on computational semantics (IWCS-5)* (pp. 410-424). Tillburg.

Habel, C. (2003). Incremental generation of multimodal route instructions. In *Natural language generation in spoken and written dialogue, AAAI Spring Symposium 2003*. Palo Alto, CA.

Heye, C. & Timpf, S. (2003). Factors influencing the physical complexity of routes in public transportation networks. *Electronic proceedings of the 10th international conference on travel behaviour research*. Lucerne.

Herrmann, T., Schweizer, K., Janzen, G., and Katz, S. (1998). Routen- und Überblickswissen - konzeptuelle Überlegungen. *Kognitionswissenschaft* 7(4):145-159.

Hirtle, S.C. (2000). The use of maps, images and "gestures" for navigation. In C. Freksa, W. Brauer, C. Habel, and K. F. Wender (eds.), *Spatial Cognition II – Integrating abstract theories, empirical studies, formal methods, and practical applications* (pp. 31-40). Berlin: Springer.

Hobbs, J.R. (1985). Granularity. In A.K. Joshi (ed.), *Proceedings of the 9th international joint conference on artificial intelligence* (pp. 432-435). San Francisco: Morgan Kaufmann

Jackendoff, R. (1990). *Semantic structures.* Cambridge, MA: MIT Press.

Jackendoff, R. (1997). *The architecture of the language faculty.* Cambridge, MA: MIT Press.

Klein, W. (1979). Wegauskünfte. *Zeitschrift für Literaturwissenschaft und Linguistik* 9(33):9-57.

Klippel, A. (2003): *Wayfinding Choremes – Conceptualizing wayfinding and route direction elements*. Universität Bremen.

Klippel, A., Tappe H., and Habel, C. (2003). Pictorial representations of routes: Chunking route segments during comprehension. In C. Freksa, W. Brauer, C. Habel, and K.F. Wender (eds.), *Spatial Cognition III. Routes and navigation, human memory and learning, spatial representation and spatial learning* (pp. 11-33). Berlin: Springer.

Klippel, A., Dewey, C., Knauff, M., Richter, K.-F., Montello, D. R., Freksa, C., and Loeliger, E.-A. (2004). Direction concepts in wayfinding assistance. In J. Baus, C. Kray, and R. Porzel (eds.), *Workshop on artificial intelligence in mobile systems (AIMS'04)* (pp. 1-8). Saarbrücken: SFB 378 Memo 84.

Klippel, A., Tappe, H., Kulik, L. and Lee, P.U. (submitted). Wayfinding choremes – A language for modelling conceptual route knowledge. (Submitted to Journal of Visual Languages and Computing).

Krieg-Brückner, B., Röfer, T., Carmesin, H.-O., and Müller, R. (1998). A taxonomy of spatial knowledge for navigation and its application to the Bremen autonomous wheelchair. In C. Freksa, C. Habel, and K.F. Wender (eds.), *Spatial Cognition. An interdisciplinary approach to representing and processing spatial knowledge* (pp. 373-397). Berlin: Springer.

Lovelace, K.L., Hegarty, M., and Montello, D.R. (1999). Elements of good route directions in familiar and unfamiliar environments. In C. Freksa & D Mark (eds.), *Spatial information theory. Cognitive and computational foundations of geographic information science* (pp. 65-82). Berlin: Springer.

Maaß, W. (1993). A cognitive model for the process of multimodal, incremental route descriptions. In A.U. Frank & I. Campari (Eds.), *Spatial information theory: A theoretical basis for GIS* (pp. 1-13). Berlin: Springer.

MacEachren, A.M. (1995). *How maps work.* New York: Guilford Press.

Miller, G.A. (1956). The magical number seven, plus or minus two: Some limits on our capacity for processing information. *Psychological Review* 63:81-97.

Montello, D.R. (in press). Navigation. In P. Shah & A. Miyake (eds.), *Handbook of visuospatial cognition.* Cambridge: Cambridge University Press.

Raubal, M. & Winter, S. (2002). Enriching wayfinding instructions with local landmarks. In M. Egenhofer & D. Mark (eds.), *Geographic information science - Second international conference GIScience 2002* (pp. 243-259). Berlin: Springer.

Richter, K.-F., Klippel, A., and Freksa, C. (2004). Shortest, fastest, — but what next? A different approach to route directions. In M. Raubal, A. Sliwinski, and W. Kuhn (eds.), *Geoinformation und Mobilität – von der Forschung zur praktischen Anwendung. Beiträge zu den Münsteraner GI-Tagen* (pp. 205-217). Münster: IfGI prints.

Schweizer, K., Katz, S., and Janzen, G. (2000). Orientierung im Raum - kognitive Grundlagen und sprachliche Realisierung. *Tourismus Journal* 4(1):79-101.

Shannon, C.E. & Weaver, W. (1949): *The mathematical theory of communication.* Urbana, Illinois: The University of Illinois Press.

Sorrows, M. & Hirtle, S.C. (1999). The nature of landmarks for real and electronic spaces. In C. Freksa & D. Mark (eds.), *Spatial information theory. Cognitive and computational foundations of geographic information science* (pp. 37-50). Berlin: Springer.

Tom, A. & Denis, M. (2003). Referring to landmarks or street information in route directions: What difference does it make? In W. Kuhn, M. Worboys, and S. Timpf (eds.), *COSIT 2003. Spatial information theory* (pp. 362-374). Berlin: Springer.

Tversky, B. & Lee, P. (1998). How space structures language. In C. Freksa, C. Habel, and K.F. Wender (eds.), *Spatial Cognition. An interdisciplinary approach to representing and processing spatial knowledge* (pp. 157-175). Berlin: Springer.

Tversky, B. & Lee, P. (1999). Pictorial and verbal tools for conveying routes. In C. Freksa & D.M. Mark (eds.), *Spatial information theory. Cognitive and computational foundations of geographic information science* (pp. 51-64). Berlin: Springer.

Wahlster, W., Baus, J., Kray, C., and Krüger, A. (2001). REAL: Ein ressourcenadaptives mobiles Navigationssystem. *Informatik Forschung und Entwicklung* 16:233-241.

Winter, S. (2003). Route adaptive selection of salient features. In W. Kuhn, M. Worboys, and S. Timpf (eds.), *COSIT 2003. Spatial information theory* (pp. 349-361). Berlin: Springer.

Investigation of Preference Between the Least-Angle Strategy and the Initial Segment Strategy for Route Selection in Unknown Environments

Hartwig H. Hochmair[1] and Victoria Karlsson[2]

[1] St. Cloud State University, Department of Geography
720 Fourth Avenue South, St. Cloud, MN 56301
hhhochmair@stcloudstate.edu
[2] University of Bremen, Cognitive Systems Group,
PO Box 330 440, 28334 Bremen, Germany
v.karlsson@home.se

Abstract. This paper presents results from a desktop experiment in which the participants' route selection behavior in an unknown street network is investigated. The participants were presented with a series of intersections in a virtual urban desktop environment in field view. Given the task to reach a distant wayfinding target that could be seen in the background, participants then had to state their preference for one of the two outgoing roads at each intersection. As the participants were unfamiliar with the environment they needed to apply a wayfinding strategy. This work analyzes the use of two wayfinding strategies with respect to the given wayfinding task, namely the *least-angle strategy* and the *initial segment strategy*. With the first strategy, the participant selects the street most in line with the target direction, whereas with the second strategy the participant prefers initially straight routes. The paper analyzes the observed preference behavior with respect to these two strategies and suggests an underlying mechanism (minimum triangle path) that explains in which situation either of the two strategies or both are applied.

1 Introduction

1.1 Route Selection Heuristics

Human wayfinding often takes place in unfamiliar environments, i.e. in situations where the navigator lacks spatial knowledge relevant for making wayfinding decisions under certainty. Under these circumstances, a wayfinding heuristics needs to be applied. Navigators who have complete knowledge about the environment with respect to the wayfinding task also use wayfinding strategies, because they help minimize the navigator's cognitive effort, but still yield satisfactory route choices (Christenfeld 1995). This paper focuses on two wayfinding heuristics, namely the *least-angle* and the *initial segment* strategies. These heuristics are used to explain the route selection behavior observed in the empirical study. An overview of further human wayfinding heuristics, for example, can be found in Løvås (1998) and Janzen et al. (2000).

C. Freksa et al. (Eds.): Spatial Cognition IV, LNAI 3343, pp. 79–97, 2005.
© Springer-Verlag Berlin Heidelberg 2005

The *initial segment strategy* (ISS) (Bailenson et al. 2000) suggests that people tend to focus disproportionately on the initial portions of the route and that they prefer routes with longer straight initial segments, regardless of what the later portions of the routes look like. This strategy is motivated by the idea that people, by turning as late as possible, try to minimize the cognitive effort required for navigation. With the *least-angle strategy* (LA) (Hochmair and Frank 2002), the navigator aims at maintaining track of the target direction throughout the trip, which is also the principle of the Compass routing algorithm (Bose and Morin 1999). Both ISS and LA are localized problem solving strategies, where the agent—as with local routing algorithms (Kranakis et al. 1999)—in an unknown environment tries to perform initial steps that minimize the difference between the initial problem state and the goal state.

Work by Bailenson et al. (1998; 2000) gives empirical evidence that humans employ ISS when planning their routes on maps, even when the selected routes—with initial long straight segments—are 50 percent longer overall than the alternatives. This effect was shown to be exaggerated when maps were regionalized and when people were under time pressure. However, a potential impact of the direction of the initial street segments on the decision maker's preference behavior is not discussed in the results.

Dalton (2001) found with wayfinding experiments in a virtual environment that the average turn angle along the routes "walked" by the subjects was closer to the minimum turn angles than to either the average or maximum turn angles at intersections. People avoided meandering routes and preferred more linear routes when given the instruction to walk to the opposite corner of the test area by the most direct possible route. The subjects were not familiar with the environment and could perceive it in the field perspective only. Standard street lengths were used to ensure that the subjects did not base their route choice decisions upon that factor. Thus, the work omits the potential effect of initially long street segments when formulating the route choice behavior.

1.2 Research Objective

In the above mentioned empirical studies, only one wayfinding strategy at the time has been investigated. We expect, however, that the decision maker considers both deviation angle and length of the initial street segment in her route choice, i.e. that both ISS and LA interfere in the decision making process. The virtual environment experiment was designed to investigate these potential interdependencies between ISS and LA empirically.

1.3 Research Method

For the experiment we used a virtual desktop environment, in which the test participants were presented a series of street intersections. At each intersection, the participants stated their preference for one of the two perceived roads, given the task to reach a distant goal as fast as possible. In the main study, the resulting preference statements were analyzed with regard to use of ISS and LA, both looking at the average results and considering differences between individuals. The observed preference behavior was further examined with respect to underlying mechanisms that may explain the interdependencies between the two strategies.

A follow-up study assessed individual choice reliability, i.e. consistency in preference statements when the same decision situation was presented several times to the individual participant.

1.4 Structure of the Paper

The remainder of the paper is structured as follows. Section 2 describes the setup of the main experiment. Section 3 presents the results and provides an analysis of the observed preference statements concerning the average behavior and differences between individuals. Section 4 describes the follow-up study which assesses individual choice consistency. Section 5 summarizes the findings and presents directions for future work.

2 Route Selection Behavior: Experiment Setup

Theories about human wayfinding and the use of navigation strategies would most effectively be tested in "real world" situations and in physical environments. This being unfeasible due to obvious practical reasons, we instead used a desktop virtual city to present intersections to the participants. Although the navigator's preference behavior found with the virtual environment may differ slightly from preference behavior for the same navigation task in the physical world, we expect that the obtained results depict a decision pattern that can be ascribed to human wayfinding behavior in general.

2.1 Participants

All test participants were either students or employees at the University of Bremen. Out of 28 persons who carried out the experiment, 10 were female and 18 were male. All participants were paid a small sum of money for their contribution.

2.2 Design

To capture the participants' preference behavior with respect to LA and ISS, the set of street segments, from which the intersections were built, contained legs with six different deviation angles at 15-degree steps between 15 and 90 degrees and three different lengths (a, 2a and 3a). This resulted in 18 different road variations (Fig. 1a) that were presented pairwise to the participant (Fig. 1b). Each street segment lead to a T-intersection. After exclusion of combinations where the preference behavior seemed predictable, 35 combinations remained. In order to optimize the combinations of leg pairs included further, a short series of pre-tests was carried out. The results of these led to the exclusion of some combinations that provided redundant information concerning the observed preference structure, but also led to the addition of others to refine preference judgment. The number of included combinations was hereby increased to 37 (Fig. 1c).

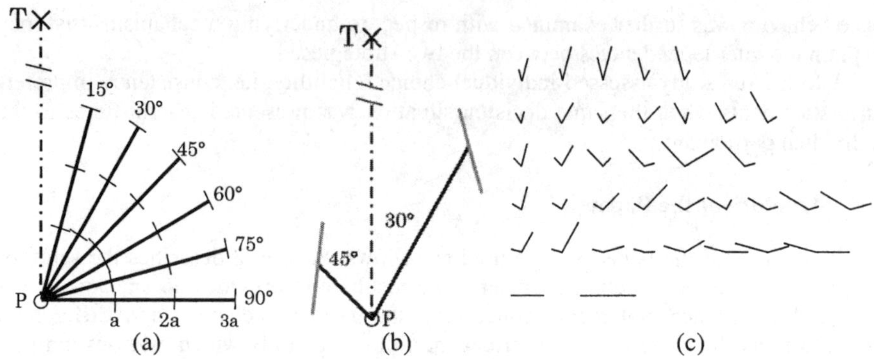

Fig. 1. Overview of all 18 road variations (a) which where combined and presented pairwise to the participant for making a binary pr eference statement (b and c)

The road constellations were visualized in a desktop 3D city environment. This environment was designed using Q3Radiant202, courtesy of Id Software Inc, and displayed using the Irrlicht-engine SDK, version 0.4, courtesy of Nikolaus Gebhardt. Q3Radiant is an editor that makes it possible to design 3D maps containing different objects, like in this case houses and streets. The Irrlicht-engine can then be used to import these maps and enable a user to move around in the environment as in a computer game. This functionality was, however, somewhat modified for the purpose of this experiment (see further below).

Each 3D city environment contained two roads positioned as a V-intersection, meeting in the point where the viewer was positioned. A watchtower, which served as the target for the wayfinding task in the experiment, could be perceived in the distance. Along the roads and in front of the target, buildings of somewhat varying size and shape were positioned. Except for deviation angle and length, the properties of the two alternative roads in each combination were kept as similar as possible: There were no intersections along any of the roads except for the T-intersections mentioned above, where the intersecting street was always oriented towards the target. Further, the type of buildings and the scenery were kept similar. We also aimed at avoiding cluttering effects (Thorndyke and Hayes-Roth 1981) and regionalization (Bailenson et al. 2000). Furthermore, the viewpoint and the distance to the target were kept constant for all combinations. One sequence of 37 scenes, randomly put together at the beginning of the experiment series, was presented to each of the participants. The screenshots of three different road combinations are shown in Fig. 2. The sequence of scenes was presented on a 19'' monitor. Using a pre-test where participants were asked to draw the deviation angles perceived on the screen on paper, we optimized the projection parameters of the Irrlicht-engine to minimize angular distortions between the physical and the simulated field of view. As a result of the pre-test we created a simulated field of view of about 80 degrees.

Using the mouse, test participants were able to pivot around the viewpoint in order to see the properties of the roads and their relation to the target more clearly. They could not, however, move forwards or backwards, since this was considered an unnecessary ability for the purpose of the experiment. The preference statements of the participants were recorded using the keys. The choices included "left road preferred", "right road preferred", and "both roads are equally preferable".

Fig. 2. Screenshots of three different road combinations in the desktop 3D city environment

2.3 Procedure

The experiment took place in a room where only the participant and the test facilitator were present. The screen, keyboard and mouse were placed on a table in front of the sitting participant, the distance to the screen being about 50 cm. Before starting the experiment, the participants were told that they would be shown a number of scenes from a desktop 3D city environment, and that each scene contained an intersection with two alternative roads. Participants were instructed to consider the properties of both roads in each combination and then to decide which alternative they would prefer, given the task to reach the distant watchtower as fast as possible. Furthermore, it was explained that there was an intersection at the end of each road, and that the intersecting road was leading in the general direction of the target. The latter explanations were made to exclude the impact of these two independent variables (number of intersections and direction of intersecting road) on the decision behavior, which allows us to analyze the decision behavior with respect to deviation angle and street length as independent variables only. Finally, the test participants were instructed on how to state their road preference by using the keys. There was no time limit for making decisions, and the participants could pivot back and forth to look and compare the choice alternatives in each scene repeatedly. They could not, however, go back to previous scenes. Three warm-up-scenes were included to give participants a chance to understand the point of the experiment and get used to pivoting with the mouse and the use of the decision keys.

3 Results and Analysis

3.1 Ranking the Choice Alternatives

For each participant we received a set of 37 binary constraints ($\succ$, ~, $\prec$) between the paired street segments. The binary relation '$\succ$' means that the person preferred the perceived left leg with deviation angle α_l and length l_l to the right leg with α_r and l_r. The binary relation '$\prec$' describes the reverse, whereas a '~' denotes the decision maker's indifference between the two options. From these 37 binary constraints a final ranking of the 18 legs was derived that reflects the decision maker's preferential structure. Formally, this ranking task corresponds to a constraint satisfaction problem (CSP). Each CSP involves a set of variables (in our case 18 leg variables), a domain of potential variables for each variable (i.e. an integer number between 1 and 18 denoting the rank), and a set of constraints, specifying which combinations of values are acceptable (i.e. 37 binary constraints). A perfect solution specifies a value to each variable that does not violate any of the constraints. A pair of values that violates a constraint is called inconsistency. In an overconstrained CSP no valid value for all variables can be found, and the CSP must be weakened, for example by removing constraints. Except for one participant, each set of 37 binary constraints recorded in the study contained inconsistencies. Figure 3 visualizes an example for a small subset consisting of three constraints that cannot be completely solved. The example is taken from a participant's binary preference statements.

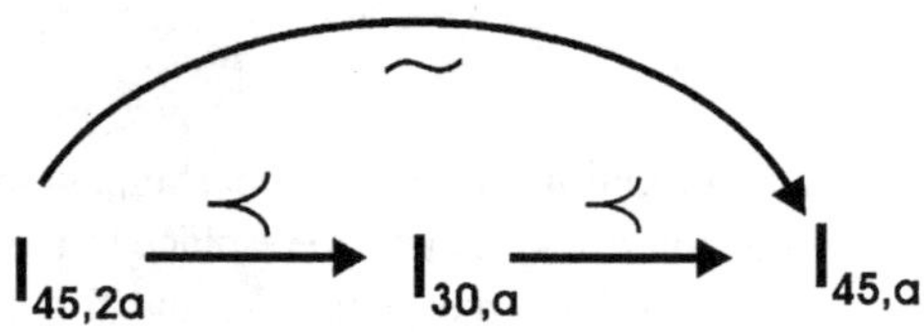

Fig. 3. Visualization of a non-satisfiable constraint system

Partial constraint satisfaction problems (PCSP) (Freuder and Wallace 1992) involve finding values for a subset of variables that satisfy a subset of the constraints, which yields a partial solution. A metric evaluates the difference between a perfect solution of a CSP and a partial solution. A metric can, among others, be expressed by the number of inconsistencies to be removed for finding a partial solution, by assigning arbitrary weights to constraints, or by introducing priorities. For our given problem we might for example use algorithms yielding a solution which satisfies as many constraints as possible (such as branch and bound or backjumping). In this case the metric would be defined over the number of constraints that cannot be satisfied. However, this strategy may yield "unnatural" results that do not reflect the decision maker's preferences, as no "semantics" is involved in the weakening process. Formalizing more complex metrics that distinguish between hard and soft constraints (Moratz and Freksa 1998; Rudová and Murray 2002) is also difficult if the participant's preferential behavior is not known in advance. As a compromise we decided to remove inconsistencies from each set of binary relations manually according to a set of intuitive rules, until the weakened CSP could be solved with a constraint satisfaction algorithm in Prolog (Poole et al. 1998). When removing constraints we tried to adhere to the following rules: Keep the number of removed binary constraints low, obtain variable values that can be interpolated from "adjacent" legs, and obtain variable values that match the general tendency of the preference pattern observed for the individual participant. On average, 4.3 out of 37 binary constraints had to be removed per participant (standard deviation = 2.5). After removing inconsistencies, the algorithm in Prolog yielded a partially ordered ranking of (α,l)-combinations from the best to the worst.

3.2 Visualization of Rankings

The domain of preference values for (α,l)-combinations contains at maximum 18 different values (Fig. 1a, section 2). Because all participants were indifferent between several (α,l)-combinations, a partially ordered ranking with fewer than 18 different preference levels (between seven and fourteen) was actually observed in all runs. For comparing the participants' judgments, the preference values of each participant were normalized to a continuous scale between 18 (most preferred) and 1 (least preferred). The numerical preference values for (α,l)-combinations found in this way are based on the rankings of the participants' statements, i.e. on ordinal and not metric data. Therefore, the normalized preference values are only an approximation for the utility associated with an (α,l)-combination. In repetitive pre-tests it was found that participants were inconsistent with their ranking order even between a small set of legs, which made direct assessment of metric preference values for the (α,l)-combinations hard to achieve, and ordinal ranking data were used instead.

After the normalization, the mean value for each (α,l)-combination was computed over all 28 participants. Fig. 4a shows a 3D scatter plot where mean preference values are plotted against the independent variables "deviation angle" and "leg length". The left axis denotes the ratio between the actual leg length l and the shortest leg a (ranging from 1 to 3), whereas the back axis denotes the deviation angle α of the leg (ranging from 15° to 90°). Fig. 4b visualizes the 3D point set as a third order polynomial regression curve.

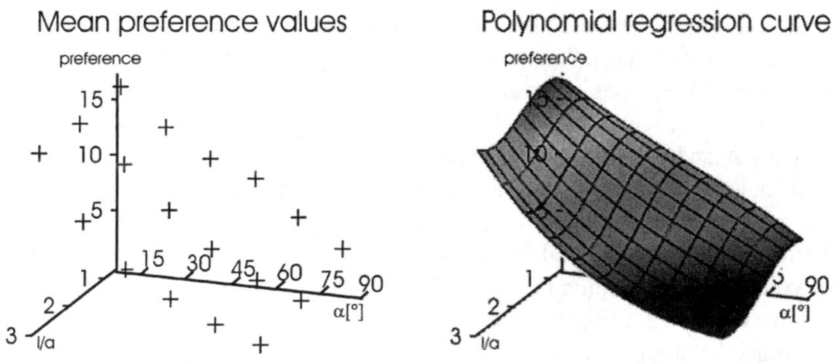

(a) (b)

Fig. 4. Mean preference values for the set of 18 street segments. 3D scatter plot (a) and third order polynomial regression curve (b)

If a horizontal cross section is made through the 3D-surface of Fig. 4b at a certain preference value, the curve that is projected onto the α-l-plane is a preferential indifference curve. That is, all legs with (α,l)-pairs which lie on the same curve share the same preference value. Fig. 5 visualizes the shape of indifference curves for integer average preference values, which are marked as numbers. The smaller the horizontal distance between the indifference curves, the higher the gradient of preference values in the corresponding direction.

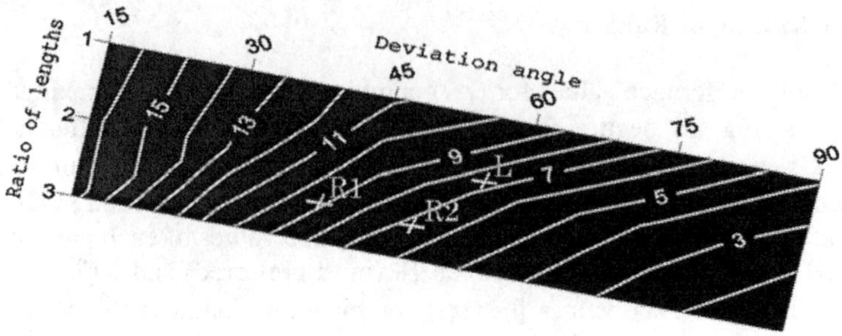

Fig. 5. Mean preferential indifference curves for the 18 legs used in the experiment

3.3 Initial Segment Strategy Versus Least-Angle Strategy: A Classification of Decision Situations

In order to clarify the analysis of the route selection behavior, we subdivided the street-leg combinations into four classes. The first class (Fig. 6a) is the group of leg-pairs, where the deviation angles for each of the two compared legs are equal, whereas the length of one leg is different from the other ($\alpha_l = \alpha_r$ and $l_l \neq l_2$). Testing such a constellation arrangement aims at finding the range of deviation angles for which ISS is actually applied. The second class is the group of leg-pairs (Fig. 6b) of equal length but different deviation angles ($\alpha_l \neq \alpha_r$ and $l_l = l_r$). Testing the street preference in such a constellation aims at finding the range of leg lengths for which the least-angle strategy is employed.

Fig. 6c and Fig. 6d visualize the more realistic situation where both parameters are different between both legs. The third class (Fig. 6c) includes leg pairs where one of the two compared legs has a larger deviation angle and a smaller initial length, i.e. $\alpha_l > \alpha_r$ and $l_l < l_r$, where the "left" and "right" indices can be swapped. According to both the LA and the ISS strategy, such a leg should be rejected. The fourth class (Fig. 6d) includes (expected) conflicting situations where one of the two compared legs has a larger deviation angle (i.e. should be rejected according to LA), but has a longer initial length (i.e. should be preferred according to ISS). Such a constellation can be written as the condition $\alpha_l > \alpha_r$ and $l_l > l_r$ where indices may be swapped.

The intersections presented to the participants were selected in a way that allowed the preference behavior concerning all four classes to be assessed from the observed choices. The results presented in this section refer to the observed mean preference values, i.e. show the general tendencies, but do not take into account preference variations between participants.

According to Fig. 5, for each angle the mean preference value increases with a shorter initial leg (i.e. when tracing the α-l-plane along lines parallel to the length-axis moving "upwards"). That is, ISS has (on average) not been applied for legs that share the same deviation angle, which describes the preference behavior related to the first class of combinations (Fig. 6a). For a deviation angle of 15° (left region in Fig. 5), however, this tendency shifts to preferential indifference between the shortest and the second shortest leg, the longest leg yet remaining the least preferred.

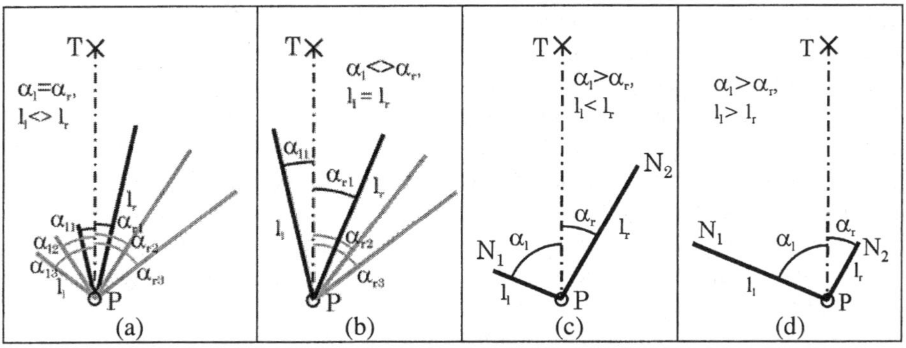

Fig. 6. Four classes of leg combinations

Moving on to the second leg combination class (Fig. 6b), Fig. 5 shows that for equally long legs, the one with a smaller deviation angle is always preferred. This can be seen when tracing the α-l-plane along lines parallel to the α-axis from right to left. Thus LA has been applied for all legs of equal initial length in the experiment.

An intersection that satisfies the geometric constraints of the third combination class (Fig. 6c) can be located in the α-l-plane as a pair of nodes, where one is located in a region more left and farther "down" than the other. No general statement about the preference behavior can be made for such a situation, which can be demonstrated by the following example: Consider the leg pairs L-R_1 and L-R_2, visualized in Fig. 5. Both leg pairs satisfy the geometric constraints for the third class. It can be seen that $R_1 \succ L$, whereas $R_2 \prec L$. Thus, the navigator makes the mental trade-off between a shorter leg and a smaller deviation angle for each situation individually. However, with an increasing difference in the deviation angles (or leg lengths) between the left and right leg the probability for choosing the leg with the smaller deviation angle (or shorter leg) increases.

Contrary to the expected conflicting situations between LA and ISS for intersections belonging to the fourth class (Fig. 6d) the leg with shorter length and smaller deviation angle is always preferred. This behavior can also be concluded from the findings about preference behavior related to Fig. 6a and Fig. 6b described above. A leg pair that satisfies these constraints can, in the α-l-plane, be located as a pair of nodes with one being located to the "upper" left of the other. The leg denoted by the upper left node will be preferred to the other.

3.4 Underlying Mechanisms: Interpretation of the Observed Preference Behavior

In the previous section we have described the observed average preference behavior along with a classification of leg combinations. In this section, we attempt to figure out a "rule" that explains the average participant's motivation for the demonstrated choice selection behavior. In other words, we seek to explain the interdependencies between the ISS and LA strategy with the help of an underlying mechanism. The first mechanism that we consider as possibly relevant denotes greedy behavior, which involves the selection of the road segment that minimizes the Euclidean distance between its endpoint and the target. A greedy algorithm works in phases and assumes that the path to the best global optimum is a series of locally optimal steps. In each phase, a decision is made that appears to be good, without regard for future consequences. The second potential underlying mechanism we looked at denotes preference for minimizing the length of a path which consists of two segments, namely the initial leg and a (fictive) leg that leads straight from the end node of the initial segment to the target (we call this mechanism "minimum triangle path"). This route thus consists of two sides of an abstract triangle, where the corners are the observer's position, the end node of the first leg, and the target.

Fig. 7 shows how the actual observed decision behavior matches the choice predictions of both theories, i.e. the greedy vs. the minimum triangle path mechanism. Fig. 7a gives an overview of the geometric situation of the experiment, assuming that the distance to the target (18a) is on average perceived as six times as long as the longest leg (3a) (see section 3.5). The lengths of the initial legs vary between a, 2a, and 3a. The Euclidean distance $c_{\alpha,l}$ of the completing legs from the end node of an initial segment to the target (which is the relevant variable for the ranking with greedy) as well as the fictive length of the triangle path ($t = l + c_{\alpha,l}$) depends on α and l, where α denotes the deviation angle and l the length of the initial segment. End nodes of initial segments are labeled $N_{\alpha,l}$. For the demonstration case, bold lines denote the initial

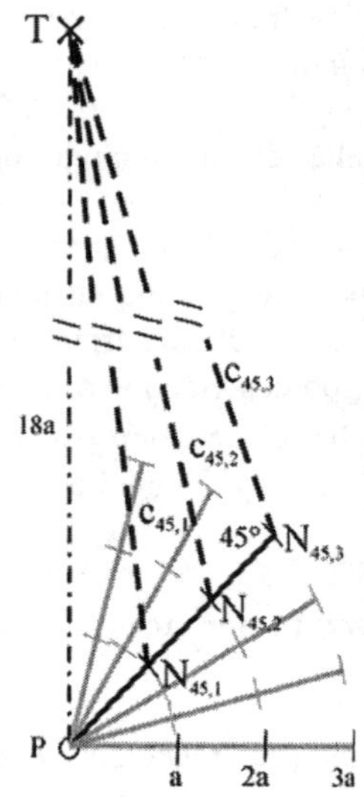

(a)

α [°]	l [a]	greedy ($c_{\alpha,l}$) [a]	greedy rank	Average rank	triangle path rank	triangle path $t = (l + c_{\alpha,l})$ [a]
15	1	17,04	8	2	1	18,04
15	2	16,08	4	1	2	18,08
15	3	15,12	1	3	3	18,12
30	1	17,14	10	4	4	18,14
30	2	16,30	5	5	5	18,30
30	3	15,47	2	6	7	18,47
45	1	17,31	11	7	6	18,31
45	2	16,65	6	9	9	18,65
45	3	16,02	3	11	11	19,02
60	1	17,52	13	8	8	18,52
60	2	17,09	9	12	13	19,09
60	3	16,70	7	14	15	19,70
75	1	17,77	15	10	10	18,77
75	2	17,59	14	15	14	19,59
75	3	17,47	12	16	17	20,47
90	1	18,03	16	13	12	19,03
90	2	18,11	17	17	16	20,11
90	3	18,25	18	18	18	21,25

(b)

Fig. 7. Greedy algorithm and minimum triangle path: Overview of geometry (a) and predicted and observed rankings (b) legs for a deviation angle of 45°, and the dotted bold lines show the completing legs for the three corresponding end nodes. Fig. 7b predicts a ranking based on minimizing the relevant variables for both discussed strategies, i.e. $c_{\alpha,l}$ and t (columns 4 and 6). Column 5 lists the average ranking for the 18 (α,l) combinations being computed from the participants' binary preference statements (section 3.1).

The scatter plots in Fig. 8 visualize the correlation between the average rankings from the experiments (Fig. 7b, column 5) and the predicted rankings for the greedy (Fig. 7b, column 4) and the minimum triangle path (Fig. 7b, column 6) algorithm.

Statistical analysis confirmed a linear correlation between the average stated rank and the minimum triangle path strategy by a Spearman's rho coefficient of 0.990 (α=0.000). Further, the average stated ranking and the greedy ranking also have a significant correlation of 0.660 (α<0.004). The results of the correlation show that the minimum triangle path fits better to the observed preference behavior than the greedy algorithm does. When following the triangle path mechanism, longer initial legs are only selected if they match with the least-angle solution, i.e. if they deviate less from the target direction than the alternative initial segment does. If participants would strictly obey ISS without taking into account the deviation angle—at least for angles up to about 60°—this would give evidence for using the greedy behavior, namely the attempt to approach the target as close as possible with the initial single step. As can be seen from Fig. 7b when comparing the best ranked greedy alternatives (the underlined values) and the corresponding average rankings, such behavior could not be observed for the average participant.

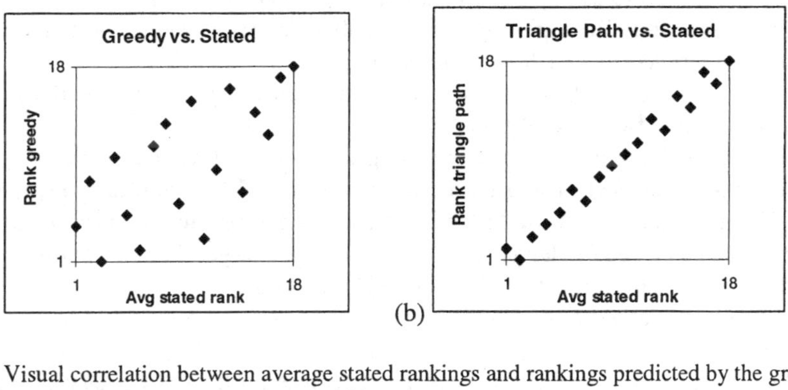

(a) (b)

Fig. 8. Visual correlation between average stated rankings and rankings predicted by the greedy (a) and the minimum triangle path strategy (b)

Although highly correlated with the observed behavior, the predictions by the minimum triangle path rule slightly deviate from the observed behavior for larger deviation angles (right half in Fig. 8b). This occurs for example in scene $leg_{60,3a}$-$leg_{90,a}$, where $leg_{60,3a} \succ leg_{90,a}$, although $t_{60,3a} > t_{90,a}$. Thus the minimum triangle rule must be slightly adopted with a weighting factor that stresses a preference for small deviation angles in addition to the preference for a small triangle path length.

Preference for shorter initial legs is further supported by participants' verbal statements made after the experiments: Participants claimed preference for the shortest leg more or less regardless of the deviation angle, as shorter initial legs would provide them with the opportunity to explore further choice options quickly at the next intersection, which in turn reduces the cost of potentially required back-tracking when compared to long initial segments. This risk-aversive behavior may explain why the ISS was rarely applied in the given scenarios, and why short initial lengths that contribute to a short triangle path length, were generally preferred to the longer lengths.

3.5 Variation in Preference Behavior Between Individuals

Although the observed average preference behavior has revealed clear patterns (section 3.3), the evaluation of participants' preference statements indicates slight variations in the preferential behavior between individuals. Some of the participants showed preference for longer initial segments at a given angle. According to the classification of decision situations provided in section 3.3, Fig. 9 visualizes the variation of preference behavior between individuals for each of the four classes.

Each bar in Fig. 9 describes a single decision situation, i.e. one street combination taken from the set of 37 intersections. The size of α and l of the two involved legs can be read from the respective end points of the bar. Each bar is split into three differently colored pieces. The relative length of each colored piece denotes the percentage of the 28 participants that prefer the corresponding street alternative: The black section of each bar denotes preference for the shorter and/or less deviating leg (Fig. 9a, b, d), whereas the white part denotes the opposite. In the decision situations referred to in Fig. 9c, which require the decision maker to mentally trade off the shorter initial distance against a smaller deviation angle, the black section denotes preference for the less deviating but longer leg. The grey sections placed between the white and black piece indicate the percentage of preferential indifference between the two legs (Fig. 9a - d).

Except for decision situations involving legs with a deviation angle of 15°, the participants' agreement on preference for shorter initial legs is clearly recognizable from Fig. 9a. The frequent use of the shorter legs led at first to the theory that the test participants perceived the longer legs to overshoot the target. In order to rule this out, a test series was performed that aimed at exploring how long the distance to the target was perceived relative to the length of the longest roads. The results showed that the average perceived distance to the target was six times as long as the longest road. Consequently, the theory that the frequent preference for the shorter legs is caused by fear of walking too far could be rejected. Agreement on preference for less deviating streets at constant leg length can be read out from Fig. 9b. In all cases of this class, the rate of agreement amounts to 50% or higher. A clear preference for the alternative that is both the shorter and less deviating one is clearly recognizable from Fig. 9d. Each of the preferred alternatives in the three discussed classes also yields the shorter triangle path. Also in the cognitively demanding class of intersections (Fig. 9c), agreement on preference for routes with the shorter triangle path can be observed. The rankings obtained from the triangle path strategy are significantly correlated with those obtained from the LA strategy (Spearman-rho coefficient: 0.893; $\alpha = 0.000$). No significant correlation could be found between ranks from the minimum triangle theory and the ISS (0.393; $\alpha > 0.10$), which means that the LA by itself fits better with the observed behavior than the ISS does. Participants did not agree upon a specific preferential behavior when the difference in the triangle path length was small between both perceived route alternatives. Examples of such intersections are $leg_{45,3}$ (t=19.02) and $leg_{75,1}$ (t=18.77) or $leg_{60,3}$ (t=19.70) and $leg_{75,2}$ (t=19.59).

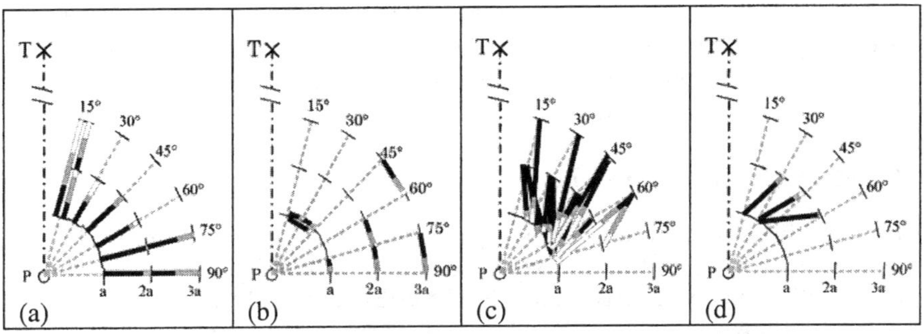

Fig. 9. Matching rates of individuals' stated preferences

4 Individual Choice Consistency

In the experiment described above, it was observed that when presenting the same intersection several times, individuals are not consistent in their choices. Numerous models of decision rules that take into account some level of uncertainty have been developed over the last decades. Two classes can be distinguished, depending on the assumptions about the source of uncertainty: Models with stochastic decision rules (Luce 1959) assume a deterministic utility, whereas random utility models (Ben-Akiva and Bierlaire 1999)—as neoclassical economic theory—assume that the decision maker has perfect discrimination behavior and that the analyst has incomplete information. This paper will neither seek the reason for observed inconsistencies in the decision makers' preference behavior nor introduce a choice selection model. Participants' potential bias for right or left is not discussed, since consideration of this aspect in the experiment would require the inclusion of many additional vertically mirrored intersections in the sequence of decision situations, which in turn would overstrain the participants' concentration abilities. This question should be addressed in a separate series of experiments within future work. However, we assume that potential directional bias would not dramatically change the findings concerning preferential behavior.

One possible method for measuring consistency is to use a metric for computing the distance between a CSP and the solved PCSP (see section 3.1). However, such a distance measure is hard to express in terms of a geometrical interpretation of the underlying inconsistency. Another disadvantage of such a measure is that the numerical result depends on the selected leg combinations actually used in the study as well as on the number of intersections compared to all possible combinations. Due to these reasons we decided to use choice repetition reliability as a substitute for a metric measure of consistency.

Internally, the experiment was split into two test series, the main study involving a group of 17, and the follow-up study involving a group of 11 participants. The evaluation of the observed statements from the first 17 participants revealed the above mentioned inconsistencies in the individuals' choice behavior. The follow-up study aimed at illuminating more details in the preference behavior by providing a reliability measure of observed decisions. To each of the last 11 participants four selected intersections (out of the 37) were shown five times each (in arbitrary order), in addition to the original set of 37 intersections from the first part. This yields a total number of 57 scenes per participant. Through this, in addition to observing the mean preference behavior, the variability of preference statements upon repeated presentation of the same intersection could be observed.

4.1 Selection of Repeated Scenes

The variability of preference statements in a repeated decision situation depends on the dissimilarity between the two involved street segments. We hypothesize that for intersections with similarly preferable legs, the choice behavior will tend towards a random decision (i.e. provide low repetition reliability), whereas for intersections with legs that are clearly discernable in terms of their assigned preference value, the choice behavior will be more reliable and show a clear preference tendency for either of the two legs. To be able to select representative scenes for the repetition task we make a simplification and treat the ordinal rank data as metric data, i.e. we assume that the preferential difference between all pairs of alternatives that are separated by one rank remains constant over the complete scale of ranks. Although this simplification may not strictly hold for each individual, it is accurate enough to find a ranking of intersections according to a measure that captures the similarity between two involved legs in an intersection. Based on this simplification we took the mean preference values for all 18 legs from the first 17 participants and computed the average of the differences of ranks (Δ rank) between the two legs involved in each of the 37 intersections, which yielded 37 Δ rank values. The absolute scores for mean Δ rank ranged between 0.2 ($leg_{15,a}$- $leg_{15,2a}$) and 7.1 ($leg_{30,a}$- $leg_{60,2a}$). The first criterion for the selection of four intersections to be used in the repetition task was that the mean Δ rank values of selected intersections should cover a wide range, thus both obvious and not so obvious decision situations should be included. Further we considered only intersections of class type c and d (Fig. 6c and d) for which the decision maker is forced to trade off mentally the deviation angle against the leg length. These cases represent more realistic conditions.

Fig. 10 shows the geometry of the four intersections (with their mean Δ rank values from the first 17 participants) that we selected for the repetition task. The left street alternative in each intersection was, according to the average results, preferred by the 17 participants in the first experiment. We expected a small repetition reliability for the left most intersection, and a tendency towards a higher reliability of statements for intersections with higher Δ rank values. The 20 additional intersections were interspersed with the 37 previously designed intersections so that participants were not able to recognize the repetition of the scenes. Thus participants were actually forced to make their decision (i.e. not just repeat previous statements) each time they were presented with one of these repeated scenes. Each of the four repeated scenes was presented five times, and no balance with left to right was made within the repetitions. Through this, a potential impact of preference for direction on the observed repetition reliability could be excluded.

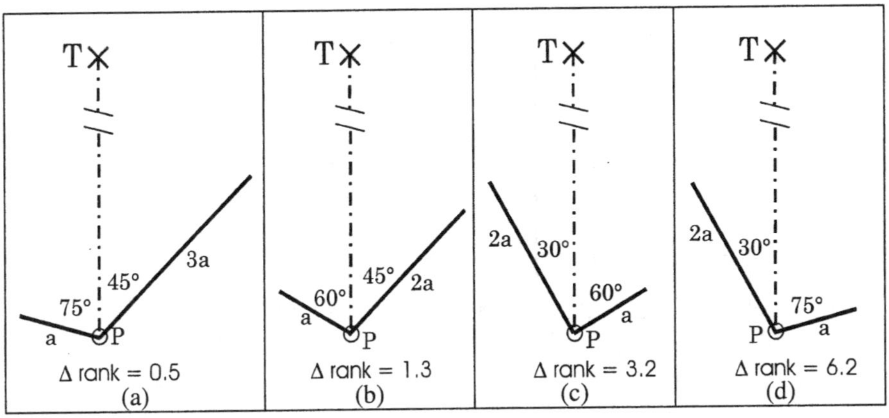

Fig. 10. The four intersections with leg pairs of increasing dissimilarity (Δ rank) used for the repetition task

4.2 Computation of Repetition Consistency

Because of variances between individual preference behavior we cannot expect the mean Δ rank value (Fig. 10) of the four chosen intersections to hold for each of the 11 participants. Thus, in a first step the Δ rank values for the four intersections had to be computed for each individual separately and then re-classified into a number of yet unknown classes. The first step yielded a list of 44 Δ rank values between 0 and 11.3 (Fig. 11). To find a representative classification of these 44 instances, we applied a cluster analysis using the Ward method with a Euclidean distance measure (Backhaus et al. 1996). The high distance coefficients at an aggregation level of 3 (marked by the bold line in the dendrogram in Fig. 11b) show that the three clustered classes capture the nature of the distribution of Δ rank values quite well. Dashed horizontal brackets in Fig. 11a indicate the groups that were finally used as the basis for the analysis of the reliability measures. The leftmost class ($0 \leq \Delta$ rank < 3.2) contains 19, the middle class ($3.2 \leq \Delta$ rank < 8.2) 17, and the rightmost class ($8.2 \leq \Delta$ rank ≤ 11.3) 8 preference statements out of 44.

Thus, each of the 44 evaluated scenes of Fig. 11a (where each scene requests five single preference statements) was assigned to one of the three clustered classes. To compute a variability measure (v) of how the preferences changed within each set of five statements we assigned numerical values to choice results (preference for left: -1; indifference: 0; preference for right: 1). We computed the variability measure by first determining the most preferred leg (left or right) for each of the 44 scenes, and by

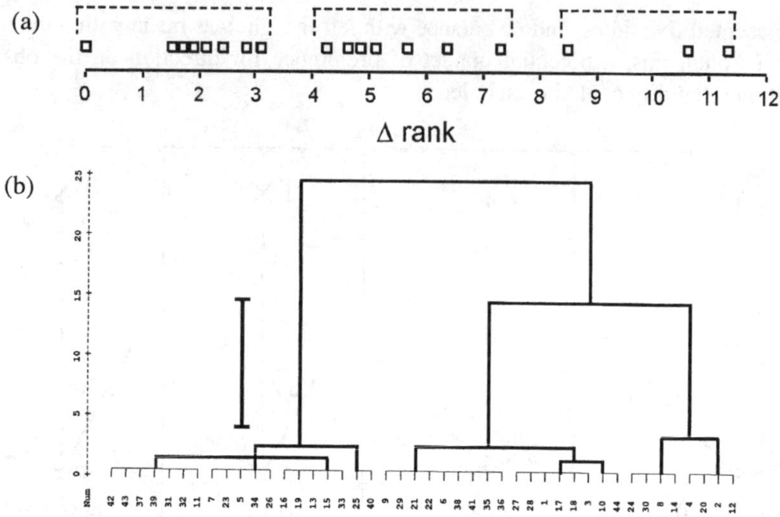

Fig. 11. Distribution of Δ rank values for the four selected repeated scenes (a) and dendrogram for hierarchical clustering of Δ rank values (b)

summing up the absolute value differences between the remaining choices in the scene and the value of the most preferred leg. Thus, the higher the variability measure for a scene, the less consistent is a participant with her preference for this particular intersection. The maximum value for v occurs with a preference statement containing the elements 1,1,0,-1,-1. For this case, the sum of differences yields 5 (through 1*(1-0) + 2*(1-(-1)), which denotes high inconsistency in the participant's preferential behavior. On the other end of the range, a statement of 1,1,1,1,1 or -1,-1,-1,-1,-1 leads to v = 0, which shows perfect consistency in the participant's decision behavior. The same (misleading) result would be received with a series of zeros (which actually means consistent indifference but not consistency in the preferential behavior at all). However, this situation did not occur in the observed data, and consequently we could apply the suggested method on our data set.

4.3 Results

Computation of the average over the variability measure in each of the three clustered classes (Fig. 11a) gives an impression of the participants' decision reliability (Fig. 12). A value of 100% denotes the maximum possible average variability (i.e. v = 5). As expected, participants were more inconsistent at intersections that involve similarly preferable street alternatives (v_{Class1} = 48.4 %), whereas the variability decreases for intersections with more discernible route segments (v_{Class2} = 28.2 %, v_{Class3} = 20.0 %). An "average" intersection for the first class would be a scene $leg_{60,a}$-$leg_{45,2a}$, for the second class a scene $leg_{15,2a}$-$leg_{45,a}$, and for the third class a scene $leg_{30,a}$-$leg_{60,2a}$.

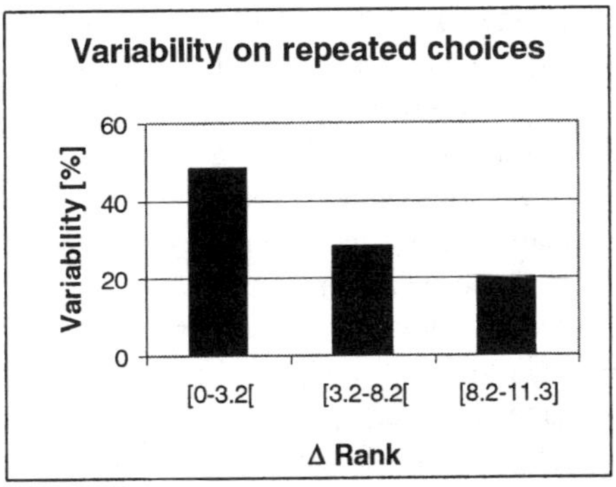

Fig. 12. Variability of choice behavior on repeatedly presented intersections

5 Conclusions and Future Work

This article has presented empirical results about human route choice behavior in an unknown environment. The following trends can be derived from the observed data: The average participant prefers the initially shorter segment, if this segment deviates equally or less from the target direction compared to the alternative (cases a and d in Fig. 6). Further, between equally long initial segments, the less deviating segment is preferred (case b). For decision situations where one choice alternative has a less deviating but longer initial leg (case c), the selection (rejection) of an alternative would mean to obey (abandon) both LA and ISS at the same time. In such a situation, the two strategies cannot be separated one from each other. The results obtained for case (c) suggest a mechanism in route choice where the decision maker tries to minimize the estimate length of the total route that would have resulted if the route had continued straight towards the goal after the initial segment (triangle path length). The minimum triangle path strategy is appropriate to explain the decision behavior for all four classes of decision situations. It highly correlates with the LA strategy.

When it comes to measuring the individual choice consistency, it has been shown that route selection behavior depends on the difference of the preference values assigned to each of the included choice alternatives. Variability in the individual route choice behavior may be caused by changing preference values assigned to an alternative over time, or due to limited cognitive ability to discriminate between two street alternatives perceived at an intersection.

The experiment gives clear evidence against the initial segment strategy in field view, although the use of this strategy on maps has been empirically proven in previous work. This leads to the conclusion that strategies observed in map-based route choice (such as the ISS) do not always generalize to view-based route choice. One of the reasons may be the higher degree of uncertainty for route choice in the real environment as compared to route choice on maps, as the whole environment can be seen on a map, which is not the case in field view (this holds especially for survey relationships between objects). The lack of information may cause the decision maker in the view-based perspective to take less risky routes, i.e. prefer short segments as opposed to long corridor like routes. A more detailed analysis of the impact of the perspective on the decision behavior is part of the future work.

The experiments further show that both the deviation angle and the initial length interfere in the decision maker's preference behavior. However, decision making in real world wayfinding situations is far more complex than this. To get a more detailed picture about human preference behavior in unknown environments, future experiments will assess the impact of additional street parameters on the decision outcome. We expect that besides deviation angle and initial leg length among others the following three geometric parameters affect preference behavior between several initial street segments: The number of intersections that are visible along the initial street segment, the direction of the intersecting streets with the initial street segment, and the width of the initial street. These parameters may affect the decision maker's judgment about how risky an alternative is with respect to time consuming detours, which in turn affects the preference for an alternative.

Acknowledgements

This research has been achieved in the framework of the project I2-[MapSpace] within the Transregional Collaborative Research Center SFB/TR8 Spatial Cognition funded by the Deutsche Forschungsgemeinschaft (DFG). Financial support (IQN grant) from the German Academic Exchange Service (DAAD) is gratefully acknowledged. The authors want to thank Elizabeth Leppman and David Wall who helped to improve the grammar and style of this work.

References

Backhaus K, Erichson B, Plinke W, Weiber R (1996). *Multivariate Analysemethoden*. Berlin: Springer

Bailenson JN, Shum MS, Uttal DH (1998). Road Climbing: Principles Governing Asymmetric Route Choices on Maps. *Journal of Environmental Psychology 18*: 251-264

Bailenson JN, Shum MS, Uttal DH (2000). The initial segment strategy: A heuristic for route selection. *Memory and Cognition 28* (2): 306-318

Ben-Akiva M, Bierlaire M (1999). Discrete choice methods and their applications to short-term travel decisions. In R Hall (Ed.), *Handbook of Transportation Science*. Kluwer, Dordrecht, pp. 5-34

Bose P, Morin P (1999). Online routing in triangulations, *ISAAC*, Vol. 1741. Springer, Berlin, pp. 113-122

Christenfeld N (1995). Choices from identical options. *Psychological Science 6*: 50-55

Dalton RC (2001). *The Secret is to Follow Your Nose*. 3rd International Space Syntax Symposium, Atlanta. A. Alfred Taubman College of Architecture and Urban Planning, University of Michigan

Freuder EC, Wallace RJ (1992). Partial Constraint Satisfaction. *Artificial Intelligence 58* (1-3): 21-70

Hochmair HH, Frank AU (2002). Influence of estimation errors on wayfinding-decisions in unknown street networks - analyzing the least-angle strategy. *Spatial Cognition and Computation 2* (4): 283-313

Janzen G, Herrmann T, Katz S, Schweizer K (2000). Oblique Angles Intersections and Barriers: Navigating through a Virtual Maze. In C Freksa, W Brauer, KF Wender (Eds.), *Spatial Cognition 2*, LNCS 1849. Springer, Berlin, pp. 277-294

Kranakis E, Singh H, Urrutia J (1999). *Compass Routing on Geometric Networks*. 11th Canadian Conference on Computational Geometry (electronic proceedings)

Løvås GG (1998). Models of Wayfinding in Emergency Evacuations. *European Journal of Operational Research 105* (3): 371-389

Luce R (1959). *Individual choice behavior: a theoretical analysis*. New York: J. Wiley and Sons

Moratz R, Freksa C (1998). Spatial Reasoning with Uncertain Data Using Stochastic Relaxation. In W Brauer (Ed.), *Fuzzy-Neuro-Systems*. Infix, St. Augustin, pp. 106-112

Poole D, Mackworth A, Goebel R (1998). *Computational Intelligence: A Logical Approach*. New York: Oxford University Press

Rudová H, Murray K (2002). University Course Timetabling with Soft Constraints. In E Burke, PD Causmaecker (Eds.), *4th international conference on the Practice and Theory of Automated Timetabling (PATAT 2002)*, KaHo St.-Lieven, Gent, pp. 73-89

Thorndyke P, Hayes-Roth B (1981). Differences in spatial knowledge acquired from maps and navigation. *Cognitive Psychology 14*: 560-589

Spatial Prepositions and Vague Quantifiers: Implementing the Functional Geometric Framework

Kenny R. Coventry[1], Angelo Cangelosi[2], Rohanna Rajapakse[2], Alison Bacon[1], Stephen Newstead[1], Dan Joyce[3], and Lynn V. Richards[1]

[1] Centre for Thinking and Language, School of Psychology, University of Plymouth, Drake Circus, Plymouth, PL4 8AA, United Kingdom
{kcoventry, ambacon, snewstead}@plymouth.ac.uk
[2] Adaptive Behaviour & Cognition Research Group, School of Computing Comms & Electronics, University of Plymouth. Drake Circus, Plymouth, PL4 8AA, United Kingdom
{acangelosi, rrajapakse}@plymouth.ac.uk
[3] Institute of Cognitive Neuroscience, University College London, Alexandra House, 17 Queen Square, WC1N 3AR, United Kingdom

Abstract. There is much empirical evidence showing that factors other than the relative positions of objects in Euclidean space are important in the comprehension of a wide range of spatial prepositions in English and other languages. We first the overview the *functional geometric framework* [11] which puts "what" and "where" information together to underpin the situation specific meaning of spatial terms. We then outline an implementation of this framework. The computational model for the processing of visual scenes and the identification of the appropriate spatial preposition consists of three main modules: (1) Vision Processing, (2) Elman Network, (3) Dual-Route Network. Mirroring data from experiments with human participants, we show that the model is both able to predict what will happen to objects in a scene, and use these judgements to influence the appropriateness of *over/under/above/below* to describe where objects are located in the scene. Extensions of the model to other prepositions and quantifiers are discussed.

1 Introduction

Expressions involving spatial prepositions in English convey to a hearer where one object (located object) is located in relation to another object (reference object). For example, in *the coffee is in the cup*, the *coffee* is understood to be located with reference to the *cup* in the region denoted by the preposition *in*. Understanding the meaning of such terms is important as they are among the set of closed class terms which are generally regarded as having the role of acting as organizing structure for further conceptual material [43]. Furthermore, from the semantic point of view spatial prepositions have the virtue of relating in some way to visual scenes being described, and therefore measurable characteristics of the world [40]. Hence, it should be possible to offer more precise semantic definitions of these as opposed to many other expressions because the definitions can be grounded in perceptual representations.

C. Freksa et al. (Eds.): Spatial Cognition IV, LNAI 3343, pp. 98 – 110, 2005.
© Springer-Verlag Berlin Heidelberg 2005

Most approaches to spatial prepositions have assumed that they only require coarse grained properties of the objects involved as constraints on their use (e.g. [28,34]. Computational models too have made the same assumption, and have focused on mapping individual prepositions onto geometric computations in the scene being described (e.g., [23,35,40,41]). Yet there is now much evidence (see Coventry & Garrod [11] for a comprehensive review) that "what" objects are influences how one talks about "where" they are. For example, Coventry, Prat-Sala and Richards [15] found that acceptability ratings of sentences such as the *umbrella is over the man* were influenced by whether the objects in the scene were shown to be fulfilling their protection (or containment) functions. For instance, with reference to the scenes shown in Figure 1, sentences were rated as being significantly more appropriate when the umbrella was depicted as protecting the man from rain (scenes in the middle row), and least appropriate when the rain was falling on the man (scenes in the bottom row). Furthermore, extra-geometric variables came into play even when the prototypical geometric constraint for the use of a term holds (i.e., effects were found even for scenes in the first column). Additionally, Coventry et al. found that function has a much bigger affect on the ratings for *over/under* than for *above/below*, and conversely that geometry (e.g., rotation of the umbrella in Figure 1) influences the ratings of *above/below* more than *over/under*.

Fig. 1. Example scenes used by Coventry, Prat-Sala and Richards [15]

Similar effects have been found across a wide range of prepositions and methodologies. For example, extra-geometric effects have been found for *in* and *on* [10,14,20,24], *above* [5,7], *over* [13], *in front of* and *behind* [6], *between* [12,48], and *near* [21]. Furthermore, the effect sizes found across these studies indicate that these effects are not minor pragmatic add-ons to geometric formulations, but rather indicate that extra-geometric variables are central to the comprehension and production of spatial terms.

1.1 The Functional Geometric Framework

Reviewing the extra-geometric evidence, Coventry and Garrod [11,12] classify these influences into two types; *dynamic-kinematic routines*, and *conceptual knowledge* regarding the specific functions associated with specific objects. Dynamic/kinematic routines implicate knowledge of what will happen to scenes over time, and the initiation of such routines is related to knowledge of *what* objects are in the scene. In particular these dynamic/kinematic routines relate to Jeannerod's [30,31] distinction between "semantic" visual representations, usually associated with visual imagery, and "pragmatic" representations associated with motor imagery. Jeannerod assumes that motor images underlie such things as preparing for an action or rehearsing an action. Furthermore he argues that the two representations, the semantic and the pragmatic, have a neural correspondence with the *what* and the *where* systems described above. Whereas "semantic" representations encode relatively detailed information about objects in a scene, "pragmatic" representations encode visual properties in relation to affordances, i.e., those visual characteristics that are important in organizing motor programs for manipulating the objects. These include information about the size, weight and shape of objects, as well as special features of those objects that are relevant for their manipulation, such as the location of handles for grasping. Empirically Freyd, Pantzer and Cheng [22] (see also [42]) carried out experiments in which they observed systematic memory errors for scenes involving the same objects in the same geometric configurations, but with different forces acting on them. Thus, in a situation where a plant pot is first seen supported by a chain then not supported, observers tend to misjudge the position of the plant pot as being lower in a subsequent memory test. In the spatial language domain, Coventry [8] and Garrod, Ferrier and Campbell [24] have demonstrated similar effects for *in* and *on*. For example, using static scenes involving pictures of ping pong balls piled high in containers with a string attached to the top ping pong ball in many scenes, they found that ratings of the appropriateness of *in* to describe such scenes was directly correlated with independent judgments of the likelihood that the ball and container would remain in their same relative positions over time should the container be moved.

In addition, a great deal of specific knowledge about objects is also required. For example, the same convex object labelled a *dish* versus a *plate* is clearly associated with the expectation of a containment versus a support relation [10]. Similarly, knowing that jugs are primarily containers of liquids has been shown to weaken *in* judgements for solids piled high in a jug as compared with *in* judgements for the same pile in a bowl with the same degree of concavity [10,14].

Coventry and Garrod [11] argue, importing terminology from Ullman [44,45], that the application of geometric and dynamic-kinematic routines underlie the comprehension of spatial prepositions. Furthermore, the application of such visual routines is driven by knowledge of the objects involved in the scene and how those objects typically interact in past learned interactions between those objects. Furthermore, just as objects are associated with particular routines, both geometric and dynamic-kinematic, prepositions themselves have weightings for these parameters. As we have seen above, the comprehension of *over/under* is better predicted by extra-geometric relations than the comprehension of *above/below*, while conversely the comprehension of *above/below* is better predicted by geometric routines than the comprehension of *over/under*. In the functional geometric framework it is how these constraints "mesh" together (cf. [2, 25]) that underpins the comprehension of spatial prepositions. The computational model we next outline implements the multiple constraint satisfaction in the functional geometric framework and maps onto new and existing datasets from human participants. The approach introduces cognitive-functional constraints by extending Ullman's [45] notion of visual routines to include operations on dynamic rather than static visual input. We next outline the components of the model, together with the experimental data used to test and validate the model.

2 Implementing the Functional Geometric Framework

2.1 Experimental Data

The model we outline shortly can deal with a range of prepositions, but here we focus on *over/under/above/below*. We conducted a series of experiments (see [9] for more details) involving three different reference objects (a plate, a dish and a bowl) pretested in a sorting task and a rating task to be the prototypical dimensions of these objects, and a variety of other objects which were all containers (e.g., a jug). Each container was presented in each of 3 x 2 positions "higher" than the other objects (representing 3 levels of distance on the x axis and two levels on the y axis from the other object). Crucially the container was shown to pour liquid such that it ended up reaching the plate/dish/bowl (the functional condition), or missed the plate/dish/bowl (non-functional condition), or liquid was not present. Figure 2 shows some example scenes. The methodology used for these experiments involved the presentation of pictures together with sentences of the form *The located object is preposition the reference object*, and the task for participants was to rate the appropriateness of each sentence to describe each picture using a Lickert scale (range from 1 = totally unacceptable to 9 = totally acceptable).

In Experiment 1 participants saw movies of the pouring scenes (or static scenes for the no liquid condition given that no movement was involved). The results showed effects of geometry and function together with interactions between these variables and *over/under* versus *above/below*, effectively replicating the results of Coventry et al. [15]. Experiment 2 compared the full movies with just the (single frame) end states, and this established that seeing the full movie makes no difference to acceptability ratings, it is what happens to the liquid that counts. Experiment 3 then compared end states to an earlier frame in the movie showing the liquid starting to protrude from the pouring container (see bottom picture in Figure 2) in order to assess whether participants predict what will happen to the liquid in order to make judgments about the appropriateness of *over/under/above/below*. Although acceptability ratings were overall lower for the predicted scenes rather than the end state scenes, effects of geometry, function and interactions between these variables and *over/under* versus *above/below* were still present, indicating that participants do predict where the liquid will go in order to ascertain the appropriateness of these prepositions. Experiment 4 confirmed this by finding a correlation between judgments of how much of the liquid will make contact with the appropriate part of the plate/dish/bowl and acceptability ratings for *over/under/above/below*.

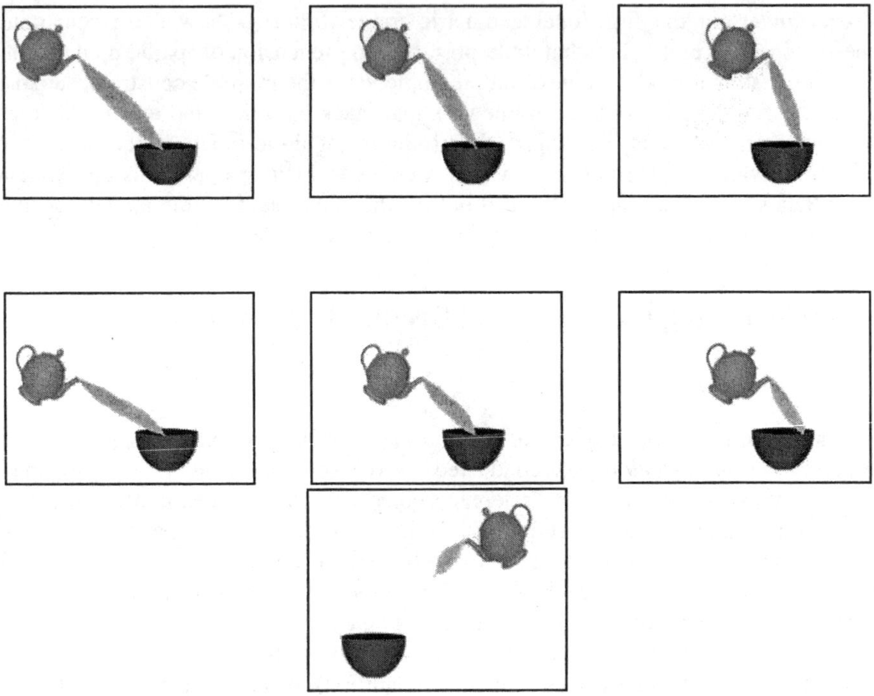

Fig. 2. Sample scenes used in the experiments. The top six pictures represent the 6 levels of geometry used. All six pictures show the functional condition where the liquid was shown to end up in the container. Non-functional scenes involved the same relative positions of teapot and container, but this time the liquid was shown to miss the container. The bottom picture shows an example of a scene where only the start state was shown

The data from these experiments indicate that participants use both information about the geometry in the scene and information about the interaction between pouring container and recipient container in the scene to assess the appropriateness of *over/under/above/below*. As has been found previously [15], the influence of geometry was stronger for *above/below* than for *over/under*, while the influence of function (whether the liquid was shown to enter or miss the recipient container, or was predicted to enter or miss the container) was stronger for *over/under* than for *above/below*.

Data from these experiments were used as a means of testing and training the model, which we outline next.

2.2 The Computational Model

The computational model for the processing of visual scenes and the identification of the appropriate spatial preposition consists of three main modules: (1) Vision Processing, (2) Elman Network, (3) Dual-Route Network (cf. Figure 3). The first module uses a series of Ullman-type visual routines to identify the constituent objects of a visual scene (reference object, located object and liquid). The Elman network module utilises the output information from the vision module to produce a compressed neural representation of the dynamics of the scene (e.g. movement of liquid flow between the reference and located objects). This compressed representation is given in input to the dual-route (vision and language) feedforward neural network to produce a judgment regarding the appropriate spatial terms describing the visual scene. We describe each of these modules and their development in turn.

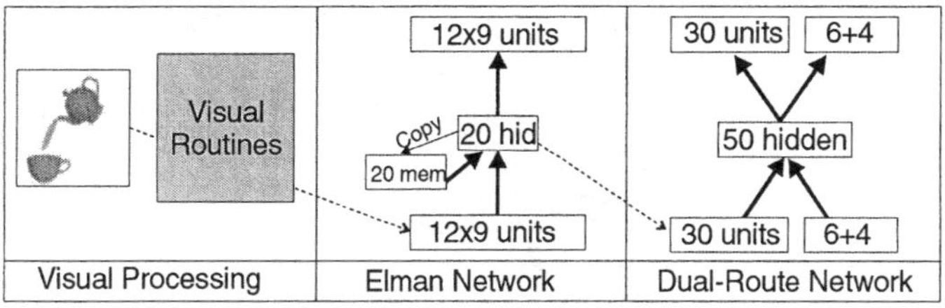

Fig. 3. Architecture of the computational model. The dotted arrows indicate functional connections between the three modules. The dual-route network has 30 visual input/output units because they copy the hidden activation of 3 different Elman networks (one with 20 hidden units, and two with 5 units each)

2.2.1 Vision Processing Module

In our computational model for spatial language, visual object recognition, spatial location and motion information are functionally necessary for the cognitive task. Beginning with the distinction between "what" versus "where" pathways (classically assumed to be the functionally segregated dorsal and ventral streams after Ungerleider and Mishkin [46]), we also needed to consider the integration of object, location and motion information when deriving a neurocomputational model. Our novel neurocomputational approach to object recognition for spatial cognition represents a compromise between the dynamic operation of the recurrent neurodynamical models of Deco and Lee [16] for selective attention, and Edelman's [17] feedforward chorus model for object recognition, and is conceptually congruent with Ballard et al's [1] model (i.e. the output of our system is a plausible deictic pointer to objects in the visual scene). Image sequences (real object images composed into moving videos) are presented to the model, which processes them at a variety of spatial scales and resolutions for object form and motion features yielding a visual buffer (functionally analogous to processing in the striate visual cortex). In addition to the basic scale representation, texture, edge and region boundary features are extracted. Motion cells (in the magnocellular pathway) are modeled as uni-directional brightness gradient-sensitive cells whose outputs are combined. This is outlined in Figure 4.

The attentional saliency map (Figure 4, Right) is a very low resolution (retinotopic) array of neurons which receive bottom-up activation from the static and motion features in the visual buffer, but which can be strongly inhibited when the region they code for is attended to or when object recognition is strong enough to require little further processing of a region. This represents information integration that might take place involving the kinds of information processed in the posterior

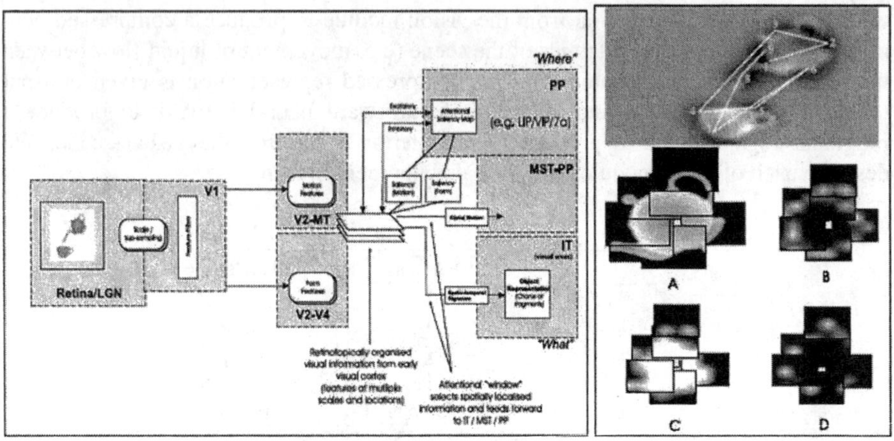

Fig. 4. Left: Constituents of the Vision Processing Module and their relationships with known neural substrates. Right (Top): Snapshots of the overall saliency map after 9 fixations. Right (Bottom): Multiple Fragments of Teapot Object (A) Full visual buffer (B) Edges (C) Region/Boundary and (D) Texture

parietal cortex. This is used to direct attention and once a region is selected (analogous to a kind of spotlight of attention), the higher-resolution information contained in the visual buffer is allowed to feedforward to the object recognition stream. Since attention selects only a windowed region of the whole visual buffer for processing in IT, our system represents a chorus of object fragments. We use Gaussian adaptive resonance models to learn the space of fragments for each object [47], leading to a probabilistic implementation.

We elaborate on the visual processing and selective attention mechanism and its role in a novel chorus of fragments framework for object recognition elsewhere [32]. We show how this may form part of a larger system for spatial language comprehension and speculatively for prefrontal cortex short term visual memory and object-place binding (via the perirhinal – entorhinal – hippocampal network), all of which further ground the understanding of the visuo-spatial processing in a computational framework.

2.2.2 Elman Network Module

This module consists of a predictive, time-delay connectionist network similar to Elman's [19] simple recurrent network, which we refer to hereafter as the Connectionist Perceptual Symbol System Network (CPSSN; [33]). Figure 3, middle image, shows the CPSSN network as an Elman SRN. As a suitable (and plausible) input representation for the CPSSN, we propose a "what+where" code (see also [18]). That is, the input consists of an array of some 9x12 activations (representing retinotopically organised and isotropic receptive fields) where each activation records some visual stimulus in that area of the visual field. This is the output information produced by the Vision module. In addition to the "field" representation, we augment a distributed object identity code. These codes were produced by an object representation system ([32]; based on Edelman's [17] theory) using the same videos. The CPSSN is given one set of activations as input which feedforward to the hidden units. In addition, the previous state of the hidden units is fed to the hidden units simultaneously (to provide a temporal context viz. Elman's [19] SRN model). The hidden units feedforward producing an output which is a prediction of the next sequence item. Then, using the actual next sequence item, back propagation is used to modify weights (see Figure 3) to account for the error. The actual next sequence item is then used as the new input to predict the subsequent item and so on. Using the coding scheme discussed, we have a total input vector of length 116 (where 8 of these 116 elements code for each object, e.g. liquid, bowl, cup etc.). The output is similarly dimensioned, and there were 20 hidden units (and 20 corresponding time-delayed hidden state nodes) to represent movement of the liquid.

The network training regime was as follows: a collection of sequences are shown to the network in random order (but of course, the inputs within a sequence are presented one after another). Each sequence contains a field and object code for the "liquid" in the videos. Multiple CPSSN networks would be required to account for the other objects in the scenes. A root-mean-square error measure is used to monitor the network's performance, and the ordering of sequences is changed each time (to prevent destructive interference between the storage of each sequence). Initially, the network is trained with a learning rate of 0.25, and after the RMS error stabilises, this is reduced to 0.05 to allow finer modifications to weights. For 6 sequences, a total of about 150 presentations are required (each sequence is therefore presented 25 times) to reduce RMS averaged over the whole training set from around 35 to around 0.4.

It is quite obvious that this network is hetero-associating successive steps in the sequence of fields, but in addition, the network is performing compression and redundancy reduction (in the hidden layer) as well as utilising the state information in the time-delayed state nodes. It is also coding for the changes between sequence items (e.g. the dynamics of how the object moves over time) rather than coding individual sequence items (which would be auto-association). The model embodies the idea that representation is inherently dynamic (cf. [22]). The network should, naturally, be able to make a prediction about a sequence given any item in the sequence. Intuitively, the network should be capable of this in the case where a cue is the first item of a sequence, since the time-delayed state is irrelevant (i.e., there can be no temporal context accumulated in the time-delay nodes). However, we propose that the network is a mechanism for implementing perceptual symbols, and therefore, a requirement is that it can "replay" the properties of the visual episode that was learned. Given a cue, the network should produce a prediction, which can be fed-back as the next input to produce a sequence of "auto-generated" predictions about a sequence (viz, a perceptual symbol). Indeed, this network is able to predict the final outcome of the visual scenes [33].

2.2.3 Dual-Route Network

The dual-route network is a feedforward neural network (3-layer perceptron) that receives in input the grounded "visual" information (hidden activations of the Elman networks) and linguistic data (name of located object, name of reference object, name of liquid + 4 spatial prepositions *over, above, below, under*). In output it must reproduce (auto-associate) the same visual data, and produce the names of objects, which are directly grounded in the input visual data. In addition, the four output units for the spatial prepositions will encode the rating values given by subjects. This architecture is directly inspired by dual-route networks for the grounding of language [3, 4, 39].

This network is trained via the error backpropagation algorithm. The training and test sets consist of the 216 scenes. These are the same as those used in the experiment on the rating of *over, above, under, below* (Experiment 1 above). Of these stimuli, 195 are used for the training and 21 for the generalisation test. The overall objective of the training is that the network must learn to produce the same average ratings for the four prepositions. We did not use the average ratings as the teaching input, because this was against the principle of mutual exclusivity [36]. During standard backpropagation training, the use of the ratings as teaching input assumes that the same scene must be simultaneously associated to the use of all four prepositions (each with an activation value proportional to the subjects' average rating). Instead, during developmental learning subjects tend to choose only one preposition to describe a scene. Naturally, the probability of choosing one preposition to describe a spatial relation is correlated to its level of appropriateness (i.e. similar to ratings). Therefore, to simulate such a learning strategy better, the original ratings of each scene-preposition pair were converted into frequency of presentation of a stimulus with an associated localist teaching input (where the output unit of the chosen preposition is 1 and the other three units are 0). To obtain such a frequency, the original average ratings were scaled and normalised within each scene and also within the whole training set. For example, individual prepositions' ratings of 7.08 (above), 7.12 (below), 3.96 (over), 4.32 (under) respectively correspond to presentation frequencies of 28, 28, 7 and 9. The conversion of ratings into preposition frequencies resulted in an epoch of 2100 stimuli.

Three networks were trained using different initial random weights and different random sets of 21 generalisation test stimuli[1]. The training parameters included a learning rate of 0.01 and momentum of 0.8, and a total number of training epochs of 500. The average final error (RMS) for the 30 vision units was 0.008 for both training and testing data, and 0.003 for the 6 output units of the object names. More importantly, for the 4 spatial preposition output units, the error was 0.044 with training data and was 0.05 with generalisation data. The error values in the preposition units were calculated off-line by comparing the actual output of the 4 preposition units and the rating data (from Experiment 1 overviewed above) converted to produce the stimulus frequencies (the actual error values used for the weight correction are always higher because they use localist teaching input). These results clearly indicate that the networks produce rating values similar to that of experimental subjects. They also indicate that the training algorithm based on presentation frequency, instead of rating teaching input, works well and provides a psychologically-plausible learning regime.

2.3 Interplay Between Experimental and Computational Work

The development of the computational model has been conducted in parallel with experimental investigations. However, in the early part of the development of the model, the experimental work has mostly influenced the model design. For example, in the previous section we explained that the training/test stimuli and the rating values were directly taken from one experiment. Later on in the development of the model, it

[1] Here we report only the data from the best simulation. Different parameters values and hidden layer sizes were tested.

was the model that directed some of the directions and objectives of the experimental investigation. In particular, new simulations produced some predictions that were subsequently tested in new experiments.

Research on the design and test of the Elman module had shown that these networks were able to predict and auto-generate the final outcome of the visual scenes, once they were given an initial cue (e.g few initial frames). The network would produce the next prediction frames, which were fed-back as the next input. To integrate such prediction ability in the overall spatial language model, the hidden activation values of these auto-generated sequences were used as visual input of the dual-route network. The model was then run as usual to produce the ratings of the 4 prepositions.

To establish if the new ratings provided by the model were consistent with those produced by real subjects, a new experiment was conducted (Experiment 4, see above). The results for this experiment, together with the results of Experiment 3, strongly suggest that subjects had to mentally "play" the visual scene and auto-generate the outcome of the scene to rate the linguistic utterance. This is very similar to what the model does, when the Elman network autogenerates the visual scene, and the dual-route network uses the Elman net's activations to produce new ratings. The Elman network used the first 3 out of 7 frames. This corresponds to the frames 0, 10 and 20 (Elman networks only see a frame every 10). The comparison of the subjects' rating data and the networks' output of the 4 prepositions resulted in an RMS error of 0.051. This is a very low error level, and confirms that the model had predicted very accurately the ratings. Overall, this result and those on the dual-route tests support the development of a psychologically-plausible model for spatial language.

3 Discussion: Extension and Links

The model we have outlined has been tested across other spatial relations as well as *over/under/above/below*, including the importance of location control for the preposition *in*. Currently we are extending the model so that it can return a description of the number of objects in the visual input scene as well as the spatial relations between objects depicted. Vague quantifiers like *a few* and *several* exhibit many of the same context effects that have been observed for spatial prepositions. For example, relative size of figure and ground objects [29,38] and expected frequency [37] have both been shown to affect the comprehension of quantifiers; *A few cars* is associated with a smaller number than *a few crumbs*, and *some people in front of the cinema* is associated with more people than *some people in front of the fire station*. These context effects appear very similar to the range of effects in evidence for spatial prepositions. Therefore the issue we are exploring is that these context effects originate from visual processing constraints such that information regarding specific numbers of objects in a scene cannot be derived very easily from visual processing of that scene.

From a theoretical perspective the functional geometric framework and the implementation of it are consonant with recent developments in the embodied cognition literature. The idea that meaning is constructed as a result of putting together multiple constraints fits with recent work by Glenberg and colleagues [25, 26] and by Barsalou [2]. Glenberg and colleagues have proposed that the meaning of a sentence is constructed by indexing words or phrases to real objects or perceptual analog symbols for those objects, deriving affordances from the objects and symbols and then meshing the affordances under the guidance of syntax. Barsalou [2] also places similar emphasis on perceptual representation for objects and nouns in his perceptual symbol systems account. For Barsalou, words are associated with schematic memories extracted from perceptual states which become integrated into what Barsalou terms *simulators* (see also [27]). As simulators for words develop in memory, they become associated with simulators for the entities and events to which they refer. Furthermore, once simulators for words become linked to simulators for concepts, Barsalou argues that words can then control simulations. We hope to be able to extend the model further by also considering interaction with objects by the model more directly (e.g., through the addition of a robotic arm), rather than simply observing interactions between objects. We hope that such developments help move embodiment arguments from the theoretical arena to showing how these ideas can be realized in a working neuro-computational model.

References

1. Ballard, D. H., Hayhoe, M. M., Pook, P. K. & Rao, R.P.N. (1997) Deictic codes for the embodiment of cognition. *Behavioural and Brain Sciences, 20*, 723-767.
2. Barsalou, L. W. (1999). Perceptual symbol systems. *Behavioral and Brain Sciences, 22(4),* 577-660.
3. Cangelosi (in press). Grounding symbols in perceptual and sensorimotor categories: Connectionist and embodied approaches. In H. Cohen & C. Lefebvre (Eds), *Categorization in Cognitive Science*, Elsevier.
4. Cangelosi A., Greco A. & Harnad S. (2000). From robotic toil to symbolic theft: Grounding transfer from entry-level to higher-level categories. *Connection Science, 12(2),* 143-162.
5. Carlson-Radvansky, L.A., Covey, E. S. & Lattanzi, K. M. (1999). "What" effects on "where": Functional influences on spatial relations. *Psychological Science, 10(6),* 516-521.
6. Carlson-Radvansky, L. A., & Radvansky, G. A. (1996). The influence of functional relations on spatial term selection. *Psychological Science, 7,* 56-60.
7. Carlson-Radvansky, L. A., & Tang, Z. (2000). Functional influences on orienting a reference frame. *Memory & Cognition*, 28, 812-820.
8. Coventry, K. R. (1998). Spatial prepositions, functional relations and lexical specification. In P. Olivier and K.-P. Gapp (Eds.) *Representation and Processing of Spatial Expressions,* (pp. 247-262). Mahwah, New Jersey: Lawrence Erlbaum Associates.
9. Coventry K.R., Cangelosi A. et al. (in submission). Spatial language and perceptual symbol systems: Implementing the functional geometric framework.
10. Coventry, K. R., Carmichael, R. & Garrod, S. C. (1994). Spatial prepositions, object-specific function and task requirements. *Journal of Semantics, 11,* 289-309.
11. Coventry, K. R. and Garrod, S. C. (2004). *Saying, Seeing and Acting. The Psychological Semantics of Spatial Prepositions.* Lawrence Erlbaum Associates. Essays in Cognitive Psychology Series.
12. Coventry, K. R., & Garrod, S. C. (in press). Spatial prepositions and the functional geometric framework. Towards a classification of extra-geometric influences. In L. A. Carlson, & E. van der Zee (Eds.), *Functional features in language and space: Insights from perception, categorization and development.* Oxford University Press.
13. Coventry, K. R. & Mather, G. (2002). The real story of 'over'?. In K. R. Coventry & P. Olivier (Eds.), *Spatial Language: Cognitive and Computational Aspects.* Dordrecht, The Netherlands: Kluwer Academic Publishers.

14. Coventry, K. R. & Prat-Sala, M. (2001). Object-specific function, geometry and the comprehension of 'in' and 'on'. *European Journal of Cognitive Psychology, 13(4),* 509-528.
15. Coventry, K. R., Prat-Sala, M. & Richards, L. V. (2001). The interplay between geometry and function in the comprehension of 'over', 'under', 'above' and 'below'. *Journal of Memory and Language, 44,* 376-398.
16. Deco, G. and T.S. Lee (2002) A Unified Model of Spatial and Object Attention Based on Inter-cortical Biased Competition. *Neurocomputing: An international journal, 44-46,* 775-781.
17. Edelman, S. (1999). *Representation and Recognition in Vision.* MIT Press, 1999.
18. Edelman, S. (2002). Constraining the Neural Representation of the Visual World. *Trends in Cognitive Sciences, 6,* 125-131
19. Elman, J.L. (1990). Finding structure in time. *Cognitive Science, 14,* 179-211
20. Feist, M.I., & Gentner, D. (1998). On Plates, Bowls, and Dishes: Factors in the Use of English 'in' and 'on'. *Proceedings of the Twentieth Annual Conference of the Cognitive Science Society,* 345-349.
21. Ferenz, K. S. (2000). *The role of nongeometric information in spatial language.* PhD Thesis, Dartmouth College, Hanover, New Hampshire.
22. Freyd, J., Pantzer, T., & Cheng, J. (1988). Representing statics as forces in equilibrium. *Journal of Experimental Psychology: General, 117,* 395-407.
23. Gapp, K. P. (1995). Angle, distance, shape and their relationship to projective relations. In J. D. Moore & J. F. Lehman (Eds.), *Proceedings of the Seventeenth Annual Conference of the* Cognitive Science Society (pp. 112-117). Mahwah, NJ: Lawrence Erlbaum Associates Inc.
24. Garrod, S., Ferrier, G., & Campbell, S. (1999). In and on: Investigating the functional geometry of spatial prepositions. *Cognition, 72,* 167-189.
25. Glenberg, A. M. (1997). What memory is for. *Behavioral and Brain Sciences, 20(1),* 1-55.
26. Glenberg, A. M., & Kaschak, M. (2002). Grounding language in action. *Psychonomic Bulletin and Review, 9(3), 558*-565.
27. Grush, R. (in press). The emulation theory of representation: motor control, imagery, and perception. *Behavioral and Brain Sciences.*
28. Herskovits, A. (1986). *Language and Spatial Cognition. An interdisciplinary study of the prepositions in English.* Cambridge, UK: Cambridge University Press.
29. Hormann, H. (1983). Then calculating listener, or how many are einige, mehrere and ein paar (some, several and a few). In R. Bauerle, C. Schwarze, & A, von Stechow (Eds.), *Meaning, use and interpretation of language.* Berlin; De Gruyter.
30. Jeannerod, M. (1994). The representing brain: Neural correlates of motor intention and imagery. *Behavioral and Brain Sciences, 17(2),* 187-245.
31. Jeannerod, M.(2001) Neural simulation of action: A unifying mechanism for motor cognition. *NeuroImage, 14,* S103-S109.
32. Joyce D., Richards L., Cangelosi A., Coventry K.R. (2002), Object representation-by-fragments in the visual system: A neurocomputational model. In L. Wang, J.C. Rajapakse, K. Fukushima, S.Y. Lee, X. Yao (Eds), *Proceedings of the 9th International Conference on Neural Information Processing (ICONP02)* IEEE Press, Singapore.
33. Joyce. D. W., Richards, L. V., Cangelosi, A. & Coventry, K. R. (2003). On the foundations of perceptual symbol systems: Specifying embodied representations via connectionism. In F. Dretje, D. Dorner & H. Schaub (Eds.), The Logic of Cognitive Systems. *Proceedings of the Fifth International Conference on Cognitive Modelling*, pp147-152. Universitats-Verlag Bamberg, Germany.
34. Landau, B., & Jackendoff, R. (1993). 'What' and 'where' in spatial language and cognition. *Behavioural and Brain Sciences, 16(2),* 217-265.

35. Logan, G. D., & Sadler, D. D. (1996). A computational analysis of the apprehension of spatial relations. In P. Bloom, M. A. Peterson, L. Nadel, & M. F. Garrett (Eds.), *Language and Space* (pp. 493-530). Cambridge, Mass.: MIT Press.
36. Markman E.M. (1987), How children constrain the possible meanings of words," In *Concepts and conceptual development: Ecological and intellectual factors in categorization.* Cambridge University Press.
37. Moxey, L. M., & Sanford, A. J. (1993). *Communicating Quantities. A Psychological Perspective.* Lawrence Erlbaum Associates; Hove, East Sussex.
38. Newstead, S. E., & Coventry, K. R. (2000). The role of context and functionality in the interpretation of quantifiers. *European Journal of Cognitive Psychology*, 12(2), 243-259.
39. Plunkett, K., Sinha, C., Moller, M.F & Strandsry, O. (1992). Symbol grounding or the emergence of symbols? Vocabulary grouth in children and a connectionist net. *Connection Science, 4(3-4)*, 293-312.
40. Regier, T. (1996). *The human semantic potential: Spatial language and constrained connectionism.* Cambridge Mass.: MIT Press.
41. Regier, T., & Carlson, L.A. (2001) Grounding spatial language in perception: An empirical and computational investigation. *Journal of Experimental Psychology: General, 130(2),* 273-298.
42. Schwartz, D. L. (1999). Physical imagery: Kinematic versus dynamic models. *Cognitive Psychology, 38*, 433-464.
43. Talmy, L. (1983). How language structures space. In H. Pick, & L. Acredolo (Eds.), *Spatial Orientation: Theory, research and application*, (pp. 225-282). New York: Plenum Press.
44. Ullman, S. (1984). Visual routines. *Cognition, 18,* 97-159.
45. Ullman, S. (1996). *High-level Vision. Object recognition and visual cognition.* Cambridge, MA; MIT Press.
46. Ungderleider, l, & Mishkin, M. (1982). Two cortical visual systems. In D. Ingle, M. Goodale, & R. Mansfield (Eds.), *Analysis of Visual Behaviour*. Cambridge, MA: MIT Press.
47. Williamson J.R. (1996). "Gaussian ARTMAP: A Neural Network for Fast Incremental Learning of Noisy Multidimensional Maps", *Neural Networks, 9(5),* 881- 897.
48. Zee, E. van der, & Watson, M. (in press). Between function and space: How functional and spatial features determine the comprehension of "between". In L. A. Carlson, & E. van der Zee (Eds.), *Functional features in language and space: Insights from perception, categorization and development.* Oxford University Press.

Reference Frame Conflict in Assigning Direction to Space

Michele Burigo and Kenny Coventry

School of Psychology, Faculty of Science, Drake Circus,
University of Plymouth, PL4 8AA, UK
{mburigo, kcoventry}@plymouth.ac.uk

Abstract. Spatial prepositions are linguistic tools to exchange information about spatial location of objects. For instance "The book is over the table" indicates that the located object (LO) is somewhere "over" the reference object (RO). Assigning direction to space (selecting a reference frame) is a necessary precursor to understanding where the LO is located. Three experiments are reported which investigated the effect of the orientation of both the LO and the RO on the acceptability of the prepositions *above/below/over/under*. We found that when the LO was not vertically aligned, the appropriateness for a given spatial preposition changes. In general scenes with the LO pointing at the RO were judged less acceptable than scenes with the LO vertically oriented. These results suggest that people generate reference frames for both LO and RO prior to assigning direction to space. Modifications to Multiple Frame Activation theory [1] are discussed.

1 Introduction

Language is a joint activity between a speaker and listener and to achieve a fruitful exchange of information it is critical that they have a common goal. Within the domain of spatial language, one of the most common goals is assumed to be to define the location of objects in space. For instance "The book is over the table" indicates that the located object ("the book") is somewhere "over" the reference object ("the table"). Prepositions like "over" and "behind" (the so-called projective prepositions) are particularly interesting as they require the selection of a reference frame before the assignment of a direction to space, specified by the preposition, can be established. Levinson [2] distinguishes between the intrinsic (object-centred), relative (or viewer-centred/deictic), and absolute (environment-centred/extrinsic) reference frames. For example, "the car is behind the house" used intrinsically locates the car in relation to the opposite wall from where the salient front of the house is (which is where the back door is). The relative use of the same expression locates the car directly behind the opposite wall to where the speaker and hearer are standing. The absolute frame locates an object with respect to a salient feature of the environment, such as the gravitational plane or cardinal directions (e.g., North, South, etc.).

Recent studies [3, 4, 5] have argued that spatial apprehension occurs in a series of stages summarised as follows; (1) identify the reference object, (2) superimpose multiple reference frames (relative and intrinsic), (3) construct spatial templates and align them to the relevant reference frames, (4) select a reference frame, (5) combine

C. Freksa et al. (Eds.): Spatial Cognition IV, LNAI 3343, pp. 111–123, 2005.
© Springer-Verlag Berlin Heidelberg 2005

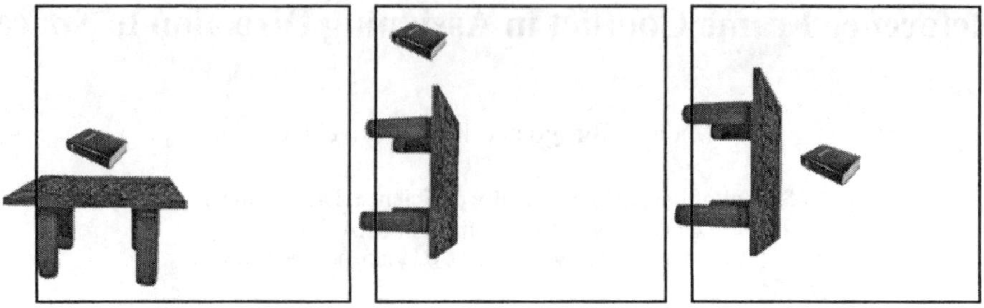

Fig. 1. Canonical absolute/intrinsic "above" (left picture), noncanonical absolute "above" (middle picture) and noncanonical intrinsic "above" (right picture)

templates into a composite template, (6) search the composite template that fits best with the located object for each position within the template, (7) calculate whether the goodness of fit measure for the located object is high (good or acceptable region) or low (bad region) (see also [6] for a computational framework). The aim of this chapter is to review some of these stages, and in particular to assess whether information regarding the reference frame of the located object, currently missing from these stages, needs to be incorporated into the spatial apprehension process.

1.1 Reference Frame Selection and Multiple Frame Activation (MFA): Empirical Evidence

Reference frames are coordinate systems that link spatial representation and perceptual representation [7, 8]. As we have seen above, the process of reference frame assignment is a fundamental stage in spatial apprehension because it is the process by which direction is given to space. Previous studies found that people usually do not use just one reference frame, but that the reference frame used to apprehend spatial prepositions is the sum of a few frames. For example experimental evidence has demonstrated that by rotating the reference object by 90° (noncanonical orientation), acceptability ratings for *above* mirror the new spatial template that is the sum of all the reference frames active in that moment [1]. The acceptability for the given spatial preposition varies as a function of the reference frame activated. Consider the scenes in Figure 1. In the canonical orientation the absolute, relative and intrinsic reference frames overlap. In the noncanonical orientation the absolute reference frame is dissociated from the intrinsic frame. This produces a lower acceptability for the given spatial preposition because a conflict emerges between all the reference frames activated at that moment [1, 9]. Further studies [10, 11] found evidence in support of the idea that multiple reference frames were activated at the same time and the evaluation of appropriateness for a spatial preposition considers all the reference frames active at that specific moment. Computational frameworks are consistent with these findings. In particular influential computational frameworks [6, 12] focus on attentional processes assuming that an attentional load (calculated as vector sum) is computed from the reference object to the located object.

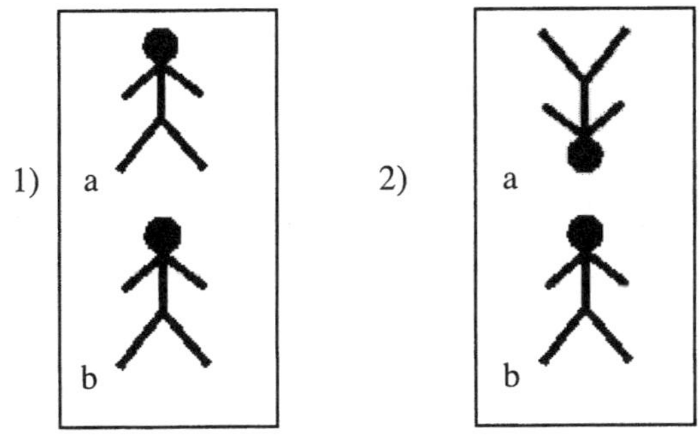

Fig. 2. Reference frame conflicts between LO and RO. The LO in 1) is in "vertical" orientation; in 2) the LO is "upside down" orientation

In summary, even if evidence has shown that reference frame activation is important, to date studies have only focused on the reference frame generated from the reference object [4, 9, 10]. Furthermore, theories of spatial language largely assume that the assignment of direction is generated from the reference object to the located object (as in the AVS model). However, further experimental evidence suggests that both objects (even distractors or those not relevant for the task) require allocation of attention to be processed [13, 14]. This suggests that both objects could play a role in the spatial apprehension process and in particular in the process of imposing reference frames and giving direction to space.

There is much evidence indicating that the LO is important in establishing the acceptability of a range of spatial prepositions (see [15] for a review). For example, Coventry, Prat-Sala and Richards [16] found that the appropriateness of a spatial preposition mirrors the degree of functional relation between located and reference object. For example, an umbrella is regarded as being more *over* a person if it is shown to protect that person from rain than when the rain is shown to hit the person. Furthermore, Coventry et al. found that the acceptability ratings for *over* and *under* were more influenced by the function of the object than by the relative positions of LO and RO, while conversely *above* and *below* were more influenced by geometry than function. Additionally, in a study which manipulated reference frame conflicts with function present (e.g., the man holding the umbrella in the gravitational plane was either upright, lying down, or upside down), Coventry et al. found that reference frame conflicts influenced the acceptability of *above* and *below* more than *over* and *under*.

However, although there is much evidence that the located object does influence the acceptability of a range of prepositions, no studies to date have examined how the located object could contribute to or affect the reference frame assignment, and hence the assignment of direction to space. Consider the scenes in Figure 2. We hypothesized that scenes where the located object (man "a" in Figure 2) was rotated past 90° would introduce reference frame conflict for the located object, and would therefore lead to a reduction in the appropriateness of spatial prepositions to describe the position of man "a" in relation to man "b". We report the results of three preliminary experiments employing an acceptability rating task where possible reference frame conflicts for both the located object and reference object are investigated.

2 Experiment 1

In this experiment we tested the hypothesis that the reference frame(s) associated with the located object would affect acceptability of *over, under, above* and *below* to describe the position of the LO in relation to a RO.

2.1 Method

Participants and Procedure
Twenty-three undergraduate students from the University of Plymouth participated in this investigation for course credit. All the participants were English native speakers. Participants had to judge the appropriateness of a spatial preposition (*above, below, over* or *under*) to describe pictures using a scale from 1 to 9 (where 1 = not at all acceptable and 9 = perfectly acceptable). All trials showed the located object in a "good" or "acceptable" location, never in a "bad" location (following Carlson-Radvansky and Logan's definitions [3]).

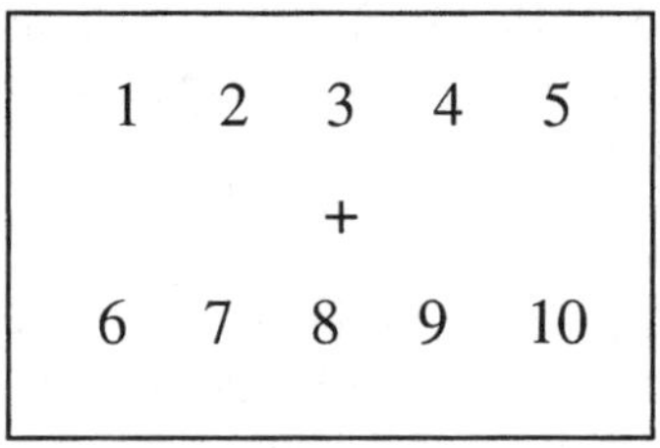

Fig. 3. Location for the located object with respect to the reference object (indicated here with a "plus")

The located object could appear in 10 different locations around the reference object (see Figure 3). The sentences were shown before the scene and in this form; <The "located object" is PREPOSITION the "reference object">. The prepositions tested were *above, below, over* or *under*. Two orientations for the located object were used: "vertical" and "pointing at" (See Figure 4 for examples). In the "pointing at" condition the axis of the located object was pointing exactly towards the center-of-mass of the reference object. Viewing distance was roughly 50 cm. The experiment was presented on a computer and was programmed and presented using the E-Prime experiment package. The experiment lasted approximately 50 minutes.

Materials
The materials consisted of three stimuli; a circle, an hourglass and a stickman. These objects were selected as the circle does not have an axis, while the hourglass has a salient axis but not an intrinsic top and bottom, and the stickman has a salient axis and an intrinsic top and bottom. We will use the following labels to classify the objects; "no axis" (circle), "ambiguous axis" (hourglass) and "intrinsic axis" (stickman). All the objects employed were presented at the same size (2.3° x 1.6° of visual degree)

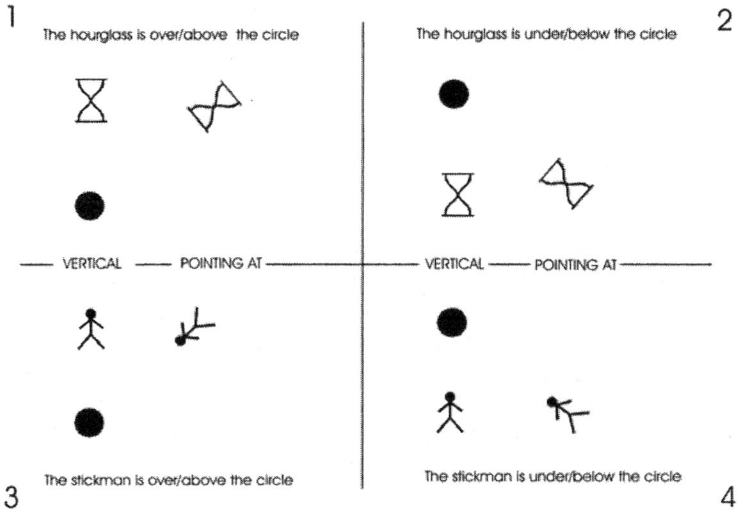

Fig. 4. This figure illustrates trials with the circle as reference object and with the located object in the "vertical" or "pointing at" orientation. Quadrants 1 and 3 illustrate superior preposition and quadrants 2 and 4 illustrate inferior preposition. Note that only one object appeared in each scene as LO

and distance from the reference object regardless of the orientation. This is because it has been found that proximity, center-of-mass orientation and distance affect the appropriateness of spatial preposition [12].The objects could appear as reference objects or as located objects, but the same object was never shown as LO and RO at the same time.

Design

The experiment consisted of 480 trials constructed from the following variables: 4 spatial prepositions X 10 locations X 6 object permutations X 2 orientations ("vertical"

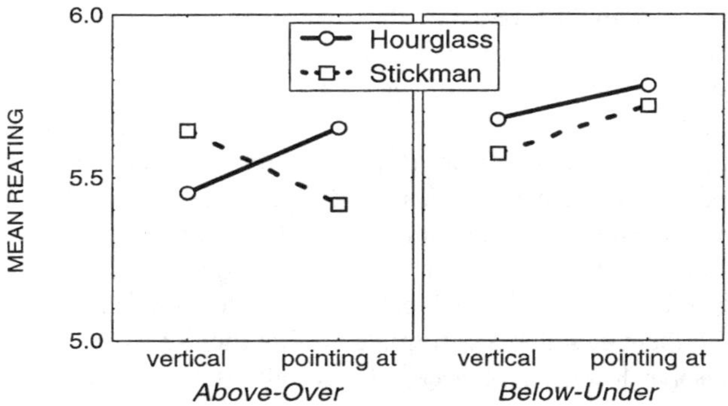

Fig. 5. 3-way interaction between superior versus inferior prepositions (above/below vs. over/under), located object and orientation of LO (collapsed over locations)

and "pointing at"). The locations were collapsed in two factors; high vs. low location (2 levels) and proximity (3 levels) as follows; far misaligned (locations 1, 6 and 5, 10 in figure 3), near misaligned (locations 2 and 4) and aligned (central location). All the trials were presented in a randomized order.

2.2 Results

A 4-way within subjects ANOVA was performed on the rating data. The variables included in the analysis were; 2 located objects (hourglass and stickman) x 2 preposition sets (above-below vs. over-under) x 2 superior versus inferior prepositions (above-over vs. below- under) x 2 orientations of LO (vertical and pointing at). The division between spatial prepositions has been employed following the Coventry et al. findings summarized above [16]. Trials with the circle as the located object were excluded from the analysis since this kind of object does not have a salient axis. Furthermore we analyzed only the trials with a circle as the reference object because it has no axis and it becomes easier to compare trials with the stickman and trials with the hourglass. A main effect of preposition set (above-below vs. over-under) was found, [$F_{(1, 22)} = 7.21$, $p < .05$]. Higher ratings were given for Above-Below ($M = 6.526$) than for Over-Under ($M = 5.192$). This is unsurprising as it known that Above-Below have larger areas of acceptability than Over-Under. No other significant main effects were found. There was a significant 3-way interaction between superior versus inferior spatial prepositions, located object and orientation of LO [$F_{(1, 22)} = 6.694$,$p < .05$], displayed in Figure 5. It is interesting to note that objects with a top/bottom orientation such as a stickman are rated less acceptable when pointing ($M = 5.42$) than when vertical ($M = 5.65$) for trials with above-over, although this was not the case for below-under ($M_{vertical} = 5.58$; $M_{pointing} = 5.72$). None of the other interactions were significant.

2.3 Discussion

An interesting difference was found between trials with the stickman and trials with the hourglass as LOs. The stickman trials generate a reference frame conflict in the pointing condition but the hourglass did not. This could be explained by a preferential assignation of a top/bottom orientation based on the vertical plane. In other words an hourglass could not be seen as upside down but always as pointing away from the reference object. Acceptability rating for inferior spatial prepositions (below-under) showed that the pointing condition was generally more acceptable than the vertical one.

These results are open to a number of possible interpretations. First, lower acceptability could be due to the activation of an intrinsic reference frame on the LO that in the case of under-below produces facilitation because it is aligned with the reference frame on the RO. In contrast, for above-over the reference frame on the LO may produce conflict because it does not match the direction of the reference frame on the RF.

Alternatively, there could be a cost in identifying the LO when it is rotated, resulting in a progressive lowering of acceptability ratings for scenes as a function of degree of rotation.

Nevertheless, the results seem to suggest that the orientation of the located object is important in establishing the appropriateness of projective prepositions. However, this experiment only used three located objects (only two were submitted to the data analysis; the hourglass and the stickman), so there is an issue regarding the extent to which the results can be generalized. For this reason the aim of the next experiment is to try to replicate the effect of the orientation of the LO using a wider range of LOs and orientations of LO.

3 Experiment 2

The second experiment utilized the same design and procedure as the first experiment, except that more materials and orientations of LO were included.

3.1 Method

Participants and Procedure
Twenty-nine undergraduate students from the University of Plymouth participated in this investigation for course credit. All the participants were English native speakers and none of them took part in the previous experiment. The procedure was the same procedure used for the previous experiment based on the acceptability rating task of the given spatial prepositions; *above, below, over* and *under*.

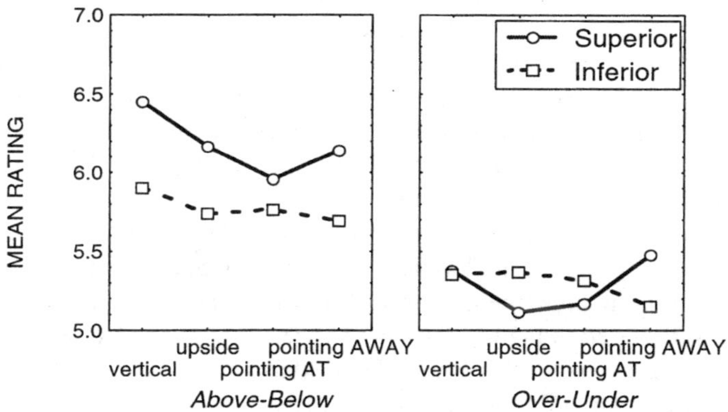

Fig. 6. 3-way interaction between orientation of LO, superior-inferior prepositions and preposition set

Materials
This experiment involved a wider number of located objects and two more orientations; "pointing away" from the reference object and "upside down". The reference object in this experiment was always a picture of a football. The located objects were picked from two sets; the first consisted of objects with a distinctive top-bottom (8 new objects "with an intrinsic axis") and the second consisted of objects with "an ambiguous axis" (7 new objects plus the hourglass). All the stimuli were hand-drawn and transformed to electronic format by a computer scanner. All the objects employed were presented at the same size (4.6° x 4.6° of visual degree) and distance from the reference object regardless of the orientation.

Design

There were 384 trials constructed from the following variables: 8 located objects X 3 locations of LO (collapsed over side) X 2 superior/inferior prepositions (above-over vs. below-under) X 2 preposition sets (above-below vs. over-under) X 4 orientations ("vertical", "upside down", "pointing at" and "pointing away"). All the trials were presented in a randomized order. We added 192 distractors where the LOs were objects without salient axes, meaning that a total of 576 trials were presented.

3.2 Results

A full factorial ANOVA was chosen to analyze the data. In this analysis we focus on trials where the LO had an intrinsic axis (following the results of Experiment 1). A significant main effect was found for preposition set (above-below vs. over-under), [$F_{(1,28)} = 15.44$, $p < .001$], with higher ratings for the above-below set. A main effect was also found for superior vs. inferior prepositions [$F_{(1,28)} = 10.72$, $p < .005$] with superior prepositions receiving higher ratings than inferior prepositions. A further main effect was found for location [$F_{(1,28)} = 80.17$ $p <.0001$]; the highest ratings were given for the central locations (3 and 8 in figure 3), with ratings decreasing as the LO moves away from being directly aligned with the RO. The last significant main effect was for direction [$F_{(1,84)} = 3.35$, $p < .05$]. Objects vertically oriented ($M = 5.75$) were judged more acceptable than the other levels of orientation. In particular the "upside down" ($M = 5.6$) and "pointing at" ($M = 5.55$) orientations produced the lowest ratings (and indeed generated the highest reference frame conflict). The analysis also revealed significant 2-way interactions between preposition set and location [$F_{(1,28)} = 10.96$, $p < .005$], between preposition set and direction [$F_{(1,84)} = 3.23$. $p < .05$] and between superior versus inferior prepositions and direction [$F_{(1,84)} = 2.82$, $p < .05$].

Finally, there was also a significant 3-way interaction between superior-inferior prepositions, preposition set and location, [$F_{(1,28)} = 5.45$, $p < .05$], and between preposition set, superior-inferior preposition and direction [$F_{(1,84)} = 3.99$, $p < .01$]. This interaction is displayed in Figure 6. As can be seen in Figure 5, the results of the orientation of LO are clearest for *above*, which exhibited a reliable difference between the vertical orientation of LO and all the other levels of LO. For *over*, pointing away from the RO is also associated with higher acceptability ratings. The results are less clear for inferior prepositions.

3.3 Discussion

The pattern of results in this second experiment confirms the hypothesis that the orientation of the located object influences acceptability ratings, although there are clear differences between prepositions. However, in the first two experiments the reference objects were objects without a salient axis. It is therefore possible that the orientation effect of LO could be restricted to scenes where the RO does not have an intrinsic axis. For this reason the next experiment tested whether the effects of the orientation of LO were present across a wider range of ROs.

4 Experiment 3

This experiment used the same basic methodology as before, but this time with a range of reference objects including ROs without a salient axis, with an ambiguous axis, and with an intrinsic axis.

4.1 Method

Participants and Procedure
Twenty-three undergraduate students from the University of Plymouth participated in this investigation for course credit. All the participants were English native speakers and they did not take part in any of the previous experiments. The procedure was the same as that used in Experiments 1 and 2.

Materials
For this experiment we used a set of 24 objects (8 "without a salient axis", 8 "with an ambiguous axis" and 8 "with an intrinsic axis"). The objects "with an ambiguous axis" and "with an intrinsic axis" were the same as those used in Experiment 2. We drew 8 new objects "without a salient axis". Thus we were able to study the effect of the reference frame activation on the located object in scenes with different kinds of reference object.

Design
The experiment was composed of 576 trials with the following factors: 8 located objects with an intrinsic axis (treated as random factor), X 3 reference objects (picked up from a set of 24 objects, 8 with no axis, 8 with an ambiguous axis, and 8 with an intrinsic axis; within subjects factor), X 2 prepositions sets (between subjects factor), X 2 superior/inferior prepositions (within subjects), X 3 locations for the probe (within subjects) and 4 directions for the located object (within subjects). This time preposition set was between subjects; half the participants received *above* and *below* and the other half received *over* and *under*.

4.2 Results

We performed two analyses; one by subjects (F^1) and one by materials (F^2). The results were similar for both analyses, so here we report the F^1 analyses alone. The means for all the conditions can be found in Table 1. Significant main effects were found for superior-inferior prepositions [$F_{(1,22)} = 18.74$, $p < .001$] with higher ratings for the above-over set. A main effect for location [$F_{(1,22)} = 69.14$, $p < .0001$] was also found; highest ratings were given for central locations (3) followed by location 2 and by location 1 respectively. The last significant main effect was for the orientation of LO [$F_{(1,44)} = 5.25$, $p < .005$]. Vertically oriented LOs were rated higher ($M = 5.69$) than "pointing at" Los ($M = 5.43$). Furthermore "upside-down" objects were rated lower ($M = 5.21$) than "pointing at" objects ($M = 5.44$).

Furthermore we found several significant 2-way interactions; between preposition set and RO [$F_{(1,44)} = 3.61$, $p < .05$], between location and RO [$F_{(1,44)} = 4.45$, $p < .05$], between superior-inferior prepositions and orientation of LO [$F_{(1,66)} = 4.93$, $p < .005$] and between location and orientation of LO [$F_{(1,66)} = 3.12$, $p < .05$].

A 3-way interaction between superior-inferior prepositions, location and orientation of LO was also significant [$F_{(1,66)} = 3.93$, $p < .05$] and a 4-way interaction between superior-inferior prepositions, location, RO and orientation of LO [$F_{(1,132)} = 2.74$, $p < .05$]. Follow-up analysis revealed significant differences in orientation between prepositions and locations, but the effects of orientation were present at all levels of RO.

4.3 Discussion

The outcomes from this experiment support the idea that the orientation of the located object affects acceptability ratings even when the reference object has an intrinsic orientation. The results for this experiment mirror the results of the previous experiment, but extend the results to show that the activation of reference frame for the LO is not restricted to cases where the RO does not provide sufficient information to cue a reference frame.

Table 1. Means for conditions across the four spatial prepositions (*above, below, over and under*)

Located Object (intrinsic)		***Reference Object***		
		No axis	Ambiguous	Intrinsic
Above	Vertical	6.281	6.307	6.375
	upside-down	5.560	5.542	5.490
	point at	5.524	5.670	5.644
	point away	6.047	6.026	5.974
Below	Vertical	5.797	6.036	6.167
	upside-down	5.411	5.604	5.453
	point at	5.786	5.823	5.754
	point away	5.387	5.536	5.578
Over	Vertical	5.419	5.084	5.479
	upside-down	5.047	4.823	5.220
	point at	5.182	5.115	5.188
	point away	5.785	5.366	5.335
Under	Vertical	5.131	5.058	5.162
	upside-down	4.691	4.901	4.889
	point at	5.335	5.073	5.156
	point away	4.698	4.693	4.901

5 General Discussion

The present investigation explored the effect of the activation of composite spatial templates assigned on both the reference object and the located object for a given spatial preposition. The results support the idea that the orientation of the located object affects acceptability ratings for projective prepositions.

The results suggest necessary extensions to the idea of Multiple Frame Activation [1] where it is indicated that in comprehending a scene multiple frames are available. However, the results we found may suggest that an additional reference frame is generated from the located object as well as from the reference object and the final template generated is influenced by its orientation. This could be also because the reference frame is activated with more difficulty when the LO orientation mismatches the direction of space defined by the RO.

In addition to the reliable effects of the orientation of LO for intrinsic objects, the results of Experiment 1 showed some interesting differences between intrinsic objects and objects such as an hourglass with a salient axis, but without an intrinsic axis. For the objects like an hourglass, the "pointing at" condition was considered more acceptable than the vertical condition. A possible explanation is that people assign a subjective top/bottom orientation to "ambiguous" objects. Thus the hourglass in trials with above-over should be seen as pointing away from the reference object instead of pointing at the RO.

The last experiment provided evidence that the conflict among reference frames emerges across a range of reference objects, including those that are more "real" with a top/bottom orientation. So the effect of the located object is not limited to cases where the RO does not activate the intrinsic frame.

But why should the orientation of the located object affect acceptability ratings when the reference frame of the RO should be sufficient to localize objects in the scene? An explanation is that people are biased in the reference frame assignment by everyday experience. For example a picture showing an object upside down[1] might not be seen as a "plausible" stereotypical mental representation. Such pictures could require consideration ("redistribution" in the attentional meaning) of more options than are actually available. In other words, an unusual pattern of objects with different orientations might activate the reference frame on the located object to maximize the chance of the success of the spatial apprehension process. Another possible explanation is based on the concept of direction of potential motion [17]. People perceive objects rotated away from the gravitational plane as falling. So a located object oriented at 90° may be perceived as moving downwards on a path to the left of/right of and away from the reference object.

5.1 Implications for Existing Models

The results found suggest modification of the key characteristics of the spatial apprehension process [1, 4, 6, 18]. We found evidence for involvement of the located object reference frame in the process of assigning direction to space. Therefore, evaluating the process of goodness of fit of a spatial preposition involves the located object as well and future studies should take this into account[2]. The finding that the located object affects the spatial apprehension process has some repercussions for computational models of spatial language as well. Models such as the Attentional-Vector-Sum

[1] Apart from polyoriented objects that do not show time differences in recognition [20].

[2] Even when no functional relationship between located object and reference object are involved.

model [12] simulate attentional processes, but thus far do not deal with attentional processing of the LO (but see [19], for a modification of AVS to deal with processing of function). It may be possible to develop the AVS model to deal with the projection of vectors from the LO to the RO as well as the other way round (see [15], for a discussion).

5.2 Limitations and Future Developments

This investigation has provided experimental evidence in support of the hypothesis that the located object, in a scene with two objects, takes part in the spatial apprehension process. However, exactly how the LO affects the apprehension process is unclear. A simple explanation is that there could be a cost in identifying the LO when it is rotated, resulting in a greater lowering of acceptability ratings for scenes as a function of degree of rotation. If this is the case, then the effects found in the studies reported should disappear when the LO is a poly-oriented object, rather than a mono-oriented object (see [21]). We are currently testing this possibility. Alternatively, it could be the case that the reference frame for the LO is indeed activated, and that this either interferes with the reference frame activation process for the RO, or that the references frames form the LO and the RO are both considered and perhaps combined. We are currently explored these alternative possibilities also.

Future investigations should attempt to ascertain the degree to which features of the LO influence the spatial apprehension process further. For example, in some contexts the LO may be more important than the RO, and vice versa for other contexts. Studies are also underway to test the conflict among reference frames using a reaction time paradigm in order to get at the time-course of processing of the LO and RO. Finally, we should consider how these findings can be implemented within frameworks such as the AVS model.

Acknowledgments

We would like to thank Lau Thiam Kok for his drawing skills.

References

1. Carlson-Radvansky, L. A. & Irwin, D. E. (1994). Reference frame activation during spatial term assignment. *Journal of Memory and Language*, 33, 646-671.
2. Levinson, S. C. (1996). Frames of reference and Molyneux's question. In P. Bloom, M. A. Peterson, L. Nadel, & M. F. Garrett (Eds.), *Language and space* (pp. 109-169). Cambridge, MA: MIT Press.
3. Carlson-Radvansky, L. A. & Logan, G. D. (1997). The influence of reference frame selection on spatial template construction. *Journal of Memory and Language*, 37, 411-437.
4. Carlson, L. A. & Logan, G. D. (2001). Using spatial terms to select an object. *Memory and Cognition*, 29 (6), 883-892.
5. Carlson, L. A., West, R., Taylor, H. A., & Herndon, R. (2002). Neural correlates of spatial term use. *Journal of Experimental Psychology: Human Perception and Performance*, 28, 1391-1408.

6. Logan, G. D. & Sadler, D. D. (1996). A computational analysis of the apprehension of spatial relations. In P. Bloom, M. A. Peterson, L. Nadel, & M. F. Garrett (Eds.), *Language and space* (pp. 493-529). Cambridge, MA: MIT Press.
7. Logan, G. D. (1995). Linguistic and conceptual control of visual spatial attention. *Cognitive Psychology*, 28, 103-174.
8. Carlson, L. A. (in press) What's up, and where is it at? Using spatial language.
9. Carlson, L. A. (1999). Selecting a reference frame. *Spatial Cognition and Computation*, 1, 365-379
10. Carlson-Radvansky, L. A. & Logan, G. D. (1997). The influence of reference frame selection on spatial template construction. *Journal of Memory and Language*, 37, 411-437.
11. Carlson-Radvansky, L. A., & Jiang, Y. (1998). Inhibition accompanies reference-frame selection. *Psychological Science,* 9, 386-391.
12. Regier, T. & Carlson, L. A. (2001). Grounding spatial language in perception: An empirical and computational investigation. *Journal of Experimental Psychology: General*, 130 (2), 273-298.
13. Lavie, N. (1995). Perceptual load as necessary condition for selective attention. *Journal of Experimental Psychology: Human, Perception and Performance*, 21(3), 451-468.
14. Lavie, N. & Cox, S. (1997). On the efficiency of visual attention: efficient visual search leads to inefficient distractor rejection. *Psychological Science*. 8(5), 395-398
15. Coventry, K. R., & Garrod, S. C. (2004). *Saying, seeing and acting. The psychological semantics of spatial prepositions.* Psychology Press; Hove and New York.
16. Coventry, K. R., Prat-Sala, M. and Richards, L. (2001). The interplay between geometry and function in the comprehension of Over, Under, Above and Below. *Journal of Memory and Language,* 44(3), 376-398.
17. Regier, T. (1996). *The Human Semantic Potential. Spatial language and constrained connectionism.* Cambridge, The MIT Press.
18. Hayward, W. G. & Tarr, M. J. (1995). Spatial language and spatial representation. *Cognition*, 55, 39-84.
19. Regier, T., Carlson, L. A., & Corrigan, B. (2004). *Attention in spatial language: Bridging geometry and function.* In L. Carlson & E. van der Zee (Eds.). Functional features in language and space: Insights from perception, categorization and development. Oxford University Press.
20. Leek, E. C. (1998). The analysis of orientation-dependent time costs in visual recognition. *Perception*, 27 (7), 803-816.

Identifying Objects on the Basis of Spatial Contrast: An Empirical Study

Thora Tenbrink

Faculty of Linguistics and Literary Sciences, University of Bremen,
Bibliothekstr. 1, 28359 Bremen, Germany
tenbrink@sfbtr8.uni-bremen.de

Abstract. In contrast to most research on spatial reference, the scenario in our human-robot experiments focuses on identifying rather than localising objects using spatial language. The relevant question in such a task is "Which" rather than "Where". In order to gain insights about the kind of language to expect in such a scenario, we collected participants' linguistic choices in a web-based empirical study. Spatial scenarios were presented that varied with respect to number, shape, and location of elements, and with respect to possible perspectives. The linguistic analysis reveals that speakers adhere to underlying principles similar to those known for non-spatial object reference. If objects only differ in spatial position, a reference system and spatial axis is chosen that is suitable for contrasting the target object from competing ones. The exact spatial location is usually not specified if there are no competing objects closeby.

1 Introduction

Much research on spatial reference focusses on the ways in which human speakers describe the spatial relation between two entities in a given context, answering a question like "Where is the object?" (e.g., the MPIP research reported in Levinson 2003; contributions in Olivier & Gapp 1998; van der Zee & Slack 2003). Spatial reference, however, is not confined to this area. In a different scenario, one of several similar objects needs to be identified on the basis of spatial location, due to the absence of other cues such as distinguishing features or perceptual salience. The relevant question in such a task is "Which" rather than "Where".

Up to now, research in spatial cognition has largely neglected this area. It has, however, extraordinary relevance in a human-robot interaction context in which the artificial communicator is not capable of making fine perceptual distinctions or interpreting pointing gestures (Moratz et al. 2003). In such a context, it is highly advantageous to be able to identify present objects on the basis of their spatial location alone, neglecting material-related differences that might be referred to by humans with more developed – and, importantly, shared – perceptual abilities. Thus, in the scenario we are interested in, there are several similar objects present together with an instructor and the instructee, usually a robot; the latter is instructed to move to one of the objects, specified on the basis of spatial position using projective terms, e.g.: "Move to the box on your right."

C. Freksa et al. (Eds.): Spatial Cognition IV, LNAI 3343, pp. 124 – 146, 2005.
© Springer-Verlag Berlin Heidelberg 2005

Identifying one of several similar objects in a spatial scenario differs in some basic respects from localising an already agreed upon object by specifying its position relative to another object. For instance, it should involve taking the overall configuration into account to a higher degree than in "Where" scenarios, in which other objects present can usually be ignored (or are simply absent in experimental settings). In contrast, "Which" questions presuppose (shared) knowledge of the situation to such a degree that the target object can be distinguished sufficiently from all other competing objects in the scenario, making the choice of spatial reference highly dependent on the number and arrangement of objects present.

Since the question of how objects are singled out of several candidates using reference to spatial position has largely been disregarded so far, there is a gap in our knowledge with regard to the linguistic variability in such a situation, and with regard to how different situational contexts and conditions influence speakers' choices. In order to fill this gap, we collected spatial instructions of the above-mentioned kind (answering a "Which" question on the basis of spatial location) in a web-based empirical study (presented in this paper) employing native speakers of English.

This study is part of the empirical work carried out in the SFB/TR 8[1] project I1-[OntoSpace] in order to investigate speakers' choices in different scenarios, supplementing various kinds of real-world experiments in human-robot interaction (HRI). On the basis of the empirical results, a flexible spatial ontology is being built in order to be employed in the HRI environment, which itself is continuously being improved by the work of other SFB/TR 8 projects. The study presented here was designed to provide insights about the range of linguistic variation (in English) to expect in a scenario that is comparable to our HRI settings as far as possible, and investigate underlying principles to which speakers may adhere.

1.1 Contrastive Spatial Reference

Spatial Localisation. A large body of research in the area of spatial cognition is devoted to the identification of spatial reference systems employed by speakers in order to localise an object in a spatial setting (e.g., Carlson 1999; Taylor et al. 1999; Bryant et al. 2000). Projective spatial expressions specifying the *reference (half-)axis* (e.g., *left, right, in front, behind*) together with a *relatum* and a *perspective* are the typical ingredients needed in order to specify a spatial relation. The perspective employed can be based on three kinds of *origin* (Herrmann 1990): speaker, listener, or a further entity (as in: "Coming out of the church, the bus station is on the right"). Likewise, speaker, listener, or a third entity (such as a landmark) available in the current context can serve as relatum. The conflation of point of view (origin) and relatum leads to the kind of reference system called *intrinsic* (e.g., by Levinson 2003), while in a *relative* reference system origin and relatum are distinct. Furthermore, *internal* vs. *external* relationships need to be distinguished (Herskovits 1986), the former specifying object parts where other objects can be located inside of other objects. Note that choice of perspective also plays a role here (cf. Retz-Schmidt 1988), since objects can be

[1] Transregional Collaborative Research Center SFB/TR 8 Spatial Cognition: Reasoning, Action, Interaction (Bremen / Freiburg, Germany), funded by the DFG.

viewed from the outside (such as a picture, where the *left side* depends on the observer's viewpoint) or from the inside (as is usually the case when talking about the internal parts of a person; but it can also be true for objects that can be occupied by a person, such as a car). This difference is often referred to as *handedness* in the literature (e.g., Levinson 2003).

If there are several objects of the same class present, it is furthermore possible to use the whole group of similar objects as a relatum. This option is called *group-based reference* by Moratz (Moratz & Fischer 2000). In contrast to situations where the relatum is an object of a different kind (here called *landmark-based relative reference system*), in a group-based relative reference system the relatum consists of one or several objects of the same kind as the target object.

Now, the question arises whether language directly reflects reference systems. For instance, it has been claimed (Eschenbach 2004) that German spatial adjectives like *link-* (*left*) can only be employed when there is at least one other object of the same class present. However, first results of our spatial human-robot interaction experiments (Tenbrink & Moratz 2003) show that more complex factors come into play. With regard to the external/internal distinction, Herskovits (1986:173f.) presents an overview of the range of linguistic options. Intuitively, for instance, *in the left/right* can only be used for the *interior* of objects, i.e., for internal object parts, while *to the left/right* is only used for external relationships. But in most cases, the language used for different reference systems is more or less the same: e.g., *on the left/right* can be used for external as well as internal relations, and *above* is used for gravity-based as well as intrinsic interpretations (Carlson 1999). In general, relative and intrinsic reference systems cannot be distinguished on the basis of the linguistic form alone. Speakers have to provide additional linguistic material if they wish to specify information about underlying reference systems, e.g., by explicitly mentioning the relatum or the point of view. However, the fact that the same linguistic form can correspond to a variety of underlying reference systems does not necessarily cause ambiguity in interpretation, as will be seen in the analysis below.

In addition to the reference systems in which projective expressions are employed, there are further options of spatial reference available: for instance, speakers can refer to *absolute* reference systems such as the earth's cardinal directions (e.g., *north* and *south*), using different linguistic material. Absolute systems, however, are usually dispreferred in small-scale (indoor) scenarios in Western countries (as opposed to some other cultures, see Levinson 2003). Furthermore, speakers can apply different (non-projective) kinds of spatial reference terms, for example, distance-related expressions such as *close* or *near*.

In order to specify the semantics of projective terms, applicability areas have been identified (e.g., Herskovits 1986; Franklin et al. 1995; Gapp 1995; Carlson-Radvansky & Logan 1997; Carlson et al. 2003). For instance, the expressions *left* and *behind* can be used straightforwardly for 90° and 180° angles, respectively. However, the more the angles between the target object and the relatum depart from these *focal axes*, the more linguistic modifications are used for specifying the spatial relation. Simple expressions are acceptable and applicable in a certain range; outside this range compounds or modifiers such as *left front* or *a little bit to the left* are more typical (Zimmer et al. 1998). Vorwerg (2001) describes in detail the graded typicality structure of spatial expressions as categories on the basis of psycholinguistic experiments in which participants are asked to point out the "best fitting" expression for a given spatial relation, to place an object on the basis of a spatial description, etc.

Task Dependency. Psycholinguistic experiments such as those designed for highlighting the graded structure of projective terms usually do not test for choices speakers make in actual discourse, which naturally depend on the level of specificity needed in a specific discourse situation. If a vague characterization of the spatial relationship is sufficient, applicability areas for simple expressions might be much larger than in a different context where a precise description is vital (as in Zimmer et al. 1998, where the interlocutor needs to find a hidden element on a screen), or where the "best fitting" expression is to be identified (Vorwerg 2001). Thus, an important factor is the *motivation* why a spatial relation should be described at all (in other words, the *question under discussion*). Many different kinds of motivations occur in actual discourse, resulting in diverging usages of spatial expressions. In route descriptions, for example, typically a goal location is described via reference to streets and landmarks which can be easily identified in the real world (e.g., Tversky & Lee 1998). There, salience and dimension of buildings play a role in the choice of landmarks; spatial relations are often sufficiently outlined via simple and vague expressions. Such a scenario differs fundamentally from psycholinguistic spatial localisation experiments where participants are asked to specify an entity's location relative to another.

Descriptions of spatial relations can also be used as a means of identifying – or finding – a target object, such as when a "Where" question is posed in order to identify the correct target object out of several possibilities. Thus, describing a non-prototypical spatial relationship via a simple expression like *left* can be sufficient in a context where there are no competing objects in the left region of the relatum. An everyday example may illustrate this: "Where is the key?" can be answered sufficiently by a vague expression like "Left of the cup" no matter what the precise spatial relationship between the objects is, if there are no other keys closeby. Ultimately, the target object can be identified by establishing sufficient contrast to competing objects, similar to scenarios where "Which" questions are asked. The latter are the target of the present research since they are specifically suited for singling out a target object in contrast to other candidates.

The present research is motivated linguistically, addressing the pragmatics of spatial communication as well as assessing the range of linguistic options for the human-robot interaction scenarios targeted in our research group. In contrast to typical psychological interests and standards we wish to leave as much freedom to our participants as possible in order to learn about their intuitive linguistic strategies (cf. Fischer 2003), since in our project we aim at enabling robots to interpret spontaneous utterances correctly. However, some important basic hypotheses are inspired by previous psycholinguistic studies on (non-spatial) object reference. These findings will be examined next.

Contrastive Object Reference. In their seminal work on object reference, Herrmann & Deutsch (1976) formulated principles of *greatest distance, dimension preference,* *redundant verbalisation,* and *partner-adapted verbalisation.* These principles capture that speakers, in choosing a reference strategy in an object identification task with several different objects, usually do at least the following:

First, they analyse the target object with respect to properties that can establish a (maximum) contrast to competing objects. Thus, if there are two black boxes of different size, the speaker chooses *size* for object reference. In case of several properties in which the objects differ, the speaker chooses the property where the distance to the competing object is most obvious. Thus, if there are two boxes, one of which is very small and dark blue, while the other is very big and black, *size* – rather than *colour* – will be chosen for reference (*greatest distance*). Individual preferences also play a role, especially if the distances are viewed as equal (*dimension preference*).

Second, the speaker encodes as many properties as needed for unambiguous object reference, but usually not more, being economic (cf. Grice's maxim of quantity, Grice 1975). But if the object reference task is complicated by the availability of multiple options, minimal differences in distance, and equal levels of dimension preference, speakers may encode more properties than needed (*redundant verbalisation*).

Third, speakers adapt to their interaction partner's view of the situation, taking into account cognitive and social distinctions, etc. (*partner-adapted verbalisation*).

Herrmann & Deutsch are exclusively concerned with object reference in non-spatial settings, designing their experiments purposively in a way that spatial reference is ruled out. In the present work, the opposite is the case. Target objects differ from competing objects only in spatial position; objects of a different class may also be present, serving as possible relata for reference. Nevertheless, the principles established by Herrmann & Deutsch can be applied to spatial scenarios, motivating hypotheses that can be approached by linguistic analyses of the data collected in the web study (though it cannot be expected, given the kind of open scenario adopted, that they will be verified conclusively). The main research question posed here is:

What principles of object reference apply when only spatial reference is available?

Herskovits (1986) noted that, although the graded structure of projective terms applies in most communicative contexts, there are situations in which an expression like *to the right*, without modification, is capable of denoting the full right side (i.e., a half-plane) with respect to the relatum (p. 182): "[T]he loosest interpretation of the preposition is adequate, provided that obvious contrasts in the context allow the expression to fulfill its function of identifying the place of the located object." This applies if there are no competing objects in the same spatial region. In case of the presence of further objects within the half-plane, unmodified projective terms can nevertheless be employed for contrastive reference. In that case, Herskovits' *shifting contrast near principle* applies (p. 81): "If two objects, A and B, are placed in a relation to a reference object in such a way that the ideal meaning of a preposition (…) is truer of A than of B, then one can use that preposition to discriminate A from B so that the locative phrase will be assumed true of A but not of B." For instance, if *to the left* is "truer" of A than of B, i.e., A is closer to the left reference axis than B is, A will be recognized as the target object even if *to the left* could also be applied to B.

Starting from these observations more concrete research questions can be formulated as follows.

How do speakers choose a reference system and point of view? If a partner is present it can be expected that participants will often choose their partner's perspective, especially if the partner is expected to act (Herrmann & Grabowski 1994:123), and that speakers will adapt their utterances to their interlocutor's in various respects (Clark 1996; Pickering & Garrod in press), for example with respect to the choice of reference systems (Watson et al. 2004). These findings further specify Herrmann and Deutsch's above principle of partner-adapted verbalisation. But the literature does not provide much evidence with respect to which reference systems are preferred if several options are available in an object identification scenario. This is so in part because such scenarios have not been in focus very often in spatial cognition research, and specifically, group-based reference has hardly been mentioned in the literature so far at all. Likewise, little can be said for the case of several options for perspective when there is no interaction partner. The present hypothesis is that the identification task – requiring reference on the basis of spatial contrast – plays a role in the decision, since situations may arise in which one kind of reference system or perspective enables a clearer contrast than the other ones available. On the grounds of Herrmann & Deutsch's principles, this means that, *in addition to adapting to their interlocutor, speakers choose a reference system and perspective that is suitable* (just as a unique object property is) *for distinguishing the target object from competing ones.*

How do speakers choose a reference axis? Within a reference system, the frontal (*front/back*), the lateral (*left/right*), and the vertical (*above/below*) axes are available for reference. With "Where" questions, the reference axis is chosen that the target object is closest to. But in contrastive reference, this may not yield unambiguous reference. It is hypothesized here that competing objects play a role in deciding about a reference axis: *A reference axis is chosen that is best suitable for distinguishing the target object from competing ones, considering the principle of greatest distance.*

How explicit are speakers about underlying reference systems and origins? The situation may offer various reference systems yielding similar results (e.g., Carlson 1999): in some situations, *to the left* could equally well be used for an intrinsic reference system conflating origin and relatum, for a group-based reference system using the other objects as relatum, and for a landmark-based relative reference system. In such situations, *to the left* can simply be used without further specification since no conflicts arise. Similar observations apply with respect to the chosen point of view (origin); if several options are available yielding no difference people do not need to provide an expression like "from my point of view". In intrinsic reference systems, the origin is often specified because it coincides with the relatum, as in "in front of me". The interesting case is when different reference systems and points of view yield different results, so that *to the left* can be interpreted in different ways. Herrmann & Grabowski (1994:132) state that speakers are usually *not* explicit with regard to the perspective used. This may be because they tacitly assume that the interlocutor will understand the intended meaning even without the additional effort, since perspective needs to be expressed by additional linguistic material and since the partner's perspective is (in many situations) conventionally preferred. But with respect to reference systems, no such conventions are known so far. Thus, it is hypothesized here that *relata – but not necessarily origins – will be made explicit in case of potential conflict.*

Under what circumstances do speakers modify and combine projective terms? With "Where" questions, speakers increasingly use modifiers and compounds as distance to the reference axis increases. According to Herskovits (1986) this is not expected in identification tasks (see above). Likewise, Herrmann & Deutsch's principles predict that speakers will not provide more information than needed, unless several options with equal properties compete. Thus, it is hypothesized that *speakers use a projective term without modification or combination with another projective term in case there are no competing objects for which the same description applies to the same degree.* If the target object is placed where it could equally well, and equally unambiguously, be referred to by two terms, such as *to the left* and *in front*, both may be combined.

2 Empirical Study

In order to address the above research questions along with the more basic aim of assessing the linguistic range of variety in settings involving contrastive spatial object reference, we collected linguistic data in a web-based empirical study. A major advantage of this approach is that large amounts of native speaker data can be collected with very little effort. On the other hand, since no clarification questions can be asked, there is a higher potential for misunderstandings. Furthermore, participants can vary factors about their participation, such as distraction, pauses between tasks, advice by other people, etc., which are controllable only in a setting involving the co-presence of experimenter and participant. They may also answer untruthfully to questions about their person regarding age, gender, language skills, etc, and they may re-start and participate several times. Further evaluation of advantages and limitations of web experiments can be found in Reips (2002). At the present stage, the advantages outweigh the disadvantages especially in light of the fact that most of the uncontrollable aspects are not considered to be central influencing factors. Nevertheless, in subsequent studies a higher degree of control is desirable, which presumably entails the participation of a smaller number of speakers.

Since the web-based study was carried out in order to establish a corpus of natural language data, in the analysis conventional methods of corpus linguistics are employed. The main focus is on the identification and description of qualitative structures in data collected in an open setting, supported by relative frequencies of usage. No statistical measures are computed because of the open setting and the large amounts of variation and potential interdependencies that render pairwise comparisons statistically less reliable. The present aim is to point to a number of systematic patterns that can be subsequently validated by more controlled experimentation.

2.1 Method

Speakers of English were asked via mailing lists and personal communication to participate in a web-based empirical study accessible at www.language-experiments.org

between September 23rd and December 31st, 2003. Participation was voluntary and not paid for. Altogether, approximately 200 self-assessed native speakers of English[2] participated. Their contributions show that there were very few problems in interpreting the instructions. Since the qualitative analysis was non-automatic and therefore time-consuming, not all of the data could be analysed exhaustively. Therefore, a target number[3] (60) of utterances for each single analysis was chosen that was considered sufficient for a fairly broad and informative exploration of the range of variety and underlying principles of speakers' choices[4]. The contributions of native English speakers were extracted at random out of the collected pool of data, annotated using the text markup tool *Systemic Coder*[5] version 4.5, and analysed linguistically. For each situation, the preferred linguistic options were identified and analysed with respect to the above research questions. Furthermore, differences between situations were examined by comparing frequencies of linguistic categories.

Each of the participants answered 15 different randomly assigned questions[6] in randomized order[7] out of a pool of 21 possible tasks which cover a range of different scenarios. A selection of the 21 tasks is analysed in this paper in order to examine the impact of specific changes in the scenario. They belong to four conditions that differ with respect to the possible perspectives on the configurations. Randomizations of tasks are only inside conditions, treating the conditions 3 and 4 as one category. In each condition, the same situations are shown (see Fig. 1 below). In the simplest scenario, S (situation) 1, three identical squares are located in a row, enabling unproblematic group-based reference. S2 shows the same scene except that the middle element is not a square but a circular element[8], providing a further option for the choice of a relatum, i.e., an element of a different kind. S3 presents only two elements in a spatial relationship that does not correspond to any of the focal axes in any kind of reference system. Thus, linguistic modifications can be expected here. Furthermore, especially in the conditions that include a viewer, it is hypothesized that the option of a group-based reference system is not available. Thus, the three configurations offer a range of spatial relationships that can be conceptualised and referred to in various

---

2 Since also non-native speakers were asked to participate the overall number of participants was much higher. Age was asked for in a questionnaire but not provided by many participants; those who stated their age were predominantly between 15 and 50 years old, which corresponds to the target of the experimental design. Since there are no hypotheses regarding the impact of age in adult speakers, age effects and distribution are not considered further.

3 As will be seen below, this number was deviated from in the analyses of S2(C4) and S2(C3).

4 Since not all participants completed all tasks, the tasks were partly completed by the same persons and partly by different ones. In treating the data as a corpus this unfortunate circumstance is regarded as marginal for the present kind of analysis.

5 Available freely at http://www.wagsoft.com/Coder/index.html.

6 The decision to limit the number of tasks for each participant to 15 was taken in order to minimise the time and effort required for participation.

7 Effects of order were also not computed in the present analysis, since they are regarded as a non-trivial additional factor that needs to be treated with specific care. Randomization guarantees a fairly even distribution of task positions.

8 This term is here used by way of contrast to the marking of the goal element by a circular line. The participants naturally used *circle* to refer to the circular element.

ways. This variability is further enhanced by the options of perspective, which are varied by the four conditions as follows.

Outside View. In the first condition, participants are presented with pictures that only contain squares and circular elements in the three configurations as depicted in Fig. 1, but without the X. In each picture, one of the elements is marked by a circle. The question to be answered by the participants is simply, "Which element of the picture is marked with a circle?" In this condition, the only available view direction is provided by the fact that the participants look at a picture on a screen (outside perspective). Concerning reference frames, one option is to use the picture itself as a relatum, applying the projective terms for the internal parts (regions) of the picture. Another option is to use a relative reference system employing either (some of) the other objects as a group relatum, or the circular element (if present) as a landmark relatum.

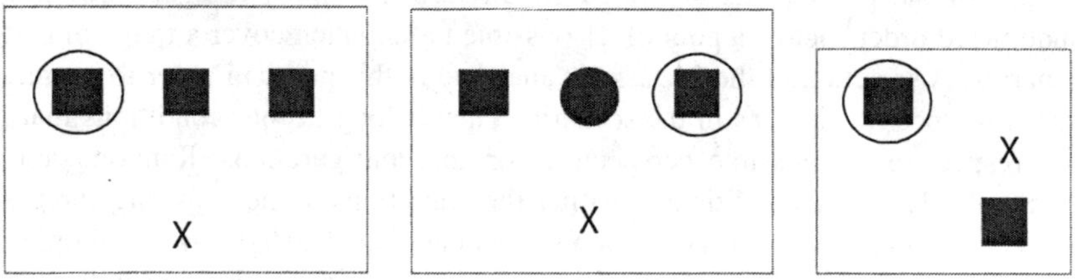

Fig. 1. Three situations S1, S2, and S3 (Condition 2: Inside view)

Inside View. In the second condition, an X appears in the picture in addition to the elements (Fig. 1). The instruction is, "Now imagine that you are looking at the figures from the position marked X. How do you describe now which element is marked with a circle?" This condition departs from the previous situation in that the participants are asked to imagine themselves *inside* the picture. Thus, the origin in their reference frames is supposed to be X directed towards the objects, but the outside perspective is still available. The reference frames available are those known from condition 1 plus intrinsic or relative reference systems specifying the target object's position relative to X, either from the imagined view direction of X or from the outside perspective.

Partner View. The third and fourth condition were designed to simulate a real world setting as much as possible, in order to represent the human-robot interaction settings that we use in our project. Now, the position of an interaction partner, Y, is added in the pictures. Additionally, both X and Y are assigned a clear view direction. In condition 3, X has the same position as in the inside view condition, and Y is positioned at a 90° angle with respect to view direction. In condition 4, the positions of X and Y are reversed. In each case, the participants read "Finally, please imagine that the figures are real world objects. You are located at X, and now your task is to instruct person Y to go to the object marked with a circle. A star ✦ shows the direction each of you is facing in." Figures 2 and 3 show the two pictures of these two conditions that are analysed in the present paper.

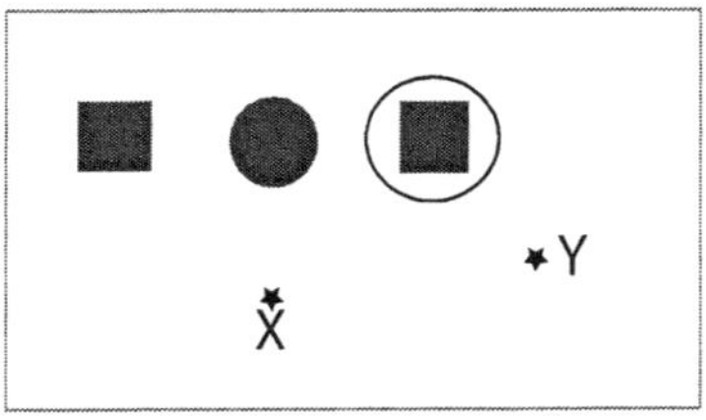

Fig. 2. S2 in cond. 3, partner view

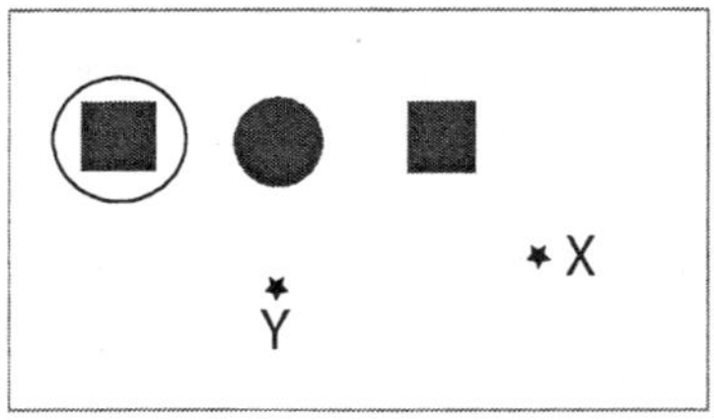

Fig. 3. S2 in cond. 4, reversed partner view

Thus, in these conditions view directions are given explicitly, and the participants are asked to imagine a dialogue situation. However, since there is no real interaction and no feedback from the interaction partner, grounding and alignment processes such as those described in Clark (1996) and Pickering & Garrod (in press) are ruled out, similar to imagined-partner experimentation as reported in Herrmann & Grabowski (1994). The instructional task differs slightly from the previous ones. This situation requires a lot of imagination by the participants; therefore, tasks in condition 3 and 4 are presented only after the first two conditions. Now, in addition to the origins and reference frames available in the previous conditions, there is the option of using Y as origin or relatum, or both.

2.2 Results

In a first examination of the data, 12 main linguistic categories were identified that occur frequently across contexts, covering the variability in utterances that refer to the goal object. These categories were used for a detailed analysis in order to examine the question how speakers choose reference terms in a variety of situations. The following list provides an overview of the categories together with the abbreviated reference term used in later sections. Note that the categories 1-11 always presuppose the presence of projective terms, which are the main focus of the present analysis, while 12 and 13 do not.

1. The projective term is a noun in a prepositional phrase and is neither modified nor further specified by providing information on the relatum, as in "The square on the right": Unmodified noun-in-pp
2. As before except that now the relatum is specified, either by another prepositional phrase, as in "to the right of the circle", or by a possessive, as in "to my right": Unmodified noun-in-pp plus relatum
3. The projective term is an unmodified adjective, stand-alone expression, preposition, or adverb (excepting handedness adjectives), e.g., "right (square)" or "the one above": Unmodified adjective / prep / adv
4. The projective term is an unmodified handedness adjective, as in "right-hand square": Handedness adjective
5. The projective term is a superlative (rightmost square): Projective superlative
6. Two projective terms of any kind are combined, as in "top left", "upper left", or "at the top left": Combination

7. The projective term is modified by a term denoting distance, as in "The one furthest to the right" or "The square on the far right": Modification-distance
8. The projective term is a noun in a prepositional phrase and is modified by counting[9] ("The third square on the right"): Modification-counting
9. The projective term is a noun in a prepositional phrase and is modified by a precisifying adverbial ("the square just to my right"): Modification-precisification
10. The projective term is part of a complex description combining several kinds of modification, as in "Go to the cube just in front of you to the right, before the sphere": Complex-description
11. A comparative relating to height is used, such as "the higher square": Comparative
12. A non-projective term is used that relates to distance, either as comparative ("the farther square") or as superlative ("farthest"): Distance-related
13. Other (mostly other kinds of non-projective expressions).

The distinctions are made on the basis of grammatical differences, principally induced by various kinds of modifications of the projective terms. Category 3 subsumes some forms because they occur only infrequently and in similar situations. This finding corresponds to the assumption that the different syntactic forms of projective terms do not necessarily reflect semantic differences (cf. Miller & Johnson-Laird 1976, Coventry & Garrod 2004). Nouns in prepositional phrases (cat. 1) are treated separately for three reasons: First, in the present data they occur frequently and in different situations than the occurrences subsumed in 3. Second, they involve a more complex linguistic construction than those in 3, containing a preposition that may itself contribute semantic content. Third, they can take an additional prepositional phrase or a possessive that specifies the relatum. This option is distinctly categorised as 2. Prepositions and adverbs such as *above* can also occur in constructions that inform about the relatum, but because they occur only very infrequently in the data this option is not assigned a distinct category. Handedness adjectives are treated separately because they appear fairly frequently in some configurations, and they may induce a semantic difference by implying a human-like origin.

The utterances are also analysed with respect to the choice of axes, perspectives and relata. However, as pointed out above it is assumed here that there is no one-to-one correspondence between linguistic surface forms and reference systems. Since utterances are in many cases underdetermined with respect to the underlying perspective and the intended relatum (as will be exemplified in some detail below), these two categories are coded with respect to all possibilities that are recognized to be compatible with the task. Since, as the present data show, participants are often creative in their conceptualisations of the spatial scenes, it is even possible that additional interpretations would yield the same results, so that the percentages listed should be regarded as tentative, only reflecting a lower limit. Note that this kind of underspecification usually does not result in ambiguity in referential identification of the target element, since this involves different interpretation processes than identification of the reference system or perspective. - Choice of axis and perspective do not

[9] This is exceptional since a short general introduction given to the participants before starting with the pictures explicitly stated that counting should be avoided.

apply for distance-based descriptions, but also there, a relatum may or may not be given explicitly.

The following situations are analysed: Condition 1 (C1): S1, S2, S3; Condition 2: S2 and S3; Condition 3: S2; Condition 4: S2. While the pool of collected data contains many more situations, this selection enables the systematic comparison of linguistic choices in situations that differ with respect to only one factor.

Overview of Results. Table 1 below summarizes the results of the linguistic analyses, providing the percentage of frequency of the various linguistic categories occurring in each situation under analysis. N is the analysed number of utterances containing projective terms in each situation. In cond. 1 and 2, 60 utterances were examined in each task under analysis, very few of which do not specify the target object directly by projective terms. In S2(C4) and S2(C3), in contrast, a high percentage of contributions do not refer directly to the goal element. These are not represented in the table but treated separately below.

Table 1. Variability in goal-related utterances (% in each task)

Task	S1(C1) (N=60)	S2(C1) (N=60)	S3(C1) (N=60)	S2(C2) (N=60)	S3(C2) (N=59)	S2(C4) (N=51)	S2(C3) (N=57)
Linguistic category							
unmodified noun-in-pp	20.0	28.3	10.0	28.3	13.6	11.8	-
unm. noun-in-pp + relat	-	8.3	-	16.7	22.0	56.9	5.3
unmod. adj., prep./ adv.	11.7	13.3	35.0	16.6	22.0	2.0	-
handedness adj.	6.7	10.0	-	6.7	-	-	-
proj. superl.	35.0	23.3	6.7	15.0	1.7	9.8	3.5
combination	-	-	40.0	13.3	20.3	2.0	-
mod.-dist.	21.6	15.0	1.7	-	-	11.8	3.5
mod.-counting	1.7	-	-	-	-	-	7.0
mod.- precisification	-	-	-	-	1.7	-	8.8
complex descr.	-	-	-	-	-	2.0	14.0
comparative	-	-	5.0	-	-	-	-
distance	-	-	-	-	13.6	3.9	57.9
other	3.3	1.7	1.7	3.4	5.1	2.0	-
Perspective							
outside view	96.7	100	100	76.7	18.6	33.3	-
X's persp	-	-	-	100	83.0	-	-
Y's persp	-	-	-	-	-	96.0	42.1
explicit	-	1.7	-	23.3	40.7	62.7	36.8

Table 1. *(Continued...)*

Task	S1(C1) (N=60)	S2(C1) (N=60)	S3(C1) (N=60)	S2(C2) (N=60)	S3(C2) (N=59)	S2(C4) (N=51)	S2(C3) (N=57)
			Spatial Axis				
only lateral	96.7	100	11.7	83.3	54.2	90.2	35.1
only frontal	-	-	-	-	1.7	-	1.8
only vertical	-	-	46.7	1.7	6.8	-	-
lat. & front.	-	-	-	10.0	6.8	2.0	5.3
vert. & lat.	-	-	38.3	3.3	10.2	2.0	-
compass dir.	-	-	1.7	1.7	3.4	2.0	-
			Relatum				
int. regions	96.7	88.3	98.3	66.7	13.6	25.5	1.8
group	96.7	88.3	88.3	66.7	13.6	27.5	1.8
landmark[10]	-	66.7	-	46.7	-	15.7	1.8
X	-	-	-	98.3	96.6	5.9	5.3
Y	-	-	-	-	-	86.3	94.7
explicit	-	11.7	6.7	25.0	49.2	72.5	83.9

In conditions 3 and 4, the perspective of Y (the "interlocutor") is consistently used, never that of X. This results in S2(C4) being more similar to S2(C2) than S2(C3) is, with regard to the origin's position. Therefore their order is reversed in the analysis. In the following, each of the above research questions is addressed separately.

Choice and Explicitness of Point of View. In condition 1, all projective expressions are based on the participant's outside view on the picture (the only available option), which is made explicit only once through all three situations examined. Note that the linguistic construction used for making perspective explicit is fairly complex, which may be a reason why it is the only such specification: "If I am viewing the picture as if it were a picture hanging on a wall, the answer is the left most square".

In S2(C2), all utterances are consistent with the perspective of X (as requested in the instruction to the participants), although the outside perspective is also available. 21.7% explicitly use this perspective by specifying X as the relatum, using an intrinsic reference system (e.g., "the one to my right"), where the relatum is identical to the origin. The other utterances are consistent both with the perspective of X and with the outside view, so that it cannot be decided which perspective is actually used. This does not, however, impede interpretation because none of the other elements could be referred to in this way in the present situation.

The second situation in condition 2, S3(C2), poses a challenge for the participants because the view direction is less clear than in the other scenarios, where all objects are placed together at one side of X. Here, in contrast, "looking at the elements" (as in the instruction) leaves some freedom for interpretation. The participants are clearly aware of this problem and therefore develop interesting strategies. One explicitly states that it is impossible to give an answer since the view direction of X is unknown;

[10] Utterances like "to the left" are coded as consistent with a landmark-based reference system because they could be expanded to "... of the circle", while "lefthand" and "leftmost" are not.

therefore, only 59 utterances are analysed. In contrast to the other tasks in conditions 1 and 2, in S3(C2) as many as 10 utterances (16.9%) are not projective; 8 of these rely solely on distance, using expressions such as "farther" or "furthest from me", which are independent of the view direction. This strategy circumvents the perspective problem. Another strategy is to avoid ambiguity via choice of axis (see below).

In conditions 3 and 4, all utterances containing projective terms are consistent with the perspective of Y, though in S2(C4) this view direction coincides with the outside view as in S2(C2) above. Therefore, in one-third of these cases perspective cannot be determined although the descriptions are not ambiguous. In the other cases, perspective is specified by mention of the relatum, as in "Go to the square on your left".

Choice and Explicitness of Reference Systems. A frequent linguistic option is the usage of a projective term as an unmodified noun in a prepositional phrase, such as "the one on the right". Here, the underlying reference system is underdetermined if the relatum is not made explicit, which is mostly the case in tasks S1(C1), S2(C1), and S2(C2). It could then be either group-based relative (in case the utterance was continued as "... of the other elements"), landmark-based relative ("... of the circle"), picture-based ("... of the picture") or intrinsic ("... of X (or Y)"). If the underlying reference system does not yield any difference in interpretation, further explication is not necessary, and accordingly does not occur at all in S1(C1). In the similar situations S2(C1,C2,C4) the relatum is mentioned increasingly often with increasing alternative possibilities. In S2(C4), as many as 72.5% explicitly use Y as relatum, as compared to 11.7% in S2(C1) and 25.0% in S2(C2). This difference is explained through the fact that there is a true alternative interpretation in S2(C4), namely X as relatum and origin, while in S2(C1,C2) all other options do not yield a different interpretation.

The circular element (landmark) is explicitly used for reference in 11.7% of utterances in S2(C1), but only in 1.7% of all utterances in S2(C2), 5.9% in S2(C4), and 1.8% in S2(C3). This leads to the hypothesis that objects of a different kind are not necessarily favoured as relata, depending on the situation.

In S3(C1), as in S1(C1), differentiating between picture-internal and group-based reference systems is not necessary, since both interpretations yield the same results. Accordingly, they are not specified except for a few utterances that refer to the picture's corner. In S3(C2), in contrast to the other tasks analysed so far, the different available reference systems yield different spatial regions. Here, the relatum is specified in half of the utterances, assisting interpretation.

Choice of Reference Axis. In S1/S2(C1), all utterances rely on the lateral axis. In S2(C2), 83.3%, and in S2(C4), 90.2% of all utterances use the lateral axis as the single axis for reference. This clear result is not surprising since the second axis is only available when using the X (in C2) or Y (in C4) as a relatum, while in group-based reference or using the picture's internal regions only one axis is available. This axis coincides with the axis that is most suited to distinguish between the objects also when X (or Y) is used as relatum, since in that case, all three elements are located *above* or *in front of* the relatum, but only one (the target object) is located *on the right.*

Thus, it is surprising that one participant wrote "the element above", since this is not informative (i.e., unambiguous) in this situation.

In S3(C1) where the target object is located diagonally between the focal axes, interestingly more than half (59.3%) of the projective utterances rely on one single axis for reference. In these cases, the axis is obviously chosen for establishing a contrast to the other element in the picture, which is not located on the same half-plane, instead of describing the exact spatial location. Whenever only one axis is used there is a clear preference for the vertical axis (80% of utterances using only one axis, as compared to 20% that use only the lateral one). This can be explained by way of Herrmann & Deutsch's principle of greatest distance: The target object is located just a trifle more clearly in the upper part of the picture, and above the other object, than it is on the left. Another explanation is that the vertical axis is more salient than the lateral, as has been claimed time and again in the literature (e.g., Bryant et al. 2000).

In S3(C2), there is a high heterogeneity in participants' choices even though the situation contains only two elements. Nevertheless, some systematic preferences can be identified: if only one axis is used, the target element is preferably referred to as "right" (44.1% as opposed to 1.7% referring to the "front") even though it is not situated close to the right focal axis with respect to X (and not at all when using the outside view). Rather, it is situated between the right and front axes of X, even closer to the front axis than to the right one. But, unlike "to the right", "in front" could, in a different interpretation, also refer to the competing object. Furthermore, since the competing object is located clearly *on the left* with respect to the lateral axis, but only *in the middle* with respect to the frontal one, the lateral axis provides a clearer contrast. Thus, as hypothesized, reference systems and spatial axes are preferred that allow for unambiguous reference in the given situation.

Modifications and Combinations of Projective Terms. Projective superlatives, e.g., "the leftmost element", can be regarded as explicit linguistic constructions with regard to contrast, since they indicate that one specific element can be singled out that is located farther in the indicated direction than any other elements. Not surprisingly, this is the most preferred option in S1(C1). In the other tasks where the elements are lined up in a row as in S1(C1), namely, S2 in all four conditions, projective superlatives are still present but the frequency decreases with each additional factor in favour of other options. In S3(C1) where there are only two elements the projective superlative is very infrequent. One utterance here states "the leftmost (leftmore?) square is", thus providing the analyst with a reason why the superlative is not used here more often: it does not seem to be appropriate in a context where there is only one competing object. Since, unlike other superlatives, there is no expression like *leftmore* in English, the participants prefer other linguistic constructions in this scenario. Note that projective superlatives do not occur at all in the literature on other kinds of elicitation of spatial expressions in which applicability areas are to be identified.

Another frequent method of establishing spatial contrast is the usage of linguistic modifiers such as "furthest (to the right)" (mod.-distance). This is especially frequent in S1(C1). The presence of a distance modifier linguistically enforces a contrast to other objects present, allowing, for example, for other elements to be "not-far to the right" (e.g., to the right but not as far as another element).

Complex descriptions combining several kinds of linguistic modifications and combinations, as in "Walk to the square object closest to you, slightly on your right", providing a fairly precise spatial description, occur only in the partner conditions, most frequently in S2(C3). Similarly, modifiers precisifying the projective term as in "Go to the square immediately in front of you" occur only in S2(C3), although also in many other cases the elements are not situated near focal axes, which could lead to the expectation that precisifying adverbials and complex descriptions should occur with much higher frequency. However, only in S2(C3) is there a need to specify the position since there are competing elements located in a similar direction. Here, the goal object is situated close to Y, enabling simple description in terms of distance, as is done by almost one third of participants. However, those that do not choose this non-projective kind of description obviously run into problems in finding a suitable description. A simple projective description that establishes sufficient contrast to competing objects does not seem to be available here.

In the other cases where the elements appear in a row, the target element is located directly on the lateral axis only in the group-based and picture-internal cases but not in the case of using X or Y for intrinsic reference in S2(C2) and S2(C4). But intrinsic reference is indeed very frequent at least in S2(C4), as indicated by the fact that in 60.8% of all goal-related utterances the relatum Y is explicitly mentioned, as in "to your left". Thus, a vague spatial description is deemed sufficient by speakers if there are no competing objects closeby. Furthermore, since modifications and combinations of projective terms are rare even in the highly problematic situation S3(C2), it can be hypothesized that the strategy of choosing a reference axis that enables unambiguous identification may in some situations have stronger consequences than the graded semantic structure of projective terms.

In the following, some contrasts between linguistic choices are highlighted by comparing situations that differ with regard to one factor.

Comparison of S1(C1) and S2(C1). These two tasks differ only in the presence of an object of a different kind in S2(C1). Interestingly, the usage of modifications including superlatives seems to be lower when a landmark is present (38.3% in S2(C1) vs. 58.3% in S1(C1)). This can be interpreted as indicating a lesser need for explicitly establishing contrast. Since the middle element in this case is of a different kind, it does not appear as a competing candidate. The only other element of the same kind is positioned at some distance, so that unmodified projective terms are more often deemed sufficient. Furthermore, participants probably refrain from employing superlatives because there is only one other object of the same class present, as in S3(C1).

Comparison of S1(C1) and S3(C1). In S1(C1), there is one clear reference axis available which is used throughout. Modifications concern position on that axis, using superlatives and modifiers denoting distance. In S3(C1), in contrast, two reference axes are equally suitable. Here, modifications concern combinations of projective terms as well as comparatives. However, unmodified adjectives are used much more frequently than in S1(C1). In S1(C1), the preferred option is the use of projective superlatives which, in turn, are clearly disregarded in S3(C1). Furthermore, in S3(C1) only one participant refers to distance, as opposed to 21.6% in S1(C1).

Figure 4 contrasts the linguistic options used in the three situations in condition 1.

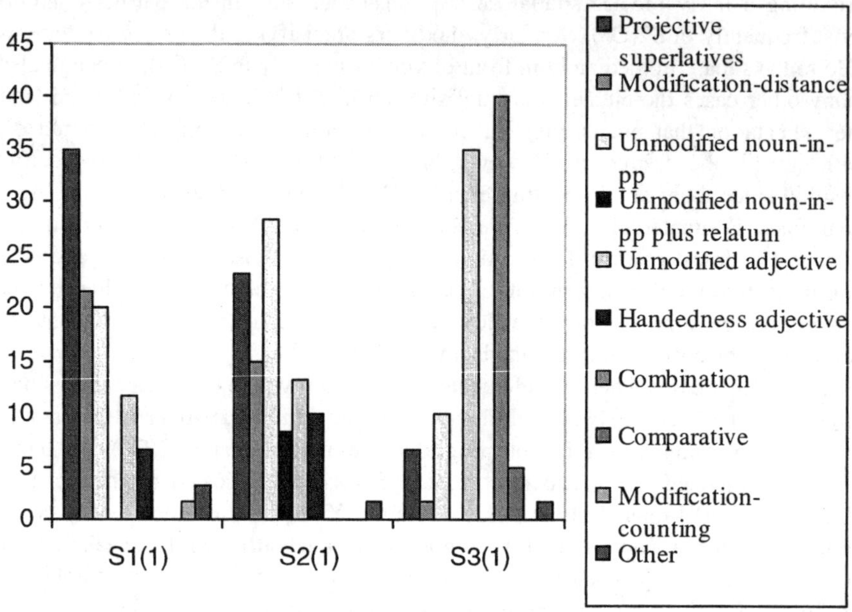

Fig. 4. The impact of the spatial configuration (without variation of perspective)

Comparison of S3 in Condition 1 and 2. The linguistic choices speakers make in S3(C2) differ considerably from those in S3(C1) although the pictures differ only with respect to the available origins and relata. Interesting differences occur, for example, with respect to the employment of a second projective term (40.0% in S3(C1) vs. 20.3% in S3(C2)), and the usage of distance-related terms (none in S3(C1) vs. 13.6% in S3(C2)). These differences point to the fact that the presence of X has a considerable impact on the conceptualisation of the situation, confirming that the participants indeed view the situations as different in spite of the fact that the outside view on the picture is still available in S3(C2), though participants are encouraged to use the perspective of X.

Condition 3 and 4. In these conditions, in addition to imagining themselves inside the picture at position X, participants are asked to instruct a person located at Y to go to one of the objects. To achieve this task, many participants do not refer directly to the goal object. The design of the study allows for free choice of instructional strategy, because the instruction to the participants only informs about the imagined persons' position and view direction, but not, for example, about the absolute scale, distances, and perceptibility of goal objects, or any other clues that could influence the choice of strategy. Therefore, I do not view utterances that are not goal-based as misinterpretations of the task, but rather as an interesting result with respect to how participants conceptualise the situation. In the present analysis (cf. Table 1), primarily goal-based utterances are targeted. However, some interesting observations can be noted with respect to the other kinds of strategies employed in conditions 3 and 4.

In S2(C4), of a total of 114 collected utterances instructing an imagined person Y to move to a goal object, only 43.0% use projective expressions referring directly to the goal, and 1.8% use distance expressions. These are the utterances that appear in Table 1 above. Another 30.7% describe path directions like left/right or angles, and 23.7% first describe the path and then refer to the goal object. In S2(C3), of a total of 125 instructions, 19.2% use projective terms pointing directly to the goal object, 25.6% rely on distance, 40.0% describe path directions like left/right or angles, and 12.0% describe the path and then refer to the goal object. This variability of instructional strategies is similar to that found in our human-robot interaction scenarios (cf. Fischer & Moratz 2001). It is depicted in Figure 5. Note that, in comparison, the utterances in the other two conditions rely almost exclusively on projective expressions, with occasional usages of distance expressions in certain situations.

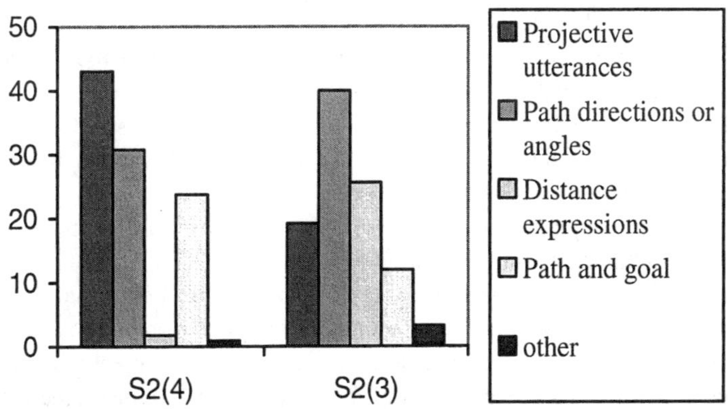

Fig. 5. Strategies in two tasks in conditions 3 and 4 differing only in perspective

The clear difference in usage of distance terms is explained through the proximity between Y and the goal object in S2(C3), especially because two competing objects are situated in the same spatial area as the target object. This option is not equally available in S2(C4) from Y's point of view, since another object is at equal distance.

3 Discussion

One of the primary aims of the present study was to investigate the variability with regard to linguistic choices in scenarios involving contrastive spatial object reference. Clearly, participants use a broad spectrum of variability, resulting in at least twelve distinct linguistic categories which could still be differentiated further (see Table 1 above). The data collected in the present study show that linguistic choices depend heavily on the spatial situation, i.e., the presence of other objects and the available perspective. Therefore, generalised predictions are difficult to formulate on a linguistic surface level. A general result, however, is the finding that *instructional strategy* changes with condition: People evidently approach a task differently when asked to instruct someone to move somewhere, than when asked to single out one target object of several competing ones. But even here, clear differences appear between spatial scenarios depending on the observer's position.

Not all of the available linguistic options are used in all situations. In spite of the high variability, however, regular patterns of usage can be identified that can be analysed in relation to the hypotheses posed in section 0 above. The systematic variations found are repeatedly explained by much the same idea, reflecting the participants' motivation to fulfill the task of providing an unambiguous – contrastive – description of the target object in a given situation. In other words, speakers intuitively (but not stringently) adhere to systematic underlying principles that could, in a first approximation, be formulated as follows.

1. A reference system is chosen that allows for unambiguous reference, i.e., that produces at least one spatial region in which an unmodified projective term can be used unambiguously (see 2), if possible at all (otherwise see 5).
2. Unmodified projective terms can be used unambiguously in the following cases, and are therefore employed there more often than in other cases:
   - if the goal object is the only one on a half-plane with respect to the reference system used (regardless of whether it is located near the prototypical axis corresponding to the projective term used or not);
   - if competing objects also situated on the same half-plane are clearly farther away from the axis than the goal object is.
3. If 2 applies for more than one spatial region, the axis is chosen that the goal object is closest to (unless 4 applies), and an unmodified projective term is used.
4. If 2 applies for more than one spatial region and the goal object is located at equal distance from two axes (e.g., left and front) then either both projective terms are combined, or one is chosen at random or via individual preferences.
5. If all regions are occupied by more than one object, modifications, combinations of projective terms, counting and/or distance expressions are used. In such a case, distance expressions are specifically likely if the goal object is clearly either the closest one or the one that is farthest away from a suitable relatum.
6. Relata are preferably mentioned explicitly in case of conflict in interpretation; otherwise they usually remain implicit. Origins are seldom given except when conflated with relata in intrinsic reference systems.

Largely, these findings are in accord with previous results of research on object reference, where it is known that speakers analyse contrast on any dimension available, with respect to the objects present in a given scenario, and use it in order to achieve unambiguous reference. In psycholinguistic studies, as well as in the present web study, the available contrastive dimensions as well as the reference area are usually clearly delimited by the experimental setting. Real world scenarios involve far greater complexity. For example, objects often differ with respect to both spatial position and non-spatial kinds of features, where under certain circumstances reference on the basis of spatial position seems to be preferred (Pobel et al. 1988). However, the factors influencing such choices and the variability with respect to related options seem to be largely unexplored. For example, in what kinds of situations do speakers deem a simple class name as sufficient for reference? Conceivably, this may be the case if a given configuration allows for unambiguous object reference via the class name. But this hypothesis does not only presuppose simplistic referability on the basis of a distinctive object name, but also a clear delimitation of the reference area (among further factors, see Freksa 1980). In real world scenarios, this may be a crucial factor influencing speakers' choices, especially in an open setting where the referential domain may be unclear. In a human-robot interaction scenario, for example, the speaker may not be informed about the limits of the robot's perception (Fischer & Moratz 2001). Brown-Schmidt & Tanenhaus (2003) show that both the form of referring expressions and their interpretation are constrained by previous specification of referential domains. Furthermore, attention focus plays a crucial role (e.g., Kessler et al. 1999). In ambiguous situations, reference is resolved with respect to a subset in focus. Attention can be directed by the speaker through focus and foregrounding on a linguistic level, influencing the listener's interpretation of spatial descriptions. The applicability of spatial terms is further influenced by functional features of the objects involved, and the relationship between them (e.g., Coventry & Garrod 2004). In natural communication, interactive processes facilitate the achievement of joint reference (Clark 1996, Pickering & Garrod in press), and previous experience (i.e., the discourse history) may influence later choices, for example in the employment of reference systems (Tenbrink & Moratz 2003). All of these major influencing factors need to be accounted for when dealing with natural conversation, in contrast to the simple scenario presented here.

Due to the open design of the study a considerably broad range of variety, and systematic patterns of choices, could nevertheless be identified, since speakers' linguistic behaviour was not governed by instructions to a high degree. The procedure adopted in the present work is in accord with the methodology adopted in our research project (cf. Fischer 2003), which ensures the production of intuitive language to a higher degree than more restricted settings would allow. The identification of intuitive strategies is vital for our research aim of enabling natural and effective human-robot interaction in spatial settings, specifically in light of the fact that current natural language interfaces are often not evaluated with respect to their effectiveness when confronted with users who are not informed about the robot's vocabulary – and if they are, this may lead to devastating results (Thrun 2004). Clearly, the results of this kind of empirical study and analysis differ in content and generalisability from results gained in more controlled psycholinguistic experiments. In the long run, it is desirable to work towards establishing methodologies combining the advantages of both approaches, as described, for instance, in von Stutterheim et al. (1993). Interdisciplinary collaboration is specifically targeted in the growing field of human-robot interaction research (Burke et al. 2004), where various research directions are combined out of necessity.

The present study can be regarded as an exploration of natural language produced in an open (though artificial) setting by unbiased speakers, which has led, on the one hand, to an assessment of the diversity in linguistic choices in a range of situations allowing for different interpretations and viewpoints, and on the other hand, to the identification of systematic principles underlying speakers' choices, which need to be confirmed by more controlled and at the same time more restricted experimentation. In addition to that, in order to gain insights with respect to a broader range of settings the validity of the hypotheses also need to be tested in other configurations, and considering further influencing factors. Furthermore, the specific strategies users develop in a human-robot interaction setting, even if the spatial situation resembles the depictions in the present study, can only be addressed in real-world experimentation with users who are not informed about the robot's capabilities.

4 Conclusion

In this paper, results of a web-based empirical study designed to collect spatial localisation utterances and test for systematic patterns of usage were presented. The linguistic analysis was carried out qualitatively and quantitatively, providing relative frequencies of participants' linguistic and spatial choices. Insights concerning linguistic variability in interplay with conditions and configurations were gained. It was shown that, out of the range of linguistic options, some were clearly preferred under specific conditions while others did not occur there at all, though favored in other situations. The qualitative analysis revealed that the systematic patterns of usage point to underlying principles speakers adhere to when establishing contrastive reference in a spatial scenario, similar to those known for non-spatial object reference and in line with previous results in spatial language research. For instance, speakers choose a reference system and spatial axis that is suitable for contrasting the target object from competing ones. The exact spatial location is usually not specified if there are no competing objects closeby. This results in the frequent usage of unmodified projective terms even if the target object is located at considerable distance from the reference axis, which contrasts from the usage of spatial terms in other kinds of tasks.

Acknowledgements

I thank the numerous anonymous participants for contributing their time and effort without reward, and the organizers of the web portal at www.language-experiments.org for hosting the study. T. Reinhardt's technical assistance in preparing the web study and data retrieval is highly appreciated, as well as K.-H. Wagner's support. Many thanks to three anonymous reviewers, M. Knauff, J. Restat, A. Stefanowitsch, and C. Vorwerg for very helpful comments on earlier versions of this paper, and the members of the SFB/TR 8 project I1-[OntoSpace] (J. Bateman, S. Farrar, K. Fischer, and R. Moratz) and other projects of the SFB/TR 8 Spatial Cognition for teamwork, cooperation, and invaluable discussions in relation to the work presented here. Funding by the Deutsche Forschungsgemeinschaft (DFG) is gratefully acknowledged.

References

Brown-Schmidt, S. & M.K. Tanenhaus. 2003. Referential domains and the interpretation of referring expressions in interactive conversation. *Proc. DiaBruck, 7th Workshop on the Semantics and Pragmatics of Dialogue, Sept. 4th-6th 2003, Wallerfangen*, pp 15-19.

Bryant, D.J., B. Tversky, & M. Lanca. 2000. Retrieving spatial relations from observation and memory. In E. van der Zee & U. Nikanne (eds.), *Cognitive interfaces: Constraints on linking cognitive information.* Oxford: Oxford University Press, 94-115.

Burke, J.L., Murphy, R.R., Rogers, E., Lumelsky, V.J., & Scholtz, J. 2004. Final report for the DARPA/NSF interdisciplinary study on human-robot interaction. *IEEE Transactions on Systems, Man and Cybernetics, Part C*, Vol. 34 No.2, pp. 103-112.

Carlson, L.A. 1999. Selecting a reference frame. *Spatial Cognition and Computation,* 1 (4), 365-379.

Carlson, L. A., T. Regier, & E. Covey. 2003. Defining Spatial Relations: Reconciling Axis and Vector Representations. In E. van der Zee & J. Slack (eds.), *Representing Direction in Language and Space.* Oxford: Oxford University Press.

Carlson-Radvansky, L.A. & G.D. Logan. 1997. The Influence of Reference Frame Selection on Spatial Template Construction. In *Journal of memory and language* 37, 411-437.

Clark, H.H. 1996. *Using Language.* Cambridge, UK: Cambridge University Press.

Coventry, K. R. & Garrod, S. C. 2004. *Saying, seeing and acting: The psychological semantics of spatial prepositions.* Psychology Press: Essays in Cognitive Psychology series.

Eschenbach, C. 2004. Contextual, functional, and geometric features and projective terms. In L. Carlson and E. van der Zee (eds.), *Functional features in language and space: Insights from perception, categorization and development.* Oxford: Oxford University Press.

Fischer, K. 2003. Linguistic Methods for Investigating Concepts in Use. In: Stolz, Th. and K. Kolbe (eds.): *Methodologie in der Linguistik.* Frankfurt a.M.: Lang.

Fischer, K. & R. Moratz. 2001. From Communicative Strategies to Cognitive Modelling. Workshop Epigenetic Robotics, Lund.

Franklin, N., Henkel, L.A. & Zangas, T. 1995. Parsing surrounding space into regions. *Memory and Cognition, 23,* 397-407.

Freksa, C. 1980. Communication about visual patterns by means of fuzzy characterizations. *XXIInd Intern. Congress of Psychology*, Leipzig, July 1980.

Gapp, K.-P. 1995. An empirically validated model for computing spatial relations. In I. Wachsmuth, C.R. Rollinger & W. Brauer (eds.): *KI-95: Advances in Artificial Intelligence. 19th Annual German Conference on Artificial Intelligence* (pp 245-256). Springer: Berlin.

Grice, H.P. 1975. Logic and conversation. In: Cole, P. and J. Morgan (eds.), *Syntax and semantics.* New York, San Francisco, London (Vol 3: 41-58).

Herrmann, T. 1990. Vor, hinter, rechts und links: das 6H-Modell. Psychologische Studien zum sprachlichen Lokalisieren. *Zeitschrift für Literaturwissenschaft und Linguistik 78.* 117-140.

Herrmann, T. & W. Deutsch. 1976. Psychologie der Objektbenennung. Bern: Huber Verlag.

Herrmann, T. & J. Grabowski. 1994. *Sprechen: Psychologie der Sprachproduktion.* Heidelberg: Spektrum.

Herskovits, A. 1986. *Language and spatial cognition.* Cambridge University Press.

Kessler, K., I. Duwe & H. Strohner. 1999. Grounding Mental Models: Subconceptual Dynamics in the Resolution of Reference in Discourse. In: G. Rickheit & C. Habel (eds.), *Mental Models in Discourse Processing and Reasoning.* Amsterdam: Elsevier.

Levinson, S.C. 2003. *Space in Language and Cognition.* Cambridge University Press.

Moratz, R. & K. Fischer. 2000. Cognitively Adequate Modelling of Spatial Reference in Human-Robot Interaction. *Proceedings of the 12th IEEE International Conference on Tools with Artificial Intelligence, ICTAI 2000*, November 13-15 2000, Vancouver, British Columbia, Canada, pp. 222-228.

Moratz, R., T. Tenbrink, J. Bateman, & K. Fischer. 2003. Spatial Knowledge Representation for Human-Robot Interaction. In Freksa, C., W. Brauer, C. Habel & K.F. Wender (Eds.), *Spatial Cognition III* (pp 263-286). Berlin, Heidelberg: Springer.

Olivier, P. & K.-P. Gapp (eds.), 1998. *Representation and Processing of Spatial Expressions.* Mahwah, New Jersey: Lawrence Erlbaum

Pickering, M. & Garrod, S. (in press). Toward a mechanistic psychology of dialogue. To be published in *Behavioral and Brain Sciences.*

Pobel, R., Grosser, Ch., Mangold-Allwinn, R. & Herrmann, Th. 1988. Zum Einfluß hörerseitiger Wahrnehmungsbedingungen auf die Überspezifikation von Objektbenennungen. Arbeiten der Forschergruppe "Sprechen und Sprachverstehen im sozialen Kontext" Heidelberg/Mannheim, Bericht Nr. 17.

Reips, U.-D. 2002. Theory and techniques of Web experimenting. In B. Batinic, U.-D. Reips, & M. Bosnjak (Eds.), *Online Social Sciences.* Seattle: Hogrefe & Huber.

Retz-Schmidt, G. 1988. Various views on spatial prepositions. *AI Magazine 9*: 2. 95-105.

von Stutterheim, C., R. Mangold-Allwinn, S. Barattelli, U. Kohlmann, & H.-G. Kölbing. 1993. Reference to objects in text production. In: J. Nuyts & E. Pederson (eds.), *Perspectives on Language and Conceptualization.* Belgian Journal of Linguistics 8., pp. 99-125.

Taylor, H.A., S.J. Naylor, R.R. Faust, & P.J. Holcomb. 1999. "Could you hand me those keys on the right?" Disentangling spatial reference frames using different methodologies. *Spatial Cognition and Computation*, 1 (4), 381-397.

Tenbrink, T. & R. Moratz. 2003. Group-based Spatial Reference in Linguistic Human-Robot Interaction. *Proceedings of EuroCogSci 2003: The European Cognitive Science Conference*, September 10-13, Osnabrück, Germany, pp 325-330.

Thrun, S. 2004. Toward a Framework for Human-Robot Interaction. *Human-Computer Interaction, Volume 19 (2004), Numbers 1 & 2*, pp. 9-24.

Tversky, B. & P.U. Lee. 1998. How Space Structures Language. In: C. Freksa, C. Habel & K.F. Wender (eds.), *Spatial Cognition. An Interdisciplinary Approach to Representing and Processing Spatial Knowledge* (pp. 157-175).

Vorwerg, C. 2001. *Raumrelationen in Wahrnehmung und Sprache: Kategorisierungsprozesse bei der Benennung visueller Richtungsrelationen.* Wiesbaden: DUV.

Watson, M.E., M.J. Pickering, & H.P. Branigan. 2004. Alignment of Reference Frames in Dialogue. *Cogsci 2004: 26th Annual Meeting of the Cognitive Science Society*, August 5-7, 2004, Chicago.

van der Zee, E. & J. Slack (eds.), 2003. *Representing Direction in Language and Space.* Oxford: Oxford University Press.

Zimmer, H.D., H.R. Speiser, J. Baus, A. Blocher, & E. Stopp. 1998. The Use of Locative Expressions in Dependence of the Spatial Relation between Target and Reference Object in Two-Dimensional Layouts. In C. Freksa, C. Habel & K.F. Wender (eds.), *Spatial Cognition* (pp. 223-240). Berlin: Springer-Verlag.

Cultural Differences of Spatial Descriptions in Tourist Guidebooks

Koshiro Suzuki and Yoshiki Wakabayashi

Department of Geography, Tokyo Metropolitan University,
Hachioji, Tokyo 192-0397, Japan
{debussy, wakaba}@comp.metro-u.ac.jp

Abstract. This paper examined spatial descriptions for guiding Japanese and American tourists from cross-cultural and geographic perspectives, based on a content analysis of 24 guidebooks to four cities in Japan and USA. Quantitative analysis of pictorial and linguistic information in guidebooks revealed that Japanese guidebooks use predominately pictorial information, whereas American guidebooks mainly depend on the linguistic one. In addition, we found a complementary relationship between the two modes of information. The contents of linguistic information were entirely influenced by socio-cultural factors rather than environmental conditions such as the street pattern regularity. In particular, difference in address systems between two countries affected the way of sorting the sites, style of maps, and the use frequency of linguistic information.

1 Introduction

The sources of spatial cognition can be divided into direct contact with the physical environment (e.g., travel behavior) and indirect one (e.g., through map and language). The latter is regarded as an information source peculiar to human beings and the only source of information about large scale environment that cannot be experienced directly. The indirect information also plays an important role when one communicates and shares spatial information with others.

The indirect sources of spatial information mainly consist of maps and languages. These are both important means of spatial information transmission, and they cannot be adequately produced by an individual who is isolated from the socio-cultural context (Downs and Liben 1993). In other words, the spatial descriptions of this type of information more or less reflect characteristics of the spatial cognition shared by the members of a given society. There have been, however, few attempts to conduct cross-cultural research in this field.

This paper is concerned with the cultural differences in the spatial descriptions for navigation with maps and languages. As Montello (1995) pointed out, the cultural difference in spatial cognition involves the ambiguity between culture-related and culturally caused. Hence it is necessary to unravel intricate aspects of cultural differences. In this study, we divide cultural differences in spatial descriptions roughly into the difference caused by socio-cultural rules or conventions (e.g., spatial languages and address systems) and the ones stemming from the characteristics of the

C. Freksa et al. (Eds.): Spatial Cognition IV, LNAI 3343, pp. 147 – 164, 2005.
© Springer-Verlag Berlin Heidelberg 2005

built environment (e.g., street pattern of the city). To examine these differences quantitatively, we analyze the contents of tourist guidebooks published in different countries for same cities. The reason for employing tourist guidebooks is that they are regarded as a material useful for cross-cultural comparison of the style of communicating spatial information.

The remaining part of this section gives a short overview of the previous studies on cultural differences in spatial descriptions with maps and languages. After providing framework and method for analyzing spatial descriptions in tourist guidebooks in the next section, we show how different are the spatial descriptions of guidebooks published in Japan and America for four sample cities in sections 3 and 4. In the last section, we discuss the role of socio-cultural factors affecting spatial descriptions and implications for future work.

1.1 Cultural Difference and Universality of Mapping Abilities

Previous studies on the cultural or social variations in mapping styles can be divided roughly into two types. One includes cultural anthropological studies that mainly noticed the spatial information transmission process (e.g., Gladwin 1970). The findings obtained in these studies, however, are difficult to generalize since they are observed in a specific environment of uncivilized society.

The other type of research is concerned with cultural difference and universality of mapping abilities. The studies of this type tried to examine the development of spatial cognition by comparing children's developmental changes in map-reading or map-drawing abilities with their age comparing different countries (e.g., Blades et al. 1998). As a result, Stea et al. (1996) pointed out that mapping behavior was investigated in one's childhood and map-like forms have been used widely since the Old Stone Age going into and existing in every culture. Consequently they demonstrated that mapping abilities are culturally universal (Blaut 1991).

However, the abilities to manipulate maps can be endowed by socio-cultural factors as well as natural factors. As Downs and Liben (1993) claimed referring to the Vygotzkian viewpoint, most of the previous studies on the development of cognitive maps focused on the individual level rather than the societal level. Hence, it is necessary to conduct research focusing on the societal level of spatial cognition by analyzing styles and contents of maps used in a given society to consider cultural context of the development of mapping abilities.

1.2 Cultural Differences in Spatial Description by Language

Though most of geographical studies on the source of spatial cognition have concerned mainly with map-like media, language is also an important means for communicating spatial information. Concerning the relationship between language and cultural diversity of spatial cognition, researchers in linguistics and related fields have argued about the reliability of the Sapir-Whorf hypothesis (Carroll, 1956). According to this hypothesis, spatial cognition is entirely determined by the structure of the language used. Hence, differences in categorization for specific languages were studied by Talmy (1983), Pederson (1995), and Bowerman (1996). Recently there have been many studies examining the usage of locative prepositions in spatial language structure (Sowden and Blades 1996; Mark and Egenhofer 1994), and differences in feature definition between languages (Mark 1993).

However, the purpose of this paper is not to grasp the structure and system of a language itself, nor to explain its role in fulfilling the relations of the space. Instead, we intend to clarify the variations in spatial descriptions with language considering socio-cultural or environmental factors in a given geographic context.

Although a large number of studies on spatial description by language have been done, little is known about the role of language in actual transmission of spatial information in a specific environment. Recently, however, several studies on this subject have been carried out. For example, there are studies that treated spatial description by language to examine the individual differences of spatial ability in use of frames of reference (Ward et al. 1986; Allen 2000), and the relationship between spatial abilities and direction giving (Vanetti and Allen 1988). Some studies also dedicated to the strategies employed in spatial description such as the order of exhibition of spatial information (Allen 2000), the relationships of spatial scale and types of referent (Taylor and Tversky 1996), the differences in route instructions between underground and urban environment (Fontaine and Denis 1999), and the usage of critical points to consider the effective navigational expression (Dennis et al. 1999).

Nevertheless, these studies did not consider whether spatial descriptions are influenced by the characteristics of built environment or by socio-cultural context of their subjects. Moreover, most researchers were mainly interested in verbal descriptions and omitted to consider the simultaneous usage of visual depictions. Although there is a similar structure between them conveying equivalent spatial information (Tversky and Lee 1999), these mediums are often used jointly and complement each other. Consequently previous studies failed to grasp the actual conditions of communicating spatial information in real-world settings. The aim of the present study is to overcome these limitations.

2 The Framework and Method

2.1 Analytical Framework

Tourist guidebooks are one of the most widely used media for traveling anywhere outside one's familiar place, containing multiple representational styles such as maps, photos and linguistic descriptions. Furthermore, they are not biased for a specific audience but are intended for general readers. Therefore, the guidebooks can be thought to reflect the properties of spatial descriptions of a given society and the environmental characteristics to a considerable extent.

To examine the effects of socio-cultural and environmental factors on the spatial descriptions in guidebooks, we created an analytical framework based on Suzuki (2003) as follows. Concerning the socio-cultural factor, we compared Japanese and American guidebooks because the difference in address systems between these countries would show distinctive features of spatial descriptions: Japan has a typical example of block-based address systems, and USA has a street-based one.

As for the environmental factor, we examined its effect by selecting cities with varied regularity of street patterns in pairs for each country (Table 1). Kyoto and Tokyo were selected as representatives of Japanese cities. Kyoto has a grid-pattern layout designed at a time in history when urban planning was based on ancient Chinese cities. In contrast, Tokyo is a city whose street pattern is less regular because the city is located on undulating plains and its urban area has spread gradually. As for the cities in America, New York and Boston were selected. The layout of Manhattan Island in New York City is famous for its regular grid form, while the street pattern of Boston is less regular because its central area is located on a peninsula.

Then, six publishers were selected using the criterion that each should publish the guidebooks for these cities in the same series. Though the Lonely Planet series is published in Australia, it was chosen because of its popularity among American tourists. Consequently, 24 guidebooks were used for the analysis (see Appendix).

Table 1. Analytical framework of this study

Guidebook	Type of information		Cities in USA		Cities in Japan	
			New York (regular street)	Boston (irregular street)	Kyoto (regular street)	Tokyo (irregular street)
American	Pictorial	Photo	American cities described by American		Japanese cities described by American	
		Map				
	Linguistic	Reference frame				
		Referent				
Japanese	Pictorial	Photo	American cities described by Japanese		Japanese cities described by Japanese	
		Map				
	Linguistic	Reference frame				
		Referent				

2.2 Hypothesis

The following hypotheses were tested to examine the influences of socio-cultural and environmental factors on the spatial descriptions for navigation in guidebooks.

First, the address system as a socio-cultural factor would influence on the spatial descriptions. Since every place is systematically numbered along street with name that is unique within local areas in USA, American people could easily find the locations of places without maps. In contrast, as addresses are not numbered consecutively along streets but based on areal place names in Japan (Longley et al 2001), Japanese people appear to heavily rely on maps to communicate locational information as exemplified by Barthes (1970). Davies and Pederson (2001) also pointed out that different cultural expectations for typical urban environments would affect the residents' mental models and behavior regarding urban wayfinding and locational knowledge. In the present study, these differences are examined by comparison between countries publishing the guidebooks.

Secondly, differences of urban environmental characteristics between two countries would also influence on the spatial descriptions included in the guidebook. Owing to narrow and irregular street pattern as well as the complicated address system as mentioned above, Japanese cities seem to be inferior in legibility to American cities. Hence people should need more information to find their way in Japanese cities. In addition, as major intersections are named instead of street name in Japanese cities, people should rely more on "nodes" (Lynch, 1960) rather than path or street in communicating spatial information. These differences can be examined through comparison between countries of the cities described.

Thirdly, street pattern of the city as an environmental factor could affect the frame of reference in spatial descriptions. As Freundschuh (1992) suggested, a regular environment with gridded road tend to facilitate the metrical configurational knowledge. Hence, absolute frame of reference would be extensively used in cities with regular street pattern, whereas intrinsic or relative frame of reference would be more used in an irregular environment. Since the street pattern depends on historical background and planning of each city, it does not necessarily correspond to the difference between countries.

Lastly, it is expected that locational information is presented in different manners between domestic and foreign cities. For example, writers of the guidebook would give detailed information to guide tourists successfully in foreign cities whereas they do not in domestic cities. This factor can be detected in the interaction effect of first and second factors mentioned above.

To test these hypotheses, we employed methods described in the subsequent sections.

2.3 Method for Analyzing Pictorial Information

Pictorial information included in guidebooks is divided broadly into the maps and the photographs. For the maps, coloration, types of legends, map grids, site arrangement, and use frequency were set up as measures. For the photographs, use frequency was assessed.

"Coloration" indicates the level of abundance of color given to the map. It can be classified into three types as full-color, two-tone color and monochrome printings. Consequently, the scoring method was employed as mono/two-colors for monochrome and two-tone color printing, full-color for full-color printing.

"Locational reference" includes two devices to aid in finding a tourism site's location. The first device is the marking of two axes on the map to form a reference grid. The second device is to attach page number relating the map to the corresponding body text for each site. Thus, the guidebooks that included both devices for the maps were scored as "exist", and the guidebooks adopting one of the two methods or with no coordinates and page numbers on maps were scored as "none."

"Site arrangement" represents the principle on which site descriptions are ordered. "Spatial" shows guidebooks that list and number all sites based on their spatial proximity. "Thematic" indicates guidebooks that list sites in categories, such as type of business.

"Map use" indicates the rate of map use for each guidebook, which is calculated by dividing the pages carrying maps by the total pages. As for photographs, "photo use" refers to the ratio of the total number of photos containing spatial information for each guidebook. Photos of cuisine or interior of building should not be categorized as spatial description since they provide no information for wayfinding, so these photographs were excluded.

2.4 Method for Analyzing Linguistic Information

Linguistic description of space in large-scale environment employs a variety of spatial expressions. To deal with them quantitatively, we must classify the elements contained in the linguistic expression. Previous studies have used a variety of classification schemes for analyzing linguistic expression, and a clear standard for classification for directional terms has not yet been established. But there is some common ground among such studies.

First we consider classification based on the referential system and referent. Vanetti and Allen (1988) divided the sentence of the spatial directional terms into environmental features and spatial relational constructs. The former indicates referents like landmarks or choice points, and the latter is indicative of spatial relational terms like direction and distance. A similar classification was also used by Allen (2000).

Second, previous studies chiefly intended to classify sentences based on verb functions. Allen (2000) classified verbs into verbs of movement and state-of-being verbs. Similar classifications were devised by Taylor and Tversky (1996) who further sorted the terms into three categories of intrinsic, relative and eclectic expressions on the basis of the agent's observing point. To classify the directional terms, it is necessary to make distinction between frame of reference and referent.

Consequently, this study reflects these classification schemata in these two points (Table 2). According to Levinson (1996), frame of reference in linguistic description can be grouped into three categories: intrinsic, relative and absolute. This corresponds to Hart's (1981) classification: egocentric, fixed and abstract. This study adopts the Levinson's categories.

Because previous studies on verbal direction-giving were mainly interested in the language structure of the sentence, taxonomy of the referent has not yet been established. But since seminal study of Lynch (1960), cognitive mapping research has addressed the question of what referent can be used as a cue. Some studies have reported that the way of using referents changed in connection with the environmental characteristics or with subjects' familiarity with the environment (Appleyard 1970; Evans et al. 1981). In this study, referents of environmental features were classified into following six types and then compiled.

The "landmark/node" referent operates as a cue to find a way to the destination, which is probably the same one as Lynch's (1960). Hence we observed another usage, in which the destination itself serves as a landmark such that a site is located in a distinctive building. When a guidebook user refers to a station as a cue and passes through it, the station takes on a hybrid function. In this study, therefore, both landmark and node were counted as in the same category.

Table 2. Typology of the element in spatial description by language

	Type	Definition	Examples
Frame of reference	Intrinsic	An explanation of routes made in the observer's point of view	Walk…, Go…, Turn …, Pass…, Down…
	Relative	An explanation in relation to a position in the environment	is located in …, is sited at…
	Absolute	Conceptual frame of reference which is not concerned with any environmental	is north, south, east, west of…
Referent	Landmark / Node (Intersection)	Environmental features which typically seen from many angles and distances and can be detected singularly (Names of the intersection themselves as a node of streets)	the red brick building, the park, the station (Hyakumanben in Kyoto)
	Path (Route)	Channels along which the observer (Names of the route of public transportations)	street, dori, avenue, (city bus No.205, metro, tramline line
	District	Names of a given two-dimensional extent recognizable as having some common, identifying character	SoHo, Higashiyama, Beacon Hill, Sibuya, names of wards
	Edge	Boundaries between two phases, linear breaks in continuity	river, railway tracks, walls

The subcategory "intersection" referent appeared predominantly in Japanese cities where most intersections are given their own names. When a guidebook described only "the intersection of street A and avenue B," it was considered to be describing two roads, rather than a specific intersection name. As with landmark/node, the term "node" as used by Lynch (1960) does not include all words in this definition although the intersection name functions as a kind of node.

"Route" gives the description of a way to destination by providing the lines of public transportation. In a functional sense, this type can be included in the "path" which is an important element for direction giving together with the "landmark/node."

"District" as used in this study corresponds closely to Lynch's definition, indicating a given area name which is recognizable as having some common, identifying character. In USA, place names such as "The Village," "SoHo," and "Midtown" for areas of New York City, and "Beacon Hill" or "Back Bay" for areas of Boston correspond to this category. In Japan, this category includes places such as "Shibuya" or "Asakusa" in Tokyo, and "Gion" or "Arashiyama" in Kyoto.

The "edge" referent almost always refers to a railroad or river. These can be distinguished in their function of blocking a path.

Using the categorization framework indicated above, for example, the sentence "my house is next to your house" can be fractionated into one relative frame of reference ("is next to") and two referents ("my house" and "your house"). All 8,698 descriptions, except those of the sites that appeared in the form of box columns, of all the tourist guidebooks selected for this study were classified into reference frames and referents, and then quantitatively analyzed.

3 Analysis of Pictorial Information

Results of the analysis for the pictorial information are shown in Table 3. Concerning the coloration, all maps in the selected Japanese guidebooks were printed in full-color which enables readers to discriminate features on maps easily, whereas the maps in American guidebooks were printed in fewer colors. Hence, Japanese maps with plenty of colors can lead to depict more information and detailed features on maps than American maps.

Table 3. Pictorial information in the guidebook

City	Guidebook	Number of sites	Coloration*	Locational reference	Site arrangement	Maps use (%)	Photos use (%)
Kyoto	JTB-K	144	F	exist	spatial	28.0	46.9
	Sho.-K	446	F	exist	spatial	6.0	123.7
	Jit.-K	159	F	exist	spatial	26.4	77.4
	Fod.-K	65	M	none	thematic	7.7	0.0
	Fro.-K	63	M	none	thematic	9.0	0.0
	Lon.-K	272	M	none	thematic	6.1	5.5
Tokyo	JTB-T	144	F	exist	spatial	16.8	84.0
	Sho.-T	511	F	exist	spatial	12.1	69.4
	Jit.-T	168	F	exist	spatial	26.3	119.8
	Fod.-T	184	M	none	thematic	9.2	0.0
	Fro.-T	294	M	none	thematic	6.2	0.0
	Lon.-T	248	M	none	thematic	6.9	12.2
New York	JTB-N	224	F	exist	spatial	9.9	86.1
	Sho.-N	256	F	none	spatial	11.1	80.2
	Jit.-N	295	F	exist	spatial	12.4	54.2
	Fod.-N	52	M	none	thematic	5.8	0.0
	Fro.-N	338	M	none	thematic	9.0	0.0
	Lon.-N	281	M	none	thematic	4.1	14.8
Boston	JTB-B	14	F	exist	spatial	7.1	133.3
	Sho.-B	22	F	none	spatial	9.1	205.3
	Jit.-B	22	F	exist	spatial	13.6	95.7
	Fod.-B	15	M	none	thematic	7.1	0.0
	Fro.-B	296	M	none	thematic	7.4	4.4
	Lon.-B	280	M	none	thematic	6.7	19.2

* F: full-color, M: mono-color or two-colors.

Note: The map/photo use represents the percentages calculated by the number of maps/photos divided by the number of pages in each guidebook.

With respect to the preciseness of the map expression, the Japanese maps tended to show the features in accurate outlines and were distinguished by colors in accordance with feature category. In contrast, American guidebooks drew most features in geometric format with arbitrary numbers (Figs. 1 and 2). Thus users could not identify a certain site without referring to the listing, which was printed separately and the sites were sometimes arranged in alphabetical order (see the site listing in Fig. 2). In other words, signs of American maps were found to be more abstract (MacEachren1995, p.259) than those of Japanese guidebooks, which call on more advanced skill for acquiring information.

Fig. 1. An example of Japanese map of Roppongi in Tokyo (Original size is13.7 cm × 7.1 cm; printed in full-color). Source: *Buru-gaido nippon 10* (1999)

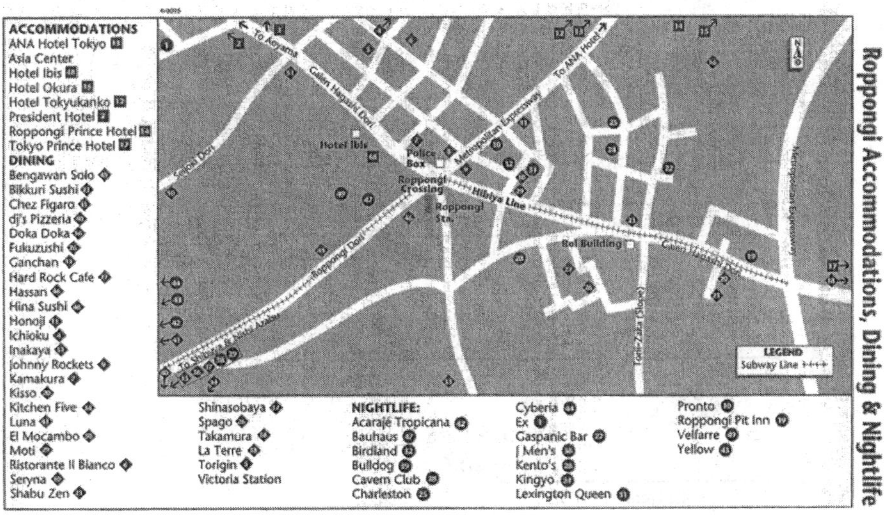

Fig. 2. An example of American map of Roppongi in Tokyo (Original size is 14.7 cm × 7.7 cm; printed in two-colors). Source: *Frommer's Tokyo* (1998)

The features on maps also showed a marked difference between two countries. As shown in Figs. 1 and 2, Japanese maps emphasized blocks and landmarks rather than streets, whereas American maps exclusively signalized streets. Moreover, the intersection name instead of street name was stressed exclusively for Japanese maps since all the streets are not named in Japanese cities and the addresses are not numbered consecutively along streets. This indicates a difference of environmental cues for navigation between the countries.

Maps in tourist guidebooks have the role of providing the positional information for each site described in the body text. Therefore, maps should be closely associated with the body text to efficiently convey the spatial information. This point is illustrated by the result of "site arrangement" in Table 3. Even here, we can confirm Japanese guidebooks contrived a way to tie a map to the text. This implies that the contents of Japanese guidebooks are entirely organized by maps.

The use frequency of maps and photographs in Japanese guidebooks significantly exceeded that of American guidebooks by as much as 4.5 times (Table 3). This was confirmed by a three-way analysis of variance (ANOVA) given in Table 4. The photos as well as maps could play a role of depicting landmarks visually, which enables readers to expect the scenery of the environment. Therefore we can conclude that Japanese guidebooks are much more likely to depend on pictorial information as a primary medium for spatial description compared with American guidebooks. This was especially true for the map use of the Japanese guidebooks in describing Japanese cities though the interaction effect between "Publisher" and "City" was not significant at 0.05 level.

These results supported the first hypothesis; the second hypothesis was partially confirmed but the third and forth ones were rejected. Consequently, the use of pictorial information can be entirely influenced by the socio-cultural background of the publishers / readers of guidebooks rather than the environmental characteristics of the cities described.

Table 4. Results of ANOVA by the types of pictorial information

Pictorial information	Factor*	df	F	p
Map use	Publisher	1	15.14	<0.01
	City	1	5.69	<0.05
	Street pattern	1	0.08	n.s.
	Publisher×City	1	3.88	n.s.
Photo use	Publisher	1	59.71	<0.01
	City	1	1.13	n.s.
	Street pattern	1	3.08	n.s.
	Publisher×City	1	0.61	n.s.

n.s: not significant at 0.05 level.

* Publisher: Country where the guidebook was published.

City: Country of the city described.

Street pattern: Street pattern regularity of the city.

4 Analysis of Linguistic Information

The results of linguistic analysis are summarized in Table 5 in which numbers in appearance of each reference frame and referent were divided by the number of sites described. At the outset of the analysis, relation between pictorial information and linguistic one was examined. Correlation coefficients between these two types of information of the guidebooks given in Table 6 indicated an inverse relationship between linguistic and pictorial types of information. This means that these two types of information complement each other in conveying route directions. Specifically, Japanese guidebooks tended to rely more on pictorial information, while American guidebooks were likely to depend heavily on linguistic information. In other words, it can be said that one of these two types of information is preferred over another according to communicative situations.

Table 5. Frequency in the use of linguistic information by types

Guidebook	Number of sites described	Frame of reference (%)			Referent (%)					
		intrinsic	relative	absolute	landmark/ node	(intersection)	path	(route)	district	edge
JTB-K	352	0.9	10.2	2.6	8.0	(0.1)	4.6	(0.1)	0.9	1.4
Sho.-K	749	2.8	35.0	9.1	34.3	(0.6)	9.1	(0.1)	2.2	3.6
Jit.-K	363	4.1	31.4	8.8	32.0	(0.4)	8.0	(0.1)	6.9	3.1
Fro.-K	107	42.1	101.9	65.4	114.0	(4.8)	103.7	(1.1)	15.0	6.5
Lon.-K	342	45.9	58.8	43.3	104.7	(1.3)	71.1	(27.3)	5.0	2.6
Fod.-K	132	65.9	51.5	49.2	134.1	(0.9)	102.2	(44.8)	19.7	4.5
JTB-T	227	0.4	9.7	0.0	8.4	(0.1)	4.0	(0.1)	1.8	0.9
Sho.-T	967	2.2	19.8	1.2	120.0	(0.1)	2.7	(0.2)	2.9	1.0
Jit.-T	415	1.3	14.0	0.4	19.5	(1.1)	8.0	(0.1)	3.7	0.9
Fro.-T	433	31.6	109.7	14.3	93.1	(8.2)	67.7	(1.1)	34.9	3.0
Lon.-T	513	48.0	96.1	11.3	105.1	(2.8)	32.9	(2.6)	32.7	1.6
Fod.-T	286	16.8	54.5	10.8	62.9	(1.5)	23.4	(2.8)	21.0	1.0
JTB-N	428	1.2	32.9	6.5	16.4	(0.0)	25.2	(0.1)	8.9	0.9
Sho.-N	319	1.6	30.1	14.1	11.3	(0.0)	19.4	(0.4)	10.7	4.1
Jit.-N	296	1.7	26.4	4.1	18.9	(0.0)	33.1	(0.4)	8.1	1.4
Fro.-N	628	1.3	29.5	1.6	15.6	(0.0)	7.0	(0.6)	15.1	0.6
Lon.-N	628	3.7	22.3	4.6	12.7	(0.0)	15.0	(1.7)	10.7	0.3
Fod.-N	205	0.5	33.7	0.5	11.2	(0.0)	18.5	(1.1)	16.6	1.5
JTB-B	43	0.0	25.6	4.7	16.3	(0.0)	9.3	(0.1)	11.6	0.0
Sho.-B	89	4.5	42.7	4.5	24.7	(0.0)	41.6	(3.5)	18.0	2.2
Jit.-B	63	6.3	49.2	27.0	46.0	(0.0)	30.2	(3.3)	7.9	3.2
Fro.-B	350	5.1	45.7	2.6	33.1	(0.0)	20.0	(1.2)	28.3	1.1
Lon.-B	633	4.3	28.0	2.4	12.8	(0.0)	13.6	(0.7)	17.7	0.6
Fod.-B	130	0.9	10.2	2.6	16.2	(0.0)	8.5	(0.9)	0.9	1.4
mean		12.2	40.4	12.2	44.6	(0.9)	28.3	(3.9)	12.6	2.0

Note: The use frequencies represent the percentages calculated by times in use of each frame of reference or referent divided by the number of sites described in each guidebook. Because one site may contain the sentences consist of more than one frame of reference or referent, the percentages may be more than one hundred percent.

Table 6. Correlation between the number of pictorial information and linguistic one

Pictorial information	Linguistic information						
	Frame of reference			Referent			
	intrinsic	relative	absolute	landmark/ node	path	district	edge
Maps use	-0.291	-0.374	-0.200	-0.251	-0.314	-0.481*	-0.054
Photos use	-0.465*	-0.370	-0.288	-0.357	-0.313	-0.400	-0.049

* significant at 0.05 level.

4.1 Frame of Reference

Among three types of the frame of reference, the relative one is most frequently used in every guidebook (Table 5). This suggests that the relative frame of reference is versatile in orienting people in large-scale environments; the other reference frames appear to be subsidiary and flexibly employed according to environmental conditions.

To examine the effects of the factors hypothesized in this study, a three-way ANOVA was carried out. Results of the analysis given in Table 7 supported three out of four hypotheses in this study; the third hypothesis concerning the effect of street pattern regularity was partially confirmed.

Table 7. Result of ANOVA by the type of reference frame

Frame of reference	Factor	df	F	p
Intrinsic	Publisher	1	34.39	<0.01
	City	1	32.07	<0.01
	Street pattern	1	1.52	n.s.
	Publisher×City	1	34.10	<0.01
Relative	Publisher	1	15.12	<0.01
	City	1	7.13	<0.05
	Street pattern	1	0.26	n.s.
	Publisher×City	1	23.18	<0.01
Absolute	Publisher	1	4.89	<0.05
	City	1	6.18	<0.05
	Street pattern	1	5.08	<0.05
	Publisher×City	1	14.84	<0.01

n.s: not significant at 0.05 level.

Concerning the first hypothesis, Table 5 exhibits a tendency for the American guidebooks to use more frames of reference in navigating tourists than do the Japanese guidebooks. This was the case for all the types of reference frames and the differences were statistically reliable at 0.01 or 0.05 level as shown in the result for "Publisher" in Table 7.

The results also varied with the cities described: more reference frames were used in describing Japanese cities than in American cities. There were significant differences in the use frequencies of reference frames between two countries as shown in the results for "City" of Table 7, which supports the second hypothesis.

Specifically, the disparity between Japanese and American guidebooks in the use frequency of reference frames was greater when they described Japanese cities. In contrast, these differences almost vanished when American cities were described. Hence, it is concluded that the American guidebooks generally rely more on linguistic information than do the Japanese ones, and this is especially true when foreign cities are described. This confirms the fourth hypothesis as demonstrated by the results for the interaction effect between "Publisher" and "City" in Table 7.

Nevertheless, the differences in street pattern regularity did not have significant effects on the usages of reference frames except the absolute one as shown in the result for "Street pattern" in Table 7. This implies that the environmental factor does not necessarily influence on the linguistic descriptions for navigation. However, the usage of absolute reference frame was exceptionally varied with the street patterns: it was used more frequently in describing cities with regular grid patterns (e.g., Kyoto and New York) than in cities with irregular street patterns (e.g., Tokyo and Boston). The reason for this is that the grid-pattern layout inherently accommodates to the description of locations with cardinal directions based on the absolute frame of reference.

4.2 Referent

Among four major types of the referent, "landmark/node" was most frequently used followed by "path" (Table 5). This appears to consistent with the result obtained in the preceding section because the relative frame of reference most frequently used in the guidebooks can lead to rely on the landmark/node. In addition, the result is consistent with the finding of Fontaine and Denis (1999) that not only landmark but also path are important element in describing route in the urban environment. The results also demonstrate that a relative frame of reference with landmark/node is the most fundamental component for giving directions.

The effects of the factors affecting the usage of each type of the referent were examined by a three-way ANOVA. The results given in Table 8 partially supported the hypotheses of this study, although strengths of the effect of each factor varied with types of the referent.

Concerning the first hypothesis, significant differences between Japanese and American guidebooks were detected for the use frequencies of the types of the referent except "edge." Specifically, the use frequencies of referents in American guidebooks exceeded those in Japanese ones. This can be due to the complementary use of pictorial and linguistic information: American guidebooks could supplement shortage of pictorial descriptions with linguistic information.

The second hypothesis about the difference between American and Japanese cities was supported only for "landmark/node" and "path". Especially, "intersection" which is included in "landmark/node" exclusively appeared in describing Japanese cities (Table 5). The reason for this is that major intersection names instead of street names would be important element for direction giving in Japanese cities.

The interaction between "Publisher" and "City" also had significant effects on the usage of all types of the referent, which confirms the validity of the forth hypothesis. This means that the guidebooks make use of more information about referents in describing foreign cities. In particular, American guidebooks tended to use more referents when they describe Japanese cities than Japanese guidebooks. This can be due to the difficulty of direction giving in Japanese cities because of the lack of systematic addresses along the streets. Especially the use frequencies of the referent for Kyoto, where the address system is complicated because of the mixture of block zoning and the combination of street names, generally exceed those for the other cities. As a result, "route" included in "path" was most frequently used in Kyoto (Table 5) whose primary means of transportation is city bus.

As for the third hypothesis, the effect of "street pattern" was not significant at 0.05 level with an exception for "edge." This means that the environmental factor does not necessarily influence on the usage of the referent, which is contrary to our expectations because it is conceivable that referents reflect the environmental characteristics more directly than reference frames. Though the "edge" was used more frequently in cities with regular streets, the result was hard to explain.

Consequently, we can conclude that the usage of referents likewise that of reference frames is entirely affected by the socio-cultural factors rather than the condition of the built environment.

Table 8. Result of ANOVA by the type of referent

Referent	Factor	df	F	p
landmark / node	Publisher	1	8.06	<0.05
	City	1	22.50	<0.01
	Street pattern	1	0.13	n.s.
	Publisher×City	1	11.19	<0.01
(intersection*)	Publisher	1	5.94	<0.05
	City	1	9.75	<0.01
	Street pattern	1	0.66	n.s.
	Publisher×City	1	5.94	<0.05
path	Publisher	1	12.04	<0.01
	City	1	5.56	<0.05
	Street pattern	1	3.48	n.s.
	Publisher×City	1	28.12	<0.01
district	Publisher	1	14.77	<0.01
	City	1	0.05	n.s.
	Street pattern	1	3.12	n.s.
	Publisher×City	1	6.05	<0.05
edge	Publisher	1	0.10	n.s.
	City	1	3.97	n.s.
	Street pattern	1	4.48	<0.05
	Publisher×City	1	5.17	<0.05

n.s: not significant at 0.05 level.
* included in landmark / node.

5 Discussion and Conclusions

The present study examined the cultural differences in communicating locational information by analyzing the tourist guidebook as a spatial information vehicle. The results showed that Japanese guidebooks are likely to depend heavily on pictorial devices such as maps and photos for communicating spatial information. In contrast, American guidebooks tend to rely more on linguistic means of communication. Moreover, we found a large difference in the proportionate use of pictorial versus linguistic description between the two countries.

The predominance of maps in the Japanese guidebooks corresponds to the way of ordering each site in the body text of the guidebook. The use of map is closely associated with the actual locations of the sites described, rather than the category of business, because maps are devices to edit the sites based on their spatial proximity or unity. To give a certain direction with maps, it is justified to classify each site based on the criterion whether they fit into a given map. Hence, Japanese guidebooks arrange spatial information primarily in order of its geographic location. This leads the maps in Japanese guidebooks to contain many kinds of information of environmental features. Therefore, the descriptions of Japanese maps need to be sophisticated and fully colored, as we found.

In contrast, American guidebooks depended basically on linguistic information. In particular, they do not necessarily depict the precise locations of each site on maps, and all the information about the sites were sorted in reference to their functional characteristics such as categories of business. Also, the maps in the guidebooks tend not to contain a variety of environmental features, and they employ symbols that are relatively simplified and highly abstract. This suggests that they emphasize route description and the sites' categories of business rather than their geographic locations. In short, the style of spatial descriptions in Japanese guidebooks resembles survey maps, while that in American guidebooks more resembles route maps.

Linguistic information giving directions are entirely influenced by the socio-cultural factors mentioned above rather than the environmental conditions such as the street pattern regularity. Specifically, the results of our study confirm the existence of marked differences of the usages of reference frames and referents between countries. There are also interactions between publishers of the guidebooks and countries of the city described. The American guidebooks, in particular, use more linguistic information when they describe Japanese cities, which is closely related to the complementary use of pictorial and linguistic information as mentioned above.

These differences of the spatial descriptions between Japanese and American guidebooks can be due to the communicative convention inherent in each country. Hence, relative advantages of the style of the spatial description probably depend on the socio-cultural context of the senders and receivers of the information. The future direction of this study is to realize which type of spatial description is useful for navigating people through unfamiliar environments in real-world settings. Clarification of this issue is necessary to examine the nature of comprehensible spatial representation for human subjects.

Acknowledgements

The authors are indebted to our sincere gratitude to Dr. Margaret Siebert for her valuable advice during the preparation of the English manuscript, and the anonymous reviewers for their profitable comments on our earlier draft.

References

Allen, G.L. 2000. Principles and practices for communicating route knowledge. *Applied Cognitive Psychology* 14: 333-359.

Appleyard, D. 1970. Styles and methods of constructing a city. *Environment and Behavior* 2: 100-117.

Barthes, R. 1970. *L'empire des signes*. Geneva, Albert Skira.

Blades, M., Blaut, J.M., Darvizeh, Z., Eluguea, S., Sowden, S., Soni, D., Spencer, C., Stea, D., Surajpaul, R. and Uttal, D. 1998. A cross-cultural study of young children's mapping abilities. *Transactions of the Institute of British Geographers, New Series* 23: 269-277.

Blaut, J.M., 1991. Natural mapping. *Transactions of the Institute of British geographers, New Series* 16: 55-74.

Blaut, J.M., McCleary Jr., G.S. and Blaut, A.S. 1970. Environmental mapping in young children. *Environment and Behavior* 2: 335-349.

Bowerman, M. 1996. Learning how to structure space for language: a cross-linguistic perspective. In *Language and space*. eds. P. Bloom, M.A. Peterson, L. Nadel and M.F., 385-436. Cambridge MA: MIT Press.

Carroll, J. B. ed. 1956. *Language, thought and reality: Selected writings of Benjamin Lee Whorf.* Cambridge, Mass.: MIT Press.

Davies, C. and Pederson, E. 2001. Grid patterns and cultural expectations in urban wayfinding. In D. Montello (ed.) *Spatial information theory: foundations of geographic information science*. Berlin, Springer, 400-414.

Dennis, M., Pazzaglia, F., Cornoldi, C. and Bertolo, L. 1999. Spatial discourse and navigation: an analysis of route directions in the city of Venice. *Applied Cognitive Psychology* 13: 145-174.

Downs, R.M. and Liben, L.S. 1993. Mediating the environment: communicating, appropriating, and developing graphic representations of place. In *Development in context; acting and thinking in specific environments*. eds. R.H. Wozniak and K. Fischer, 155-181. Hillsdale NJ: Lawrence Erlbaum Associates.

Evans, G.W., Marrero, D.G. and Butler, P.A. 1981. Environmental learning and cognitive mapping. *Environment and Behavior* 13: 83-104.

Fontaine, S. and Denis, M. 1999. The production of route instructions in underground and urban environments. In *Spatial information theory: cognitive and computational foundations of geographic information science*. eds. C. Freksa and D.M. Mark, 83-94. Berlin, Springer.

Freundschuh, S. M. 1992. Is there a relationship between spatial cognition and environmental patterns. In *Theories and Methods of Spatio-Temporal Reasoning in Geographic Space*. eds. A.U. Frank, I. Campari and U. Formentini, 288-304. Berlin: Springer Verlag.

Gladwin, T. 1970. *East is a big bird: Navigation and logic on Puluwat Atoll.* Cambridge MA: Harvard University Press.

Hart, R.A. 1981. Children's spatial representation of the landscape; lessons and questions from a field study. In *Spatial representation and behavior across the life span*. eds. L.S. Liben, A.H. Patterson and N. Newcombe, 194-233. New York: Academic Press.

Levinson, S.C. 1996. Frames of reference and Molyneux's question: Crosslingustic evidence. In *Language and space*. eds. P. Bloom, M.A. Peterson, L. Nadel and M.F. Garrett, 109-169. Cambridge MA: MIT Press.

Longley, P.A., Goodchild, M.F., Maguire, D.J. and Rhind, D.W. 2001. *Geographic information systems and science*. New York: John Wiley & Sons.

Lovelace, K. L., Hegarty, M., and Montello, D. 1999. Elements of good directions in familiar and unfamiliar environments. In *Spatial information theory: cognitive and computational foundations of geographic information science*. eds. C. Freksa and D.M. Mark, 65-82. Berlin, Springer.

Lynch, K. 1960. The image of the city. Cambridge MA: MIT Press.

MacEachren, A.M. 1995. *How maps work: representation, visualization and design*. New York: Guilford Press.

Mark, D.M. 1993. Toward a theoretical framework for geographic entity types. In *Spatial information theory: A theoretical basis for GIS*. eds. A.U. Frank and I. Campari, 270-283. Berlin : Springer.

Mark, D.M. and Egenhofer, M.J. 1994. Modeling spatial relations between lines and regions: combining formal mathematical models and human subjects testing. *Cartography and GIS* 21: 195-212.

Montello, D. 1995. How significant are cultural differences in spatial cognition?. In *Spatial information theory; a theoretical basis for GIS*. eds. A.U. Frank and W. Kuhn, 485-500. Berlin: Springer.

Pederson, E. 1995. Language as context, language as means: spatial cognition and habitual language use. *Cognitive Linguistics* 6(1): 33-62.

Pocock, D.C.D. 1976. Some characteristics of mental maps: an empirical study. *Transactions of the Institute of British Geographers, New Series* 1: 493-512.

Rovine, M.J. and Weisman, G.D. 1989. Sketch-map variables as predictors of way-finding performance. *Journal of Environmental Psychology* 9: 217-232.

Sowden, S. and Blades, M. 1996. Children's and adult's understanding of the locative prepositions 'next to' and 'near to'. *First Language* 16: 287-299.

Stea, D., Blaut, J.M. and Stephens, J. 1996. Mapping as a cultural universal. In *The construction of cognitive maps*, .ed. J. Portugali, 345-360. Dordrecht: Kluwer Academic Publishers.

Suzuki, K. 2003. A comparative study of the spatial descriptions in tourist guidebooks. *Geographical Review of Japan* 76(5): 249-269.

Talmy, L. 1983. How language structures space. In *Spatial orientation: theory, research and application*. eds. P. Herbert and L. Acredolo, 225-282. New York: Plenum Press.

Taylor, H.A. and Tversky, B. 1996. Perspectives in spatial description. *Journal of Memory and Language* 35: 371-392.

Tversky, B. and Lee, P.U. 1999. Pictorial and verbal tools for conveying routes. In *Spatial information theory: cognitive and computational foundations of geographic information science*. eds. C. Freksa and D.M. Mark, 51-64. Berlin, Springer.

Vanetti, E.J. and Allen, G.L. 1988. Communicating environmental knowledge: the impact of verbal and spatial abilities on production and comprehension of route directions. *Environment and Behavior* 20: 667-682.

Ward, S.L., Newcombe, N. and Overton, W.F. 1986. Turn left at the church, or three miles north: a study of direction giving and sex difference. *Environment and Behavior* 18: 192-213.

Appendix: List of the Guidebooks

City	Abbreviation	Title	Year	Publisher (country*)
Boston	JTB-B	*JTB no poketto gaido 115*	2001	JTB Press (J)
	Sho.-B	*Earia gaido 103*	1990	Shobun-sha (J)
	Jit.-B	*Buru-gaido wa-rudo 20*	1999	Jitsugyononihon-sha (J)
	Fod.-B	*Forder's New Edition USA*	2001	Fodor's Travel Publications (U)
	Fro.B	*Frommer's Boston 2001*	2001	IDG Books Worldwide Inc. (U)
	Lon.-B	*Lonely Planet Boston*	2000	Lonely Planet Publications (A)
New York	JTB-N	*JTB no poketto gaido 139*	2000	JTB Press (J)
	Sho.-N	*Earia gaido 115*	1994	Shobun-sha (J)
	Jit.-N	*Buru-gaido wa-rudo 20*	1999	Jitsugyononihon-sha (J)
	Fod.-N	*Forder's New Edition USA*	2001	Fodor's Travel Publications (U)
	Fro.-N	*Frommer's 99 New York City*	1999	IDG Books Worldwide Inc. (U)
	Lon.-N	*Lonely Planet New York City*	1997	Lonely Planet Publications (A)
Kyoto	JTB-K	*JTB poketto gaido 42*	2000	JTB Press (J)
	Sho.-K	*Earia mappu tabi oukoku 29*	2001	Shobun-sha (J)
	Jit.-K	*Buru-gaido nippon 27*	2000	Jitsugyononihon-sha (J)
	Fod.-K	*Forder's updated edition*	2000	Fodor's Travel Publications (U)
	Fro.-K	*Frommer's Japan*	2000	IDG Books Worldwide Inc. (U)
	Lon.-K	*Lonely planet Kyoto*	1999	Lonely Planet Publications (A)
Tokyo	JTB-T	*JTB poketto-gaido 15*	2000	JTB Press (J)
	Sho.-T	*Earia mappu tabi oukoku 12*	2000	Shobun-sha (J)
	Jit.-T	*Buru-gaido nippon 10*	1999	Jitsugyononihon-sha (J)
	Fod.-T	*Forder's updated edition Japan*	2000	Fodor's Travel Publications (U)
	Fro.-T	*Frommer's Tokyo*	1998	IDG Books Worldwide Inc. (U)
	Lon.-T	*Lonely Planet Tokyo*	1998	Lonely Planet Publications (A)

* J: Japan, U: USA, A: Australia.

Reasoning About Consistency with Spatial Mental Models: Hidden and Obvious Indeterminacy in Spatial Descriptions

Georg Jahn[1,2], P.N. Johnson-Laird[3], and Markus Knauff[1,2]

[1] University of Freiburg, Centre for Cognitive Science,
D-79098 Freiburg, Germany
[2] Max-Planck-Institute for Biological Cybernetics,
D-72076 Tübingen, Germany
{georg.jahn, markus.knauff}@tuebingen.mpg.de
[3] Princeton University, Department of Psychology,
Princeton, NJ 08544, USA
phil@princeton.edu

Abstract. The assessment of whether a statement is consistent with what has gone before is ubiquitous in discourse comprehension. One theory of the process is that individuals search for a mental model of a situation in which all the statements in the discourse are true. In the case of spatial descriptions, individuals should prefer to construct models, which retain the information in the description. Hence, they should use strategies that retain information in an efficient way. If the descriptions are consistent with multiple models then they are likely to run into difficulties. We report some preliminary results of experiments in which the participants judged the consistency of spatial descriptions. The participants made more errors when later assertions in the description conflicted with the preferred models of earlier assertions.

1 Reasoning About Consistency with Mental Models

Sometimes, it is hard to make sense of a description. Imagine, say, that you receive a report about a car accident. One car hit another at a complex crossing. First, you get a description of the crossing, then you are told about the movements of the two cars. But, as you imagine the layout of the crossing and the trajectories of the cars, it seems that they would never have collided. If you have a written report, you can step back and see whether at any point it was possible to interpret the report differently. However, if you listened to the report and did not take notes, it is nearly impossible to consider alternative possibilities. You probably cannot remember exactly what was said. There might be an alternative interpretation of the description in which the cars did collide. Then the whole report would be consistent, even though you had difficulty in understanding it. But sometimes, descriptions just do not make sense. The author of the report may have accidentally used "left" instead of "right" at some point, and so the

C. Freksa et al. (Eds.): Spatial Cognition IV, LNAI 3343, pp. 165–180, 2005.
© Springer-Verlag Berlin Heidelberg 2005

report in fact was inconsistent. In general, the comprehension of a description and the evaluation of its consistency calls for a search for a possible interpretation. In this article, we briefly summarize what is known about the processing of ambiguous spatial descriptions. Then, we report preliminary results of two experiments that tested predictions about what determines the difficulty of judging the consistency of a spatial description. Finally, we discuss the consequences of our findings for accounts of indeterminacy effects in human spatial reasoning.

Spatial descriptions are often consistent with more than one spatial layout. In everyday discourse, this indeterminacy is usually not a problem, because either the exact spatial relations are unimportant, or common knowledge and conventions allow the indeterminacy of a description to be resolved. Hence, even if a description yields alternative interpretations, different recipients usually arrive at compatible interpretations. One account of the interpretation of spatial descriptions is provided by the theory of mental models (Johnson-Laird, 1983; Byrne & Johnson-Laird, 1989)[1, 2]. In terms of the theory of mental models, when individuals understand a spatial description, they construct a mental model of a state of affairs that is consistent with the description. They begin the model as soon as they receive information, and they integrate further information building on the model as it has been constructed up to this point. It follows that the initial mental model should have a crucial influence on the subsequent interpretation of the discourse. A direct consequence of the updating of mental models is that a difficulty arises when new information is not consistent with the model constructed so far. There might be another model of the information presented so far that is consistent with the new information, but this model may be hard to find. The model theory accordingly predicts that the evaluation of the consistency of a set of assertions should be more difficult if a mental model of earlier assertions is contradicted by a later assertion. In this case, individuals need to search for an alternative model consistent with the complete description (Legrenzi, Girotto, & Johnson-Laird, 2003; Johnson-Laird, Legrenzi, Girotto, & Legrenzi, 2000)[3, 4].

What determines which will be the initial mental model if there are several possibilities? In everyday discourse, prior knowledge and conventions may determine the choice among alternative interpretations. In the example about the car accident, individuals take for granted that the familiar regulations governing driving hold. They may imagine, for instance, that drivers keep to the left-hand side of the road. Hence, it would be difficult for them to construct a model of a crash in which one of the drivers turned left. In a more formal task with schematic descriptions, similar phenomena can influence the choice of the initial mental model and consequently can determine whether or not individuals detect an inconsistency.

1.1 Preferences for Certain Sorts of Spatial Models

When a spatial description is indeterminate, individuals tend to make one sort of interpretation rather than another. They prefer one sort of spatial model to another. For indeterminate descriptions of spatial intervals, such as:

The green interval overlaps the blue interval from the left
The blue interval overlaps the red interval from the left
individuals show a consensus in the sort of model that they envisage (Knauff, Rauh, & Schlieder, 1995; Knauff, 1999)[5, 6]. The two assertions above result in a spatial model, in which the green interval ends before the red interval starts. The other possible models - in which the green interval meets or even extends into the red interval - are systematically neglected by the participants. This example illustrates a more general principle that characterizes preferences for models of spatial intervals. As far as possible, individuals try to keep the linear ordering of start points (green, blue, red in the example) also for the linear ordering of endpoints in models of spatial intervals (linearization). In this way, they minimize the amount of information that they have to keep in working memory.

In everyday descriptions, indeterminacy arises in various ways. The following description:
The hammer is to the left of the saw
The saw is to the right of the drill
is consistent with two arrangements:
hammer drill saw
drill hammer saw
Presumably, there is a preference that biases towards <hammer drill saw>. Individuals seem to place the two objects that are mentioned in the second assertion, saw and drill, adjacent to each other in the model. We assumed this preference for adjacency, which has also been implemented in computer simulations of reasoning with spatial mental models (Johnson-Laird, 1983; Payne, 1993)[1, 7], and constructed a subset of the experimental reasoning tasks accordingly.

There are also spatial prepositions that convey indeterminacy more directly, because they are inherently indeterminate. "Next to" can be used to present an indeterminate description in a single assertion. If "The hammer is next to the saw" refers to a horizontal layout, it is consistent with <hammer saw> and with <saw hammer>. We are going to report evidence that these two possibilities are not equally probable to be chosen as an initial mental model. We will show that <hammer saw>, the ordering that matches the order of mention in the premise, is the preferred alternative. Likewise, indeterminacy can be conveyed with "between". The following assertion:
The hammer is between the drill and the saw
is consistent with two possible horizontal layouts:
drill hammer saw
saw hammer drill
Only <drill hammer saw> reflects the order of mention of drill and saw in the assertion and is the preferred mental model as will be shown.

1.2 Reasoning Problems with Contradicted Preferred Models

In the following experiments, we built on the two supposed preferences: order of mention and adjacency. We constructed reasoning problems that should be difficult, because a preferred mental model is first endorsed in a sequence of assertions, but gets contradicted later on. The experimental task is to evaluate whether the set of assertions is consistent, that is whether there is a one-dimensional layout for which all assertions are true. In the following, we refer to the reasoning problems as Cons1A, Cons1B, Cons2A, Cons2B, Cons3A, Cons3B, and Incons1A, Incons1B, Incons2A, Incons2B. The prefixes "Cons" and "Incons" indicate consistent or inconsistent descriptions, respectively. "A" and "B" refer to the two versions of each problem, for consistent problems this indicates hidden (A) and obvious (B) indeterminacy (see below). Consider the sequence of assertions in problem Cons3A in Table 1. The third column shows the supposed mental model after reading the respective assertion, the fourth column shows the possible layouts that are all consistent with the assertions presented up to this point.

The first assertion in Cons3A, "C is next to D", should yield CD as the preferred mental model. The second assertion, "B is between D and A", can be integrated without changing the order CD and yields CDBA, but the third asssertion, "D is to the right of A", contradicts this preferred mental model. In the problems Cons1A and Cons2A it is the fourth assertion that contradicts the preferred mental model. Presented in a different sequence, the same sets of assertions should be easier to recognize as consistent.

1.3 Reasoning Problems with Obvious Indeterminacy

The supposedly easier sequences are listed as problems Cons1B, Cons2B, and Cons3B in Table 1. They should be easier, because the alternative possibilities are more obvious in the first two assertions. Furthermore, the first two assertions can be represented in a way that supports reorganizing the mental model. Consider problem Cons3B that starts with "B is between D and A" and "C is between D and A". The assertions match with regard to the relation term "between" and the arguments D and A. They invite to construct the relation "being between D and A" ad hoc and to represent that it applies to both, B and C. In terms of relational complexity theory (Halford, Wilson, & Phillips, 1998; Birney & Halford, 2002)[8, 9] this ad hoc simplification of relations is a case of strategic chunking. In this way, individuals reduce the number of variables in relational assertions to be used in steps of the reasoning process (reduction of dimensionality). By means of reducing the number of variables, humans cope with demanding relational reasoning problems when this is possible.

In terms of the theory of mental models, the integrated model of the first two assertions of problem Cons3B consists of one pair of object tokens in between another pair of object tokens and an annotation that the order in both object pairs is not fixed. The parentheses in the third column of Table 1 have been inserted to convey the supposed structuring of the model: D(BC)A. With this model, it is easy to integrate the third assertion, which fixes the order of the outer pair, and also the fourth assertion, which fixes the order of the inner pair.

Table 1. Consistent and inconsistent problems. Assertions of consistent problems occur in two sequences. In A-sequences (Cons1A, Cons2A, Cons3A) a preferred mental model gets contradicted by a later assertion, in B-sequences indeterminacy is obvious and integration of assertions is easier

Problem	Premises	Mental Model	Possible Layouts
Consistent Problems			
Cons1A	C right of B	BC	1: BC
	D right of C	BCD	1: BCD
	D right of A	BCAD	3: BCAD BACD ABCD
	B right of A	ABCD	1: ABCD
Cons1B	D right of C	CD	1: CD
	D right of A	(CA) D	2: CAD ACD
	C right of B	(BC A) D	3: BCAD BACD ABCD
	B right of A	ABCD	1: ABCD
Cons2A	C between D and A	DCA	2: DCA ACD
	B right of A	DCAB	4: DCAB ACDB ABCD ACBD
	D next to C	DCAB	3: DCAB ACDB ABCD
	B next to C	ABCD	1: ABCD
Cons2B	C between D and A	DCA	2: DCA ACD
	B next to C	D (BC) A	4: DBCA DCBA ABCD ACBD
	B right of A	A (BC) D	2: ABCD ACBD
	D next to C	ABCD	1: ABCD
Cons3A	C next to D	CD	2: CD DC
	B between D and A	CDBA	4: CDBA DCBA ABCD ABDC
	D right of A	ABCD	2: ABCD ABDC
	C between D and A	ABCD	1: ABCD
Cons3B	B between D and A	DBA	2: DBA ABD
	C between D and A	D (BC) A	4: DBCA DCBA ABCD ACBD
	D right of A	A (BC) D	2: ABCD ACBD
	C next to D	ABCD	1: ABCD
Inconsistent Problems			
Incons1A	A right of B	BA	1: BA
	C between D and B	DCBA	4: DCBA BCDA BDAD BACD
	D right of C	BACD	3: BACD BCDA BCAD
	B right of A	inconsistent	0
Incons1B	A right of B	BA	1: BA
	C between D and B	DCBA	4: DCBA BCDA BDAD BACD
	D left of C	DCBA	1: DCBA
	B right of A	inconsistent	0
Incons2A	B between A and C	ABC	2: ABC CBA
	D next to C	ABCD	4: ABCD ABDC DCBA CDBA
	B next to D	ABDC	2: ABDC CDBA
	C next to A	inconsistent	0
Incons2B	B between A and C	ABC	2: ABC CBA
	D between B and C	ABDC	2: ABDC CDBA
	B next to D	ABDC	2: ABDC CDBA
	C next to A	inconsistent	0

1.4 Predictions from Model Preferences and Obvious Indeterminacy

Several predictions can be derived from the mental models account of evaluating consistency and from the supposed model preferences. First, if indeterminacy that is introduced in the first assertion of a spatial description cannot be easily represented in a mental model, a single mental model is constructed sequentially. Probably, the single model is the one that conforms with the supposed model preferences. If a later assertion contradicts the preferred model, individuals should have difficulty to find an alternative model that would be consistent with all premises. Therefore, the A-sequences of the problem types Cons1-3, which first support the preferred mental model and contradict it later on, should be difficult.

Second, if indeterminacy is obvious in the first assertion and can be easily represented in a mental model with associated and movable tokens, it should be easier to find a model that is consistent with all assertions. Therefore, the corresponding B-sequences Cons1B, Cons2B, and Cons3B should be less difficult.

And finally, the contrast between consistent problems with preferred models that get contradicted (consistent A-sequences) and problems with obvious indeterminacy (consistent B-sequences) should be more pronounced, if the premises are presented serially. If all premises are presented together, it is possible to start anew with an alternative model and the order of assertions presumably has less effect.

We tested these predictions in two experiments. We report a brief summary of these experiments and selected preliminary results. A full report will be published after additional experiments have been completed. In both experiments, participants evaluated the consistency of consistent and inconsistent premise sets. Each of three consistent sets of assertions was presented in a sequence that contradicted a preferred model (A-sequence) and in a sequence leading to obvious indeterminacy (B-sequence). The two inconsistent sets of assertions were also presented in two variants, but the assertions in the A- and B-sequences of inconsistent problems differed as can be seen in Table 1.

In Experiment 1, assertions were presented serially and reading times were collected to see which assertions are harder to integrate and when difficulty arises. In Experiment 2, assertions were presented in parallel and participants were allowed to draw sketches. Drawing was recorded on video and provided information on the possibilities that were considered first. We expected participants to start with the possibilities that correspond to the presumed preferred models.

2 Experiment 1 - Evaluating the Consistency of Serially Presented Sets of Assertions

In Experiment 1, participants judged the consistency of the problems listed in Table 1 and of analogue problems which contained "left of" instead of "right

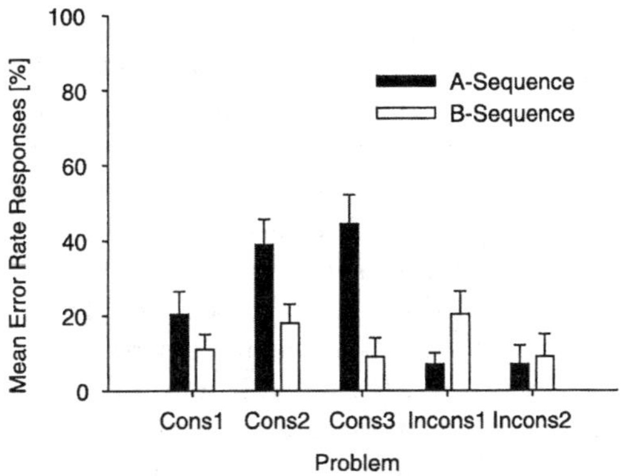

Fig. 1. Mean error rates of consistency judgments for consistent and inconsistent problems in Experiment 1 ($N = 27$); error bars denote the standard error

of". Each consistent set of assertions was presented in two sequences, one that should yield an initial mental model that got contradicted later on (consistent A-sequences) and a second one that should ease the representation of indeterminacy when reading the first assertions (consistent B-sequences). There were consistent and inconsistent sets of assertions. For consistent problems, we expected the A-sequences to cause more errors, the consistent B-sequences should lead to less errors.

The problems were presented self-paced on a LCD screen. Each trial began with the presentation of the first two assertions, each assertion in a separate line. With the space-key, participants could request the third and then the fourth assertion, which were presented separately. Participants were instructed to judge with response keys labeled "Yes" and "No" whether there was a layout for which all four assertions were true.

As expected, error rates showed that consistent problems were harder, if later assertions contradicted a preferred model of earlier assertions (consistent A-sequences). Mean error rates for consistency judgments are shown in Figure 1. For all three consistent problems, the error rates are higher for A-sequences.

For inconsistent problems, error rates were lower than for consistent problems. For the Incons1 problems, error rates were lower for the A-sequence, in which the preferred model gets contradicted. Presumably, the contradiction of the preferred model caused participants to more often judge the set of assertions correctly as inconsistent. For the Incons2 problems, in which such a contradiction was not expected for either sequence, there was no significant difference in response error rates.

Reading times for the first two assertions provided evidence on whether the advantage in representing the information in the first two assertions of consistent B-sequences occurred as predicted. The first two assertions were presented together and their information had to be retained. Participants had to stay aware of possible alternatives. They may have tried to retain the assertions verbally. But more probably, they constructed one possible model and tried to remember in addition how this model could be changed so that it still would be consistent with the information in the premises. The expected advantage for B-sequences was confirmed for Cons1 and Cons3 problems. The reading times of the first two assertions of B-sequences were around 8 s shorter than for A-sequences. For Cons2 problems the advantage was only small (2 s).

The B-sequences were constructed to ease the encoding of multiple possibilities. Matching relations and terms in the first two assertions made it possible to construct simplified relations ad hoc as proposed by Halford et al. (1998)[9]. In terms of the theory of mental models, matching assertions made it easier to construct suitable annotated models. Presumably, in initial models of B-sequences, indeterminacy was represented by inner and outer pairs of object tokens with the annotation that the order within the pairs was unspecified. This was also possible for Cons2B, although the relations in the first two assertions did not match. Verbally, participants may have used the integrated relation "Both, C and B are between D and A".

Experiment 1 confirmed the predicted difficulty of consistent A-sequences and showed that participants had the expected advantage in coping with indeterminacy in consistent B-sequences. In order to collect more direct evidence for the preferences that presumably caused the difficulty of consistent A-sequences, we observed participants' sequential consideration of layouts in Experiment 2. Participants saw all premises of a problem at once and were encouraged to draw sketches of the layouts they considered while evaluating a set of assertions.

3 Experiment 2 - Parallel Presentation and Sequential Drawing

In Experiment 2, participants evaluated the same consistent problems as in Experiment 1. However, instead of reading sequentially presented assertions on a screen, they received all four assertions of a problem at once on paper. They were encouraged to think aloud and to draw sketches. We recorded their drawing on video. Participants should start with the layouts that correspond to the presumed preferred models in Experiment 1. Furthermore, error rates should be diminished with parallel presentation of assertions, because participants could refer to all the assertions during the entire solution process.

The A- and B-sequences of Cons1, Cons2, Cons3, and Incons2 problems in Table 1 were used. Each participant evaluated eight problems. Problem sheets were prepared with the four assertions of one problem printed on the top and free space below for drawing sketches. The experimenter handed the problem sheets to participants one sheet at a time. Participants were encouraged to draw sketches and were prompted to comment on what they did. A video camera was positioned above the desk and recorded the drawing and participants' verbal comments. The purpose of this procedure was explained to the participants and they gave their consent before the experiment started.

The coding of video recordings and answer sheets yielded frequencies for preferred interpretations of assertions. We were interested in how often the layouts considered first were the ones consistent with the order of mention in the first premise and how often adjacency was reflected in the drawings. The first assertion of Cons3A states a "next to" relation and 70.8% of participants' initial drawings that could be coded for Cons3A conformed with the order of mention in the assertion. For first assertions stating a "between" relation (Cons2, Cons3B, and Incons2), 80.2% of initial drawings were consistent with the order of mention. Therefore, the analysis of initial drawings confirmed that order of mention determined the preferred interpretation of first assertions that stated "next to" and "between" relations.

Adjacency should take effect in Cons1A problems. The layouts considered first after reading the third assertion of Cons1A problems were relevant for the question of whether participants preferred adjacency in integrating the third assertion. Only 28.0% of the drawing recordings unequivocally showed that the layout predicted by the adjacency assumption was considered first. However, the variety of participants' strategies that were obvious in the drawings and video recordings might have prevented that a preference for adjacency showed up.

A preference for adjacency was shown, if the first layout that participants considered for Cons1A after the third assertion "D is to the right of A" was the one in which A is adjacent to D, that is BCAD rather than BACD or ABCD (Participants did not use "A, B, C, D" but instead the initial letters of the objects in the presented assertions). The drawings often could not reflect adjacency, because participants' drawing strategies evoked by the parallel presentation of assertions counteracted a possible adjacency effect. When reading the third assertion, their drawings up to this point often differed from a simple notation of BCD, which is the single possible layout consistent with the first two assertions. For example, 3 of the 25 participants first wrote separate pairs of letters for each assertion of Cons1 problems in a vertical alignment. Then they constructed the final layout in a single line. In addition, even if simply BCD was drawn, several participants considered the third and the fourth assertion ("D is to the right of A" and "B is to the right of A") before adding the fourth letter. Therefore, no drawn layout could be observed for the integration of the third assertion alone. For example, five participants just wrote one line while integrating the assertions sequentially and left space initially to fill in the letters for later assertions. These participants might have considered the adjacency layout first, but their drawings were not informative on this point.

For the other problems there was also a variety of strategies reflected in the drawings. Many participants never or seldom layed out all possibilities systematically, but instead added information and corrections to a single layout or started anew to draw the solution.

As expected, the parallel presentation of assertions reduced error rates. They lay between 8% and 32% for consistent problems. Inconsistency was always detected. The video recordings showed that participants usually checked their solution against the assertions and therefore changed some wrong solutions before giving the final answer.

Experiment 2 was conducted to collect evidence for the order of mention and adjacency preferences. The analysis of video recordings confirmed that the layouts most frequently considered first corresponded to the supposed preferred models for "next to" and "between" relations. As expected, with problems that began with "next to" or "between", participants more often started with the layouts in which the order of objects matched the order of mention in the first assertion.

Although all assertions of a problem were accessible at once in Experiment 2, participants usually considered the assertions one after the other in the sequence in which they were printed on the problem sheet. Consequently, with A-sequences they were initially biased toward a layout that preserved the order in which objects were mentioned in the first assertion. They had to change this preferred model to arrive at consistency with later assertions. With B-sequences, obvious indeterminacy as in Experiment 1 counteracted the order of mention preference and increased participants' awareness of alternative possibilities.

Our experiments provided indirect evidence for the adjacency preference by showing that Cons1A was more difficult than Cons1B. However, more direct evidence can be expected from an experiment, in which only the first three assertions are presented and participants are instructed to draw a consistent layout. Such an experiment is currently under way.

4 Model Preferences and Strategies in Spatial Relational Reasoning

In the reported experiments, we tested effects of model preferences in evaluating the consistency of sets of spatial assertions. According to the theory of mental models, humans evaluating the consistency of a set of assertions search for a consistent model. As soon as they encounter the first assertion, they start to construct a model to represent the information contained in the assertion. The initial model is then extended and changed to account for information in later assertions. As a direct consequence of this process, the search for a consistent model can go astray, if earlier assertions support one possibility, but a later assertion contradicts the model constructed so far and requires to consider alternative interpretations of earlier assertions. In Experiment 1 the sequential consideration of assertions was ensured by serial visual presentation. Sequences, in which later assertions contradicted the preferred model of earlier assertions, were harder to evaluate as predicted. In contrast, the same sets of assertions were more often evaluated correctly, if they were presented in sequences whose first assertions revealed indeterminacy and could be more easily represented in a way that allowed participants to stay aware of alternative interpretations.

For problems with contradicted preferred models, we postulated two model preferences. The preference for the order of mention was confirmed in both experiments. If the first assertion was of the form "A is next to B", participants thought of AB rather than BA, and if the first assertion had the form "B is between A and C", participants thought of ABC rather than CBA. They preferred the order that matched the order in which objects were mentioned in the assertions. The reason for this preference is not entirely clear, but most likely it reflects a convention for spatial descriptions that is rooted in the culturally determined habit to visually scan from left to right.

4.1 Causes Underlying the Order of Mention Preference

Imagine, someone would utter a sentence similar to a "next to"-assertion to describe a spatial layout of two objects to somebody else who cannot see the layout or is facing the layout from the same side. The object next to which an assertion locates the other object is the reference object. For instance, in "The hammer is next to the saw", the saw is the reference object. If no object in particular qualifies as reference object, the speaker is not constrained by the conventions regarding the choice of reference objects (e.g., salient, inanimate, stable). Rather, the speaker is influenced by the order in which one encounters the objects if the layout is visually scanned. In Western cultures, people visually scan, read, and write from left to right. In a recent study, participants were asked to formulate a question referring to one of two identical looking objects that they should imagine to be in front of them both equally distant. The objects and the participants' imaginary location were depicted in a diagram (Mainwaring, Tversky, Ohgishi, & Schiano, 2003, Experiment 3)[10]. Participants could freely choose to which object they referred with their question. A majority of participants from the US referred to the left object, however Japanese participants tested in Japan more frequently referred to the right object. This suggests that the direction of visual scanning (left to right in the US, whereas right to left in Japan) determines the order in which equivalent objects are mentioned in a spatial description (prior to cultural experience, the preference seems to be left to right as studied by Tversky, Kugelmass, & Winter, 1991)[11]. The strategy to take recipients on a "gaze tour" is a known strategy in describing, for example, the interior of rooms (Linde & Labov, 1975)[12]. The convention to describe a layout in the order in which it is visually scanned may induce the preference to interpret indeterminate spatial assertions with a preference for the order of mention.

Even if participants did not expect conventional descriptions, visually scanning from left to right might have caused model construction proceeding from left to right (see also Huttenlocher, 1968)[13]. If participants entered the objects in mental models from left to right as they were mentioned, this resulted in a order of mention preference. In a study, in which participants were asked to draw a map according to a description they had learned before, they drew the locations in the map in the order that matched the order of mention in the text (Taylor & Tversky, 1992; see also Spivey, Tyler, Richardson, & Young, 2000)[14, 15]. Presumably, participants in our experiments, even in Experiment 2 with parallel presentation, similarly not only considered assertions one after the other in the presented order, but also the objects mentioned in an indeterminate assertion.

Some predictions can be derived from the order of mention effects that are worth to be tested. First, the difficulty of sequences of assertions in which the preferred model is only induced by order of mention (Cons3A) should not exist for participants who read texts and scan images from right to left. Second, with vertically oriented layouts, there should be no differential effects of the habitual direction of visual scanning and difficulty from the order of mention preference should be found independently of cultural background. And finally, there should also be no differential effect of the direction of visual scanning, if the problems are formulated with relations from non-spatial domains, for example temporal relations.

4.2 Causes Underlying the Adjacency Preference

The second supposed preference in the reported experiments was a preference for adjacency in model construction. If a "left of"- or "right of"-assertion related a new object to an object already in the model, we expected participants to place the new object adjacent to the reference object. Such a bias for adjacency might reflect another convention in spatial descriptions. Speakers describing a layout of objects by means of "left of" or "right of" would choose an adjacent object as reference object (if no other object is a more salient reference object). The reported experiments provided first evidence for an adjacency preference, because predictions on difficulty from adjacency were confirmed. However, participants' drawings were not informative with regard to an adjacency preference for reasons explained in Section 3.

4.3 Representing Indeterminacy

The order of mention and adjacency preferences caused difficulty in detecting consistency, because they biased participants toward a model that was different from the single consistent solution. Participants knew that they needed to find just one consistent layout to answer "consistent". They also knew that to answer "inconsistent" they had to check alternative interpretations of earlier assertions, if later assertions did not fit the interpretation that they had chosen earlier. Therefore, they attempted to retain the information in assertions in addition to a single possible interpretation.

We have designed the problems to yield the representation of indeterminacy more or less difficult. With problems that should induce preferred models, the representation of indeterminacy was difficult, because either multiple mental models were necessary or the necessary propositional annotations to a single model were complex (Johnson-Laird, 1983)[1]. Both, multiple mental models as well as multiple propositional annotations are difficult to retain in working memory. It would have been also difficult to retain the assertions verbally while integrating assertions.

Several terms in the literature convey the idea that indeterminacy may be represented in mental models, for example, by *annotations* (Johnson-Laird, 1983; Vandierendonck, Dierckx, & De Vooght, in press)[1, 16], *isomeric models* (Schaeken, van der Henst, & Schroyens, in press)[17], or *mental footnotes* (Rauh, 2000)[18]. Strategies for representing indeterminacy probably differ between participants. The short description of drawing strategies in the discussion of Experiment 2 may have given an idea of the variability to expect. Despite the variety of participants' strategies, we succeeded in demonstrating that certain sequences of assertions support humans in representing indeterminacy.

The spatial relations that have been most extensively used in experiments on human spatial reasoning are relations such as "left of" or "in front of". They fix the order in a layout and project a direction into a spatial scene. The interpretation of those relations requires the consideration of reference frames (left of the speaker, or left of the recipient, or left of an object with a defined front side, or left of an object with a defined front seen from the viewpoint of the recipient; e.g., Levinson, 1996)[19]. It is true that with these relations any attempt to annotate a mental model in order to represent indeterminacy will exceed working memory after a few assertions. However, an effective annotation was possible for indeterminate relations as they were used in the reported experiments.

The relations "next to" and "between" can be interpreted independently of a reference frame (topologically). In the context of a one-dimensional layout, they directly convey indeterminacy even with the order of mention preference biasing towards one possibility. "Next to" allowed to treat the two related objects as one chunk and consequently as one token in a model as long as no asserted relation such as "left of" fixed their order. Moreover, because participants knew that the layouts consisted of four objects, "between" made further chunking possible. If a combination of premises identified the inner and outer pairs of objects, a model with two tokens for the pairs and annotations specifying which was the outer pair and that the order in both pairs is unspecified was sufficient to correctly represent all four possible layouts. Such chunking and ad hoc reduction of complexity may be seen as an instance of the general strategies for reducing demands that are postulated in relational complexity theory (Halford et al., 1989)[9]. Given the domain of spatial reasoning, participants presumably have constructed spatial representations beyond orderings on one dimension. Human visuo-spatial abilities are manifold (Barsalou, 1999)[20]. Participants did not have to restrict themselves to one-dimensional models and detached tokens. Rather, we suppose that participants used models with combined tokens and a representation of the possible changes of the model conceivable as flipping or rotating pairs of objects.

Chunking and strategic representation of indeterminacy was induced by the demands of Experiment 1, in which participants were not allowed to draw sketches or to take notes of sequentially presented assertions. The effort of strategic chunking was mainly invested in Experiment 1 when it was necessary to reduce processing load. However, in Experiment 2, the strategies we observed in the video recordings of participants' drawing were straightforward notations of possible orders in layouts.

With the use of the indeterminate spatial relations "next to" and "between" we were able to demonstrate effects of model preferences and of strategic reductions of relational complexity in deductive reasoning with mental models. Human spatial relational reasoning has been mainly studied with relations such as "left of" that determine the order of two arguments on a single dimension. One reason for this is that those (asymmetric) relations correspond to common relations in other domains (for example, better, heavier, later). Hence, results obtained with those spatial relations probably generalize to relational reasoning in other domains. Effects that are known for those relations were surely effective in the present experiments, too. Integration of assertions with relations such as "left of" is easier, if the term with which an assertion refers back to the previous assertion has the role of the reference object (relatum = given) and if this term is mentioned first (given-new) as has been shown in a recent study that successfully joined several earlier findings (Hörnig, Oberauer, & Weidenfeld, in press)[21]. However, these order effects cannot explain the large sequence effects in the reported results. Those resulted from model preferences and the support for representing indeterminacy.

We have shown that indeterminate relations such as "next to" and "between" are suitable and practical to study how humans cope with indeterminacy in relational reasoning. In the temporal domain, the equivalent relations are "at the same time as" and "between". In other domains "as R as" is not uncommon and "between with regard to Rness" may be uncommon in language, but seems to be used by individuals in relational reasoning. Therefore it is worthwhile to use indeterminate relations more often in studies of deductive reasoning. In the spatial domain this would also meet the prevalence of indeterminate relations in everyday spatial language. For example, if given the choice between referring to a layout with "left/right" or "near", "near", which is independent of reference frames, is usually chosen (Mainwaring et al., 2003)[10].

5 Conclusions

In summary, our preliminary results demonstrate the effects of indeterminate spatial descriptions. When individuals interpret such descriptions, they have distinct preferences for certain sorts of spatial models. If they then encounter an assertion that is inconsistent with this model, though not with the discourse itself, then they have difficulty in finding a consistent interpretation. In other words, the need to consider multiple possibilities creates a special difficulty if individuals' preferences bias them toward a model of the wrong possibility. Indeterminacy usually increases difficulty in deductive reasoning, because individuals need to represent multiple possibilities. As we have shown, support for detecting and efficiently representing indeterminacy can effectively reduce indeterminacy effects that usually impair human deductive reasoning.

The reported results have also implications for cognitive ergonomics. Systems supporting users in schematic spatial reasoning should anticipate model preferences, which may be culturally determined. As has been shown, the order of presentation affects how humans integrate information. Furthermore, humans can handle indeterminacies better if they are represented in an obvious and familiar way. One- or two-dimensional diagrams may induce suboptimal mental representations even if those layouts are the explicit task content. Users should be supported in detecting indeterminacy and, if possible, also in representing indeterminacy.

Acknowledgements

This research was supported by grants from the Deutsche Forschungsgemeinschaft (DFG) within the Transregional Collaborative Research Center SFB / TR 8 Spatial Cognition (www.sfbtr8.uni-bremen.de) to Markus Knauff, and by an NSF grant (BCS-0076287) to Johnson-Laird to study strategies in reasoning. Markus Knauff is also supported by a Heisenberg Award from the DFG.

References

1. Johnson-Laird, P.N.: Mental models. Harvard University Press, Cambridge, MA (1983)
2. Byrne, R.M., Johnson-Laird, P.N.: Spatial reasoning. Journal of Memory and Language **28** (1989) 564–575
3. Johnson-Laird, P.N., Legrenzi, P., Girotto, V., Legrenzi, M.S.: Illusions in reasoning about consistency. Science **288** (2000) 531–532
4. Legrenzi, P., Girotto, V., Johnson-Laird, P.N.: Models of consistency. Psychological Science **14** (2003) 131–137
5. Knauff, M., Rauh, R., Schlieder, C.: Preferred mental models in qualitative spatial reasoning: A cognitive assessment of Allen's calculus. In: 17th annual conference of the Cognitive Science Society, Mahwah, JN: Erlbaum (1995) 200–205
6. Knauff, M.: The cognitive adequacy of Allen's interval calculus for qualitative spatial representation and reasoning. Spatial Cognition and Computation **1** (1999) 261–290
7. Payne, S.J.: Memory for mental models of spatial descriptions: An episodic-construction-trace hypothesis. Memory & Cognition **21** (1993) 591–603
8. Birney, D.P., Halford, G.S.: Cognitive complexity of suppositional reasoning: An application of the relational complexity metric to the knight-knave task. Thinking and Reasoning **8** (2002) 109–134
9. Halford, G.S., Wilson, W.H., Phillips, S.: Processing capacity defined by relational complexity: Implications for comparative, developmental, and cognitive psychology. Behavioral and Brain Sciences **21** (1998) 803–864
10. Mainwaring, S.D., Tversky, B., Ohgishi, M., Schiano, D.J.: Descriptions of simple spatial scenes in English and Japanese. Spatial Cognition and Computation **3** (2003) 3–42
11. Tversky, B., Kugelmass, S., Winter, A.: Cross-cultural and developmental trends in graphic productions. Cognitive Psychology **23** (1991) 515–557
12. Linde, C., Labov, W.: Spatial networks as a site for the study of language and thought. Language **51** (1975) 924–939
13. Huttenlocher, J.: Constructing spatial images: A strategy in reasoning. Psychological Review **75** (1968) 550–560
14. Taylor, H.A., Tversky, B.: Spatial mental models derived from survey and route descriptions. Journal of Memory and Language **35** (1992) 371–391
15. Spivey, M.J., Tyler, M.J., Richardson, D.C., Young, E.E.: Eye movements during comprehension of spoken scene descriptions. In: 22nd Annual Conference of the Cognitive Science Society, Mahwah, NJ: Erlbaum (2000) 487–492
16. Vandierendonck, A., Dierckx, V., De Vooght, G.: Mental model construction in linear reasoning: Evidence for the construction of initial annotated models. Quarterly Journal of Experimental Psychology: Section A, Human Experimental Psychology (in press)

17. Schaeken, W., van der Henst, J.B., Schroyens, W.: The mental models theory of relational reasoning: Premise relevance, conclusion phrasing and cognitive economy. In Schaeken, W., Vandierendock, A., Schroyens, W., d'Ydewalle, G., eds.: The mental models theory of reasoning: Extensions and refinements. Erlbaum, Mahwah, NJ (in press)
18. Rauh, R.: Strategies of constructing preferred mental models in spatial relational inference. In Schaeken, W., De Vooght, G., Vandierendonck, A., d'Ydewalle, G., eds.: Deductive reasoning and strategies. Erlbaum, Mahwah, NJ (2000) 177–190
19. Levinson, S.: Frames of reference and Molyneux's question: Crosslinguistic evidence. In Bloom, P., Peterson, M., Nadel, L., Garrett, M., eds.: Language and space. The MIT Press, Cambridge, MA (1996) 109–169
20. Barsalou, L.W.: Perceptual symbol systems. Behavioral and Brain Sciences **22** (1999) 577–660
21. Hörnig, R., Oberauer, K., Weidenfeld, A.: Two principles of premise integration in spatial reasoning. Memory & Cognition (in press)

Spatial Principles in Control of Focus in Reasoning with Mental Representations, Images, and Diagrams

Dominik Engel, Sven Bertel, and Thomas Barkowsky

SFB/TR 8 Spatial Cognition, Universität Bremen, Germany
{dengel, bertel, barkowsky}@sfbtr8.uni-bremen.de

Abstract. The effective control of attentional focus is an essential requirement in mental reasoning based on mental models and mental images, as well as in the interaction with external diagrams. In this paper, we argue for spatial organization principles common to various mental subsystems that entail a non-centralistic control of focus. We give a brief overview of mental spatial reasoning and present a review of psychological findings related to cognitive control. We review existing modeling approaches that realize control of focus in imagery, scene recognition, and mental animation. Based on these foundations, we identify basic spatial organizing principles that are shared by the diverse subsystems collaborating in mental spatial reasoning. We discuss the implications of these principles in the framework of a computational modeling approach and give an outline of the conception of control of focus in our computational architecture *Casimir*.

1 Introduction

This paper addresses issues of computational modeling of mental image-based reasoning with spatial configurations. The work is based on the assumption that representational and procedural aspects of cognitive systems come together, and are but two sides of a single coin. As a matter of fact, all computational attempts to model cognitive phenomena are based on the hypothesis that computational metaphors such as data and storage, information and processing, function and system provide adequate concepts for understanding and describing cognitive phenomena. Different approaches differ in which of the metaphors they follow and to what extent; yet, the basic dual abstraction into representations and processes persists.

For the computer scientist, it can hardly come as a surprise that representations and processes should be related (neither should it surprise the psychologist, cf. Palmer, 1978); they only make sense when seen as pairs. On a representation theoretic level, representations in fact set the standards for processes, and vice versa. On a practical level, however, it can be worthwhile to focus on properties of one *given* the other. This is especially true where our knowledge of a system is incomplete. Such is often the case in the modeling of cognitive phenomena, as the level of detail of what is postulated for mental representations frequently differs from that of the corresponding processes. The investigation into visual mental imagery provides a prime example of a debate that for a long time has been guided in particular by considerations of various representational formats.

C. Freksa et al. (Eds.): Spatial Cognition IV, LNAI 3343, pp. 181–203, 2005.
© Springer-Verlag Berlin Heidelberg 2005

We argue that spatial properties in mental knowledge representations influence mechanisms in the mental processes which operate on such representations. Second, we propose that there exist basic spatial organizing principles which are common to different types of mental representations and processes, and that it is in part based on those principles that the different types are related or interact. Specifically, we will address the role that mechanisms resulting from spatial or spatio-analogical structures play in the control of focus in reasoning with mental models, visual mental images, and external diagrams. To this end, the next section gives a minimalist review of mental spatial reasoning, followed by a section providing a synopsis and discussion of selected psychological findings on attentional control. Section 4 presents an overview of existing approaches that give a range of functional accounts of the processes involved. Subsequently in Section 5, we will identify basic spatial organizing principles and mechanisms that are common across different mental representations and processes. In Section 6, implications of these principles and mechanisms for the development of a specific computational model of mental image-based reasoning will be discussed. Section 7 concludes the paper and gives an outline of future work.

2 Mental Spatial Reasoning

This section gives an overview of the topic of mental spatial representations, reasoning with mental images, and the role of externalization and diagrams from the perspective of spatial reasoning.

2.1 Mental Representations of Space and Visual Mental Images

As we move through the world, a variety of sensory inputs are continually presented to the brain. Motor and sensory information are used to construct mental representations of the space in which we move. Studies on deficits following from parietal lobe lesions suggest that multiple mental representations are constructed (cf. Colby & Duhamel, 1991), for instance differing in the frame of reference, but that some of these representations are truly intermodal (e.g. in that they receive input through visual as well as somatosensory streams; cf. Duhamel et al., 1998).

As a bottom line, mental representations of the surrounding space are in many ways analogous to the space. Sometimes the analogies lie in distinct spatial properties (e.g. topological ones, Sereno et al., 2001), in accessibility to manipulation (e.g. Shepard & Metzler, 1971, for mental rotation of images, or Hegarty, 1992, for mental animation of mechanical systems), or in properties with respect to reasoning (Kosslyn et al., 1978, for image scanning; Moyer, 1973, for size judgments). The construction of mental images and mental image-based reasoning can be seen as rather extreme cases in which a plethora of analogies between mental and external representations can be drawn. These analogies have led to vivid debates about the actual representational format of the mental representations (cf. Tye, 1991; Kosslyn & Thompson, 2003; Pylyshyn, 2003).

Much is known about human mental conceptions of large-scale spaces; there exist a range of theories that detail their properties and structure in relation to developmental issues, to learning, to different classes of space (e.g. to geographic or environmental spaces), or to a variety of spatial reasoning tasks. Over the past decades, the increase in knowledge along various lines of research has been reflected by the introduction of metaphors such as cognitive maps (Tolman, 1948), spatial images (Lynch, 1960), cognitive atlases (Kuipers, 1982), geographic information systems (Hirtle, 1998), or cognitive collages (Tversky, 1993). None of these metaphors must be taken literally, since, for example, mental knowledge about geographic spaces is found to be frequently distorted, fragmentary, incomplete (cf. Montello, 1992; Tversky, 1993), or potentially contradictory. In addition, mental representations of spatial knowledge are often organized hierarchically (cf. Stevens & Coupe, 1978) or chunked together to more complex structures (e.g. as visual chunks; cf. Kosslyn & Pomerantz, 1977).

Sometimes, spatial information is associated with mental representations of non-spatial information (i.e. in the form of spatial tags attached to semantic representations). A use of locational indexing procedures has been revealed in memory tasks for which location should be irrelevant (Richardson & Spivey, 2000), again confirming the elemental role of external (spatial) structures for inner mechanisms of an embodied cognitive system (cf. Wilson, 2002; Lockhead & Pomerantz, 1991). Functionally, using spatial indexes can be computationally efficient as they relieve working memory load (Ballard et al., 1997).

2.2 Externalization and the Spatial Properties of Diagrams

The perceptual and cognitive advantages of external diagrams over sentential representation have been frequently stressed, both with respect to representational and procedural properties; consequent to these are computational advantages. It is because of the way locality and indexing are realized in diagrams that, for instance, information which is needed at the same time is displayed in groups, and that correspondences between diagram parts are established without the need to introduce explicit labels (Larkin & Simon, 1987). Also, the drawing of direct "perceptual inferences" is permitted by a close coupling between (bottom-up) processes in visual perception and (top-down) processes in mental imagery (e.g. Kosslyn & Sussman, 1995). Further computational advantages arise from the specificity of information required by the diagrammatic representation format (Stenning & Oberlander, 1995); in this respect, diagrams have similar advantages over propositional formats as mental images have over isolated knowledge fragments. The use of diagrams supports processes of creativity and reflection (Goldschmidt, 1995), fosters synchronized communication (Healey et al., 2002), and introduces structure (Purcell & Gero, 1998) into problem solving processes.

With respect to the spatial properties and effects discussed in this paper, we argue that diagrams as accessed through visual perception can be compared to mental representations of spatial knowledge, such as in long-term memory (LTM), in working memory, and in particular in visual mental images. The near relationship between images and diagrams is further supported by findings of similar patterns of eye movements in visual perception of a diagram and consequent re-instantiations as a mental image, as well as of similar functional roles of such patterns (Laeng & Teodorescu, 2002). Accordingly, spatial effects of scanning, sequentialization, neighborhoods or locality, and of grouping in reasoning with external diagrams can be related to those in reasoning with mental representations of space.

3 Control of Mental Visuo-Spatial Processes

After having reviewed some issues of mental reasoning about space, in this section we discuss a selection of psychological findings relating to cognitive control of the processes involved. In cognitive psychology, the issue of control has traditionally been discussed in the context of the research on *attention*. Within the multitude of contributions from cognitive science that have been made to the topic of attention, we discuss three issues that are fundamental to the phenomena we are interested in.

3.1 Control of Visual Attention and Selectivity

Traditional research on attentional control centered around the *selection* of information from perceptual data. Broadbent's (1958) filter theory suggests a bottleneck in the processing system located at the transition from a parallel, high-capacity system to a linear, limited-capacity system. Much research followed this tradition (e.g. Treisman, 1964; Moray, 1973) and focused on the question where this bottleneck is located. Johnston and Dark (1986) categorize the theories in Broadbent's tradition as *cause* theories. For cause theories, they distinguish between two domains of processing, *Domain A* and *Domain B*, which they abstract from various labels used in causal theories about attention: "nonconscious and conscious, automatic and controlled, peripheral and central, intraperceptual and extraperceptual, preattentive and attentive, and passive and active". What they subsume under *Domain A* then is the "large-capacity, non-conscious, and passive system that is responsible for encoding environmental stimuli", whereas *Domain B* is characterized as "a relatively small-capacity, conscious, and active system that is responsible for controlling various forms of information processing including selective attention. [...] Domain B is ... an attentional mechanism or director, a cause of selective processing" that selects portions of the data offered by *Domain A*.

The problem they identify for causal theories is that "Domain B has all the characteristics of a processing homunculus" and thus the question how an individual pays attention has to be asked how this "attentional director" pays attention, resulting in an infinite regress (Johnston & Dark, 1986). More recently, Allport (1993) criticizes the set of assumptions that the causal theories often take for granted; some of these assumptions seem at least questionable from the perspective of more recent neuropsychological and neurophysiological findings. In many causal theories, it is assumed that mental processing is taken as a "linearly ordered, unidirectional sequence from sensory input to overt output" (Allport, 1993). Furthermore, processing of nonsemantic, especially spatial, features is assumed to be done prior to semantic categorization.

If these assumptions do not hold, however, then the discussion of early vs. late selection loses its basis. Regarding the location of attentional selection, Allport points out that the mutually exclusive distinction between early and late selection presumes a single locus of attentional selection, and thus also a unitary computational process. The most problematic assumption is the postulation of such a unitary central system that is the sole component responsible for attentional control. Monsell and Driver (2000) address the same problem and remark that even with advances in neuropsychology the problem of postulating a homunculus in attentional control has not been sufficiently resolved.

A possible answer to the problem of cause theories are approaches that view attention as "the consequence of natural priming effects" (Johnston & Dark, 1986). Johnston and Dark fail to report what exactly these effects are. However, the idea of attention emerging from a set of effects is reflected in more recent psychological theories (cf. Monsell & Driver, 2000; Posner, 1993), and it is supported by findings that suggest that control of attention involves distributed systems, both on the functional and on the neural level (Nobre et al., 2004; Ishai et al., 2000; Hommel et al., 2004; Allport, 1993; Posner, 1993).

In the light of these results, another traditional assumption on attention is to be questioned: the view of attention as a kind of limited resource that forms a bottleneck in the processing system and induces selectivity in information processing. In evolutionary terms, the idea that processes developed *on* the available resources is more sensible than the idea that processes developed independently of the (limited) resources, and then had to get by with these resources through selectivity. While the idea of selectivity remains important, recent evidence points out the perspective of attention as a set of distributed control mechanisms that are not only concerned with selection but also cover management, scheduling, and communication tasks (cf. Kieras et al., 2000). Together with the idea of attentional control as an emergent, distributed phenomenon, these theories have interesting implications for a computational model.

3.2 Control of Focus in Mental Imagery and Visual Perception

Traditionally, research on attention has been concerned with external sources, i.e. perception of objects and events in the extrapersonal world. Only recently, attention in the domain of internal mental representations is moving into the focus of research programs (e.g. Nobre et al., 2004; Griffin & Nobre, 2003). Nobre and colleagues (2004), for example, performed an array of brain imaging studies that try to link the orienting of spatial attention to extrapersonal objects and events with the orienting of spatial attention on internal representations held in working memory (WM). They report evidence that may indicate that "orienting of attention in the perceptual and working memory domains share common substrates". They also present data that suggest "that shifting and zooming the spatial focus of attention in the absence versus presence of a memory context recruits highly overlapping but not coextensive systems". In their brief discussion of the relevance of their experiments to mental imagery, Nobre and colleagues point out the link between mental imagery and internal orienting of attention, based on work by Ishai and Sagi (1995) and Ishai and coworkers (2000).

It seems plausible that some of the mechanism at work in visual attention can also be postulated for attention in mental imagery. Not only is there evidence suggesting strong overlap between the two systems, both on the functional and on the neural level (cf. Michelon & Zacks, 2003; Ishai & Sagi, 1995) for higher level cognitive processing, but there are also findings that support the view that imagery and perception systems are interfaced at an early processing stage (Ishai & Sagi, 1997). In this view, mechanisms of attentional control induced by mental imagery and visual perception are likely to exhibit a similar coupling.

3.3 Space as Structure in the Control of Focus

As Allport (1993) points out, "spatial location is far from being 'a simple characteristic' so far as its coding in the brain is concerned". Space is fundamental in providing structure during the control of focus. This has been shown, for example, in studies with disorders of spatial attention, which also affect internal, non-sensory, mental representation (cf. Bartolomeo & Chokron, 2002), for instance during mental imagery. There is abundant research on the topic of selection based on spatial attention. Posner (1984) suggests three internal mental operations in the covert orienting of spatial attention: disengagement of attention from its current focus, moving attention to the target focus, and engaging attention to the new target. In the context of mental imagery, Hazlett and Woldorff (2004) subdivide the shifting of spatial attention further into separable stages of planning and executing. As Eimer and colleagues (2003) argue, functional imaging studies have shown that the processes involved in control of spatial attention are also distributed among different attentional networks, possibly employing different spatial reference frames.

In Kosslyn's (1994) model of mental imagery, the issues of space as structure and control of focus come together in the formation of *multipart images* (i.e. images that consist of more than one component). According to Kosslyn's theory, mental images can be formed in a piecemeal manner with several activations from long-term memory and subsequent visualization of the activated information in the visual buffer. The locations of the visualization of the image parts are determined by the position of the attention window. In order to construct multipart images, the window of attention is moved to the right positions by the attention shifting system, based on spatial information from long-term memory. The same system is also responsible for moving the window of attention to the sequence of scanning positions during image inspection.

Noton and Stark (1971) introduce the spatial notion of the *scanpath* as the sequence of eye movements with which a subject scans a picture and organizes its features in a linear order. They suggest that the scanpath, i.e. the sequence of eye movements, could be part of the mental representation of a visual scene (cf. Stark & Choi, 1996).

Brandt and Stark (1997) show that the scanpath during mental imagery reflects the content of the imagined scene. Laeng and Teodorescu (2002) present evidence that suggests that eye scanpaths during mental imagery play a functional role. They argue that scanpaths might provide a spatial index to the parts of a mental image, a position that is embraced by Mast and Kosslyn (2002). In this view, the scanpath can be conceived of as a part of mental representations that is abstracted from the actual eye movements, and that plays a functional role in general relative spatial indexing. The frames of spatial reference in this respect still remain subject to research (cf. Eimer et al., 2003).

3.4 Summary

In summary, the psychological evidence discussed above supports the view of attentional control as a set of distributed processes rather than just as a resource. Control of attentional focus is a cognitive effect realized by distributed executive processes that are based on fundamental mechanisms. In the case of spatial cognition, there is evidence that spatial structures and representations lead to spatial mechanisms in the control of focus.

4 Existing Approaches

The idea that specific spatial knowledge representations are used to control focus-related processes in the mind has been used with respect to several partial aspects in cognitive modeling approaches so far. Examples are structural encodings that are used to control attention shifts in mental image construction, representations that guide eye movements employed for detecting salient features in visual scene analysis, or control of attention based on functional implications in understanding diagrams that convey dynamic operations.

4.1 Control of Focus in Mental Image Processing

In the construction of visual mental images, control of attention is an essential requirement to combine several image parts in a meaningful manner. Mental images are formed from well-structured components that are retrieved from long-term memory. In *multi-part images*, control of focus is used on the visual buffer representation structure to detect where a new image part has to be integrated into a partially constructed image. In the implementation of his early mental imagery model, Kosslyn (1980) proposes a *cathode-ray tube metaphor* to illustrate the successive, focus-oriented construction of the mental image on the *surface representation* (the visual buffer structure that holds the image proper). The content to be visualized on the surface representation is retrieved from the *deep representation*, which conceptually corresponds to human long-term memory.

In the deep representation structure, there are two types of representations: *literal* or *perceptual* representations that encode what entities look like (i.e. shape representations); and *discursive* representations that propositionally encode compositional information related to mental images (i.e. general knowledge about part-whole relationships, rough size information, or information about object categories). The literal image representation comes in two types: as *skeletal encodings* and as *individual encodings*. Skeletal encodings contain the overall shape and structure of an image, whereas individual encodings provide further detail. When a complex image is to be constructed in the visual buffer, a rough overall image based on a skeletal image is generated first. The encoding of the skeletal image contains propositional descriptions of details that may further specify the image. With respect to the control of focus, the positions of the detailing image parts are given as sequences of descriptions of abstract search procedures that are performed on the image in the visual buffer until the proper location is detected.

The mechanisms realized in this implementation have also been proven to be plausible through the later, neuropsychologically motivated work by Kosslyn (1994). According to his later conception, in a first processing step a global image is retrieved from associative memory, which afterwards is further detailed. Subsequent parts are arranged according to spatial descriptions retrieved from associative memory (the mechanisms employed here are the same that are also used in top-down hypothesis testing in object recognition during visual object recognition). The spatial descriptions retrieved are used directly to position the attention window on the visual buffer.

In a more technical application inspired by Kosslyn's mental imagery research, similar strategies are used in the *computational imagery* system by Glasgow and Papadias (1992): propositional descriptions of spatial relationships are used to generate a spatio-analogical representation on a symbolic array structure. Also, the same propositional descriptions are used to control a region of attention on the array structure that, for instance, can be used to reason about spatial motion planning.

4.2 Control of Eye Movements in Visual Scene Analysis

It has been shown that the mental imagery mechanisms described above can be regarded as being functional in vision processes, for instance, when a complex object or an entire scene has to be inspected and analyzed (cf. Kosslyn & Sussman, 1995). In a technical field of application, these ideas have been adopted in the *active perception* paradigm in computer vision (see Aloimonos, 1993, for a review).

A cognitively motivated computational system that integrates spatial representations of objects and visual scenes with sensorimotor representations has been developed by Schill and co-workers (1998, 2001). These integrated knowledge representations are used directly for the control of saccadic eye movements during visual scene analysis tasks. The basic representation scheme used in this approach is a feature triple of the form: *current sensory feature – eye movement – target sensory feature*.

The basic sensorimotor features (e.g. polygon patterns representing the characteristics of a polygon's vertices) are organized in hierarchical structures that represent entire visual scenes. The overall architecture of the system consists of a visual preprocessing component that extracts salient two-dimensional features, and a reasoning component that operates on the basis of belief measures according to Shafer (1976). The system adapts itself to possible scenes using a supervised learning component that learns to relate scene concepts and sensorimotor features for subsequent processing. At each processing step the system analyzes the current visual feature (i.e. the feature that currently is in the focus); based on this feature and knowledge gained from already performed eye movements, the system determines a new target feature (i.e. its relative location) that promises to provide the maximum information gain towards the goal of identifying the scene under consideration. The corresponding shift of focus (modeling eye movement in natural cognitive systems) is performed and, the whole process starts over again. This cycle is repeated until a certain threshold of belief is reached and the scene is claimed to be identified.

4.3 Control of Attention in Diagram Understanding

Aiming at describing and modeling the mental processes involved in understanding external diagrammatic depictions is related to both fields discussed above, and thus it exhibits specific requirements with respect to control of focus: on the one hand, diagrams are conveyed through external media and are therefore related to object and scene recognition; on the other hand they are combined from interpreted and well-organized components that mentally need be dealt with in a specific manner, thus requiring extensive use of corresponding mental imagery processes.

Diagram understanding is especially demanding when the objective of the diagram is to convey dynamic behavior, for instance of a chain of events or of some mechanical device. Research on *mental animation* (e.g. Hegarty, 1992) aims at explaining how people manage to infer dynamic motion from static diagrammatic depictions. Since mental animation relies on complex reasoning processes in working memory, mental capacity limits make an efficient allocation of mental resources necessary. The prevailing mental strategy employed is by decomposing the task and by solving partial problems subsequently (*piecemeal* strategy). This decomposition is done according to the causal chain of events that characterizes the system's behavior.

Through dual task experiments, it has been shown that mental animation is highly related to visual mental processes: there are strong interferences between visuo-spatial working memory and mental animation, whereas verbal tasks do not interfere with animation tasks (Sims & Hegarty, 1997). Moreover, there is an immediate correlation between eye fixations of specific image components and the entities in the causal chain that need to be analyzed to investigate specific motion properties in the diagram.

Hegarty (1992) presents a production system that models mental animation of pulley systems that account for a strong successiveness in the reasoning chain (i.e. motion of a specific part can only be inferred based on immediately neighboring parts) as well as for a limited working memory capacity with respect to the number of image parts that can be simultaneously animated. So, in this model, the focus of the reasoning process successively moves from one entity in the chain to the next one, which propagates the inferred direction of motion through the pulley system. Being a production system, however, the model does not account for the analogical aspects of mental animation, i.e. for mental animation being performed in visuo-spatial working memory.

5 Spatial Representations Lead to Spatial Mechanisms

In the last two sections, we have presented an overview of some existing approaches and psychological models to get hands on a number of phenomena of attentional control and focus. In the following, we will identify basic spatial organizing principles and mechanisms across different mental representations and the processes associated with these representations.

In an editorial, Hommel et al. (2004) make the case for a highly distributed view of executive control of human behavior; with respect to the processes involved, they conclude that "most if not all of these processes may turn out to be disappointingly common and it may be their concert that creates the emergent property of being 'executive'". With respect to attentional control and focus in mental imagery, the rationale for the current paper is to motivate that basic spatial organizing principles and basic spatial processing mechanisms can account for a good part of those processes – at least from a computer science point of view. The argument for which we will try to provide grounds in the following is that (a) spatial properties in mental knowledge representations lead to spatial mechanisms in the mental processes which operate on such representations, and that (b) there exist basic spatial organizing principles and mechanisms which are common to different types of mental representations and processes, and which play an important role for relating representations of the different types. The interplay of spatial mechanisms resulting from spatial properties of the underlying representations and the way in which mechanisms in one domain can trigger mechanism in another domain can form the basis for an abstracted, functional model view of control of focus. This model view may avoid the problem of causal theories and may fulfill the requirements put forward by recent psychological evidence as discussed in Section 3.

In its postulation of basic (spatial) organization schemes and mechanisms common to different types of mental representations, our approach finds an ally in Cowan's embedded-process model of working memory (Cowan, 1988, 1999). His model posits a set of basic mechanisms of activation and attention which are universal to all or most working memory components. It is the comparable functional nature of mental procedures across different memory subsystems that should be considered rather than their respective and potentially differing implementations. Cowan states: "there is no single separate theoretical entity that I would call working memory; that is a practical, task-oriented label. What are potentially more meaningful in a theoretical sense are the basic mechanisms proposed to underlie this complex system, including activation of memory contents of an attentional process, and the contextual organization of memory" (Cowan, 1999, p.88).

In the following, we will specifically discuss spatial knowledge held in three mental faculties: representations in long-term memory, (spatial) mental models in working memory, and visual mental images. In addition, we will consider depictions in external diagrams as a fourth kind of representation. Figure 1 illustrates these four representational domains of mental spatial reasoning along with the connections we propose on the basis of spatial mechanisms and organizing principles.

5.1 Origins of Spatial Properties in Mental Representations of Space

There are two ways in which we can consider spatial properties of mental representations: First, there exist properties that are induced by the spatial structure inherent to the entities that are represented. Second, there exist spatial properties that are induced by the organization of the mental representation structures. Richardson and Spivey (2000), for example, present evidence that "spatial indexes are being employed by the cognitive system, even in a memory task where location is irrelevant".

An example for the first type are mereological properties of the represented entities that are reflected in mental representation by contiguity, leading to subsequent activation. In spatial configuration tasks we find another example in the way in which spatial ordering information is reflected in mental models as proposed by Johnson-Laird (1983). In the case of external diagrams that show causal connections, for example the working of a system of pulleys, the spatial organization of the external diagram is reflected in the sequence of reasoning steps (Hegarty, 2000).

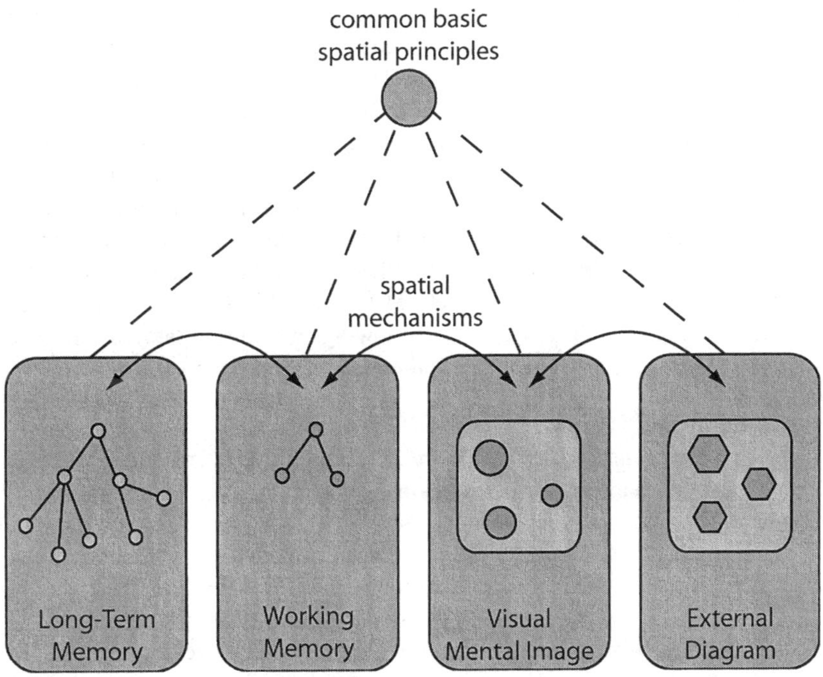

Fig. 1. Domains of spatial mechanisms and principles

Mental images provide good examples for the second type, i.e. representations in which spatial properties are induced by the organization of the representation structure: their construction involves an instantiation of spatial knowledge that exists in a non-visual mental format into a visuo-analogical form. Representational properties of this visuo-analogical form are largely determined by the representation structure (the cognitive substrate, if you will) that holds the mental image. Along with an increase in specificity that necessarily goes along with the instantiation process (i.e. through graphical constraining, Scaife & Rogers, 1996), it is the specific representation format that allows to read off novel bits of information. In this property, mental images exhibit many similarities to external diagrams: through analogical representation and knowledge contained implicitly they facilitate reasoning (Shimojima, 1996). With respect to attentional focus during mental image inspection, the quasi-pictorial structure of the image has a strong influence on focus shifts. Spatial properties also get induced by spreading activation in mental representations that integrate individual knowledge fragments. Examples are integrated semantic structures in long-term memory as well as mental models in working memory. Effects of hierarchical organization of spatial knowledge (such as observed in the classical experiments of Stevens and Coupe, 1978) might be explained by activation spreading in semantic nets.

5.2 Spatial Principles Resulting from Effects of Spatial Structures

We have discussed two types of origins for spatial structures in mental representations, and we have pointed out that spatial properties are more than just another characteristic for a range of representations. This is true for visual mental images and external diagrams, but also for non-visual mental representations attributed to long-term memory and working memory. This subsection addresses the role that spatial effects and mechanisms, resulting from this spatial structure, play for attentional control and focus in reasoning with working memory representations, visual mental images, and spatial knowledge in long-term memory. The number of discussed effects have to be seen as an exemplary selection, rather than an exhaustive account. We thus focus on principles that are representative for our claims and that span all four representational domains (Fig. 1). In the following, we concentrate on the issues of *grouping* and *chunking*, *zooming* and *scanning,* and *sequentialization*.

5.2.1 Grouping and Chunking

Organizing pieces of information into larger, meaningful units is a universal cognitive principle. It can be found in all four domains of interest. If, throughout different representational formats, one conceptualizes spatial knowledge in terms of individual knowledge fragments, a number of different organizational schemes can be found. In the context of spatial reasoning, the different schemes find a unifying characteristic in that they all relate to the spatial properties of the representations involved.

In long-term memory, knowledge may be organized semantically (e.g. such as described by metaphors of semantic networks) or by temporal coherence in terms of episodes. Models of activation spreading on networks are frequently used to describe aspects of activation flow through those structures (e.g. Kokinov, 1997), and connection strengths between individual pairs of fragments in the structure may vary significantly. Individual fragments are largely described by propositional metaphors; however, certain information of shapes and configuration, certainly with respect to human faces, may be encoded otherwise (McDermott et al., 1999). The interlinking of the fragments frequently is conceptualized in the term of networks, possibly exhibiting a hierarchical structure (Stevens & Coupe, 1978). While long-term memory remains the domain that is dominated least by spatial mechanisms, the characteristics of these network conceptualizations are at the transition from the purely propositional to the spatial, paving the way from semantically encoded knowledge to spatio-analogical representations in spatial reasoning.

For spatial reasoning in working memory, knowledge fragments may be organized in terms of mental models which fulfill roles as problem instantiations, mental theories, or are employed in mental simulations. The number of fragments integrated into one model is usually small, especially when compared to the structures that exist in long-term memory from which knowledge fragments in WM are typically derived. However, the degree of integration between fragments is high in mental models. With respect to computational characterizations of structures of spatial knowledge in WM, annotated graphs have often been suggested (e.g. by Barkowsky, 2002), where spatial entities serve as graph nodes and are connected by (mostly binary) topological, directional, distance, or other relations.

Visual mental images, on the other hand, are mental constructions in working memory that adhere to a specific representational format. The prevalent imagery paradigm stresses the relationship between mental images as representations and the visual buffer as the images' main representational substrate (Kosslyn, 1994; Kosslyn & Thompson, 2003). Mental images are constructed from mental models, and the individual actual organization of image parts into substructures largely matches that found in the corresponding mental model. Early work by Kosslyn and Pomerantz (1977) reported chunking in mental images, and the chunks reported seem to relate to a basic spatial structure underlying the mental model, as well (e.g. in the schematized figure of a human body, visual parts that make up an arm are chunked together). Content in mental images is frequently described to be interpreted (Logie, 2001), since, for most of its parts, it stems from semantic structures in LTM. The assumption of underlying spatial and semantic organization is in line with findings that structural reinterpretations of figures which are easy in perception are often hard in imagery (Chambers & Reisberg, 1992; Verstijnen et al., 1998): images already possess such a spatial and semantic grouping, where visual percepts yet have to be assigned one.

When visual mental images are externalized into diagrams, the groupings present in the image are preserved in the external diagram. Conversely, grouping effects during inspection, i.e. internalization of external diagrams, are reflected in the mental image formed, for instance according to Gestalt laws (Rock & Palmer, 1990).

Grouping, then, is one of the effects resulting from spatial organization principles. Although the actual mechanisms are different in each of the representational domains, they are connected on a functional level by their spatial grounding, and thus contribute to cognitive control during spatial reasoning: groupings are not only reflected in the various representations, they also entail one another on the basis of a common principle.

5.2.2 Zooming and Scanning

Inspection of visual mental images provides a good illustration of how the assumption of a common basic spatial organization underlying spatial knowledge in LTM representations, mental models, and mental images can help explain mechanisms of retrieval or maintenance. Spatial operations during inspection of a visual mental image, like *zooming*, i.e. focussing on a detail of an image, or *scanning*, i.e. bringing another part of the image into focus, have direct effect on the image representation. Zooming might be conceptualized in terms of concentration whereas scanning might be conceptualized in terms of translation. They both lead to a change of focus on the image, or, in Kosslyn's (1994) terms, to a shift of the attention window. As with grouping, the effects are not limited to the single domain of mental imagery, but have direct influence in control in the other domains. For example, a scanning operation on a mental image may trigger retrieval processes that lead to activation of associated long-term memory content that thus becomes part of working memory.

In information-rich LTM structures, zooming can be conceptualized in terms of different extents of activation across structural parts: On a coarser level of inspection, only some distributed parts of the representation are fairly activated. Zooming in on one specific part leads to higher activation of its neighborhood along the structural connections, thus making neighboring components part of working memory (cf. the mechanisms of activation and attention suggested by Cowan, 1999).

The scanpath, as discussed in Section 3.3, is an example for how LTM content can influence operations in mental imagery: during the construction of multipart images, the scanpath is retrieved from long-term memory and enters working memory along with the stored knowledge fragments. During visualization, the spatial information of the scanpath directs the construction of multipart images, and thus has direct influence of the control of focus. The same can be suggested for the externalization processes of the content of the visual mental image into an external diagram: the scanpath might provide the spatial index during the piecemeal inspection and externalization. This feature brings us to the principle of sequentialization.

5.2.3 Sequentialization

We have identified the control of focus as a phenomenon resulting from distributed processes. In order to jointly achieve computational processing goals in reasoning with mental images, these processes need to work on a succession of representations. We identify the principle of sequentialization, i.e. the ordering of salient features in a complex structure, as a mechanism making the emergence of goal-directed behavior possible from distributed processes.

Sequentializations work in – and across – all four representational domains as introduced above. We have already identified one of the prime structures involved in sequentialization: the scanpath. During the inspection of external diagrams, eye movements, i.e. overt shifts of attention, organize the salient features of the perceived input into a meaningful sequence. As discussed above, this is the basis for the model of saccadic eye control, devised by Schill et al. (2001). According to the evidence presented in Section 3, this external scanpath can be conceived of as being transformed into a sequence of relative spatial indices that can be stored along with the perceived information in long-term memory. The scanpath in this apprehension is an abstracted spatial index. In this view, it can be attributed as a sequencing mechanism to all of the four representational domains of spatial mental reasoning. As with the other mechanisms, the actual implementations might differ in the domains, but the principle stays the same.

During spatial reasoning with mental images, spatial indexing plays a crucial role. It structures the parts of the working memory mental model to be visualized as a mental image in the spatio-analogical representation format. The sequence also controls the succession of activations from long-term memory for the generation of multipart images and the position for new parts of a multipart image to be integrated in an existing image.

During image inspection, the scanpath influences the chain of positions for the attention window, that in turn determine the succession of readings from the visual mental image. As an aside, the spatial sequentialization of the scan points of visual mental image could also be held partially accountable for the fact that mental images are harder to reinterpret than visual percepts (Verstijnen et al., 1998): the sequentialization of the scanpath provides a commitment to an interpretation that is not easily overruled. Only by re-externalization and re-inspection of the content can there be new sequentializations and thus new interpretations.

Finally, sequentialization is a necessary principle in the externalization of mental images. While many processes are engaged in the inspection of the mental image, the transformation to motor actions, and the feedback from eye movements and other sensory input in a highly parallel manner, the sequentialization of spatial indices provides the thread that gives structure to the adequate collaboration between the processes involved.

5.3 Summary

At the examples of grouping, scanning, and sequentialization, we have shown that on an abstracted level, there exist basic principles in reasoning with mental images that stem from spatial properties inherent in the representations involved. These principles cross the different representational domains of mental spatial reasoning, in each of which their actual implementation might differ. Through this universality, they guide the control of focus and aid goal-directed behavior emerging from autonomously working processes.

6 Implications for a Computational Model

In this section, we show how the insight of the previous sections can be reflected in a computational model. We discuss the implications of spatial principles on the one hand and distributed, non-central control mechanisms on the other, in the control of focus with respect to the *Casimir* model (computational architecture, specification and implementation of mental-image based reasoning). In the present discussion of cognitive control, we will concentrate on aspects of Casimir's underlying conceptual model.

6.1 Overview of the Conceptual Model

Figure 2 shows an exemplary processing cycle of imagery-based reasoning in the conceptual model of *Casimir*. We will use this high-level flow to discuss the workings of five major subsystems: long-term memory activation, image construction, image inspection, memory update, and diagram inspection. As can be seen, these subsystems span all the four representational domains discussed in the previous section.

6.1.1 Long-Term Memory Activation

Long-term memory representations are accessed based on the representation of a (propositionally stated) spatial problem. Such a problem could be, for example, to decide upon the relative spatial orientation of two geographic locations. The problem representation could contain a query regarding the spatial relations that holds between two entities. The access process activates spatial knowledge fragments, i.e. structures that bring *n* entities into an *n*-ary spatial relation in long-term memory. Through this process, the knowledge fragments are transferred to working memory where they are integrated into a mental model representation.

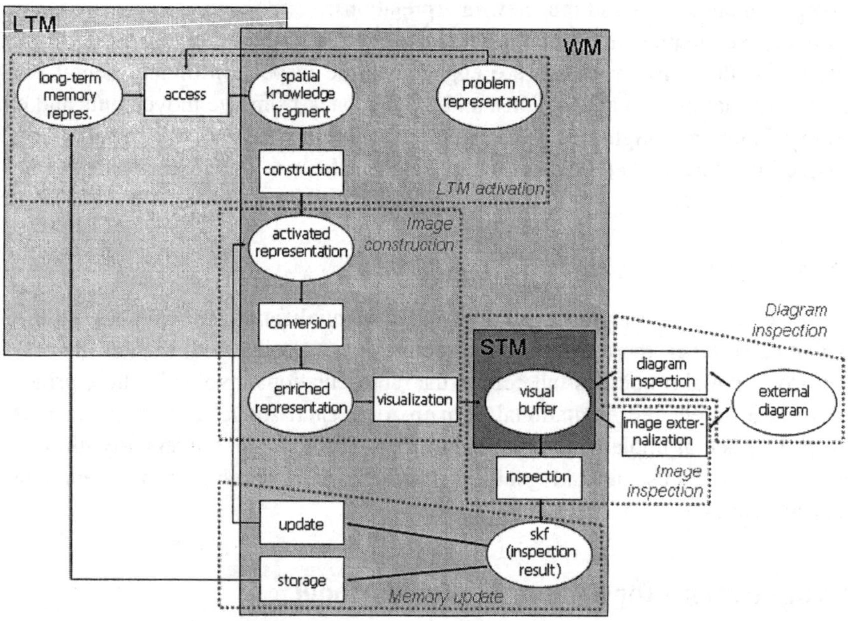

Fig. 2. Exemplary processing cycle and division into subsystems in the conceptual model of *Casimir*

6.1.2 Image Construction

The knowledge contained in this initial representation is usually underdetermined; it might be incomplete or too coarse to be visualized in a mental image. The *conversion* process in working memory enriches the representation by filling gaps in specificity with default assumptions, assigning ontological types and completing relations where necessary. The enriched mental model representation is the basis for the visualization process that activates parts of the visual buffer to produce a spatio-analogical representation, i.e. a mental image.

6.1.3 Image Inspection

The mental image held in the visual buffer is subject to inspection. The processes involved in inspection employ scanning and zooming on the mental image in order to read off previously undetected knowledge that can aid in reasoning and solving the problem at hand. The inspection processes thus may produce new spatial knowledge fragments, which can either serve as input to the memory update system for storage or further scrutiny during a subsequent processing cycle, or which might be passed on to the externalization processes, leading to the construction of an external diagram.

6.1.4 Memory Update

Regarding the further use of spatial knowledge fragments passed on by the inspection system, the memory update system has a twofold responsibility: On the one hand, these fragments can serve as an input to memory update processes in working and long-term memory, where long-term memory update is conceptualized as storage. The inspection result may also be passed to further cognitive functions, for example it may be verbalized or result in a motor action being taken.

6.1.5 (External) Diagram Inspection

The processes and structures responsible for interpreting perceptual input from external diagrams are grouped in the external diagram inspection subsystem. This system can receive perceptual input as well as input from working memory systems, reflecting the overlap between bottom-up perceptual and top-down imagery processes discussed in Section 3.

6.2 Control of Focus in *Casimir*

In line with the psychological findings we presented in Section 3, the flow of control in our model is conceived of as a distributed phenomenon of the interplay between autonomous components. In this collaboration, each of the components aims at achieving a local goal specific to that component, thus furthering the convergence of the system towards a global goal that is derived from the initial problem representation. The same is true on a macro-level for the individual subsystems: each of the systems triggers actions that refine a representation to match a local goal, for example the image construction subsystem works towards a representation that is suitable for visualization in the visual buffer. The global flow of control in *Casimir* thus results from the hierarchical composition of local goals on different granularities. The structure of the representations involved plays an important role in this process, as do unifying spatial principles. *Grouping*, for example, is a mechanism that is employed to transcend hierarchies of organization, whereas *sequentialization* is used to implement causal relationships on the same level of hierarchy; both principles can serve to provide action triggers that spread control of focus over subsystems and representational domains. Thus, on the finest granularity level, local goals are causal relationships, i.e., they are triggered from other goals on the basis of the structure of the representations they work with. On coarser granularity levels, the goals gradually assume a more global character, up to the goal of solving the initial problem.

The spatial principles are used to connect different subsystems and domains, both vertically and horizontally in terms of organizational hierarchy. A shift of focus in one domain can entail a shift of focus in a neighboring domain. This is best illustrated with a metaphoric conception of the transcendence of representations between the representational domains discussed above. Metaphorically, the representational domains can be conceived of as rotating disks on which the mental representations lie (Fig. 3). Consider the example of working memory content being visualized in a mental image. The trigger to activate visualization can be seen as a rotation of the "mental image disk", as illustrated by the rightmost arrow in Fig. 3. The "rotation" of more content into the mental image domain inevitable leads to other parts of working memory to come into focus and to the activation of associated long-term memory content, illustrated by the rotation of the leftmost arrow that is triggered by the rotation of the middle disk. While the "rotation" of the disks, i.e. the actual mechanisms, might differ in implementation in the different domains, there are principles that work across the domains and facilitate more global goals.

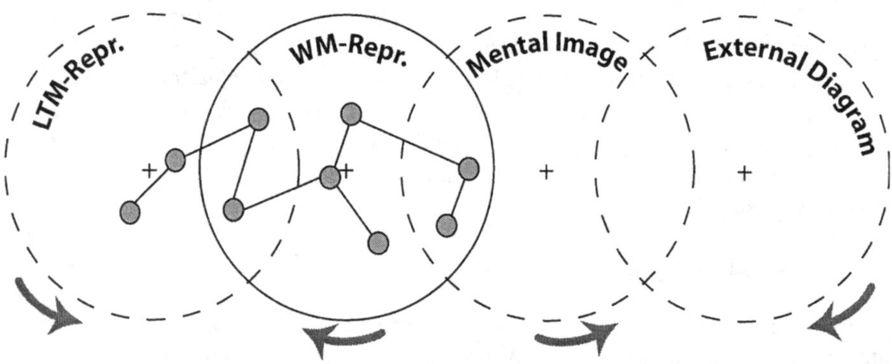

Fig. 3. *Rotating disks* metaphor for computational control mechanisms

As an example, consider zooming on a mental image as illustrated in Fig. 4. Based on activated long-term memory, i.e., content that is integrated into a working memory mental model ①, a first mental image is visualized in the visual buffer ②. According to the goals of local inspection processes, a part of the mental image is subjected to zooming ③. This triggers a shift of focus in the mental model representation ④ and leads to further retrieval from long-term memory ⑤, to provide a more detailed model ⑥. The changed focus in the mental model leads to an adaptation of focus in the mental image, which provides the frame for a detailed visualization ⑦. As this rough sketch shows, changes in the flow of control in any one of the representational domains entail a change in focus in the other domains and may trigger actions in other domains that serve to achieve local goals, for instance acquiring more detail for the entity in focus, regardless of the implementation of the parallel processes of the next lower level.

The aim to achieve locally defined goals and the connection of representations over different representational domains result in dynamic collaboration between neighboring system components. This can facilitate the construction of collaborative networks in which components join processes, capacities, and resources on the basis of the connecting principles. A possible result could, for example, be that *access*, *construction*, and *conversion* processes are joined in a special-purpose network to collaborate in solving a simple spatial configuration problem with only little help from other components.

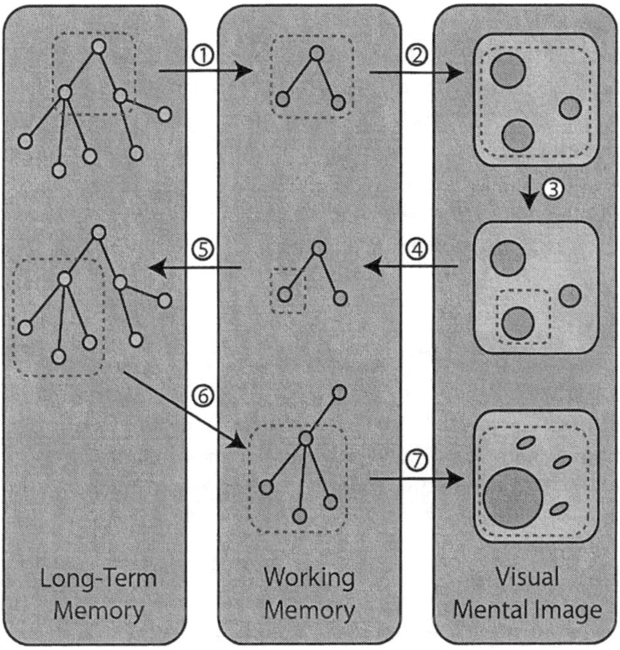

Fig. 4. The effect of zooming in different domains

7 Conclusion

We have reviewed psychological evidence relating to cognitive control and mental spatial reasoning with representations in four domains: long-term memory and working memory representations, visual mental images, and external diagrams. After having discussed the origins of spatial structures in all four of these domains, we have presented arguments that basic spatial organizational principles exert direct influence on the control of focus during mental spatial reasoning. We have outlined possible implications for a specific computational model.

All of the above has been conducted on a relatively high level of abstraction, with the aim of pointing to the existence of unifying principles across representational domains that have an influence on cognitive control. In further work, we will address these principles in more detail and look into the involved mechanisms in each of the domains. As regards the application in a computational model, we are currently refining the metaphors we presented in this paper and extract implications and structures for the implementation of *Casimir*.

With respect to the application perspective in an assistance scenario, for example in spatial configuration tasks, the presented work provides the foundation for more research into principles effecting cognitive control for providing users with adequate assistance with regard to (a) representations in external diagrams, (b) process flows, and (c) cognitive load. Sequentialization operations, for example, might be applied to external representations for highlighting important features (cf. the notion of *aspectualization* as introduced in Bertel et al., 2004). The insights of connecting principles across different domains of representations regarding shifts of focus could be reflected in the design of work flows and process flows; thus, navigation through dynamic configurations may be facilitated. Finally, the insights into mechanisms of cognitive control of focus in the *Casimir* model may serve to give a rough estimate of the cognitive load (induced by mechanisms associated with cognitive control) that a specific spatial task imposes on a human reasoner.

Acknowledgments

This paper presents work done in the project R1-[ImageSpace] of the Transregional Collaborative Research Center SFB/TR 8 Spatial Cognition. Funding by the Deutsche Forschungsgemeinschaft (DFG) is gratefully acknowledged. We also thank three anonymous reviewers for their constructive comments and suggestions.

References

Allport, A. (1993). Attention and control: have we been asking the wrong questions? A critical review of twenty-five years. In Meyer, D. E. & Kornblum, S. M. (Eds.), *Synergies in experimental psychology, artificial intelligence, and cognitive neuroscience*, Vol. XIV of *Attention and Performance*. Cambridge, MA: MIT Press.

Aloimonos, Y. (Ed.) (1993). *Active Perception*. Mahwah, NJ: Lawrence Erlbaum.

Ballard, D. H., Hayhoe, M. M., Pook, P. K., & Rao, R. P. N. (1997). Deictic codes for the embodiment of cognition. *Behavioral and Brain Sciences, 20* (4), 723–767.

Barkowsky, T. (2002). *Mental representation and processing of geographic knowledge – A computational approach*. Berlin: Springer.

Bartolomeo, P., & Chokron, S. (2002). Orienting of attention in left unilateral neglect. *Neuroscience and Biobehavioral Reviews*, *26*, 217–234.

Bertel, S., Freksa, C., & Vrachliotis, G. (2004). Aspectualize and conquer in architectural design. In J. S. Gero, B. Tversky, & T. Knight (Eds.), *Visual and Spatial Reasoning in Design III* (pp. 255–279). Key Centre of Design Computing and Cognition, University of Sydney.

Brandt, S. A., & Stark, L. W. (1997). Spontaneous eye movements during visual imagery reflect the content of the visual scene. *Journal of Cognitive Neuroscience*, *9*, 27–38.

Broadbent, D. E. (1958). *Perception and Communication*. New York: Pergamon.

Chambers, D., & Reisberg, D. (1992). What an image depicts depends on what an image means. *Cognitive Psychology*, *24*, 145–174.

Colby, C. L., & Duhamel, J. R. (1991). Heterogeneity of extrastriate visual areas and multiple parietal areas in the macaque monkey. *Neuropsychologia, 29*, 517–537.

Cowan, N. (1988). Evolving conceptions of memory storage, selective attention, and their mutual constraints within the human information processing system. *Psychological Bulletin, 104*, 163–191.

Cowan, N. (1999). An embedded-process model of working memory. In P. Shah, & A. Miyake (Eds.), *Models of Working Memory: Mechanisms of Active Maintenance and Executive Control* (pp. 62-101). New York: Cambridge University Press.

Duhamel, J. R., Colby, C. L., & Goldberg, M. E. (1998). Ventral intraparietal area of the macaque: Convergent visual and somatic response properties. *Journal of Neurophysiology, 79*, 126–136.

Eimer, M., Forster, B., & van Velzen, J. (2003). Anterior and posterior attentional control systems use different spatial reference frames: ERP evidence from covert tactile-spatial orienting. *Psychophysiology, 40*, 924–933.

Glasgow, J., & Papadias, D. (1992). Computational imagery. *Cognitive Science, 16*, 355–394.

Goldschmidt, G. (1995). The designer as a team of one. *Design Studies, 16*, 189–209.

Griffin, I. C., & Nobre, A. C. (2003). Orienting attention to locations in internal representations. *Journal of Cognitive Neuroscience, 15* (8), 1176–1194.

Hazlett, C. J., & Woldorff, M. G. (2004). Mechanisms of moving the mind's eye: Planning and execution of spatial shifts of attention. *Journal of Cognitive Neuroscience, 16* (5), 742–750.

Healey, P. G. T., Swoboda, N., Umata, I., & Katagiri, Y. (2002). Graphical representation in graphical dialogue. *International Journal of Human-Computer Studies, 57*, 375–395.

Hegarty, M. (1992). Mental animation: inferring motion from static displays of mechanical systems. *Journal of Experimental Psychology: Learning, Memory, and Cognition, 18* (5), 1084–1102.

Hegarty, M. (2000). Capacity limits in diagrammatic reasoning. In M. Anderson, P. Cheng., & V. Haarslev (Eds.), *Theory and application of diagrams* (pp. 194–206). Berlin: Springer.

Hirtle, S. C. (1998). The cognitive atlas: using GIS as a metaphor for memory. In M. Egenhofer & R. Golledge (Eds.), *Spatial and temporal reasoning in geographic information systems* (pp. 267–276). Oxford University Press.

Hommel, B., Daum, I., & Kluwe, R. H. (2004). Exorcizing the homunculus, phase two: Editors' introduction. *Acta Psychologica, 115*, 99–104.

Ishai, A., & Sagi, D. (1995). Common mechanisms of visual imagery and perception. *Science, 268*, 1772–1774.

Ishai, A., & Sagi, D. (1997). Visual imagery facilitates visual perception: Psychophysical evidence. *Journal of Cognitive Neuroscience, 9* (4), 476–489.

Ishai, A., Ungerleider, L. G., & Haxby, J. V. (2000). Distributed neural systems for the generation of visual images. *Neuron, 28*, 979–990.

Johnson-Laird, P. N. (1983). *Mental models*. Cambridge, MA: Harvard University Press.

Johnston, W. A., & Dark, V. J. (1986). Selective attention. *Annual Review of Psychology, 37*, 43–75.

Kieras, D. E., Meyer, D. E., Ballas, J. A., & Lauber, E. J. (2000). Modern computational perspectives on executive mental processes and cognitive control: Where to from here? In S. Monsell, & J. Driver, (Eds.), *Control of Cognitive Processes*, Vol. XVIII of *Attention and Performance* (pp. 681–712). Cambridge, MA: MIT Press.

Kokinov, B. (1997). Micro-level hybridization in the cognitive architecture dual. In R. Sun, & F. Alexander (Eds.), *Connectionist-symbolic integrations: From unified to hybrid architectures*. Hilsdale, NJ: Lawrence Erlbaum Associates.

Kosslyn, S. M., & Pomerantz, J. P. (1977). Imagery, propositions and the form of internal representations. *Cognitive Science, 9*, 52–76.

Kosslyn, S. M., Ball, T. M., & Reiser, B. J. (1978). Visual images preserve metric spatial information: evidence from studies of image scanning. *Journal of Experimental Psychology: Human Perception and Performance, 4*, 47–60.

Kosslyn, S. M. (1980). *Image and mind.* Cambridge, MA: Harvard University Press.

Kosslyn, S. M. (1994). *Image and brain – The resolution of the imagery debate.* Cambridge, MA: MIT Press.

Kosslyn, S. M., & Sussman, A. L. (1995). Roles of imagery in perception: Or, there is no such thing as immaculate perception. In M. S. Gazzaniga (Ed.), *The cognitive neurosciences* (pp. 1035–1042). Cambridge, MA: MIT Press.

Kosslyn, S. M., & Thompson, W. M. (2003). When is early visual cortex activated during visual mental imagery?. *Psychological Bulletin, 129* (5), 723–746.

Kuipers, B. (1982). The 'map in the head' metaphor. *Environment and Behavior, 14* (2), 202–220.

Laeng, B., & Teodorescu, D. (2002). Eye scanpaths during visual imagery reenact those of perception of the same visual scence. *Cognitive Science, 26*, 207–231.

Larkin, J. H., & Simon, H. A. (1987). Why a diagram is (sometimes) worth ten thousand words. *Cognitive Science, 11*, 65–99.

Lockhead, G. R., & Pomerantz, J. R. (1991). *The perception of structure.* Washington, D.C.: American Psychological Association.

Logie, R. H. (2001). Working memory: A mental space for design and discovery. In J. S. Gero, B. Tversky, & T. Purcell (Eds.), *Visual and spatial reasoning in design II* (pp. 223–235). University of Sydney, Australia: Key Centre of Design Computing and Cognition.

Lynch, K. (1960). *The image of the city.* Cambridge, MA: MIT Press.

Mast, F. W., & Kosslyn, S. M. (2002). Visual mental images can be ambiguous: Insights from individual differences in spatial transformation abilities. *Cognition, 86*, 57–70.

McDermott, K. B., Buckner, R. L., Peterson, S. E., Kelley, W. M., & Sanders, A. L. (1999). Set- and codespecific activation in frontal cortex: an fMRI study of encoding and retrieval of faces and words. *Journal of Cognitive Neuroscience, 11* (6): 631-40.

Michelon, P., & Zacks, J. M. (2003). What is primed in priming from imagery? *Psychological Research, 67*, 71–79.

Monsell, S., & Driver, J. (2000). Banishing the control homunculus. In S. Monsell, & J. Driver (Eds.), *Control of Cognitive Processes*, Vol. XVIII of *Attention and Performance* (pp. 3–32). Cambridge, MA: MIT Press.

Montello, D. R. (1992). The geometry of environmental knowledge. In A. U. Frank, I. Campari, & U. Formentini (Eds.), *Theories and methods of spatio-temporal reasoning in geographic space* (pp. 136–152). Berlin: Springer.

Moray, R. S. (1973). Attention in dichotic listening: Affective cues and the influence of instructions. *Quarterly Journal of Experimental Psychology, 9*, 56–90.

Moyer, R. S. (1973). Comparing objects in memory: Evidence suggesting an internal psychophysics. *Perception & Psychophysics, 8*, 228–246.

Nobre, A. C., Coull, J. T., Maquet, P., Frith, C. D., Vandenberghe, R., & Mesulam, M. M. (2004). Orienting attention to locations in perceptual versus mental representations. *Journal of Cognitive Neuroscience, 16* (3), 363–373.

Noton, D., & Stark, L. (1971). Eye movements and visual perception. *Scientific American, 224*(6).

Palmer, S. E. (1978). Fundamental aspects of cognitive representation. In E. Rosch & B. B. Lloyd (Eds.), *Cognition and categorization* (pp. 259–303). Hillsdale, NJ: Lawrence Erlbaum.

Posner, M. I., Walker, J. A., Friedrich, F. J., & Rafal, R. D. (1984). Effects of parietal injury on covert orienting of attention. *Journal of Neuroscience, 4*, 1863–1874.

Posner, M. I. (1993). Attention before and during the decade of the brain. In D. E. Meyer, & S. M. Kornblum (Eds.), *Synergies in experimental psychology, artificial intelligence, and cognitive neuroscience*, Vol. XIV of *Attention and Performance*. Cambridge, MA: MIT Press.

Purcell, A. T, & Gero, J. S. (1998). Drawings and the design process. *Design Studies*, *19*, 389–430.

Pylyshyn, Z. W. (2003). Return of the mental image: Are there really pictures in the brain? *Trends in Cognitive Sciences, 7*, 113–118.

Richardson, D. C., & Spivey, M. J. (2000). Representation, space and Hollywood squares: looking at things that aren't there anymore. *Cognition*, *76* , 269–295.

Rock, I., & Palmer, S. (1990). The legacy of Gestalt Psychology. *Scientific American*, *263* (6), 84–90.

Scaife, M., & Rogers, Y. (1996) External cognition: How do graphical representations work?, *International Journal of Human-Computer Studies*, *45*, 185–213.

Schill, K., Umkehrer, E., Beinlich, S., Zetzsche, C., Deubel, H., & Pöppel, E. (1998). A hybrid system for the scene analysis with saccadic eye movements: Learning of feature relations. *Perception, 27*, 6.

Schill, K., Umkehrer, E., Beinlich, S., Krieger, G., & Zetzsche, C. (2001). Scene analysis with saccadic eye movements: top-down and bottom-up modeling. *Journal of Electronic Imaging, Special Issue on Human Vision and Electronic Imaging, 10* (1), 152–160.

Sereno, M. I., Pitzalis, S., & Martinez, A. (2001). Mapping of contralateral space in retinotopic coordinates by a parietal cortical area in humans. *Science*, *294*, 1350–1354.

Shafer, G. (1976). *A mathematical theory of evidence*. Princeton, NJ: Princeton University Press.

Shepard, R. N., & Metzler, J. (1971). Mental rotation of three-dimensional objects. *Science*, *171*, 701–703.

Shimojima, A. (1996). Operational constraints in diagrammatic reasoning. In Allwein, & Barwise (Eds), *Logical reasoning with diagrams* (pp. 27–48). Oxford University Press.

Sims, V. K., & Hegarty, M. (1997). Mental animation in the visual-spatial sketchpad: Evidence from dual-task studies. *Memory & Cognition, 25*, 321–332.

Stark, L. W., & Choi, Y. S. (1996). Experimental metaphysics: The scanpath as an epistemological mechanism. In W. H. Zangemeister, H. S. Stiehl, & C. Freksa (Eds.), *Visual Attention and Cognition*. Amsterdam: Elsevier.

Stenning, K., & Oberlander, J. (1995). A cognitive theory of graphical and linguistic reasoning: logic and implementation. *Cognitive Science*, *19*, 97–140.

Stevens, A., & Coupe, P. (1978). Distortions in judged spatial relations. *Cognitive Psychology, 10*, 422–437.

Tolman, E. C. (1948). Cognitive maps in rats and men. *The Psychological Review, 55* (4), 189–208.

Treisman, A. M. (1964). Monitoring and storage of irrelevant messages and selective attention. *Journal of Verbal Learning and Verbal Behavior*, *3*, 449–459.

Tversky, B. (1993). Cognitive maps, cognitive collages, and spatial mental models. In A. Frank, & I. Campari (Eds.), *Spatial information theory – A theoretical basis for GIS* (pp. 14–24). Berlin: Springer.

Tye, M. (1991). *The imagery debate*. Cambridge, MA: MIT Press.

Verstijnen, I. M., van Leeuwen, C., Goldschmidt, G., Hamel, R., & Hennessey, J. M. (1998). Creative discovery in imagery and perception: Combining is relatively easy, restructuring takes a sketch. *Acta Psychologica, 99*, 177–200.

Wilson, M. (2002). Six views of embodied cognition, *Psychonomic Bulletin and Review*, *9*, 625–636.

Perceptually Induced Distortions in Cognitive Maps

Alexander Klippel[1], Lothar Knuf[2], Bernhard Hommel[2,3], and Christian Freksa[4]

[1] Cooperative Research Centre for Spatial Information,
Department of Geomatics, University of Melbourne, Australia
aklippel@unimelb.edu.au
[2] Max Planck Institute for Psychological Research, Munich, Germany
[3] Department of Psychology, Leiden University, The Netherlands
hommel@fsw.leidenuniv.nl
[4] SFB/TR 8 Spatial Cognition & Cognitive Systems Group,
University of Bremen, Germany
freksa@sfbtr8.uni-bremen.de

Abstract. Cities on a map that are directly connected by a route are judged closer than unconnected cities. This route effect has been attributed to memory distortions induced by the integration of map information with high-level knowledge about implications of route connections. However, depicted routes also connect cities visually, thereby creating a single visual object—which implies a perceptual basis of the route effect. In this article we show that the effect does not depend on whether a map is presented as a map or as a meaningless pattern of symbols and lines (Experiment 1), and that the effect occurs even if spatial judgments are made vis-à-vis a permanently visible configuration (Experiment 2). These findings suggest that the distorted spatial representation is a by-product of perceptual organization, not of the integration of abstract knowledge in memory by given organization principles.

1 Introduction

When people use or try to remember knowledge from spatial representations of the environment, such as geographic maps, they produce systematic errors. This suggests that their cognitive representations of those maps and the represented knowledge, respectively, are systematically distorted. A well-known demonstration of such a distortion is what we will refer to as the route effect, reported by McNamara et al. (1984). These authors had participants estimate distances between cities whose locations were previously memorized from a map. When comparing estimates for location pairs of equal Euclidean distance, they found those estimates to depend on whether or not a given pair was directly connected by a route: Connected cities were judged closer than unconnected (or not directly connected) cities.

These and related observations have been taken to suggest two types of conclusions (for overviews, see McNamara, 1991; Tversky, 1991). First, cognitive representations of spatial layouts do not seem to be mere mental copies or pictures of the represented arrays but, rather, integrated and highly organized knowledge structures, i.e. processed according to cognitive principles (e.g., McNamara, 1986; Stevens &

C. Freksa et al. (Eds.): Spatial Cognition IV, LNAI 3343, pp. 204 – 213, 2005.
© Springer-Verlag Berlin Heidelberg 2005

Coupe, 1978). Second, in the process of being integrated and organized the represented information is merged with, enriched and sometimes even modified by (pre) knowledge of the representing individual (e.g., Merrill & Baird, 1987; Tversky & Schiano, 1989). The route effect, for instance, is commonly attributed to the interaction of information about spatial distance with knowledge about the functional implications of route connections (e.g., McNamara et al., 1984). These conclusions receive considerable support from available findings. Indeed, judgments of spatial relations are not only affected by route distance but also by geographical (Stevens & Coupe, 1978), political (Maki, 1981), and semantic (Hirtle & Mascolo, 1986) relations between locations, suggesting that spatial information is integrated with both nonspatial and nonperceptual information (but see our conclusions for a possible perceptual interpretation).

However, evidence of information integration and of knowledge-based effects does not mean that all distortions of map representations result from background knowledge. Given that visual maps are often rather complex configurations, the way they are perceived and perceptually organized may shape the emerging cognitive representation some time before processes of memory storage or retrieval come into play. Accordingly, Tversky (1981) has argued that at least some distortions of cognitive maps may reflect principles of perceptual organization, i.e., Gestalt laws (e.g., Coren & Girgus, 1980). Indeed, spatial memories and relation judgments have been found to be affected by manipulating Gestalt factors, such as grouping by proximity (Tversky, 1981), closure (McNamara, 1986), symmetry (Tversky & Schiano, 1989), and similarity (Hommel et al., 2000). As to be expected from perceptually based effects, the impact of Gestalt factors do not only show up in memory tasks but in perceptual tasks as well, that is, in judgments of spatial relations between currently perceived locations (Baylis & Driver, 1993; Hommel et al., 2000). Thus, at least some demonstrations of distorted spatial memories may reveal rather the influence of perceptual organization on memory than effects of memory processes as such.

In the present study, we asked whether the same logic may apply to McNamara et al.'s (1984) route effect, hence, whether even this standard spatial-memory effect might be of perceptual origin. Consider our slightly simplified version of the map used by McNamara et al. in Figure 1. Take pair E-F as an example of a connected pair and M-N as one of a distance-matched unconnected pair. If the visual configuration is taken to represent a road map, it is obvious that E and F are, in a sense, "closer together" because the direct route makes it easier to get from E to F, or vice versa, than from M to N. However, not only are E and F functionally linked—someone can travel directly with no other stop from E to F, they also have a perceivable visual connection. Connecting visual elements is likely to affect their perceptual organization in creating a single perceptual object to which these elements then belong (Baylis & Driver, 1993; Humphreys & Riddoch, 1992)—the Gestalt law of connectedness. If so, judging a relation between E and F would represent a within-object judgment and judging M and N a between-object judgment, which is known to be more difficult (Baylis & Driver, 1993). In other words, the finding of McNamara et al. (1984) may be better explained by a (perceptually based) line effect rather than by a (memory based) route effect.

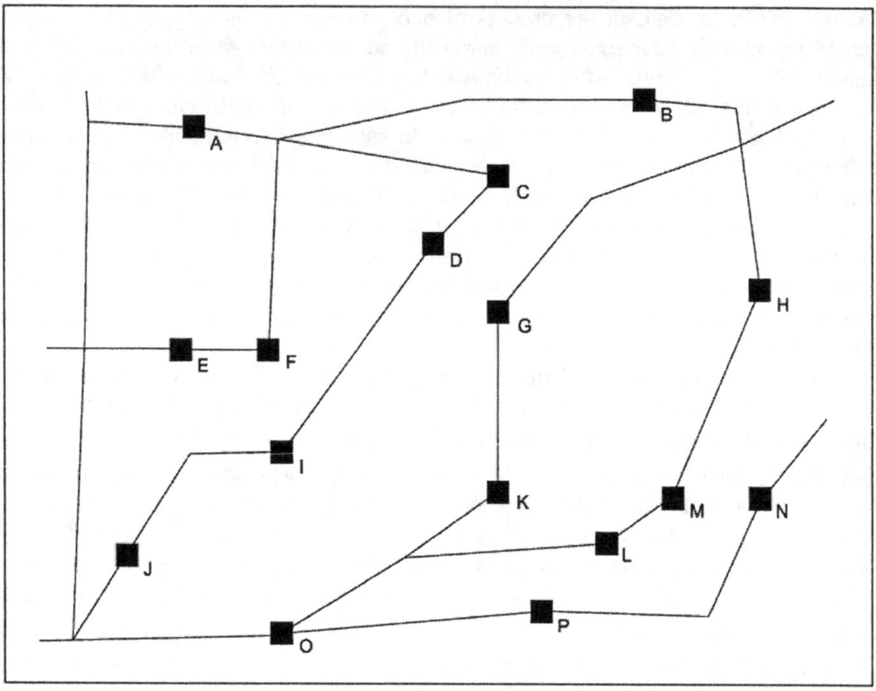

Fig. 1. Stimulus layout used in Experiments 1 and 2. The letters indicating locations were not presented; instead each location was identified by a nonsense name appearing at the bottom right corner of the corresponding square. The layout was presented in two orientations (rotation angle 0 and 180), balanced between participants

2 Experiment 1

Attributing the route effect to the knowledge-based (re-)organization of spatial information presupposes that respective knowledge (here: about implications of route connections) is not only available but is also actually used. That is, to produce a route effect participants would not only have to know about the fact that roads "bring cities together"; they also would need to interpret a particular layout as a road map. Otherwise it would be hard to see why road-related knowledge should come into play.

Experiment 1 tests this idea by introducing the stimulus layout shown in Figure 1 as either a road map or a meaningless visual pattern. Like in the study of McNamara et al. (1984) participants performed a memory task, i.e., they estimated distances between locations of the previously acquired layout. Under map instruction we expected a normal route effect, that is, estimated distances should be shorter for connected than unconnected locations. Under pattern instruction, however, a knowledge-based account would predict no route effect, whereas a perceptual account would expect the same effect as with map instruction. Hence, if the route effect is really a route rather than a line effect, it should depend on interpreting the stimulus as a map (rather than an arbitrary graph).

2.1 Method

Participants

Thirty-two paid adults (mean age 24.5 years, 24 female) participated in single sessions of about 90 minutes; they were unaware of the purpose of the study and reported having normal or corrected-to-normal vision. Sixteen of the participants received the map instruction and 16 the pattern instruction (see below).

Procedure and Design

Data acquisition was controlled by a standard PC. Stimuli were projected via a video projector (BARCODATA 800) onto a vertical 144 x 110 cm surface, about 200 cm in front of the seated participants. The stimulus layout was a visual black-on-white configuration of lines and squares (see Fig. 1), which was introduced as either a map showing cities and connecting roads (map instruction) or a meaningless graphical layout of squares and lines (pattern instruction). In the pattern group any hint to the semantics of the pattern or its elements was carefully avoided. Indeed, when asked after the experiment, none of the participants of this group reported to have recognized or imagined a map. As Figure 1 shows, the map/pattern consisted of partially connected square symbols (representing the cities under map instruction) of 34 x 34 mm, each individually named by a consonant-vocal-consonant nonsense syllable chosen to avoid any obvious phonological, semantic, or functional similarities. Names appeared at the bottom right corner of the squares. Two versions of the configuration were balanced between participants: the one shown in Figure 1 and a copy rotated by 180 degrees.

Eight location pairs were chosen for distance estimation. Half of them were composed of squares directly connected with a line (E-F, D-C, G-K, O-P) and the other half consisted of squares that were not (directly) connected (M-N, P-L, I-O, G-H). As can be taken from Figure 1, for each of the connected location pairs there was a corresponding unconnected pair with an identical Euclidean distance and orientation: item1 = E-F vs. M-N (135 mm); item2 = D-C vs. P-L (145 mm); item3 = G-K vs. I-O (260 mm); item4 = O-P vs. G-H (400 mm). A small set of vertical and diagonal pairs was used as fillers. On basis of the critical items a set of 160 judgments was composed, consisting of eight repetitions for each location pair and two orderings of presentation within the pair (A—B, B—A), plus 32 judgments on filler pairs that were not further analyzed.

At the beginning each session, participants were shown the display and were asked to memorize the locations of the "cities" or "squares", respectively. After a 2-minute study period, the display was replaced by a "road" or "line" grid that no longer showed squares and their locations (a procedure also used by McNamara et al., 1984). In each of 16 randomly ordered trials, an empty frame of a square's size then appeared in the upper center of the display, together with the "name" of a display element. Using a computer mouse, participants were then to move the frame to the correct position of the square with that name and to confirm their choice by pressing the left mouse button. After completing 16 trials, correct locations were superposed on the judged locations and the experimenter pointed out any errors the subject may have made. If in a sequence a square was misplaced by more than 15 mm, the whole procedure was repeated until a participant positioned an entire sequence correctly.

Following this acquisition phase, participants judged distances between pairs of labeled squares. The 160 pairs were displayed, one pair at a time, in the upper center of a projection surface. Labels were displayed in adjacent positions, separated by a hyphen. A horizontal line of 110 cm in length was shown above the names and participants were explained that this line would represent a line of 70 cm (half of the width of the whole projection surface). Thus, a scaling factor of 11:7 was applied to the represented length. A vertical pointer of 5 cm in length crossed the horizontal line. This pointer could be moved to the left or right by pressing a left or a right response key, respectively. For each pair of squares, participants were required to estimate the distance between the corresponding squares (center to center) by adjusting the location of the pointer accordingly. Then the participants had to verify their estimation by pressing a central response key. They were instructed to take as much time as they needed for their decisions, but not longer. The latencies of distance estimations were recorded as well.

2.2 Results

Over all conditions, the real average distance of 235 mm was overestimated (328 mm), with the relative magnitude of overestimation decreasing with actual distance between location pairs: item1 = 201 mm; item2 = 215 mm; item3 = 410 mm; item4 = 485 mm (actual distances were 135, 145, 260 and 400 mm, respectively).

Table 1. Mean estimated Euclidean distances (in millimeters) between symbol pairs in Experiments 1 and 2 as a function of symbol relation (judged pairs connected or unconnected), instruction (Experiment 1: road map or pattern), and actual connectedness (Experiment 2: pairs actually connected by lines or not). Real mean distance was 235 millimeters

	Pairs		
	Connected	Unconnected	Δ
Experiment 1			
Map Instruction	328	359	31
Pattern Instruction	292	336	44
Experiment 2			
Symbols & Lines	356	376	20
Symbols Only	371	372	1

Mean estimated distances (in mm) were computed for each participant and condition by collapsing across the four distances used (see Table 1), so that comparisons could be made between the pooled estimates of the connected location pairs E-F, D-C, G-K, and O-P and the pooled estimates of the unconnected location pairs M-N, P-L, I-O, and G-H. A three-way mixed ANOVA (analysis of variance) was run with the within-participants factor symbol relation (connected/unconnected) and the between-participants factors stimulus display (original/rotated) and instruction (map/pattern). The analysis revealed only a significant main effect of symbol relation, $F(1,28) = 15.872$, $MSE = 1227.124$, $p < .001$, whereas the interaction of instruction and symbol relation was far from significant ($p > .3$). Thus, distances were judged shorter between connected pairs than unconnected pairs under either instruction. If anything, the connectedness effect was stronger under pattern than under map instruction (η^2's = .499 and .223, respectively).

The latencies of distance estimations were also analyzed. A three-way mixed ANOVA revealed a significant main effect of symbol relation, $F(1,28) = 30.827$, $MSE = 456072$, $p < .001$, indicating that the participants spent less time estimating connected than unconnected pairs (7.667 vs. 8.604 s), this paralleling the estimation results.

2.3 Discussion

The results show a clear route effect: Distances between directly connected location pairs were estimated shorter than between unconnected pairs, and the estimation latencies, likewise, were shorter for connected pairs than for unconnected pairs. Both the estimation and latency patterns fully replicate the findings of McNamara et al. (1984) and demonstrate the robustness of the route effect. However, the effect also occurred if the stimulus display was introduced as a meaningless pattern of lines and symbols, i.e. under conditions that made the employment of route- or map-related knowledge at least less likely. Moreover, there was no indication that the pattern interpretation might have weakened the effect; on the contrary, the effect was even stronger in the pattern group. Thus, Experiment 1 provides first evidence that the route effect might have a perceptual origin.

3 Experiment 2

Although the outcome of Experiment 1 is consistent with a perceptual account of the route effect, there are two reasons to search for further, converging evidence. First, there was no way to check to which degree our instruction manipulation really worked. True, none of the participants in the pattern group reported about perceiving the pattern as a map. But even if some of the participants did perceive the pattern as a map, the route effect should have been at least somewhat reduced. Nevertheless, we do not know whether the self-reports were correct and we do not know whether members of the map group might have failed to actually perceive the layout as a map.

Second, taking evidence from a memory task to conclude on a perceptual effect is still rather indirect. Indeed, if connecting lines do affect the perception of location arrays we should be able to demonstrate such effects in a perceptual task, that is, in a task performed vis-à-vis the stimulus array. This is what we did in Experiment 2. Here we presented one group of participants (the symbols-and-line group) with the pattern condition of Experiment 1, except that distance estimations were performed in front of the permanently visible stimulus configuration. To control for possible perceptual Gestalt effects apart from the connecting lines we further investigated another group (the symbols-only group). This group worked with a display version where all lines were omitted. According to a perceptual account, a route effect was expected in the symbols-and-line group but not in the symbols-only group.

3.1 Method

Participants
Thirty-two new paid adults (mean age 24.5 years, 22 female) were recruited, 16 in each of the two groups.

Procedure and Design
The method was as in the pattern group of Experiment 1, except that the acquisition phase was omitted and participants performed the distance-estimation task in front of the constantly visible stimulus display. In the symbols-and-line group the same stimulus layout as in Experiment 1 was used, whereas in the symbols-only group all lines were omitted.

3.2 Results

On average, distances again were overestimated in both the symbols-and-line group (366 mm) and the symbols-only group (372 mm). Pooled estimates were entered into a three-way ANOVA including one within-participants factor, symbol relation, and two between-participants factors, connectedness and stimulus display (for means, see Table 1). Although treated as an orthogonal factor, connectedness had a different meaning in the two groups: In the symbols-and-line group it distinguished locations that were directly connected by a line from those that were not (i.e., E-F, D-C, G-K, and O-P vs. M-N, P-L, I-O, and G-H). In the symbols-only group the location pairs were sorted in exactly the same way, even though there were no actual lines.

The main effect of symbol relation was highly significant, $F(1,28) = 11.015$, $MSE = 164.618$, $p < .001$, as was the symbol relation x connectedness interaction, $F(1,28) = 8.192$, $MSE = 164.618$, $p < .01$. Planned paired comparisons showed a highly significant effect of symbol relation in the symbols-and-line group, $t(15) = 3.627$, $p < .001$, but not in the symbols-only group, ($p > .4$; always one-tailed). Comparable patterns were observed in estimation latencies. A reliable interaction of symbol relation and connectedness, $F(1,28) = 5.502$, $MSE = 275043$, $p < .05$, and corresponding t-tests indicated that actually connected pairs were estimated faster than unconnected pairs (8.877 vs. 9.294 s), $t(15) = 2.431$, $p < .05$, whereas the same pairs produced the same results when not actually connected (10.482 vs. 10.250 s, n.s.).

3.3 Discussion

As predicted by a perceptual account, the symbol-and-line group replicated the findings from Experiment 1 in all detail, even though here participants estimated in front of a visible display: Distances between connected location pairs were estimated shorter than between unconnected pairs, and a comparable pattern showed up in the estimation latencies. In contrast, no effects were obtained in the symbol-only group, demonstrating that the connecting lines, not the configuration were responsible. That is, a "route" effect can be obtained even in the absence of any routes and even in a perceptual task, implying that the route effect is actually an effect of connectedness.

4 Conclusions

Altogether, our findings demonstrate that spatial distortions are not only present in the memory representation of map-like configurations but in their perceptual representation as well. In principle, distortions in perception and memory—as well as their underlying causes—may be independent and may co-exist. However, it seems more reasonable and parsimonious to assume that the latter simply reflects the former, hence, memory distortions may be a by-product of perceptual organization (Hommel et al., 2000).

On one hand, this raises the question of whether other phenomena attributed to post-perceptual integration are actually of perceptual origin. For instance, take another classical finding of Maki (1981) that judging spatial relations between cities of the same country (e.g., Alamo and Burlington, North Dakota) takes less time than comparing cities of different countries (e.g., Jamestown, North Dakota, and Albertville, Minnesota). It may well be that effects of this sort reflect the (apparently hierarchical) way spatial information is organized in memory as proposed by McNamara (1986) and others. Nevertheless, this very organization may not be a memory-specific characteristic but it may merely mirror the way this information has been perceptually organized in the acquisition process, i.e., in map-reading (Tversky, 1981). Indeed, the same logic applies to other classical findings, as those of Stevens and Coupe (1978), Thorndyke (1981), or Wilton (1979). So, the structure of (parts of) our spatial memory may be perceptually derived.

On the other hand, though, it may be farfetched to attribute all effects on spatial memory to processes of perceptual organization. For instance, Hirtle and Mascolo (1986) had participants memorize map locations falling in two functional clusters, recreational facilities and city buildings. When later judging inter-location distances, participants showed a tendency to underestimate distances between places belonging to the same functional cluster as compared to pairs belonging to different clusters. Again, this is an indication of hierarchical memory organization—but in this case without an obvious perceptual basis. Similarly, Hommel and Knuf (2003) found that participants are faster in verifying spatial relations between objects that previously had been associated with the same action than between objects associated with different actions. As the authors argue, cognitive codes of the actions may be integrated into object representations, thereby functionally linking the codes of objects belonging to the same action (Hommel & Knuf, 2000; Hommel et al., 2001). This leads us to conclude that perceptual organization is only one of perhaps several types of processes shaping the structure, and in part even the content, of spatial memory. However, the present findings suggest that perceptual organization plays a powerful role.

Acknowledgments

The research reported in this paper was supported by grants of the Deutsche Forschungsgemeinschaft (HO 1430/6-1/2 and FR 806/8-1/2) and by the Max Planck Institute for Psychological Research in Munich. We are grateful to Edith Mueller, Melanie Wilke, and Kathrin Jepsen for collecting the data.

References

Baylis, G. C., & Driver, J. (1993). Visual attention and objects: Evidence for hierarchical coding of location. *Journal of Experimental Psychology: Human Perception and Performance*, 19, 451-470.

Coren, S., & Girgus, J. S. (1980). Principles of perceptual organization and spatial distortion: The Gestalt illusions. *Journal of Experimental Psychology: Human Perception and Performance*, 6, 404-412.

Hirtle, S. C., & Mascolo, M. F. (1986). Effect of semantic clustering on the memory of spatial locations. *Journal of Experimental Psychology: Learning, Memory, and Cognition*, 12, 182-189.

Hommel, B., Gehrke, J., & Knuf, L. (2000). Hierarchical coding in the perception and memory of spatial layouts. *Psychological Research*, 64, 1-10.

Hommel, B., & Knuf, L. (2000). Action related determinants of spatial coding in perception and memory. In C. Freksa, W. Brauer, C. Habel, & K. F. Wender (Eds.), *Spatial cognition II: Integrating abstract theories, empirical studies, formal methods, and practical applications* (pp. 387-398). Berlin: Springer.

Hommel, B. & Knuf, L. (2003). Acquisition of cognitive aspect maps. In C. Freksa, W. Brauer, C. Habel, & K. F. Wender (Eds.), Spatial cognition III: Routes and navigation, human memory and learning, spatial representation and spatial learning (pp. 157-173). Berlin: Springer.

Hommel, B., Müsseler, J., Aschersleben, G., & Prinz, W. (2001). The theory of event coding (TEC): A framework for perception and action planning. Behavioral and Brain Sciences, 24, 849-878.

Humphreys, G. W., & Riddoch, M. J. (1992). Interactions between object and space systems revealed through neuropsychology. In. D. E. Meyer & S. Kornblum (eds.), *Attention and performance XIV: Synergies in experimental psychology, artificial intelligence, and cognitive neuroscience* (pp. 143-162). Cambridge, MA: MIT Press.

Maki, R.H. (1981). Categoriziation and distance effects with spatial linear orders. *Journal of Experimental Psychology: Human Learning and Memory*, 7, 15-32.

McNamara, T. P. (1986). Mental representation of spatial relations. *Cognitive Psychology*, 18, 87-121.

McNamara, T. P. (1991). Memory's view of space. *The Psychology of Learning and Motivation*, 27, 147-186.

McNamara, T. P., Ratcliff, R., & McKoon, G. (1984). The mental representation of knowledge acquired from maps. *Journal of Experimental psychology: Learning, Memory, and Cognition*, 10, 723-732.

Merrill, A. A., & Baird, J. C. (1987). Semantic and spatial factors in environmental memory. *Memory and Cognition*, 15, 101-108.

Stevens, A., & Coupe, P. (1978). Distortions in judged spatial relations. *Cognitive Psychology*, 10, 422-437.

Thorndyke, P. W. (1981). Distance estimation from cognitive maps. *Cognitive Psychology*, 13, 526-550.

Tversky, B. (1981). Distortions in memory for maps. *Cognitive Psychology*, 13, 407-433.

Tversky, B. (1991). Spatial mental models. *The Psychology of Learning and Motivation*, 27, 109-145.

Tversky, B., & Schiano, D. J. (1989). Perceptual and conceptual factors in distortions in memory for graphs and maps. *Journal of Experimental psychology: General*, 118, 387-398.

Wilton, R. N. (1979). Knowledge of spatial relations: The specification of the information used in marking inferences. *Quarterly Journal of Experimental Psychology*, 31, 133-146.

Characterizing Diagrams Produced by Individuals and Dyads

Julie Heiser and Barbara Tversky

Department of Psychology, StanfordUniversity, Stanford, CA 94305-2130
{jheiser, bt}@psych.stanford.edu

Abstract. Diagrams are an effective means of conveying concrete, abstract or symbolic information about systems. Here, individuals or pairs of participants produced assembly instructions after assembling an object. When working individually, nearly all participants used a combination of text and diagrams. Those high in spatial ability produced the step-by-step action diagrams that in later studies were rated higher by all and improved performance of low ability participants. In a second experiment, pairs of participants assembled the object and produced instructions jointly. Pairs assembled the object faster and more accurately than individuals. Surprisingly, in the instructions produced, fewer than half the dyads used diagrams, and dyads produced fewer of the more effective diagrams. We speculate that the social verbal nature of the interactions of pairs encouraged verbal instructions.

1 Introduction

Designers, computer scientists, psychologists, and educators alike are interested in diagrammatic communication. Designing effective diagrams, whether for instructional, educational or computational purposes, is not simply a matter of realism. Effective diagrams abstract the essential information, and omit information that is irrelevant to the problem or task at hand. In addition, the spatial structures of diagrams provide a familiar foundation for spatial and conceptual inferences based on proximity, similarity, grouping, and more [e.g., 9, 14].

For conveying how to operate systems or how systems operate, diagrams are especially effective. Diagrams can convey structural information directly by depicting parts of a system in their spatial relations [20]. To convey dynamic or conceptual information, diagrams can be enriched with extra-pictorial devices such as arrows [7, 22]. Yet, diagrams are not always produced even in situations where they are most useful. For example, in a study of way-finding, most informants expressed a preference for using maps, yet most people writing down directional information provided words rather than sketch maps [24].

Benefiting from diagrams depends in part on the mental processes or resources the problem solver has to work with, whether expertise or ability [6, 10]. In particular, individuals low in prior knowledge or spatial ability often have more difficulty extracting relevant information and making inferences from diagrams [6, 7, 9].

C. Freksa et al. (Eds.): Spatial Cognition IV, LNAI 3343, pp. 214–226, 2005.
© Springer-Verlag Berlin Heidelberg 2005

Specifically, it seems they have difficulties in "mentally animating" a system they are less familiar with, which could inhibit them from making correct inferences about the behavior of a system [6]. People are generally good at making perceptual inferences about the structure of a system, but spatial ability or prior knowledge are often needed to make inferences about motion, behavior, or causality from a static diagram. Often, in this case, verbal descriptions may be more helpful [8].

Diagrams are all too often poorly designed for all learners, high and low ability alike. A survey of thousands of visual instructions revealed many that were misleading, ambiguous, confusing, sometimes downright incomprehensible, causing frustration and error [8, see also 15]. An example appears in Figure 1. Our focus here is on assembly instructions, because they are so common, because they are often poorly designed, because they entail conveying both structural and functional information. As such, they are representative of a large class of diagrams meant to convey systems from how the heart works to how to pass a law. Assembling an object requires understanding both the structure of the system, that is, how the parts are to be configured, and the operation of the system, that is, the sequence of actions needed to put the parts together. Because assembly is both visual and spatial, diagrams are essential to effective instructions.

Clues to effective diagram design for all types of problem solvers can be obtained through user testing and empirical investigations. For example, Novick & Morse [13] found that in a complicated origami task, users needed step-by-step instructions. Participants in that study were unable to infer the intermediate steps from diagrams of the initial and final states. Maps provide another example. They have been refined by use of way-finders all over the planet for many generations. The natural process of iterative testing and refinement can be brought into the laboratory to serve as an empirical way to discover design principles. Informed participants are asked to produce instructions that will allow others to carry out the instructions or to understand the system [1, 8, 21]. Their productions can be rated and tested by users. Some evidence suggests that further benefits can be obtained by having participants work in pairs [e.g. 17]. Presumably, the iterative processes of producing and comprehending occur within the pairs, facilitating the refinement of instructions.

We are involved in a project to generate visualizations on demand, currently for assembly instructions. The aim is to create algorithms that instantiate empirically revealed cognitive design principles [see 1, 8]. In addition, because the process of assembly requires spatial transformations, often imaginal, we investigate the effects of spatial ability on production of diagrams and performance of an assembly task. Here we extend that project to collaborations of dyads in both object assembly and diagram production.

We report two projects on production of visual instructions by individuals and dyads. We chose a simple object assembly task, construction of a television stand, because it can be completed in a typical laboratory session and because it is representative of more complex tasks that rely on visual instructions and diagrams for instructional or learning purposes. To assure expertise in assembling the object, participants first assembled the TV stand using a photograph of the assembled TV stand as their only guide. Then they produced instructions. In the first experiment, participants worked individually; in the second experiment, they worked in pairs, or dyads. In addition we review the outcome of previous experiments where the quality of diagrams were rated and later tested by new participants so that the critical features of successful assembly diagrams could be extracted.

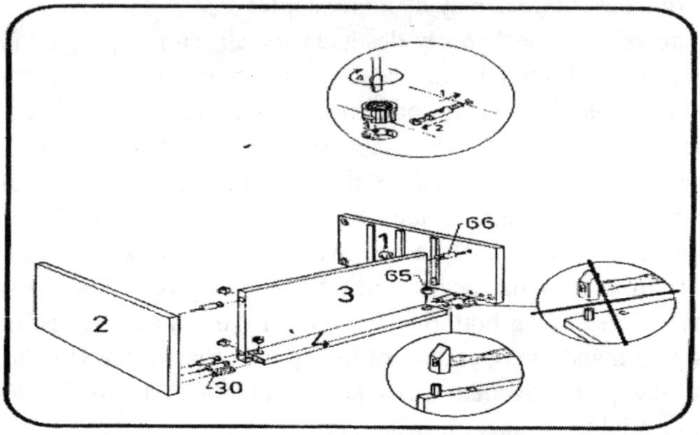

Fig. 1. This figure is a diagram illustrating how to assemble a drawer. This diagram contains several steps, incorporating several different parts and connector pieces, and also insets with more detailed instructions. In addition, there is no indication of order. According to Heiser et al., (2003), this would not be an effective representation for novice assemblers

2 Method

Participants completed the same task in Experiment 1 & 2. In Experiment 1, individuals completed it, whereas in Experiment 2, dyads completed the task.

2.1 Assembly and Writing Task

The object we chose to be assembled in Experiments 1 and 2 is a basic television stand, a standard build- your-own piece of modular furniture (see Figure 2). To participate in the experiment, participants could not have previously assembled this model or similar models of furniture. Assembling the TV stand is relatively simple: it consists of 5 major pieces (excluding wheels) and 2 types of connector parts, screws and pegs.

Participants were given a picture of what the assembled TV Stand looks like, and were given no other instructions as to how to assemble it. Figure 3 is a step-by-step schematic of the assembly process. In its most abstract form, the process consists of 5 steps.

Upon completing the assembly, participants were asked to create instructions to assemble the TV stand. They were told to use information they thought was necessary so that a novice assembler could efficiently and effectively assemble the TV stand, using diagrams and or text to convey this information.

Fig. 2. Picture of assembled TV Stand used in Experiments 1 & 2. The picture on the box, shown on the left, is the only picture participants in Experiment 1 & 2 had to assemble the TV stand

2.2 Individual Difference Measures

In both Experiment 1 and 2, participants completed a questionnaire about their prior experience with assembling or building objects, such as model airplanes, Legos, dollhouses, or other toys.

Participants also completed 2 tests of spatial ability, the Vandenburg and Kuse [23] test of mental rotation and the Money Spatial Navigation Task [12], a 1-minute test that evaluates egocentric perspective transformations. In the rest of the paper, we will be referring only to results from the Mental Rotation task [23] in terms of spatial ability. This test is a stronger predictor of performance (relevant to Experiments 1 & 2) than the Money Spatial Navigation Task.

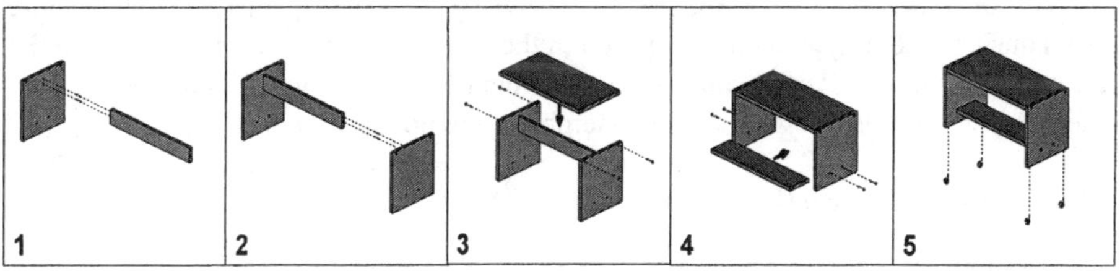

Fig. 3. Schematic depiction of steps to assemble the TV Stand (see Agrawala, et al, 2003, for origin of these instructions)

3 Experiment 1: Individuals Assembling and Creating Instructions

3.1 Participants

Forty-five Stanford University undergraduates participated for pay in individual sessions. The data of two participants were eliminated as they had participated more than once. Gender of participants was roughly equal in the final sample.

3.2 Procedure

Participants were tested individually. Each session began with a short interview assessing participants' prior experience with the TV stand, assuring that experience would not influence their performance. As described in more detail in Section 2, participants then assembled the TV Stand without instructions, only a picture of what the assembled TV stand looks like. Upon successful assembly, participants wrote instructions for assembling the TV stand.

4 Results: Experiment 1

Participants' scores on the spatial ability task were coded and participants were divided into high and low spatial categories using a median split, yielding 21 low and 22 high spatial participants. Participants had to perform below average to be included in the low spatial category, and above average to be categorized as high spatial. Both performance on the assembly task, and an analysis of the instructions produced (focusing on the diagrams) will be presented in the following sections. Performance and instructions were both highly correlated with spatial ability of the participant, and strong patterns were found in high and low categories, thus the results will be presented with respect of spatial ability scores.

4.1 Assembly Performance

All participants were able to assemble the TV stand without instructions. On the average, participants took 10.1 minutes (SD = 3.9) to assemble the TV Stand. Low spatial participants took 12.7 minutes (SD = 3.56 min) to assemble the TV stand, while high spatial participants completed the assembly on an average of 7.3 minutes (SD = 2.09 min), $F(1,41) = 36$, $p < .01$. Low spatial participants also made more errors during assembly, which manifested in the instructions produced (reported in the following section). Participants in Experiment 1 were not videotaped during assembly, so records of errors during assembly were not analyzed.

4.2 Analysis of Instructions

Even though participants had just completed the assembly task, nearly half of participants included an error in their assembly instructions. 86% of low spatial participants included an error of an "impossible action," such as putting the support board in (Step 2) after the top board was connected to both sideboards (Step 3) (see Figure 3). 12% of instructions produced by high spatial participants had such errors, $t(1,41) = 5.9$, $p < .01$.

The average number of assembly steps in the instructions produced by participants was 5.44 (SD =1.64) steps, which corresponds well with the steps portrayed in Figure 2. 42/43 (98%) of participants in Experiment 1 included some type of visual representation or diagram in the instructions they created. 26/42 (62%) of the diagrams represented information that was redundant with the text, and of these, all the diagrams were integrated into the text as tools for reference.

The diagrams used in these instructions (for both Experiment 1 & 2) can be categorized into 3 types of representations. First, people drew diagrams of parts, demonstrating the way parts look, sometimes to help differentiate 2 parts, and often times just used as sort of a part "menu." Second, people drew "structural" diagrams. A structural diagram is defined as 2 or more parts in configured position (see Figure 4). Structural diagrams could be used to show a step that has just been completed, or perhaps a demonstration of what your object should look like at a given point. Third, people drew "action" diagrams. Action diagrams are diagrams that represent, for example, 2 parts moving together, demonstrating the action between 2 structures. Note that action diagrams also contain structural information. Sentential representations of an action diagram, for example, would be "Put A into B, using a peg," or "Place A on top of B" (see Figure 4).

Differences between high and low spatial participants appeared in the sketches drawn in the instructions (see Figure 5 for examples of representative instructions). High spatial participants produced 2.67 action drawings per instruction set on average. By contrast, low spatial participants produced less than 1 (.64) action drawings per instruction set, $F(1,41) = 16$, $p <.01$. Conversely, low spatial participants included an average of 1.45 drawings that depicted the structure of the system, but high spatial participants produced only .81 structural drawings per instruction, though this difference was not significant due to high variance. Action diagrams necessarily depict structure, so the majority of drawings produced by the high spatial participants depicted both action and structure. Low spatial participants were more likely to include sketches of parts on their own (low had mean of 4.14 compared to high mean of 2.19, $F(1,41) = 5$, $p < .05$). In addition, high spatial participants were more likely to include diagrams with multiple perspectives, with information about depth and shading. Importantly, high spatial participants also made effective use of diagrammatic elements, such as guidelines and arrows to indicate placement or direction.

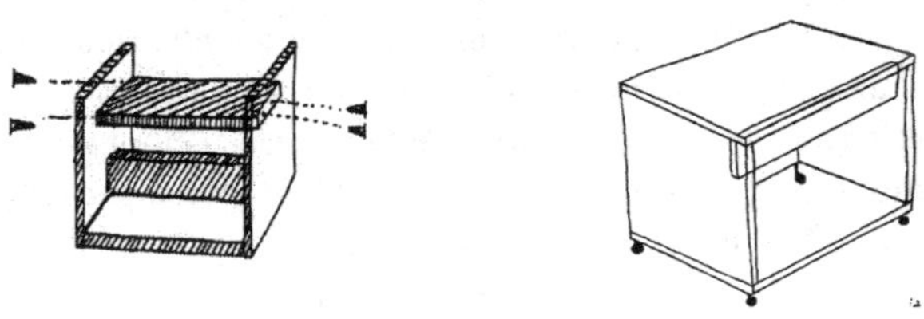

Fig. 4. Diagrams produced by participants in Experiment 1. The diagram on the left, with the shading, is an example of an action diagram, as it shows screws connecting 2 boards together. The diagram on the right is an example of a structural diagram, as it shows parts in configuration

5 Experiment 1: Conclusions

Diagrams are an integral part of instructions for an object oriented, visual and spatial tasks such as assembly. Participants, both more and less experienced, agree that diagrams are important as shown by the high number of participants that include

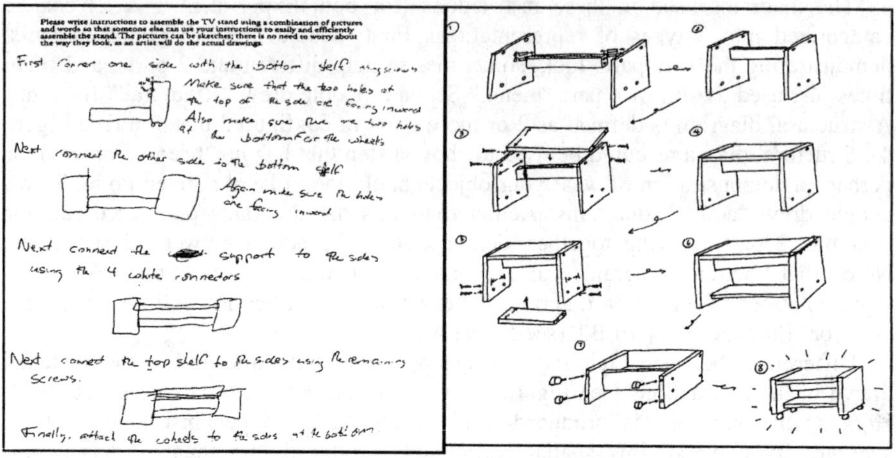

Fig. 5. The figure of the left is an example of instructions from low spatial participant. The figure on the right is an example of instructions from high spatial participant.

them in the instructions (98% of participants). An analysis of the types of diagrams that users produce can aid in revealing design principles for effective instructional visualizations.

In follow up studies (discussed in more detail in [1,8]) we had participants rate the instruction sets produced by participants in Experiment 1 and extracted the factors of those instructions that influenced high ratings. These factors included but were not limited to step-by-step illustrations of the assembly process (see Figure 3 for an example of instructions produced algorithmically according to empirically determined cognitive design principles; from [1]), clear and explicit order of assembly operations, showing mode of attachment and relevant parts being attached, action diagrams instead of structural, and consistent and effective use of diagrammatic elements such as guidelines and arrows to convey actions. When these factors guide the design of assembly instructions, they improve performance of low spatial participants [8].

Besides educating the design of instructions, the results from Experiment 1 raise the important issue of individual differences in diagram comprehension and production. The differences in diagrams produced by high and low ability participants were striking. Participants low in spatial ability produced diagrams of part menus or structure, in contrast to the step-by-step action perspective diagrams produced by high ability participants. There are several possible interpretations, not mutually exclusive. Low ability participants may be uncertain how to depict action in static diagrams. Depicting action, perspective, and even structure may depend on mental rotation ability, on facility in holding complex figures in the mind and imagining them from other points of view. For low spatial individuals, language may be an easier way to conceptualize action [7].

The variation across individuals found in Experiment 1 motivated the design of Experiment 2, having dyads assemble the TV stand and create instructions. Would dyads create more effective visualizations than individuals, as Schwartz [17] found? Would collaboration compensate for effects of ability? Collaboration requires reconciliation of different points of view, which has the potential to yield better visualizations. Collaboration may also reduce error, as participants' errors may be independent, and they may catch each other's errors.

6 Experiment 2: Dyads Assembling and Creating Instructions

The method of Experiment 2 was identical to Experiment 1, except the assembly task and the instructions were done with pairs of new participants.

6.1 Participants

Participants were 34 students in an Introductory Psychology course at Stanford University participating to fulfill a course requirement. Each participant signed up with a person they did not know personally. This created 17 dyads. Overall, there were 22 men, and 12 women: 1 Female-Female dyad, 7 Male-Male dyads, and 9 Male-Female dyads.

6.2 Procedure

Participants were tested in pairs. Each session began with a short interview assessing participants' prior experience with the TV stand, assuring that experience would not influence their performance. Participants were told to work together to assemble the TV Stand without instructions, only a picture of what the assembled TV stand looks like. Upon successful assembly, participants worked together to write one set of instructions for assembling the TV stand. Generally, one of the participants did the writing while the other talked through it, but a few dyads switched off the writing task.

7 Results: Experiment 2

Participants' scores on the spatial ability task were coded and participants were divided into high and low spatial categories using a median split, yielding 15 low and 19 high spatial participants. Participants had to perform below average to be included in the low spatial category, and above average to be categorized as high spatial. Performance on these tasks allowed us to categorize the dyads in terms of spatial ability, giving us 5 Low-Low dyads, 5 High-Low dyads and 7 High-High dyads. There were no gender difference in performance and hence, none are reported.

7.1 Assembly Performance

Participants took an average of 6.6 min (SD = 1.8) Assembly times across spatial ability groups did not differ significantly. Only 2/18 (11%) of participants made an error in assembly (explained in section 4.1) that was reflected in the instructions they produced. Of the 2 dyads that made an error, one was a Low-Low dyad, and one was a High-High dyad.

7.2 Analysis of Instructions

The average number of steps to assemble the TV stand dyads included in their instructions was 6.4 (SD = 2.7). 9/17 dyads (53%) included one or more of the 3 types of diagrams in their instructions, parts, structural or action diagrams (see 4.2 for explanation). Thus, almost half of the instructions written by dyads only included text descriptions. 5/9 of the instructions with diagrams included action diagrams, either step by step or exploded diagrams. The remaining 4/9 included structural diagrams, and 2/9 participants included part menu, neither of which were from low spatial dyads. There were no significant differences in the instructions written across spatial ability groups.

8 Comparing Individuals and Dyads

8.1 Assembly Performance

Dyads assembled the TV stand more efficiently (M = 6.6 min) than individuals (M = 10.1), This is not surprising given that assembling is much smoother and faster when one person can stabilize the whole as another attaches parts. Because participants had only one screwdriver, parallel work was limited. For dyads, one participant could plan the next step while the other was performing an assembly step.

8.2 Analysis of Instructions

Only two of the 17 dyads made an error in their instructions. This contrasts with the instructions produced by individuals, where 20 out of 43 made an error in instructions. Fully 86% of the low spatial participants made errors in instructions in the first study. For accuracy, two heads were indeed better than one, especially for low spatial participants.

The dyads' improvement in instruction accuracy and in assembly performance was not mirrored in the quality of the instructions dyads produced compared to individuals. There was a sharp decrease in number of diagrams participants included in their assembly instructions. Ninety-eight percent(42/43) of individuals writing instructions in Experiment 1 included a diagram in their instructions whereas only 53% (9/17) of dyads did. Moreover, of the 9 dyads who included diagrams, only 5 included 1 or more action diagrams. Of the remaining 4, half included only a "menu" of parts, and half included a structural diagram. Only one person out of 43 individuals in Experiment 1 used only text in their assembly instructions, whereas, 8/17 dyads in Experiment 2 used only text. The omission of diagrams is significant because in other studies using the same task, users rated instructions with diagrams higher than instructions with text, and low ability assemblers benefited from instructions with clear diagrams [8].

9 Discussion

Design of effective instructions and explanations can be informed by testing the creations of experienced users. A classic example is maps, which have been used by cultures all over the world for many purposes.

Route maps, for example, have become highly refined to convey a route as a sequence of lines and nodes, with minimal embellishment. The refinement occurs as people produce and use maps with varying degrees of ease and success. The refinement of visualizations, then, occurs in a community of users, and the processes parallel those of establishing common ground in language [3]. The iterative design processes can be brought into the laboratory in order to uncover principles of effective instructional design.

A critical feature of many instructions and explanations, as for maps, is visualization. Diagrams use elements and relations on paper to convey elements and relations of instructions and explanations. Users can then understand diagrams by interpreting elements and spatial relations in the diagrams as elements and spatial relations in a broader spatial or abstract space. Effective diagrams convey only the essential elements and relations, removing irrelevant clutter. Because instructions and explanations are communicative, creating them in a communicative setting, by dyads instead of individuals, is expected to improve design of instructions and explanations. Schwartz [17] found that junior high school students working in pairs produced more effective diagrams in several scientific domains. Dyads' diagrams were more abstract and contained less idiosyncratic, often decorative rather than useful, information. However, even when diagrams are acknowledged as effective, they are not always produced [24].

Here, we compared individuals and pairs in an instruction design task. Participants first assembled a TV stand using the photograph on the package as a guide. Participants high in spatial ability assembled the TV stand faster and with fewer errors than those of low ability. After assembling the TV stand, participants produced instructions they thought would be sufficient for a novice assembler to complete the task. Nearly all the instructions created by individuals contained both diagrams and text. The effectiveness of the diagrams varied considerably, from simple menus of flat parts to a sequence of step-by-step perspective drawings that showed the actions required for assembly and used extra-pictorial devices such as arrows and guidelines to convey assembly. The more sophisticated drawings in the instructions were produced by participants that were high in spatial ability. In other research, the instructions produced by individuals were evaluated by new participants [8]. The step-by-step perspective drawings showing assembly actions were rated higher by participants of all ability levels. The lowest ratings were given to instructions where text dominated and diagrams were minimal. In a third study, low spatial ability participants benefited from more effective instructions, assembling the TV stand faster and with fewer errors [8]. Successful explanations, therefore, rely on diagrams more than text, and rely on sequential perspective diagrams that convey function or action as well as structure. For participants of high ability, the quality of instructions made no difference; in fact, they were hardly used, as the photograph on the package was sufficient.

Participants working in pairs assembled the TV stand with fewer errors than participants working individually. Although there were pairs where both individuals were of high or low ability as well as mixed pairs, spatial ability had no effect on assembly. The improved performance of even low ability pairs suggests that working together on this type of task can compensate for ability. Similarly, for dyads, spatial ability had no effect on quality of diagrams. Nevertheless, the improvement in assembly performance did not translate into creating more effective instructions. The surprising result is that only half the dyads included diagrams in their instructions, sometimes only a single diagram. What's more, dyads included fewer of the more effective kinds of instructions, those kind that showed action as well as structure. These results only seem to contradict those of Schwartz [17]; his participants were instructed to construct diagrams, and ours were free to invent the format of the instructions.

Why should dyads produce fewer and less effective diagrams than individuals? There are several ways to approach explaining this surprising finding, and more than one factor may be at work. One key reason may be that the dyads communicated between themselves by language. The natural extension is then to continue the task in language. This is a form of entrainment, a familiar process in establishing common ground, where cooperative collaborators take up each other's formulations, language, and gesture [e.g., 4]. In addition, in the present situation, the dyads did not test the instructions they had written on themselves or on others, so they had no feedback on the efficacy of their productions. It is natural to think of the design processes of individuals as a conversation between the designer and whatever the designer places on paper [e. g., 16, 5, 19]. The designer may put diagrams on paper instead of language for several reasons. The design task is about something visualizable, and it is natural to translate something visualizable to a depiction. Then, thinking about something visualizable in order to refine it is easier from a diagram than from language, which needs to be visualized, an extra step. In addition, for the most part, dyads talk as equals; when they construct instructions, this symmetry is broken, as one partner typically dictates, and the other records. Turning diagrams into talk takes great effort, something dyads may avoid by extending talk into instructions. This analysis is consonant with the "Principle of Least Collaborative Effort" [2] according to which individuals sacrifice their "individual cognition" to facilitate the collaborative effort. We speculate that in the dyadic situation, the conversation does not take place over the pieces of paper that will constitute the instructions. Instead the conversation takes place in language, and prior to putting something down on paper. So the design is more likely to be in language, put down after the design conversation as a final product. For individuals, the design thinking, conversation if you will, takes place over the markings on paper, in this case, diagrams, which are a more direct mapping of assembly than language.

What is clear from the differences between individuals and dyads is that the effects are due to the dynamics of collaboration. The reasonable assumption is that participants have the same cognitive representations and procedures whether working alone or in pairs. If these were the only factors, then the outcomes of the dyads would be similar to those of individuals, perhaps comparable to the best performer in the dyad. An alternative account, one compatible with the discrepant performance of individuals and dyads, is that group cognition is distinct from individual cognition and the outcome is not equivalent to the average, sum, or best of the members' cognition. The social component of group cognition influences the dynamics between the individuals collaborating, which in turn influences the outcome of the collaboration. Further analyses of on-line creation of instructions should reveal the ways individuals and dyads interact with their own creations in design.

Acknowledgments

Special thanks to Christina Vincent, Mario Scussel, Rachel Mackenzie, Heesoo Kim and Vincent Pham for help in data collection and analysis. This research was funded by ONR grants NOOO14-PP-1-O649, N000 14-01-1-0717, and N00014-02-1-0534.

References

Agrawala, M., Phan, D., Heiser, J., Haymaker, J., Klingner,J., Hanrahan, P., and Tversky, B. (2003). Designing effective step-by-step assembly instructions. *SIGGRAPH* 2003.

Clark, H. and Wilkes-Gibbs, D. (1986). Referring as a collaborative process. *Cognition*, 22, 1-39.

Clark, H. H. (1996). *Using language*. Cambridge: Cambridge University Press.

Garrod, S. and A. Anderson (1987). Saying what you mean in dialogue: A study in conceptual and semantic co-ordination. *Cognition*, 27: 181-218.

Goldschmidt, G. The dialectics of sketching. *Creativity Research Journal*, Vol. 4, No 2, (1991) pp. 123-143.

Hegarty, M. and Sims, V. K. (1994). Individual differences in mental animation during mechanical reasoning. *Memory & Cognition*, 22, 411-430.

Heiser, J & Tversky B. (2002). Descriptions and depictions in acquiring complex systems. In the *Proceedings of the 24th annual meeting of the Cognitive Science Society.*

Heiser, J., Tversky, B., Agrawala, M., and Hanrahan, P. (2003). Cognitive Design Principles for Visualizations: Revealing and Instantiating. In the *Proceedings of the 25th annual meeting of the Cognitive Science Society.*

Larkin, J. & Simon, H. (1987). Why a diagram is (sometimes) worth ten thousand words. *Cognitive Science*, 11:65-99.

Mayer, R. E., & Gallini, J. K. (1990). When is an illustration worth ten thousand words? *Journal of Educational Psychology*, 82(4), 715-726.

Mijksenaar, P. & Westendorp, P. (1999). *Open here: the art of instructional design*. London: Thames and Hudson.

Money, J., and Alexander, D. (1966). Turner's syndrome: further demonstration of the presence of specific cognitional deficiencies. *Journal of Medical Genetics*, 3, 47-48.

Novick, L. and Morse, D.L. (2000). Folding a fish, making a mushrooms: The role of diagrams in executing assembly procedures. *Memory & Cognition*, 28(7), 1242-1256.

Pinker, S. (1990). A theory of graph comprehension. In R. Freedle (Ed.), *Artificial intelligence and the future of testing*. Pp. 73-126. Hillsdale, N. J.: Erlbaum.

Richards, C. (2000). Getting the picture: Diagram design and the information revolution. *Information Design Journal, 9(2).*

Schon, D. A.: *The Reflective Practitioner*. Basic Books, New York. (1983)

Schwartz, D. (1995) The emergence of abstract representations in dyad problem solving. *Journal of the Learning Sciences*, 4, 321-354.

Shepard, R. & Cooper, L. (1982). *Mental Images and Their Transformations*. Cambridge, Mass: The MIT Press.

Suwa, M., Tversky, B. (1997). What do architects and students perceive in their design sketches?: A protocol analysis. *Design Studies*, 18, 385-403.

Tversky, B. (1995). Cognitive origins of graphic conventions. In F. T. Marchese (Ed.). *Understanding images*. (pp.29-53). New York: Springer-Verlag.

Tversky, B. & Lee, P.U. (1999). Pictorial and verbal tools for conveying routes. In C. Freksa, D.M. Mark (eds.) *Spatial information theory* (pp. 51-64). Berlin: Springer

Tversky, B, Zacks, J., Lee, P. U., & Heiser, J. (2000). Lines, blobs, crosses, and arrows: Diagrammatic communication with schematic figures. In M. Anderson, P. Cheng, and V. Haarslev (Editors). *Theory and application of diagrams*. Pp. 221-230. Berlin: Springer.

Vandenburg, S.G. and Kuse, A.R. (1978). Mental rotations. A group test of three-dimensional spatial visualization, *Perceptual Motor Skills*, 47, 599-604.

Wright, P., Lickorish, A., Hull, A, and Ummelen, N. (1995). Graphics in written directions: Appreciated by readers but not by writers. *Applied Cognitive Psychology*, 9, 41-59.

Sketch Map Analysis Using GIS Buffer Operation

Kohei Okamoto, Kei-ichi Okunuki, and Toshibumi Takai

Nagoya University, Department of Geography
464-8601 Nagoya, Japan
h44540a@nucc.cc.nagoya-u.ac.jp, nuki@lit.nagoya-u.ac.jp,
takainu@ss.iij4u.or.jp
http://www.geogr.lit.nagoya-u.ac.jp

Abstract. We developed a method to analyze sketch maps by GIS, and applied it to an actual case study. We found that analysis using buffer operation was more effective for sketch map analyses than other methods, such as the entire road length method and area method. After modeling the buffer method, an experimental study of the micro-genetic cognitive process was conducted on sketch maps from Japanese students and Brazilian residents in Japan.

1 Introduction

Increasing mutual affects are seen between spatial cognition research and GIS, and integration of the two has been attempted from various viewpoints (Kuhn et al., 2003). After GIS research evolved in the 1990s to GISc (Geographic Information Science), which examines basic problems concerning the use of GIS, interest in the spatial cognition of man as the GIS user has increased (Wakabayashi, 2003). Schuurman (2000) called this movement a 'cognitive turn' in GIS research.

GIS researchers were originally concerned with cognitive aspects in human-computer interaction; in other words, operation of GIS (Nyerges et al., 1995). Spatial cognition researchers, on the other hand, were seeking progress in systematic understanding of human spatial knowledge (Golledge, 1993; 1995). These directions led to polarization in the two fields of study; applied research such as for the development of visualization tools on the one hand, and conceptual studies such as classification of geographical knowledge on the other. Basic studies on how GIS can improve methods in spatial cognition study were left behind. This paper proposes a method for sketch map analysis, and examines its applicability in an actual analysis.

In previous studies on spatial cognition, having subjects draw sketch maps was a popular method to elicit environmental knowledge. Sketch maps have been regarded as external representations of cognitive maps (e.g. Matthews, 1992), although there has been some discussion as to whether or not sketch maps are an appropriate means for revealing cognitive representations (e.g. Blades, 1990).

Recently, sketch map research using GIS is seen (e.g. Forbus et al., 2003; Blaser and Egenhofer, 2000). The principal object of many of these studies is the

C. Freksa et al. (Eds.): Spatial Cognition IV, LNAI 3343, pp. 227–244, 2005
© Springer-Verlag Berlin Heidelberg 2005

development of a visual tool or software to express and analyze sketch maps on GIS. We present a method of analyzing sketch maps by using an existing GIS operation, and apply this method to the task of distinguishing between route-type sketch maps and survey-type sketch maps. The hypothesis that people's cognitive maps simply develop from the route-type to the survey-type as their land experience increases has been challenged recently (e.g. Montello, 1998). We do not go deep into this issue, but focus on how to analyze external representations in sketch maps.

2 Sketch Maps in Spatial Cognition Research

The analytical methodologies used with sketch maps have consisted of analysis of contents (what is drawn in the map) and analysis of form (how the map is drawn). For the latter, there are three major subjects of analysis.

Subject 1: The drawing style of sketch maps (whether the drawing is linear or planar; in other words, whether it is sequential or spatial.)
Subject 2: The range and elaboration of sketch maps (the extent to which real geographical space is drawn, and the density of cognitive elements such as paths, nodes, and landmarks (Lynch,1960) drawn in the sketch maps).
Subject 3: Spatial relationships among the elements (distance and directions among the elements)

Of these, Subject 3 has been studied quantitatively since before the development of GIS. One example is the examination of distortions in cognitive maps by comparing the cognitive configurations with the actual ones (Wakabayashi, 1994: Lloyd, 1997). Analysis of Subjects 1 and 2 has been performed nonobjectively, relying mainly on visual judgment.

Subject 1 has been studied in terms of its relationship to the development of spatial cognition. It was suggested that as one grows older, one's sketch maps change from route-type (linear drawings) to survey-type (planar drawings) maps. Many studies have shown that such changes occur both in ontogenetic development, that is, the development that accompanies the aging process from infancy (Hart, 1981), and in micro-genetic development, which is the development that comes with learning about a new place after migration (Lee and Schmidt, 1988; Humphreys, 1989). However, the criteria for route maps and survey maps in those studies were always ambiguous, and sketch maps were classified based on researchers' visual estimates or general impressions. The classic work by Appleyard (1970) was not an exception, in that the method was conceptual in effect and quite difficult to apply to empirical studies.

Subject 2 has been understood in terms of its relation with spatial behavior. If one's daily activity covered a wide area, the geographical space drawn in one's sketch map would be large. The extent of action space has been discussed in relation with the actor's mobility, which is affected by income or occupation, or the geographic size of the actor's community, which is influenced by ethnicity. Orleans (1973) for instance, compared the sketch maps drawn by different ethnic and socio-status groups, namely Hispanic residents and wealthy white residents in the same urban area, and showed that the sketch maps by wealthy whites were more elaborate and covered a wider area, while those by Hispanic residents consisted of quite limited areas.

The range of a sketch map is also affected by development of spatial cognition. In general, the space represented by a survey map is more extensive than the space represented by a route map. Thus, it is important to assess the geographical space drawn in a sketch map. However, previous studies rarely quantified the range of sketch maps, only roughly estimating it in terms of the distribution of elements. Although a few studies attempted to measure the area, the methodology employed was not advanced.

In this paper we suggest a method using GIS for quantitative analysis of sketch maps. In particular, we focus on (1) analysis of the areas drawn in sketch maps in reference to actual geographical space, and (2) distinction of the forms of sketch maps.

3 Analysis of Sketch Maps Using GIS

As mentioned above, previous studies often focused on the range of sketch maps. Suppose, for instance, there was a sketch map where a person drew his/her neighborhood. On this sketch map, we may find the drawer's home, stations, bus stops, parks, supermarkets, schools, railways, etc. We call each of these a geographical element. In order to understand the drawer's cognitive map, we could consider the distribution of all elements on a sketch map. From the geometric viewpoint, each element is either a point-like element, a line-like element, or a polygon-like element. A building on a sketch map is a point-like element, a road network is a line-like element, and a park is a polygon-like element.

In order to employ GIS for the analysis, it is necessary to conceptualize the space where these elements are distributed. In the research field of spatial analysis, researchers have proposed various methods to analyze a distribution of point-like elements (Bailey and Gatrell, 1996); however, there has been little success in developing methods to analyze a distribution of line-like elements or polygon-like elements. There has been no general method to understand a space where different types of element (for instance, point elements and line elements) are distributed. In this chapter, we will discuss such a method using GIS.

One critical issue here is how to consider the space where geographical elements are distributed. In the real world, when we move between two points, we can not usually move as the crow flies, but we move along roads. Thus, a space where elements are distributed is like a road network spread over a plain. One approach to analyzing the distribution of geographical elements, therefore, is to consider them in a linear network space.

Another method to analyze the distribution of elements is to assume that the space is Euclidean and one can move along a straight line between any two points. In other words, the distribution of geographical elements is analyzed in a planar space.

In the following, we will describe how GIS is applied to the analysis of sketch maps with these two methods.

3.1 A Method to Analyze Distribution in Linear Space: Entire Road Length Method

One method to quantify the extent of distribution is to measure the entire length of the roads drawn in sketch maps by using GIS. The area recognized is considered to be larger as the entire length is longer.

However, this method is insufficient for distinction between a route-type sketch map and a survey-type sketch map. For instance, the same length could be measured for the case in which one long road was drawn (route-type) and that in which many roads were densely drawn in a small area (survey-type). Although it is possible to analyze the distribution of geographical elements in a linear space by measuring entire road length by GIS, this method is not yet able to distinguish the forms of sketch maps.

3.2 A Method to Analyze Distribution in Planar Space: Area Method

The simplest way to analyze distribution in a planar space is to measure the area within the range where the drawn elements are distributed in the map, i.e. to measure the *drawn area*. However, the drawn area of a sketch map is usually ambiguous and needs to be defined. For that we now suggest the following method. The drawn area is defined as a polygon that is enclosed by the lines connecting landmarks and nodes at the edge of the sketch map drawing. A dashed line in Fig.2 indicates the drawn area, for example. By measuring the area of this polygon in actual space, we can describe the planar range of distribution quantitatively. It can be considered that the larger the area of the polygon, the larger is the acknowledged space.

However, this method holds one shortcoming. In the area method, the space within the polygon is all defined as a cognitive space. Even domains not actually recognized may be interpreted as being within the cognitive space.

3.3 Combination of Linear Space and Planar Space: Buffer Method

Although both the entire road length method and area method can analyze the range of the distribution quantitatively, there are problems with both. To overcome these problems, we propose a new method to analyze sketch maps. To explain this new method, we introduce the concept of *buffered regions*. The buffer regions are obtained by applying a buffer operation, which is implemented in GIS. The buffer operation selects only the parts of a map or those features that lie within a certain distance of a point, a set of points, a line, or an area (Clarke, 2002). We refer to the parts of a map as buffered regions and define these regions mathematically as follows. Let P be a point or a set of points, and a line or an area consist of the set of points. $B(P, h)$ is then the buffered region in buffer distance h from all the points in P. The buffered region is described as $B(P, h)=\{q|d\,(p, q)\leqq h, \text{ for all } p\in P\}$, where $d(p, q)$ is Euclidean distance between any two points p and q. The new method we propose here is to obtain the area of the buffered regions within buffer distance h of all roads drawn on a sketch map. We call this method the *buffer method.*

Using the buffer method, it is possible to distinguish roughly between route-type and survey-type maps. In route-type maps, the area of the buffered region increases at the same rate as the buffer distance is increased. In the survey-type map, when the buffer distance is also increased, the rate of increase in the area of the buffered regions should become smaller after a certain point. Because the buffered regions will begin to overlap each other, the increase in the entire buffered region will be slowed. Based on this difference, the two forms can be distinguished. Also, the misunderstanding of unrecognized space for recognized space can be avoided, because this method identifies only the region a certain distance from a road (See 5.2 for theoretical explanation).

4 A Sample Analysis of Sketch Maps Using GIS

In this chapter, we show a case study in which the above-described entire road length method, area method, and buffer method are attempted. Sketch maps of the neighboring areas around one's home were drawn by 35 Japanese and 19 Japanese-Brazilians living in Japan (e.g., Fig. 1). The Japanese were undergraduate students at the Hamamatsu campus of Shizuoka University. Their residences were

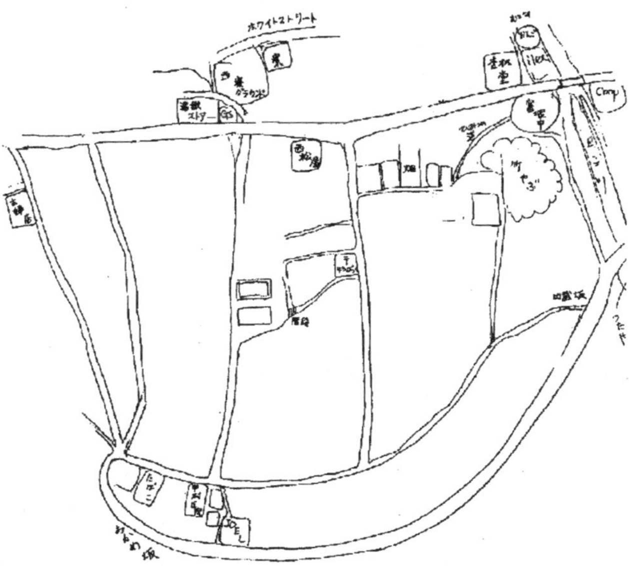

Fig. 1. A sketch map (Sketch map 1). The number of the sketch map is same as in Table 1

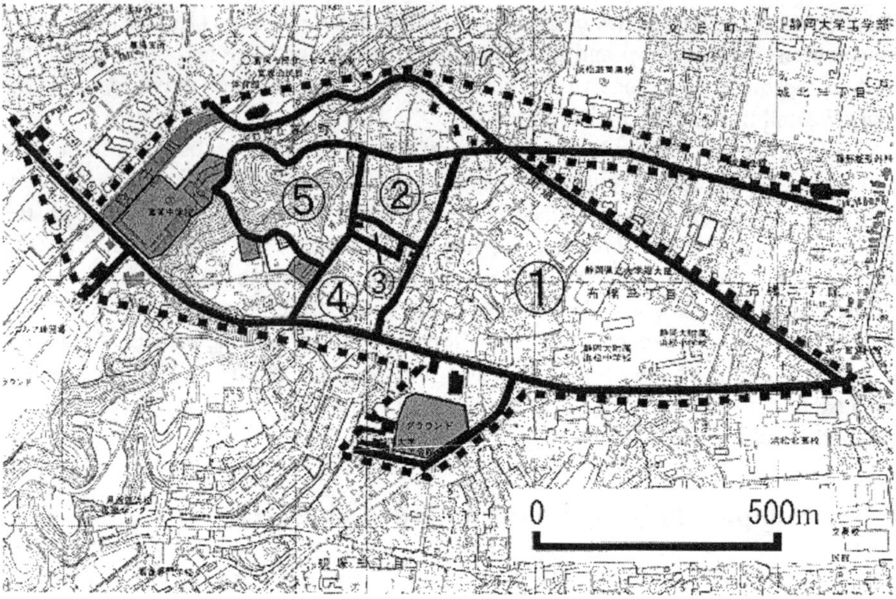

Fig. 2. A base map on which the elements of Fig.1 are traced. Drawn from a topographical map of Hamamatsu on a scale of 1/10,000. ①～⑤ are circuits, and the dash line indicates the drawn area

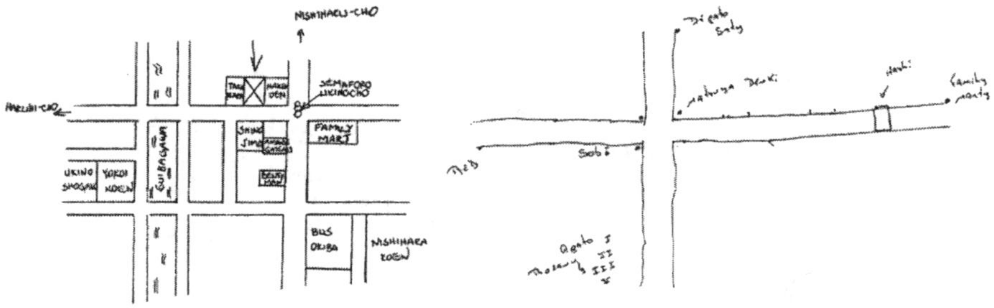

Sketch map 43 Sketch map 47

Fig. 3. A sketch map where the forms differ despite having the same entire road length. The numbers of the sketch maps are same as in Table 1

distributed throughout Hamamatsu city and its vicinity. The Japanese-Brazilians were mostly workers in automobile factories who lived in Nagoya, Toyota, and Toyohashi in Aichi prefecture and Hamamatsu in Shizuoka prefecture. They are descendants of migrants from Japan to Brazil. In 1990, the immigration control law was revised, enabling foreign nationals of Japanese ancestry to work as manual laborers. The population of Japanese-Brazilians then increased rapidly in Japan. Among the Brazilians who drew the maps, some spoke Japanese fluently, but very few read the Chinese characters used in Japan (Takai, 2004).

Table 1. Results of analysis of sketch maps

	Sketch Map No.	length of residence (year)	entire road length (km)	drawn area (km^2)	N of circuit	area of buffered region (km^2)							A	N
						0	10	20	30	···	390	400		
Japanese	1	1	6.9	0.76	5	0	0.13	0.26	0.38	···	2.93	3.00	0.94	8.83
	2	2	6.7	0.71	10	0	0.13	0.25	0.36	···	2.56	2.62	0.74	11.82
	3	14	12.7	0.69	42	0	0.24	0.43	0.59	···	2.49	2.54	0.70	53.33
	4	2	8.9	0.91	14	0	0.17	0.33	0.48	···	3.18	3.25	1.08	18.03
	5	2	6.2	0.55	9	0	0.12	0.23	0.33	···	2.37	2.43	0.64	11.95
	6	1	9.0	0.82	14	0	0.17	0.33	0.48	···	3.27	3.34	1.13	16.41
	7	2	4.0	0.33	3	0	0.08	0.15	0.22	···	2.14	2.19	0.53	2.84
	8	2	4.4	0.47	3	0	0.09	0.17	0.25	···	2.25	2.31	0.58	9.36
	9	1	3.2	0.29	2	0	0.06	0.13	0.18	···	1.86	1.91	0.40	4.04
	10	2	7.2	0.78	8	0	0.14	0.28	0.41	···	2.88	2.94	0.91	8.39
	11	2	5.0	0.44	4	0	0.10	0.18	0.26	···	2.37	2.43	0.65	8.26
	12	20	14.6	2.95	14	0	0.29	0.56	0.83	···	7.35	7.48	3.84	19.34
	13	21	6.9	0.40	18	0	0.13	0.24	0.34	···	2.00	2.05	0.46	23.58
	14	20	4.5	0.42	7	0	0.09	0.17	0.24	···	2.01	2.06	0.47	10.04
	15	2	3.7	0.25	2	0	0.07	0.14	0.21	···	2.00	2.05	0.46	6.58
	16	2	3.9	0.45	8	0	0.08	0.15	0.21	···	1.83	1.88	0.38	8.12
	17	2	7.5	0.90	3	0	0.15	0.29	0.43	···	4.51	4.62	1.91	4.79
	18	20	9.2	0.80	15	0	0.17	0.32	0.45	···	2.69	2.76	0.81	31.46
	19	20	8.3	0.40	24	0	0.15	0.26	0.36	···	2.01	2.06	0.47	27.78
	20	2	4.6	0.55	3	0	0.09	0.18	0.27	···	2.60	2.66	0.76	3.80
	21	2	5.0	0.64	2	0	0.10	0.20	0.29	···	2.85	2.92	0.90	2.65
	22	2	4.5	0.48	9	0	0.09	0.17	0.24	···	1.67	1.71	0.31	10.20
	23	2	12.9	2.42	11	0	0.25	0.49	0.72	···	5.81	5.92	2.76	16.16
	24	16	6.6	0.63	9	0	0.13	0.25	0.35	···	2.86	2.93	0.90	11.95
	25	2	5.7	0.50	13	0	0.11	0.20	0.29	···	1.98	2.04	0.46	16.78
	26	18	15.5	1.80	20	0	0.30	0.59	0.85	···	5.04	5.14	2.24	28.97
	27	4	17.0	2.13	14	0	0.33	0.64	0.94	···	6.77	6.90	3.43	23.33
	28	19	3.0	0.26	5	0	0.06	0.11	0.16	···	1.50	1.54	0.25	5.53
	29	20	7.1	0.79	4	0	0.14	0.28	0.41	···	3.53	3.61	1.29	5.38
	30	21	1.6	0.16	4	0	0.03	0.06	0.08	···	1.08	1.12	0.10	5.93
	31	2	0.9	0.11	0	0	0.02	0.04	0.05	···	0.99	1.03		
	32	20	7.1	0.54	16	0	0.14	0.26	0.36	···	2.53	2.59	0.73	20.09
	33	2	1.4	0.04	5	0	0.03	0.05	0.07	···	0.86	0.90	0.05	5.36
	34	3	9.5	0.80	13	0	0.18	0.36	0.52	···	3.27	3.35	1.14	14.96
	35	3	7.9	1.28	6	0	0.15	0.30	0.45	···	3.44	3.51	1.23	8.86
Brazilian	36	8	5.1	1.03	1	0	0.10	0.20	0.30	···	3.74	3.84	1.43	2.59
	37	1	1.9	0.38	0	0	0.04	0.08	0.12	···	1.90	1.96		
	38	7	1.1	0.35	0	0	0.02	0.04	0.06	···	1.16	1.20		
	39	1	0.6	0.05	0	0	0.01	0.03	0.04	···	0.94	0.98		
	40	7	0.4	0.03	0	0	0.01	0.02	0.03	···	0.73	0.76		
	41	7	0.6	0.03	0	0	0.01	0.02	0.04	···	0.78	0.81		
	42	8	5.3	0.70	0	0	0.11	0.21	0.32	···	4.23	4.33		
	43	8	1.6	0.03	3	0	0.03	0.05	0.07	···	0.91	0.95	0.06	6.03
	44	6	1.8	0.07	1	0	0.04	0.07	0.11	···	1.54	1.59	0.27	2.88
	45	9	1.9	0.11	0	0	0.04	0.08	0.11	···	1.80	1.86		
	46	10	1.2	0.18	0	0	0.02	0.05	0.07	···	1.32	1.37		
	47	8	1.8	0.38	0	0	0.04	0.07	0.11	···	1.74	1.80		
	48	2	1.6	0.12	0	0	0.03	0.06	0.10	···	1.56	1.61		
	49	2	1.6	0.12	0	0	0.03	0.06	0.10	···	1.56	1.61		
	50	8	3.6	0.32	2	0	0.07	0.14	0.20	···	1.87	1.92	0.40	3.66
	51	7	1.7	0.05	2	0	0.03	0.06	0.09	···	1.41	1.46	0.21	2.95
	52	2	0.9	0.03	0	0	0.02	0.03	0.05	···	0.85	0.88		
	53	2	1.8	0.15	0	0	0.04	0.07	0.11	···	1.69	1.74		
	54	2	1.0	0.05	0	0	0.02	0.04	0.06	···	1.05	1.09		

Since the sketch maps like that in Fig. 1 cannot be directly analyzed by GIS, it is necessary to trace the roads and landmarks drawn in Fig. 1 onto a 1:2,500 scale digital map, as shown in Fig. 2. The polygons consisting of line segments among the landmarks and roads on the edge of traced map are subject to measurement by the area method. In the figure, ①~⑤ indicate the domains surrounded by roads. We simply call this the domain *circuit*.

After tracing all the sketch maps on digital maps, the entire length of roads, area of drawn range, number of circuits, and domain areas for each buffer distance are measured for each map (Table 1). Using these data, the entire road length method, area method, buffer methods are applied as follows.

4.1 Entire Road Length Method

This method is effective in quantitative analysis of linear space when the entire length of the drawn roads is measured. The longer the entire length, the larger the cognitive space is assumed to be. However, even if the entire road length was the same in two maps, it does not necessarily mean that the forms are the same. For example, the entire road lengths are almost the same in sketch map 43 and sketch map 47 in Fig. 3. However, map no. 43 is a survey-type map with the roads drawn densely in a small area, while no. 47 is a route-typ map with two roads simply drawn crossing (Fig. 3). As is shown here, the entire road length method is unable to distinguish the forms of sketch maps.

4.2 Area Method

The area method is effective in quantitative analysis of planar space. It is natural to assume that the drawn area of a planar survey map is larger than that of a linear route map. However, even if only a small area is drawn, the drawing style can be planar. For example, no. 43 in Fig. 3, as described in the previous section, is a survey-type map with quite small area. No. 47 is a route-type sketch map, but the drawn area tends to be overestimated as two roads cross perpendicularly. Thus, the area method is also unsuitable to distinguish route-type and survey-type maps.

4.3 Buffer Method

As mentioned before, both the entire road length method and area method can be used to analyze sketch maps quantitatively, but they both have problems. As a means to overcome these problems, we show a case study employing the buffer method.

We denote a set of all points on the roads drawn on a sketch map by P. Forty buffered regions are obtained by drawing buffered regions $B(P, 10\text{m})$, $B(P, 20\text{m}), \cdots$, $B(P, 400\text{m})$ in buffer distance h=10m, 20m,$\cdots$, 400m using GIS (because buffering is performed in 10m intervals between 10m and 400m). We calculate the area of each region, and obtain a scatter-plot graph where h is on the X axis, and areas of $B(P, h)$ are on the Y axis. The reason for changing the buffer distance in 10m intervals is that most of the blocks in the actual urban areas are larger than 10m square. So, even if we use the intervals shorter than 10m, the scatter-plot pattern is unlikely to differ.

Following the above procedure, we worked out graphs based on the sketch maps by Japanese and Brazilian subjects. Table 1 shows some of the areas of buffered regions that were obtained by moving buffer distance from 10m to 400m in 10m intervals. Figure 4 is a graph of sketch maps by 35 Japanese resulting of the above procedure. Each line corresponds to an individual sketch map by Japanese. The same procedure was applied to the sketch maps by the 19 Brazilians (Fig. 5).

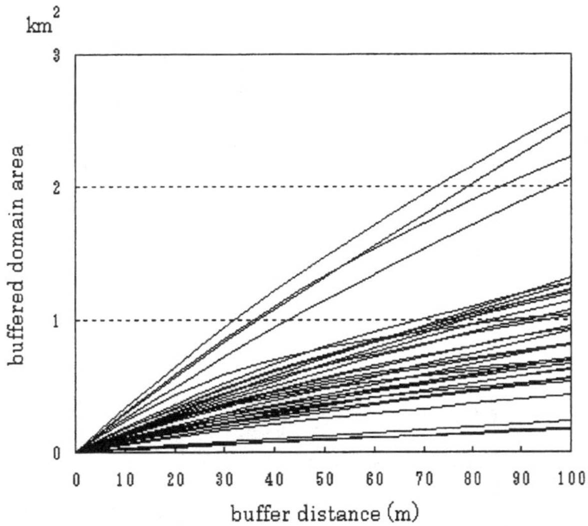

Fig. 4. The transition of buffered areas in each sketch map (Japanese)

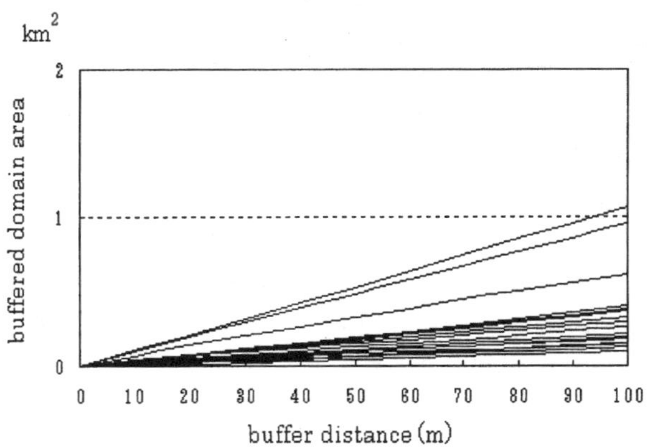

Fig. 5. The transition of buffered areas in each sketch map (Brazilian)

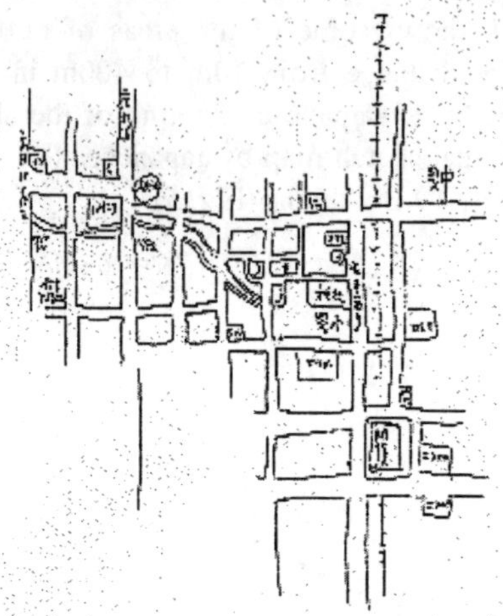

Sketch map 27

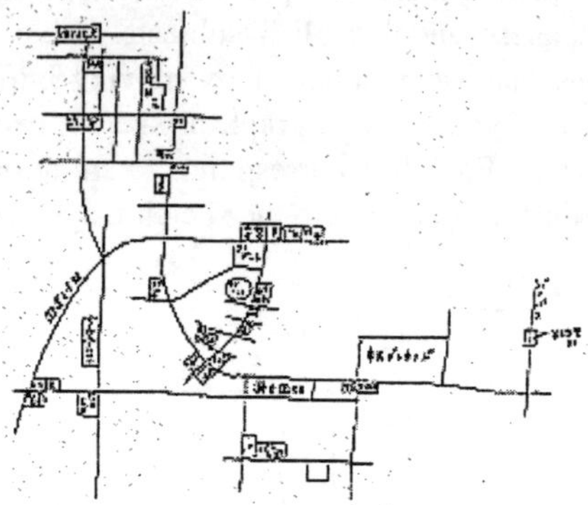

Sketch map 12

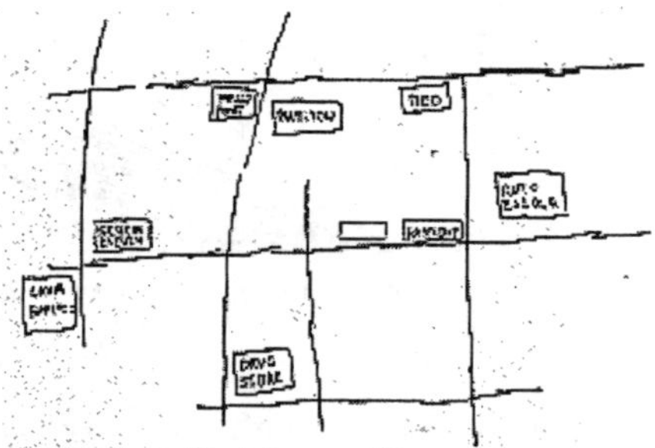

Sketch map 50

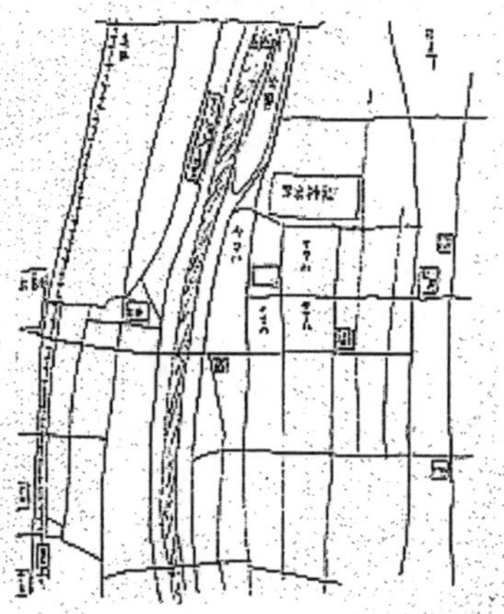

Sketch map 19

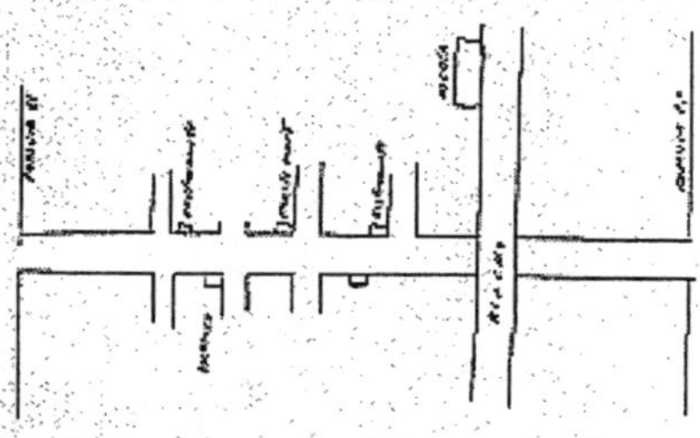

Sketch map 45

Fig. 6. Typical sketch maps that produce graph pattern 1-5. The numbers of the sketch maps are same as in Table 1

On these graphs, we can see visually how the areas of the buffered regions increase with an increase in buffer distance. We related every line on the graph with the forms of sketch maps, paying special attention to the inclination and the buffer distance that gives the greatest impact on the inclination, and found several patterns in the forms of graph showing the transition of the areas of buffered regions.

Pattern 1: The inclination at the origin is relatively large. The graph shows a narrow curve when the buffer distance is short. Sketch map that produce this form are survey-type maps, having large drawn area, and containing many circuits (Sketch map 27 in Fig. 6, for example).

Pattern 2: The inclination at the origin is in the mid-range. The graph curves moderately with increase in buffer distance. Sketch maps that produce this pattern of graph are survey-type maps, having large drawn area, but relatively few circuits (Sketch map 12 in Fig. 6, for example).

Pattern 3: The inclination at the origin is in the mid-range. The graph shows a narrow curve when the buffer distance is short. The drawn area in the sketch maps that produce this type of graph is small, but there are many circuits (Sketch map 19 in Fig. 6, for example).

Pattern 4: The inclination at the origin is relatively small. The graph curves moderately as the buffer distance increases. The drawn area in the sketch maps that produce this pattern is small, and the circuits are few (Sketch map 50 in Fig 6, for example)

Pattern 5: The graph extends in almost at same rate, or at a slightly increasing rate like a convex function. Sketch maps that produce this type of graph are route-type maps with no circuit (Sketch map 45 in Fig. 6, for example).

As shown the above, the forms of graphs can be criteria to roughly distinguish between survey-type or route-type sketch maps. Graph that trace an arc like a concave function (Pattern 1~Pattern 4) indicate a survey-type map, and graphs with a little change in inclination that trace an arc like a convex function (Pattern 5) indicate a route-type map. Also, with survey-type maps, once the graphs are distinguished as one of Pattern 1~Pattern 4, the forms of sketch maps can be classified more strictly.

Although the method is more objective compared with classification directly by subjective impression of sketch maps, it still retains some ambiguity in that it gives visual references, such as the forms of graphs. In order to develop a more objective method, mathematical consideration of the relationships between the interval of buffer distance and the increase rate of the area of buffered region (buffered area) is necessary. In the following chapter, we will attempt a basic mathematical consideration.

5 Modeling the Buffer Method

5.1 Relationship Between Buffer Distance and Buffered Area (Route Map)

A route map is a representation structured by movement along a road, and a survey map is a collective representation consisting of relationships among multiple locations. Also, a survey map is not an ego-centered representation (Hart and Moor, 1973). Given these characteristics, it is possible to define a representation without circuits as a route map, and a representation with circuits as a survey map. There is only one route between two locations on a network with no circuit, whereas there are plural routes on a network with one or more circuit, and plural routes produce survey knowledge. Therefore, we define a sketch map with no circuits as a route-type map, and a sketch map with one or more circuits as a survey-type map.

In the modeling of buffer method, buffering on a line with length *k*, which is the simplest route map, is considered here. In this case, the buffered area *Y*(*X*) in buffer distance *X* is

$$Y(X) = \pi X^2 + 2kX . \tag{1}$$

This is a quadratic curve, and its rate of increase always increases for $X >= 0$. Therefore, as described in the previous section, the graph draws an arc of a convex function.

Fig. 7. A route which winds vertically by a move in length *k*

However, when a route turns vertically *m* times with every move in length *k*, as shown in Fig. 7, the buffered area in buffer distance *X* is

$$Y(X) = (\frac{\pi - 4}{4} m + \pi) X^2 + 2k(m+1)X . \tag{2}$$

In this case, a coefficient of X^2 is negative for *m*>=15, when the graph traces an arc of a concave function. Consequently, the form of graph does not identify whether the sketch map is a route map. We avoid this problem by setting a condition with a route map in which the number of circuits is 0.

5.2 Relationship Between Buffer Distance and Buffered Area (Survey Map)

Here a sketch map such as that in Fig. 8 is considered. This map consists of streets in a lattice-like form in which *a* length square is divided into *n* x *n* grids. *a* indicates the range of the drawn area, and n indicates the density of a drawn road network; that is, elaboration of the sketch map. a^2 corresponds to the area of the drawings, and n^2 corresponds to the number of circuits. When n=1, the number of circuits is 1, and the figure becomes a square loop road without inner branches.

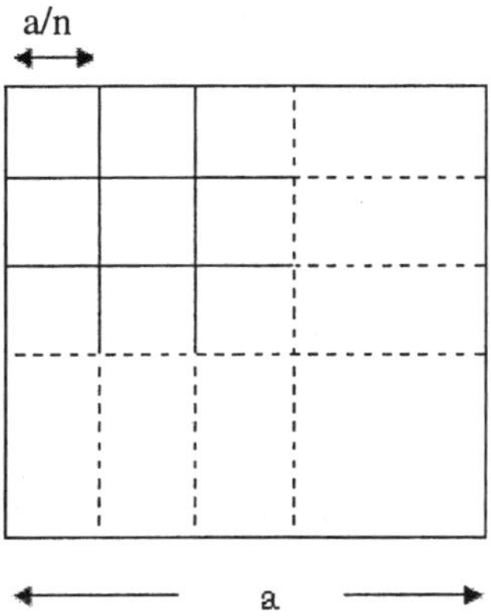

Fig. 8. Survey map with n×n grids

Buffering is performed on these roads. When the buffer distance is *a*/2*n,* the sections between the grids are fulfilled. Therefore, the area of buffered region *Y(X)* is written as follows:

$$Y(X)=\begin{cases} -(4n^2-\pi)X^2+4a(n+1)X, & \text{for } X \le a/2n \\ \pi X^2+4aX+a^2, & \text{for } X \ge a/2n. \end{cases} \tag{3}$$

The first derivative of *Y(X)* is

$$\frac{dY(X)}{dX}=\begin{cases} -2(4n^2-\pi)X+4a(n+1), & \text{for } X \le a/2n \\ 2\pi X+4a, & \text{for } X \ge a/2n. \end{cases} \tag{4}$$

And the second derivative is

$$\frac{d^2Y(X)}{dX^2}=\begin{cases} -2(4n^2-\pi), & \text{for } X \le a/2n \\ 2\pi, & \text{for } X \ge a/2n. \end{cases} \tag{5}$$

Since n is a natural number in the above, $(4n^2-\pi)$ >0 is maintained. From equation (3), we find that *Y* (*X*) changes as X increase as in Fig. 9.

The typology described in chapter 4 is supported by equations (4) and (5). That is,

when $\frac{dY(0)}{dX}=4a(n+1)$, $X : a/2n$, we have $\frac{d^2Y(X)}{dX^2}=-2(4n^2-\pi)$.

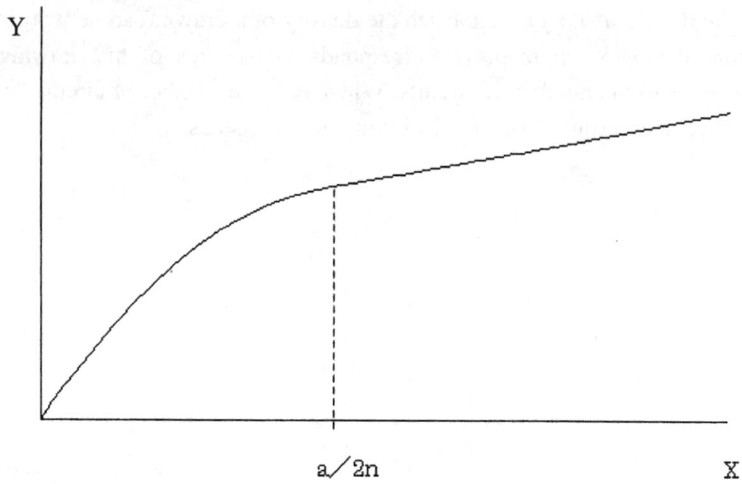

Fig. 9. Relationship between buffer distance and buffered area in survey map

Therefore, when both a and n are large, the inclination at the origin is relatively large. This produces a graph that curves steeply as buffer distance increases (Pattern 1). When a is large and n is small, the inclination at the origin is moderate, and the graph curves moderately as buffer distance increases (Pattern 2). When a is small and n is large, the inclination at the origin is moderate, and the graph curves teeply while buffer distance is small (Pattern 3). When both a and n are small, the inclination at the origin is relatively small, and the graph shows a moderate curve as buffer distance increases (Pattern 4).

5.3 Expected Values for Area, Number of Circuits, and Entire Road Length

From formula (3), the following equation is obtained when buffer distance X is large enough:

$$a^2 = (8-\pi)X^2 + Y(X) - 4X\sqrt{(4-\pi)X^2 + Y(X)}. \tag{6}$$

Therefore, the following index is obtained:

$$A(X, Y(X)) = (8-\pi)X^2 + Y(X) - 4X\sqrt{(4-\pi)X^2 + Y(X)}. \tag{7}$$

When buffer distance is small in equation (3),

$$n^2 = -\frac{1}{8}\frac{d^2Y(X)}{dX^2} + \frac{\pi}{4}. \tag{8}$$

Therefore, the following index is also obtained:

$$N\left(\frac{d^2Y(X)}{dX^2}\right)=-\frac{1}{8}\frac{d^2Y(X)}{dX^2}+\frac{\pi}{4}. \tag{9}$$

A and *N* are indices for a^2 and n^2 in Fig. 1 respectively. Thus, *A* indicates the drawn area (the extent of cognitive space) and *N* indicates the number of circuits.

A and N correspond to the expected values for an area and number of circuits obtained by the buffer method.

5.4 Relationship Between Expected and Actual Values

Using the data for buffer distance and buffered areas in Table 1, the expected values of the drawn areas and the number of circuits are obtained by (7) and (9) for each sketch map:

$$A(400,Y(400))=(8-\pi)\times 400^2+Y(400)-4\times 400\sqrt{(4-\pi)\times 400^2+Y(400)}\,, \tag{10}$$

$$\begin{aligned}N\left(\frac{d^2Y(X)}{dX^2}\right)&=-\frac{1}{8}\frac{d^2Y(X)}{dX^2}+\frac{\pi}{4}\\&\approx-\frac{1}{8}\frac{1}{\Delta X}\left(\frac{dY(X+\Delta X)}{dX}-\frac{dY(X)}{dX}\right)+\frac{\pi}{4}\\&\approx-\frac{1}{8}\frac{1}{\Delta X}\left(\frac{Y(X+2\Delta X)-Y(X+\Delta X)}{\Delta X}-\frac{Y(X+\Delta X)-Y(X)}{\Delta X}\right)+\frac{\pi}{4}\\&=-\frac{1}{8}\frac{1}{\Delta X^2}(Y(X+2\Delta X)-2Y(X+\Delta X)+Y(X))+\frac{\pi}{4}.\end{aligned} \tag{11}$$

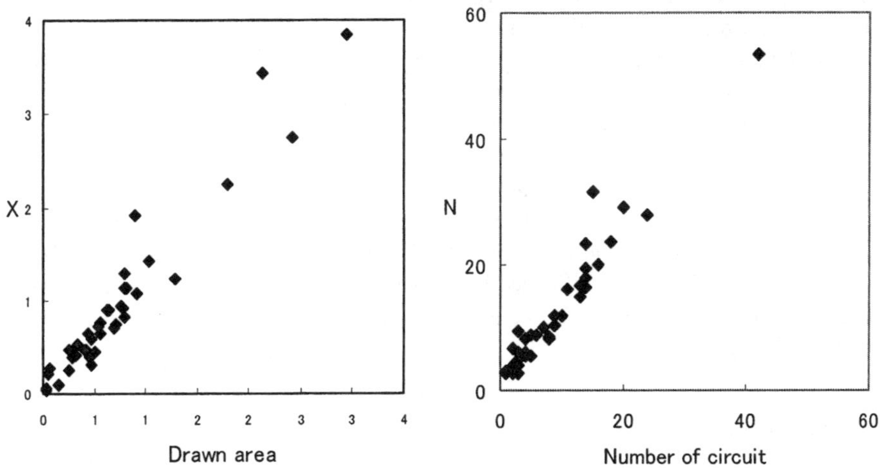

Fig. 10. Relationship between actual values and expected values. Left: drawn area and A. Right: number of circuit and N

When we depict the expected values obtained above and the actual values of areas and number of circuits that were measured directly from the sketch maps onto a correlation diagram, strong correlations are seen for two values (Fig. 10). This suggests the effectiveness of the buffer method, and that the model in Fig. 8 appropriately typifies a wide variety of survey-type sketch maps. Parts of the procedure shown in Fig. 2 as a measure of calculating the drawn area remain ambiguous; however, the buffer method can measure the drawn areas more objectively.

6 Examination of Micro-genetic Development

In this chapter, we examine micro-genetic development by looking into the relationships between the forms of sketch maps and the length of current residence of those who drew the maps.

The first subject for inquiry is the development from a route map to a survey map. Based on the definition that the number of circuits is 0 in a route map, only one of 35 sketch maps by Japanese corresponded to a route map. On the other hand, 14 out of 19 sketch maps by Japanese-Brazilians are route-type, and only 5 are survey-type. The reasons many Brazilians do not obtain survey-type knowledge are considered to be a limited proficiency in reading the Chinese characters used in Japan, a lack of landmark knowledge attributed to difficulties in learning the characteristics of Japanese buildings, and inexperience with the Japanese address indication system (Takai, 2004). An additional reason may be that their area of activity is small and daily travel patterns have relatively little variation compared to Japanese university students.

The Japanese-Brazilians can be roughly divided into two groups: those who have resided in Japan less than 3 years and those who have resided 6 or more years (Table 1). All the individuals in the former group drew route-type maps, while 5 out of 12 individuals in the latter drew survey-type maps with one or more circuits. This shows a tendency for micro-genetic development.

The Japanese subjects can be divided into a group that has resided in the current location for less than 4 years, that is, the student group who moved to Hamamatsu city to go to university, and a group with residence in the city for 10 or more years. The only student who drew a route-type map belonged to the former group. For the remaining 34 individuals, we examined the difference in the forms of survey maps for the short residence group and long residence group. A Wilcoxon rank-sum test for A and N as in section 5.4. indicated no significant difference in A, but a significant difference in N at a 5% significance level. This result suggests that the drawn area does not become larger but the contents become more elaborate with increase in length of residence.

7 Conclusion

We herein proposed a method to analyze sketch maps by GIS, and applied it in an actual case study. In addition to the entire road length method for analysis of linear space and the area method for analysis of planar space, we introduced a buffer method to analyze sketch maps to which the former two methods are not applicable. GIS analysis of sketch maps was practiced by applying each method to the analysis of the neighborhood sketch maps drawn by Japanese and Japanese-Brazilians. As a result, classification of route maps and survey maps, and description of their characteristics including drawn range and elaboration was done according to number of circuits and the values of A and N obtained by buffer method. The findings indicate that the buffer method with GIS is effective as a method for sketch map analysis, which enables automatic distinction of the forms of sketch maps. A specific advantage of the buffer method was the ability to obtain easily and objectively the approximate value of the area drawn in survey map. We used these values to analyze the micro-genetic development shown in sketch maps.

In this study, we traced the roads drawn in the sketch maps onto a digital map, and applied buffering on the roads. This method focused especially on paths among other elements of a cognitive map. Further examination of the method by applying it to other elements such as landmarks and edges is needed. In order to develop a more versatile model, which is applicable to a greater variety of relationships in buffer distance and buffered areas, further mathematical consideration is also needed.

References

Appleyard, D.(1970): Styles and methods of structuring a city. *Environment and Behavior*, 2, pp.100-117.

Bailey, T.C. and Gatrell, A.C. (1996): *Interactive Spatial Data Analysis.* Longman Scientific & Technical, Harlow.

Blades. M.(1990): The Reliability of Data Collected From Sketch Maps. *Journal of Environmental Psychology,* 10, pp 327-339.

Blaser, A. and Egenhofer, M. (2000): A visual tool for querying geographic databases. Di Ges, V., Levialdi, S. and Tarantini, L. eds.: *AVI2000--Advanced Visual Databases*, Salerno, Italy, pp. 211-216.

Clarke, K.C.(2002): *Getting Started with Geographic Information Systems, 4th edition*, Prentice Hall, Upper Saddle River.

Forbus, K., Usher, J. and Chapman, V. (2003). Qualitative spatial reasoning about sketch maps. *Proceedings of the Fifteenth Annual Conference on Innovative Applications of Artificial Intelligence*, Acapulco, Mexico.

Golledge, R.G.(1993): Geographical perspectives on spatial cognition. Garling, T. and Golledge, R.G. eds.: *Behaviour and Environment: Psychological and Geographical Approaches.* Elsevier Press, pp.16-46.

Golledge, R.G.(1995): Primitives of spatial knowledge. Nyerges, T.L., Mark, D.M., Laurini, R., and Egenhofer, M.J. eds.: *Cognitive Aspects of Human-Computer Interaction for Geographic Information System (NATO ASI Series D83).* Kluwer Academic Publishers, pp.29-44.

Hart, R.A.(1981): Children's spatial representation of the landscape: lessons and questions from a field study. Liben, L.S., Patterson, A.H. and Newcombe, N. eds.: *Spatial Representation and Behavior across the Life Span*, Academic Press, pp.195-233.

Hart, R.A. and Moore, G.T.(1973): The development of spatial cognition: a review. Downs, R.M. and Stea, D. eds.: *Image and Environment: Cognitive Mapping and Spatial Behavior.* Ardine, pp.246-288.

Humphreys, J.(1989): Place Learning and spatial cognition: a longitudinal study of urban newcomers. *Tijdschrift voor Economische en Sociale Geografie,* 81, pp.364-380.

Kuhn, W., Worboys, M. and Timpf, S. eds.,(2003); *Spatial Information Theory: Foundations of Geographic Information Science. Conference on Spatial Information Theory (COSIT) 2003.* Springer: Berlin.

Lee, Y. and Schmidt, C.(1988): Evolution of urban spatial cognition: patterns of change in Guangzhou, China. *Environment and Planning* A 20, pp.339-351.

Lloyd,R.(1997): *Spatial Cognition: Geographic Environments.* Kluwer Academic Publishers.

Lynch, K.(1960): *The Image of the City.* MIT Press.

Matthews, M.H.(1992): *Making Sense of Place: Children's Understanding of Large-Scale Environments.* Harvester Wheatsheaf.

Montello, D.R.(1998): A new framework for understanding the acquisition of spatial knowledge in large-scale environments. Egenhofer,M.J. and Golledge. R.G. eds.: *Spatial and Temporal Reasoning in Geographic Information Systems.* Oxford University Press, pp.143-154.

Nyerges, T.L., Mark, D.M., Laurini, R., and Egenhofer, M.J.(1995b): Cognitive aspects of HCI for GIS: an introduction. Nyerges, T.L., Mark, D.M., Laurini, R., and Egenhofer, M.J. eds.: *Cognitive Aspects of Human-Computer Interaction for Geographic Information System (NATO ASI Series D83).* Kluwer Academic Publishers, pp.1-8.

Orleans, P.(1973): Differential cognition of urban residents: effect of social scale on mapping. Downs, R. and Stea, D. eds.: *Image and Environment: Cognitive Mapping and Spatial Behavior* Aldine, pp.115-130.

Schuurman, N.(2000): Critical GIS: Theorizing an emerging science. *Carto- graphica Monograph*, 53, pp.1-107.

Takai, T.(2004): Spatial cognition of Japanese-Brazilians in Japanese urban space. *Geographical Review of Japan*, 77, pp.523-554. (in Japanese)

Wakabayashi, Y.(1994): Spatial analysis of cognitive maps. *Geographical Reports of Tokyo Metropolitan University*, 29, pp.57-102.

Wakabayashi, Y.(2003): Spatial cognition and GIS. *Geographical Review of Japan*, 76, pp.703-724. (in Japanese)

Imagined Perspective–Changing Within and Across Novel Environments

Marios N. Avraamides[1] and Jonathan W. Kelly[2]

[1] Department of Psychology, University of Cyprus, P.O. Box 20537, 1678 Nicosia, Cyprus
mariosav@ucy.ac.cy
[2] Department of Psychology, University of California, Santa Barbara, CA 93106-9660, USA
j_kelly@psych.ucsb.edu

Abstract. Results suggesting that changing perspective and switching across spatial environments held in memory are processes that take place in parallel were obtained from a task-switching experiment. Participants learned layouts of objects in two virtual rooms and then were asked to use their memories to locate the objects from various imagined viewing perspectives. Results revealed that, even after experiencing multiple perspectives, participants maintained viewpoint-dependent memories for the layouts, and that the latencies for changing perspective within and across environments followed a different pattern depending on whether participants imagined adopting the preferred view.

1 Introduction

Everyday tasks, such as giving and taking route directions, often require imagined navigation within spaces that are not perceptually available. In order to comprehend route directions, for example, we often perform a mental simulation of our movement, imagining ourselves passing through various neighborhoods and intersections that are held in memory.

Past research on spatial memory has suggested that we represent the world in our memory in a hierarchical fashion. That is, we parse the world into smaller meaningful units, creating thus a collection of separate representations (Hirtle & Jonides, 1985; McNamara, 1986). A hierarchy of representations is then created by linking these representations to higher-order representations (e.g., the separate representations for the rooms of a building are connected to a representation for the building).

Supporting accounts of hierarchical organization of spatial memory, a number of studies (e.g., Stevens & Coupe, 1978) have shown that our spatial memory is often biased by the use of information stored at a superordinate level in the hierarchy. Furthermore, other studies have provided evidence for the presence of multiple separate spatial representations. McNamara (1986), for example, has demonstrated that spatial judgments about targets in the same spatial region are faster than judgments about targets in different spatial regions, suggesting that the knowledge is organized in separate spatial representations.

C. Freksa et al. (Eds.): Spatial Cognition IV, LNAI 3343, pp. 245 – 258, 2005.
© Springer-Verlag Berlin Heidelberg 2005

The hierarchical organization of spatial memory also predicts that mentally switching from one spatial representation to another entails some cognitive effort, since switching across representations requires activating a new spatial representation and possibly inhibiting the previous one.

Indeed, past research has demonstrated this cost for switching between mental representations of different environments. In a study using the cued task-switching paradigm, Brockmole and Wang (2002) had college professors locate, from memory, objects in two nested environments (their office and the building in which their office was located). The subjects were physically located in a separate environment, and their imagined perspective was held constant. Trials probed unpredictably one or the other environment, thus producing a sequence of trials in which the probed environment was the same or different from that of the previous trial while the imagined global perspective was held constant. Results revealed that participants were slower on trials that required mentally switching to a new environment.

However, performing a spatial task such as imaginally following route directions requires not only mentally switching across environments, but also mentally shifting perspective within them. The present study examines the relation between the processes of mentally switching from one imagined environment to another and that of changing imagined perspective.

Prior research on imagined navigation has shown that, even in the absence of landmarks, changing perspective is time demanding (e.g., Avraamides & Carlson, 2003). This is, of course, expected if mental operations are believed to evoke the perceptual-motor mechanisms that apply to real situations (Jeannerod, 1995; Kosslyn, 1994; Wexler, Kosslyn, & Berthoz, 1998). Furthermore, when responses such as locating objects from imagined perspectives are mandated, additional cognitive effort -- and hence time -- is needed. In order to be able to localize a target after having moved (physically or mentally) to a new perspective, one needs to update that target's position relative to one's self. A number of studies have established that doing so is harder when the movement is imagined rather than real (e.g., Avraamides, 2003; Klatzky, Loomis, Beall, Chance, & Golledge, 1998; Presson & Montello, 1994). Furthermore, Rieser (1989) has shown that while the latency for locating objects after imagined rotations of the viewer increases monotonically with the extent of the rotation, it remains constant with physical rotations.

While both perspective-changing and environment-switching are associated with temporal costs, it is unclear whether those costs are independent of each other. If, for example, one adopts a new imagined perspective in a new environment, is the total temporal cost simply the sum of the two costs? This should be expected if there are two processes that take place sequentially. Alternatively, if switching environment and changing perspective can be carried out in parallel, then the total time needed to do both should depend on the time needed to complete the more difficult of the two. To be more precise, a race model (e.g., Logan, 1988) would predict that the time needed to perform both mental actions should be, on average, somewhat longer than the duration of the longest action, given that the distributions of latencies for the two actions are likely to overlap. Finally, a third possibility is suggested by Brockmole and Wang (2003): the cost for changing perspective could be smaller when switching environments than when remaining in the same environment.

Evidence for the latter possibility was provided by Brockmole and Wang (2003). Using a similar procedure to their earlier study, they had participants judge from memory the relative locations of targets from differing imagined perspectives in two familiar environments. In a within-environment block, trials probed targets from only one of the environments but randomly from two different perspectives. This produced a string of trials in which the perspective could be the same or different from the preceding trial. In an across-environment block, each of the two environments was fixed to a specific (but unique) perspective. Trials probed objects from the two environments in a random manner, thus producing a sequence of trials in which the environment (and, as a result, the perspective) could have either remained the same or changed from trial to trial.

Results revealed that the cost for changing perspective (operationalized as the latency difference between same and different perspective trials in each environment condition) was smaller in the across-environment than in the within-environment condition. The authors interpreted this finding as evidence for distinct processes governing perspective change within and across environments and proposed an interference account to explain it. Specifically, they argued that when changing perspective within a single environment, conflicts occur between the targets' current and previous relative locations. Such conflicts do not exist when changing perspective across environments (because the targets are different in the current and previous trial), making it faster to change perspective across environments.

The results of Brockmole & Wang (2003) could have important implications for studies of spatial updating. Spatial updating refers to the process of computing the changed egocentric locations of objects that result from observer or object movement. While spatial updating is rather effortless during physical movement, this process is cognitively demanding and is carried out in a backward manner (i.e., after the movement is completed) when the movement is only imagined (e.g., Avraamides, 2003; Rieser, Guth, & Hill, 1986). The inferior spatial performance associated with responding from perspectives adopted after imagined movement is typically attributed to the absence of vestibular feedback, proprioceptive information, and/or efferent copy during imagined movements. Recently, an alternative account has been proposed to explain why spatial performance is inferior during imagined compared to physical movement. Specifically, May (1996, 2004; see also Avraamides, Klatzky, Loomis, & Golledge, 2004, and Wraga, 2003) has suggested that the difficulty (or at least part of the difficulty) when responding from imagined perspectives might be due to conflicts between the physical and imagined egocentric locations of objects. When an observer attempts to localize a target from an imagined position that differs from his/her actual physical position, s/he must only consider object locations relative to his/her imagined position and disregard their locations relative to his/her physical position. As the results of May (2004) and Avraamides et al. (2004) show, suppressing the physical egocentric locations of targets is not always an easy task.

The findings of Brockmole and Wang's (2003) suggest another explanation for the inferior performance evidenced during imagined movements. Specifically, their interference account posits that the problems with reasoning from imagined perspectives might be due (completely or partially) to interference from the objects' previous imagined locations. Indeed, if a spatial updating experiment requires that the participant adopt a different imagined perspective in every trial then an imagined target-object will have a different relative location on every trial. If the imagined location of a target-object is subjected to interference from its imagined location in the previous trial or trials, then performance in the experiment overall will be hindered. This interference account is similar to that proposed by May (1996; 2004) but differs from it in that the proposed conflicts are between the previous and current imagined locations of objects and not between the objects' current physical and imagined locations.

While this interference account is plausible, it is based solely on the study by Brockmole and Wang (2003), which had one important design limitation: because each environment in the across-environment block was fixed to a specific yet unique perspective, a sequence of same perspective trials could only occur if the environment also remained unchanged. In essence, there were no true same-perspective/switch-environment trials.

Furthermore, one finding from that experiment is particularly noteworthy: Latencies were substantially smaller in the across-environment than the within-environment block for both perspective conditions. The interference account cannot explain why maintaining the same perspective was faster in the across condition than the within condition. Maintaining the same perspective in the same environment should not produce any conflicts and it is also free of the temporal cost of environment switching. Intuitively, one should predict this to be the fastest condition.

The present study reexamined the relation between changing perspective and switching environment using a design that allows for the independent manipulation of the two variables. Participants first learned two environments and then performed target-localization trials which probed targets, in a random fashion, from the two environments and four different perspectives. Because spatial updating studies are typically carried out in novel settings (e.g., Farell & Robertson, 1998; Mou, McNamara, Valiquette, & Rump, in press; Loomis, Lippa, Klatzky, & Golledge, 2002), and because our interest is to evaluate the interference hypothesis in the context of spatial updating, the present study uses novel scenes rather than highly familiar ones.

2 Method

2.1 Participants

A total of 20 students participated in the experiment. Nineteen (8 male) participants were students of introductory psychology classes at the University of California, Santa Barbara who participated in exchange for course credit. One female participant was a graduate student in psychology who volunteered to participate.

2.2 Stimuli and Apparatus

The experiment was divided into two phases: a learning phase during which participants studied two virtual rooms, and a testing phase in which their spatial performance was tested.

2.2.1 Learning Phase

Participants experienced the virtual rooms via a Virtual Research V8 head-mounted display (HMD; a stereoscopic display with dual 680 × 480 resolution LCD panels that refresh at 60 Hz). The projectively correct stereo display was rendered by a 2.2 GHz Pentium 4 processor computer with a GeForce 4 graphics card using the Vizard (www.worldviz.com) software package. The simulated viewpoint was continually

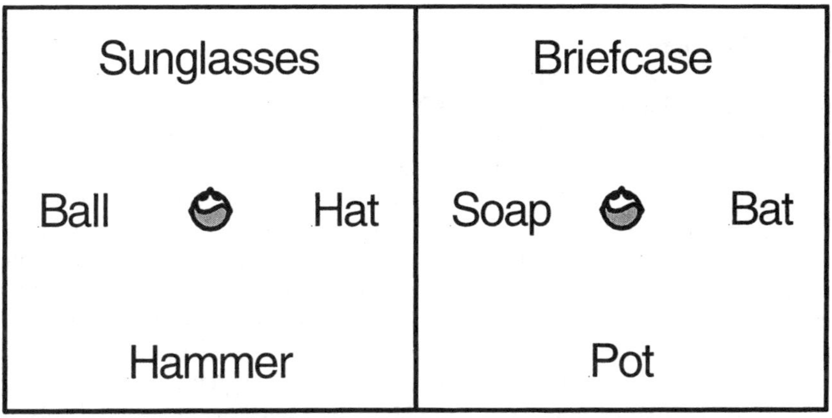

Fig. 1. Schematic diagrams of the layouts of objects in the two virtual rooms

Fig. 2. A view of the brick-walled room

updated by the participant's head movements. The orientation of the participant's head was tracked by a three axis orientation sensing system (Intersense IS300), and the location of the participant's head was tracked three-dimensionally by a passive optical position sensing system (developed in-house and capable of measuring position with a resolution of 1 part in 30,000, or approximately 0.2 mm in a 5 m square workspace). The system latency, or the amount of delay between a participant's head or body motion and the concomitant visual update in the HMD, was 42 ms maximum.

Two virtual rooms, each with unique wall textures and objects, were used in the learning phase of the experiment (Figure 1). Each virtual room contained four objects that were placed on 1.2 m tall pillars. The four objects were spaced evenly around the subject (every 90°).

The names of the objects appeared printed on the front of the pillars. One of the rooms appeared to be made of bricks (Fig. 2) and the other appeared to be made of wood (Fig. 3). This was done to help subjects form distinct representations of the two rooms.

2.2.2 Testing Phase

The testing phase was carried out on a laptop computer, also running the Vizard software package. Participants were presented with statements of the form: "Face: x, Find y", where x and y were objects from the same room, and were asked to imagine facing object x and report the location of object y relative to their facing direction. Responses were collected using four keys on the computer labeled with "F", "B", "L", and "R", corresponding to front, back, left, and right. Timing began with the presentation of the statement and ended when participant pressed a key to indicate an answer. The probed object was actually never directly in front of the imagined perspective, so the "F" response was always incorrect. Therefore, there were three object locations that could be probed from each imagined perspective in each environment. Each participant completed 72 trials total, comprised of 3 repetitions of the 12 possible object-perspective combinations, for the 2 different environments. Trials were presented in a different random order for each participant thus producing trials in which the probed room and/or the facing direction were the same or different from the immediately previous trial[1]. Data from the first trial were considered practice and were discarded.

Fig. 3. A view of the wood-walled room

[1] Because participants experienced the two rooms from the same standpoint and initial facing orientation, we assume that they have noted the correspondence between the two rooms (e.g., left in one room corresponded to left in the other).

2.3 Design

The experiment was a 2 (environment: switch, no switch) × 2 (perspective: same, different) within-subjects factorial design.

2.4 Procedure

Participants stood at a fixed location and orientation in the laboratory and wore the HMD. They were placed into the first of the two virtual rooms (the order of room presentation was counterbalanced across participants) and were given unlimited time to study and memorize the four objects that were visible around them. They were instructed to rotate in place in order to examine the room but not to move to a new position in it. Once participants indicated that they had learned the layout, the experimenter tested their memory. To do so, all objects except one were removed from the room and participants were asked to face the only visible object and then point to each of the remaining three objects as probed by the experimenter. This was repeated until participants pointed correctly to all objects from all possible perspectives. Then, participants were asked to adopt their initial facing direction and were placed into the second room and the same procedure was repeated. Upon completion, they were given the option to revisit the first room and refresh their memory for it.

Once participants reported that they knew both virtual rooms very well, they were led to a different laboratory for the testing phase. They were seated in front of a computer at a random physical facing orientation. This phase of the experiment was conducted on a laptop computer, and it was therefore quite easy to manipulate their physical facing direction. After reading the instructions for the testing phase, they began the 72 test trials. As soon as they completed each trial the next one followed with no delay; that is, as soon as they had pressed a key to respond to a statement, the statement for the next trial appeared. Participants were asked to perform the task as fast as they could but without sacrificing accuracy for speed.

3 Results

Data from one participant were eliminated from all analyses because of low accuracy (<40%). Because accuracy was extremely high in all conditions (92%), latency data for correct trials are the primary focus.

Data from trials in which response latency deviated 3 or more standard deviations from each subject's cell mean were considered outliers and were therefore discarded from all analyses. This resulted in eliminating 1.5 % of correct-response trials.

A preliminary analysis of latencies revealed that memories for the two environments were viewpoint-dependent. Participants were 598 ms faster when judging statements that probed targets from the first view they received upon entering each of the rooms (hereafter referred to as *initial viewpoint*[2]) in the learning phase than from

[2] We use the term *viewpoint* whenever we refer to the distinction between the first view of the room during the learning phase vs. the remaining views. The term *perspective* is used instead when distinguishing trials based on whether they probed the same or different view (compared to the previous trial) during the test phase.

the remaining three perspectives[3] (hereafter referred to as *novel viewpoint*), t(18)=4.16, p<.001. More importantly, the pattern of latencies for switching environment and perspective were different for the initial and novel viewpoints. Therefore, separate repeated-measures analyses of variance (ANOVAs) were performed for each viewpoint condition using environment and perspective as factors.

3.1 Initial Viewpoint

First, an ANOVA was carried out using the data from trials in which participants imagined adopting the viewpoint that coincided with the first view they got upon entering each of the virtual rooms (the initial viewpoint). This analysis revealed that participants performed the task faster when they maintained the imagined perspective of the previous trial (i.e., when the current and previous trials both tested the initial viewpoint) than when they shifted to a new perspective (i.e., when the current trial tested the initial viewpoint and the previous trial tested a novel viewpoint), F(1,15)=7.36, MSE=13.98, p<.05. They were also faster when they remained in the same environment than when they switched, but the difference was not statistically significant.

As seen in Figure 4, changing perspective took longer than switching environment. Furthermore, when participants had to change both environment and perspective, latencies were slightly longer than simply changing perspective, the longer of the two processes.

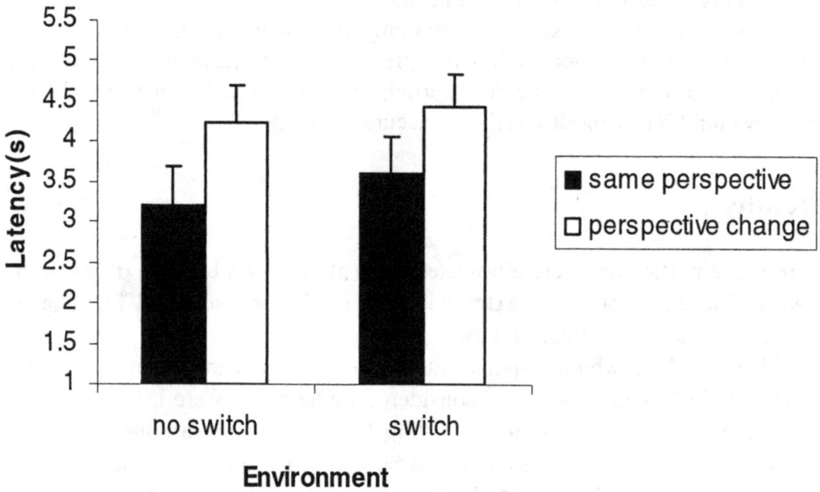

Fig. 4. Initial Viewpoint. Mean latency by perspective and environment condition.Error bars indicate standard errors

[3] Latencies for the remaining three perspectives did not differ significantly from each other and were therefore averaged to form a single novel viewpoint mean.

3.2 Novel Viewpoint

An ANOVA was also carried out for the trials in which participants imagined adopting a view that differed from their first view of the rooms. Results showed that participants were overall faster when they remained in the same environment than when they switched to the other, $F(1,18)=27.67$, $MSE=14.47$, $p<.001$. As in the initial viewpoint case, they were faster when they maintained the perspective of the previous trial than when they adopted a different one, $F(1,18)=13.13$, $MSE=6.58$, $p<.01$. However, this effect was only present for the same-environment condition. This was supported by a significant environment × perspective interaction, $F(1,18)=3.26$, $MSE=6.21$, $p<.05$. In contrast to the results from the initial viewpoint analyses, participants were faster changing perspective within than across environments, $t(18)=-2.59$, $p<.05$. As seen in Figure 5, switching environment was the longest process, and changing both environment and perspective took slightly longer than only switching environment.

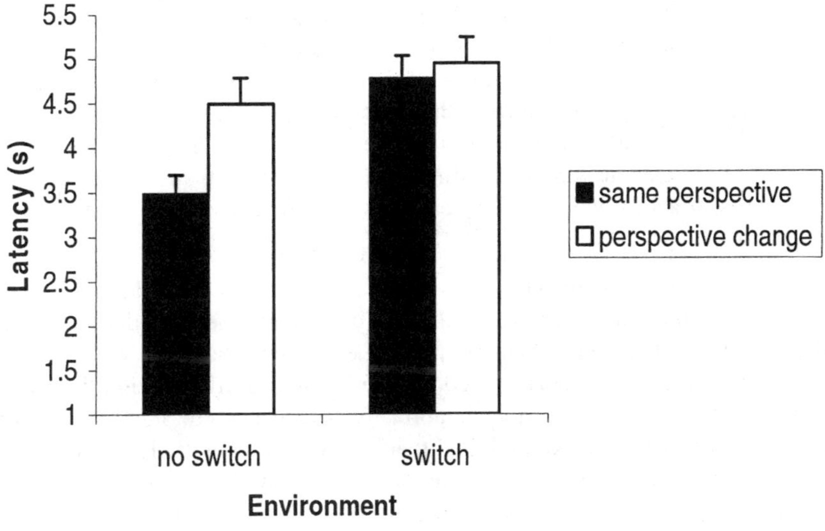

Fig. 5. Novel Viewpoint. Mean latency by perspective and environment condition. Error bars indicate standard errors

4 Discussion

The present experiment examined the relation between mentally switching between imagined environments and changing perspective within imagined environments. The results are most compatible with the race model (a parallel-processing model), which predicts that when one has to both switch environment and change perspective, the total time needed is slightly longer than the duration of the longer of the two. When participants mentally adopted the viewpoint with which their spatial memory was aligned to (i.e., the initial viewpoint), changing perspective was the longest process. In contrast, when they adopted a novel viewpoint, switching environment was longer. In both cases there is clear evidence that the costs associated with perspective changing and environment switching are not additive. Instead, results suggest that changing perspective and switching environment are processes that can take place in parallel.

A striking result in the present experiment was that the latency for maintaining the same perspective across environments was substantially longer when people adopted a novel viewpoint than the original. We believe that this was the case because participants represented the scenes in memory from the initial viewpoint and always referred to these representations whenever they had to either change perspective or switch environment. If that was the case, one should also expect the following:

1. Maintaining the same perspective and changing to a new one across environments should not differ when the novel viewpoint is tested because both conditions would entail consulting the initial viewpoint. Our results clearly support this hypothesis (see Figure 3).

2. Changing perspective should be faster when the previous trial tested the initial viewpoint. That is, there should be an advantage not only when people change perspective to adopt the initial viewpoint, but also when they change perspective away from it. This, however, should be the case only when the two trials probe the same environment. The results from the initial-viewpoint case (Figure 2) are compatible with this hypothesis: same-perspective trials (that is, trials in which both the current and previous viewpoint were the initial) were faster than change-perspective trials (that is, trials in which the current viewpoint was the initial but the previous one was novel).

To test the hypothesis with novel viewpoints we performed an additional analysis. We coded each perspective-change trial based on whether the immediately preceding trial tested a novel or the initial viewpoint. Results revealed that, within the same environment, the mean latencies for both these conditions differed significantly from the same perspective mean[4]. More importantly, latencies were significantly shorter when the previous trial tested the initial than a novel viewpoint[5]. This result was not obtained when there was an environment switch. This additional analysis further supports the hypothesis that participants maintained viewpoint-dependent representations for the two layouts and that these preferred orientations mediated their performance.

It should be pointed out that this was the case despite the fact that during the learning phase participants had experienced the scenes from four different viewpoints and had even performed localization training trials from all four. Despite this extensive experience with novel viewpoints, participants created and maintained mental representations that were aligned with the initial viewpoint. This finding is consistent with McNamara's (2003) theory of spatial memory which posits that people interpret spatial scenes by assigning a reference frame intrinsic to the layout and update their mental representation only if a later view provides a chance for superior encoding

[4] p's$<.01$

[5] $t(18)=2.06$, $p<.05$

(e.g., Mou & McNamara, 2002; Shelton & McNamara, 2001). Extending previous work on viewpoint-dependency (e.g., Christou & Bülthoff, 1999; Richardson, Montello, & Hegarty, 1999), we show here that even elaborate experience with a scene from novel viewpoints does not override the initial encoding of the scene.

Despite the evidence for viewpoint-dependent encoding of the spatial layouts, our data do not speak to the question of what reference frame participants used to organize their memories. Our results are compatible with a number of different hypotheses that have been proposed by other researchers. One possibility is that participants have used an egocentric reference frame to encode the location of objects in relation to their bodies as in the case of "spatial frameworks" (e.g., Franklin & Tversky, 1990), An alternative account is that participants have encoded the allocentric relations of objects using a reference frame that was intrinsic to the scene but was oriented with respect to the first viewing experience (McNamara, 2003). A further possibility is that participants had formed allocentric representations to organize their memory when learning the scenes and imposed an egocentric reference frame when retrieving each spatial relation during the testing phase (Easton & Sholl, 1995).

In summary, our results replicated those of Brockmole and Wang (2002) showing that when remaining in the same perspective people are faster when they also remain in the same environment. However, our findings differed from those of Brockmole and Wang (2003) which showed faster performance for changing perspective across than within environments.

Many experimental aspects differed between the present study and that of Brockmole & Wang (2003), all of which could have produced the discrepant results. For example, while Brockmole and Wang (2003) used environments that were highly familiar to their participants, the present study used environments that were learned by participants just prior to the target-localization trials. It could be the case that interference is greater for familiar than novel environments. There is some evidence in the literature that memories for familiar environment are organized in a different manner than those for novel environments (but see McNamara, 2003 for a critique). The classic study by Evans & Pezdek (1980), for example, has shown that university students had formed viewpoint-invariant representations for their campus, possibly as a result of having experienced their campus many times from multiple perspectives. In contrast a group of students from a different school who learned the campus only via a map exhibited the typical viewpoint-dependent performance found in the present and many other studies using novel environments (e.g., Presson & Montello, 1994).

Moreover, in Brockmole & Wang (2003) the two test environments represented two distinct levels of hierarchical encoding. Participants in that study were tested on their memory for objects located in their office and the building in which their office was located. In contrast, the environments in our study were not nested. It is possible that switching across two presumably unrelated environments is not comparable to switching across two environments that are held in memory hierarchically.

Despite these and other methodological differences between the two studies, the interference account should have, in theory, applied in the present experiment. If it did, this would have cast serious doubts on the conclusions of many studies that use target-localization from imagined perspectives in novel environments. Although our results cannot rule out the presence of interference from the previous relative (mental) locations of targets, they can at least suggest that such interference (if any) is not severe.

In summary, our results show that there are costs for both mentally switching across environments in memory and changing perspective within them, and that these costs are not additive when one needs to mentally switch both environment and perspective.

In our view, the cost for environment-switching results from the need to shift the focus of attention, as defined by Cowan's (1999) theory of working memory, from one spatial representation to the other. Due to the nature of our task (i.e., cascaded trials probing unpredictably one or the other environment), we believe that our participants maintained two separate spatial representations active in their working memory. Nevertheless, only one of them could be at the center of the mental focus at a given point in time. Hence, we propose that the cost for switching from one environment to the other represents the time needed to move the attentional focus.

Additionally, we posit that the cost for changing perspective results from updating the stored view (i.e., the initial viewpoint) — which is activated when the environment is brought into the focus of attention — to the view that is probed by the trial instruction. Updating in this case is not effortless as it is with physical movement. Rather, changing imagined perspective demands cognitive resources as it requires an explicit computation of how the relative locations of objects change as a result of adopting the new imagined perspective.

Acknowledgments

We thank Roberta Klatzky, Jack Loomis, and Frances Wang for valuable comments and suggestions during the preparation of this paper. We are grateful to Jack Loomis for letting us use his laboratory.

References

Avraamides, M. N. (2003). Spatial updating of environments described in texts, *Cognitive Psychology, 47, 402-431.*

Avraamides, M. N., & Carlson, R. A. (2003). Egocentric organization of spatial activities in imagined navigation. *Memory & Cognition, 31*, 252-261.

Avraamides, M. N., Klatzky, R. L., Loomis, J. M., & Golledge, R. G. (2004). Use of cognitive vs. perceptual heading during imagined locomotion depends on the response mode. *Psychological Science, 15*, 403-408

Brockmole, J. R., & Wang, R. F. (2003). Changing perspective within and across environments. *Cognition, 87*, B59-B67.

Brockmole, J. R., & Wang, R. F. (2002). Switching between environmental representations in memory. *Cognition, 83*, 295-316.

Christou, C.G., and Bülthoff, H.H. (1999). View dependence in scene recognition after *active learning. Memory and Cognition, 27, 996-1007.*

Cowan, N.(1999). An embedded-process model of working memory. In A. Miyake, & P. Shah (Eds.), *Models of Working Memory: Mechanisms of Active Maintenance and Executive Control* (pp. 62-101). New York: Cambridge University Press.

Easton, R. D., & Sholl, M. J. (1995). Object-array structure, frames of reference, and retrieval of spatial knowledge. *Journal of Experimental Psychology: Learning, Memory, & Cognition, 21*, 483-500.

Evans, G. W., & Pezdek, K. (1980). Cognitive mapping: Knowledge of real-world distance and location information. *Journal of Experimental Psychology: Human Learning and Memory, 6*, 483-500.

Farell, M. J., & Robertson, I. H. (1998). Mental rotation and the automatic updating of body-center spatial relationships. *Journal of Experimental Psychology: Learning, Memory, & Cognition, 24*, 227-233.

Franklin, N., & Tversky, B. (1990). Searching Imagined Environments. *Journal of Experimental Psychology: General, 119*, 63-76.

Hirtle, S. C., & Jonides, J. (1985). Evidence of hierarchies in cognitive maps. *Memory & Cognition, 13*, 208-217.

Jeannerod, M. (1995). Mental imagery in the motor context. *Neuropsychologia, 33*, 1419-1432.

Klatzky, R., L, Loomis, J., M, Beall, A., C, Chance, S., S, & Golledge, R., G. (1998). Spatial updating of self-position and orientation during real, imagined, and virtual locomotion. *Psychological Science, 9*, 293-298.

Kosslyn, S. M. (1994). *Image and brain: The resolution of the imagery debate*. Cambridge, MA: MIT Press.

Logan, G. D. (1988). Toward an instance theory of automatization. *Psychological Review, 95*, 492-527.

Loomis, J. M., Lippa, Y., Klatzky, R. L., & Golledge, R. G. (2002). Spatial updating of locations specified by 3-D sound and spatial language. *Journal of Experimental Psychology: Learning, Memory, & Cognition, 28*, 335-345.

May, M. (2004). Imaginal perspective switches in remembered environments: Transformation versus interference accounts. *Cognitive Psychology, 48*, 163-206.

May, M. (1996). Cognitive and embodied modes of spatial imagery. *Psychologische Beiträge, 38*, 418-434.

McNamara, T. P. (2003). How are the locations of objects in the environment represented in memory? In C. Freksa, W. Brauer, C. Habel, & K. Wender (Eds.), *Spatial cognition III: Routes and navigation, human memory and learning, spatial representation and spatial reasoning* (pp. 174-191). Berlin: Springer-Verlag.

McNamara, T. P. (1986). Mental representations of spatial judgments, *Cognitive Psychology, 18*, 87-121.

Mou, W., & McNamara, T. P. (2002). Intrinsic frames of reference in spatial memory. *Journal of Experimental Psychology: Learning, Memory, & Cognition, 28*, 162-170.

Mou, W., & McNamara, T. P., Valiquette, C. M, & Rump, B. (2003). Allocentric and egocentric updating of spatial memories. *Journal of Experimental Psychology: Learning, Memory, and Cognition. 28*, 162-170.

Presson, C. C, & Montello, D. R. (1994). Updating after rotational and translational body movements: Coordinate structure of perspective space. *Perception, 23*, 1447-1455.

Richardson, A. E., Montello, D. R., & Hegarty, M. (1999). Spatial knowledge acquisition from maps and from navigation in real and virtual environments. *Memory & Cognition, 27*, 741-750.

Rieser, J. J. (1989). Access to knowledge of spatial structure at novel points of observation. *Journal of Experimental Psychology: Learning, Memory, & Cognition, 15*, 1157-1165.

Rieser, J. J., Guth, D. A, & Hill, E. W. (1986). Sensitivity to perspective structure while walking without vision. *Perception, 15*, 173-188.

Shelton, A. L., & McNamara, T. P. (2001). Systems of spatial reference in human memory. Cognitive Psychology, 43, 274-310.

Stevens, A., & Coupe, P. (1978). Distortions in judged spatial relations, *Cognitive Psychology, 10*, 422-437.

Wexler, M., Kosslyn, S. M., & Berthoz, A. (1998). Motor processes in mental rotation. *Cognition, 68*, 77-94.

Wraga, M. (2003). Thinking outside the body: An advantage for spatial updating during imagined versus physical self-rotation. *Journal of Experimental Psychology: Learning, Memory, & Cognition, 29*, 993-1005.

Through Diagrams: Discovery in Game Playing

Susan L. Epstein

Department of Computer Science, Hunter College and The Graduate School of The City University of New York, New York, NY 10021, USA
susan.epstein@hunter.cuny.edu

Abstract. Diagrams often play an important role in human problem solving. A diagram can make a task more difficult, however, when it obscures important problem features, or requires repeated effort to interpret what it represents. Moreover, the nature and origin of diagrams can be as important as their exploitation during problem solving. This paper chronicles the complex interleaving of visual cognition with high-level reasoning in three subjects. Their diagrams and subsequent verbal protocols offer insight into human cognition. The diversity and richness of their response, and their ability to address the task via diagrams, provide an incisive look at the role diagrams play in the development of expertise. This paper recounts how their diagrams led, and misled, them; how their diagrams both explained and drove explanation; and how a spatial approach reported here can lead to deeper understanding of other simple games.

1 Introduction

Spatial information can play important roles in problem solving, learning, and discovery. The centerpiece of this paper is an experiment in complex cognition that explores this idea. Three subjects investigated the simple game described in Figure 1. Each of them invented diagrams to cope with the task's complexity and reasoned from that spatial information. The primary results of this work are the subjects' preference for one-dimensional representation in thinking about a two-dimensional space; the power of a diagram to displace the original reasoning context; a new diagrammatic convention and associated algorithm that provide insight into cyclic, two-person games; and

Pong hau k'i

- **Two-player game**
- **5 locations**
- **First to move plays black**
- **Second to move plays white**
- **Move = slide to the next empty location**
- **Initially, the board is as shown**
- **Mover who cannot slide loses**

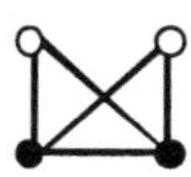

Fig. 1. As presented to the subjects, the game of pong hau k'i [21]. The mover slides her piece along a line to the single empty location on the board. The goal is to *trap* the other contestant, so that on her turn she cannot slide. Contestants take turns until a mover loses because she is trapped, or until a draw is declared because the mover and the location of all the pieces have been repeated more than three times

C. Freksa et al. (Eds.): Spatial Cognition IV, LNAI 3343, pp. 259 – 282, 2005.
© Springer-Verlag Berlin Heidelberg 2005

a demonstration of the synergy between visual and high-level processes during abstract reasoning. The focus here is on how spatial information supports, and interferes with, high-level reasoning.

This work addresses cognition about movement in an abstract, rather than a realistic, two-dimensional space. Although there are studies of how individuals change representation to facilitate the solution of one-person problems [1], and studies of how individuals respond to a variety of representations for the same two-person game [22], to the best of this author's knowledge there are no studies of how individuals change representations to facilitate learning a two-person game. This paper is in the tradition pioneered in [2], particularly with respect to the third subject. The first section of this paper provides terminology and an overview of related work. Subsequent sections describe the experiment, the subjects' solutions, and the processes used to reach them. Finally, the results are discussed, along with suggestions for future work.

1.1 Representing Problem Solving and Search Spatially

The classical AI approach to behavior in a *domain* (problem area) is to specify *states* (ways the world can appear) and *actions* that transform one state into another. Given a set of states and actions defined on them, a *problem* is a *start state* (description of the initial state of the world) and a description of one or more desired *goal states,* to which a sequence of actions should lead. *Problem solving* can thus be modeled as *search,* finding a sequence of actions that transform the start state into a goal state.

The set of all possible states in a domain is its *state space*; it represents all the states within it, and how they relate to one another A *state space diagram* represents a state space spatially, as a set of *nodes* (each of which depicts a state), some pairs of which are linked by *edges*. An edge from one node to another indicates the existence of a problem-solving action that transforms the first state into the second. In a state space diagram, problem solving can be envisioned as finding a path along edges through the diagram, from the start state to a goal state. To construct a state space diagram, one must describe both the individual states and their interrelationships.

A *game tree* is a state space diagram specialized for two-person games. An example of part of a game tree appears in Figure 2. Note the horizontal alignment of the nodes there into *levels*, and the labels to the left that note whose turn it is to move on that level. A state that ends play in a game tree is *terminal*, and is associated with an

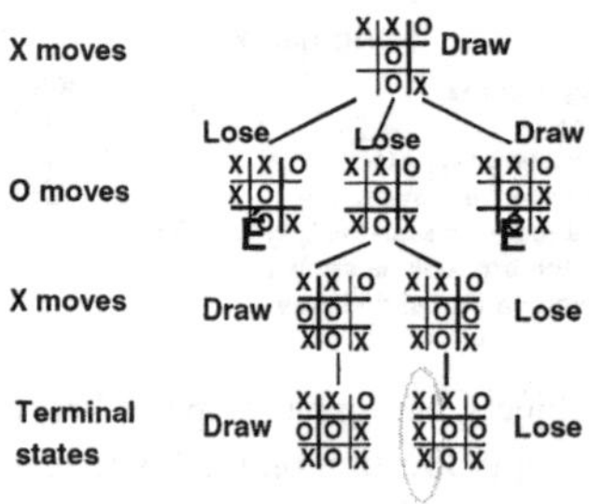

Fig. 2. Part of a game tree for another game, lose tic-tac-toe, where the object is to avoid three in a row, column, or diagonal

outcome (e.g., win, loss, or draw). *Move generation* is the enumeration of every possible action from a game state, and can itself be a complex task; some subjects in this experiment developed diagrams as aids to move generation.

For a game like that in Figure 2, where a piece, once placed, is fixed to a single location on the board, a game tree describes possible lines of play clearly: a contest moves downward from the start state along one edge at a time until it reaches a terminal state. In the game described in Figure 1, however, where a move may return the board to a previously visited state, the game tree must be adapted, either by repeating states (so the tree becomes infinite) or by introducing a *cycle* (a path of edges returning to an already-represented state).

1.2 Cognitive Issues in State Representation

A *representation* for a state describes it with some degree of detail. One might, for example, represent the state in Figure 1 as follows: "it is black's turn, the pieces are made of wood, black's pieces are on the two lower locations, white's pieces are on the two upper locations, and the center location is empty." A good representation captures the salient features of a state, without irrelevant details. In pong hau k'i, the material from which the pieces are made has no impact on problem solution. Similarly, their size and shape are irrelevant.

Once the salient features of a problem state are identified, there are still many possible ways to represent an individual state. For example, the pong hau k'i start state could be represented in any of the ways shown in Figure 3. In some games (e.g., tic-tac-toe), it is possible to determine whose turn it is to move merely by examining the board. In pong hau k'i and many other games, however, one must identify the *mover* (the contestant whose turn it is to move) as part of the state description. Figures 3(a) and 3(b) are *visualizations*, representations that convey information by the spatial arrangement of elements of the display itself [10]. Figure 3(a) is a *complex visualization* because it suggests animation (sliding pieces along the lines).

A human problem solver has at least two representations of a problem: the *external representation* provided by the environment and the *internal representation* constructed within the subject's memory. The external representation, although both explicit and implicit, is for the most part under experimental control; the internal representation can only be glimpsed through a subject's behavior. Furthermore, a problem solver may construct additional implicit representations, and, within the restrictions of the experiment, additional explicit ones.

A good state representation speeds solution. Kaplan and Simon's work on the mutilated checkerboard problem showed that cues (e.g., colors, or words that are

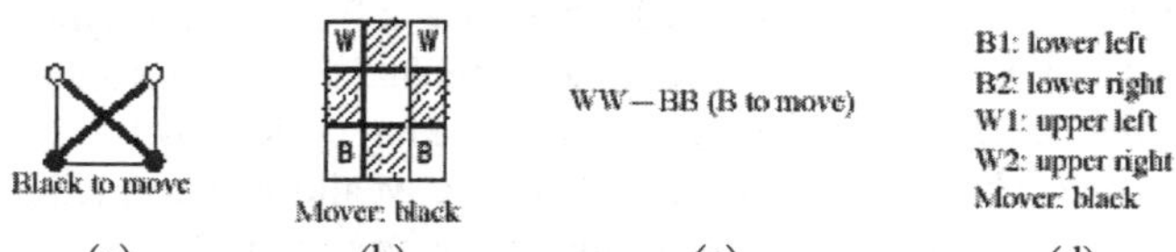

(a) (b) (c) (d)

Fig. 3. Some representations for the start state in pong hau k'i commonly paired with each other) inherent in a representation can suggest a particular line of reasoning to people [11]. They also showed that hints to subjects who had difficulty solving the problem could help them change to a representation (make a *representational shift*) that incorporated such cues. Their subjects required motivation to make that shift, however, and cues proved constructive only once subjects were dissatisfied with their progress in the current representation. Their best subjects could detect the salient features of the space, and relied heavily on perceptual details. They reported that a good representation produces a smaller space with an effective generator that relies on *invariants*, things unchanged by an action.

Representational shifts sometimes increase expertise. Amarel's work on the missionaries and cannibals problem, for example, argued the increasing power of a sequence of different representations, progressing from natural language to a compact mathematical symbolism [1]. On the 3-disk tower of Hanoi problem, Anzai and Simon described how a single, untrained individual performed a sequence of representational shifts that led to better performance and to a deeper understanding of the problem [2]. Deliberately obscured representations of the 3-disk problem (involving monsters and globes), however, were far more difficult for humans than the original problem [12]. Careful analysis showed that most of those subjects' time was devoted to exploration of the possible states, followed by rapid detection of a correct sequence of actions. Fast, accurate move generation from the problem description proved essential to expertise; the ability to construct simple two- or three-move plans was a further asset. Memory load was also a consideration: the more a subject was forced to recall (e.g., imagining the monsters, the size of their globes, and changes to their globes), the more difficult it was to plan. Practice on repeated problems seemed to support planning and learning. An external representation (e.g., a diagram) also helped, but objects in more than one location, actions entailing spatial changes, or more than one imaged entity still made solution more difficult.

1.3 Search, Spatial Information, and Learning

Spatial information can reduce search effort by providing problem-solving cues. For example, a program that constructed simple plane geometry proofs from an input problem and an accompanying diagram was able to construct more complex proofs when it could take metric cues from the diagram [8]. Spatial information can also encourage the substitution of (faster) perceptual judgments for logical inference. Such a judgment is a recognition that the elements in a diagram are spatially organized in a way that supports problem solving. DiBS (Display-Based Solver) used list-based problem input to construct its own set of one-dimensional diagrams [14]. These diagrams captured relationships (coded as "below") that described the goal state and represented subgoals as objects that blocked other objects from their intended final locations. The program could solve several quite different, simple, one-person problems (the 3-disk tower of Hanoi, simple algebra, and making coffee) by transformations to its diagrams alone. This approach demonstrates the wealth of information a diagram can store, the cues it can offer a solver, and the power of an automated representational shift. DiBS, however, requires careful encoding of domain knowledge and of the problem itself, and is intended for single-agent problems with a single relationship category. Moreover, DiBS requires the description of goal states in advance, although part of some tasks is to identify goal states for the first time. DiBS also assumes that the order in which actions occur is irrelevant, and that there is a single agent, the results of whose actions are certain. The rules of two-player games, however, are quite strict about sequence, and a single decision may not determine the next state in which one makes a choice.

Domain expertise can be pre-specified, or it can be learned. From an AI perspective, *learning* is a change in one's problem solving behavior based on experience. The most challenging form of learning is *discovery*, where the learner has no external teacher. In game playing, for example, learning to play merely by playing, without expert instruction, constitutes discovery.

Spatial information can support learning in a variety of ways. The construction of a state space diagram helps identify the states that actually arise during problem solving. Some spaces also lend themselves to a kind of organization that prevents repetition and supports move generation (e.g., the tic-tac-toe game tree). For game playing, the paths through a state space diagram can delineate opportunities (possibilities to win) and dangers (possibilities of loss) during play. Furthermore, a diagram can support generalizations about the paths in the state space, generalizations readily converted into heuristics for efficient, effective problem solving. Examples of this abound in the central experiment described here.

The importance of spatial information in human scientific discoveries is well-documented [9, 15, 17]. One way to validate a theory about the discovery of scientific laws is to automate the process and replicate the results. Most such AI artifacts begin with numeric data that simulate the results of hypothetical experiments, and induce rules from them [7, 13]. HUYGENS, however, first transformed the data into a set of one-dimensional diagrams, and then sought perceivable regularities within those diagrams, using heuristics to focus attention [4]. HUYGENS could extend its initial diagrams during search, so that it not only reasoned about diagrams but also constructed new ones. The program was limited, however, to domains where one-dimensional diagrams could represent salient features.

Diagrams can also make learning more difficult. People learning to play isomorphs of tic-tac-toe experienced varying degrees of difficulty with eight disguised versions of the game [22]. Each isomorph was accompanied by a diagram that offered or obscured invariants ordinarily perceivable without high-level reasoning. Zhang showed that perceptions about those invariants supported or undermined one's ability to learn to play this simple game, to explore the space, and to discover structure within it. Tic-tac-toe has a large state space: 5746 reachable distinct states, and 120 different possible paths to a terminal state. Zhang argued that, if accessible, the symmetric invariants made search in so large a space more manageable, and that the winning invariants elicited a particular human bias about the world: more is better. His work is an elegant argument, at least in this specific domain, for *representational determinism*, the idea that a representation both guides and constrains what a problem solver can achieve.

2 The Experiment

Pong hau k'i was an informal assignment for students at the 2001 Cognitive Science Summer School at the New Bulgarian University. During a lecture, they were given the rules for this old Korean game, with the slide in Figure 1. None of the students knew the game. They were told that it was a very simple, even boring, game, played between black and white. They were asked to "play it, take protocols on the development of your own expertise, and capture any representational shifts," defined as "a different way of seeing things." (No example of a representational shift was provided.) They were encouraged to obsess over it and to report back on their experience.

The same lecture introduced state space search, including the slide in Figure 2. Later in the lecture, the students were asked, on additional slides, to "draw out the state space for pong hau k'i, label [it] with outcomes, produce a rule set to play it, ...and produce a good heuristic for it." No guidance on representation for repeating states was provided, and no representation for the pong hau k'i states was specified. The author expected that the students would use the representational conventions of Figures 1 and 2, but, as we shall see, the students quickly abandoned them.

This task began as a teaching device, not as a planned experiment. As a result, it lacks the timed photographs, videos, and recordings that lend rigor to some other work in this field, [5, 16, 18, 19]. Once several students began to respond in such great depth, however, documentation was as extensive as possible. The students of the Cognitive Science Summer School are a diverse lot, drawn from many countries and many disciplines. The Summer School is conducted in English, but the students' written materials were idiosyncratic, and included notation in both German and Polish. For uniformity, in what follows this author has made minor spelling corrections, inserted clearer or more grammatical language [in brackets like these], and substituted B for "black," W for "white," and "–" for an empty location. Any italics are the subjects' own, and all names have been omitted. During the experiment, no subject described here saw the protocols of the others.

2.1 Pong Hau k'i

By game-playing standards, pong hau k'i is very simple: there are only 60 states in its search space. Moreover, pong hau k'i is a *draw game*, that is, if both sides play correctly, play always ends in a draw. Even if one is not familiar with the game, it ought to be relatively easy to play well.

Compared to other simple games (e.g., tic-tac-toe), however, pong hau k'i is actually rich and challenging, for several reasons. The pieces are permitted to move about the board; this encourages competitors to envision "movies" rather "snapshots" of the board, and places a high load on their spatial memory [12]. It is also impossible to tell from the board alone whose turn it is to move; it could be black's or white's from any state. Finally, there is only symmetry across the vertical axis on the board; diagonal and horizontal symmetries are invalid.

2.2 Subject 1 — An Automated Approach

Subject 1, a graduate student in psychology, ostensibly drew no diagrams at all. At lunch after the lecture, several students made pieces from bread (the crusts against the bread centers) and began to play. Although he professed disinterest, eventually the fun drew Subject 1 in, and he began to play against Subject 3. "After a while the game became dull, because neither [of us]... understood the point of it.... We started to discuss, that there must be some good strategy, and we must find it.... [Subject 3 began] to draw a tree-like decision graph, where every branch represented a possible move. ...The only thing I remember of his drawing was that he was thinking of a good labeling of the [locations]. We agreed that numbers are good indicators of the locations, but we didn't talk about [how to assign numbers to locations]. He wanted to count how many [distinct states there were]. But it was really time to go back to [class]."

Subject 1 is not a trained programmer. Nonetheless, alone later that evening he wondered "why people should bother with counting [moves] so much ... if computers are much more suitable.... Therefore I decided to build a program that can play this game." Although his original plan was to play against the computer and have it count the states for him, he subsequently realized that the program might also serve as an opponent. "At this point, I did not think of any strategy, only the rules [of the game]... I remembered the lunchtime games, and [Subject 3's] idea that a good labelling is needed. Without much thinking I came up with [Figure 4]. The reason I chose this labelling is because if you make a 'line' from the original shape by 'pushing' from the top and from the bottom the numbers will increase from 1 to 5."

Two hours later Subject 1 had a program that reproduced the pong hau k'i game board, but with red and blue (rather than black and white) circles for the playing pieces. The code, however, did not employ Subject 1's numbering in Figure 4. The order in which the circles were drawn on the screen on each turn, and the order in which they were stored, using his numbering from Figure 4, was 2, 1, 3, 5, 4.

Subject 1 now began to play against his program, which made random legal moves. "This was the first time, I wanted to understand the strategy." Subject 1 soon won against his program, played "10 –30" more contests, and remained convinced that he was correct. Still later the same evening, Subject 3 appeared in Subject 1's room. "He [Subject 3] ... seemed very tired. He became delighted that" another person was working on pong hau k'i so intensely. "So we quickly started to play," but Subject 3 warned that he was undefeatable. "After 5 – 8 steps we gave up, because I couldn't lose [either]. Tie game. By that time, we realised that both of us [had become experts at pong hau k'i]. He did his way theoretically, I [did] mine practically."

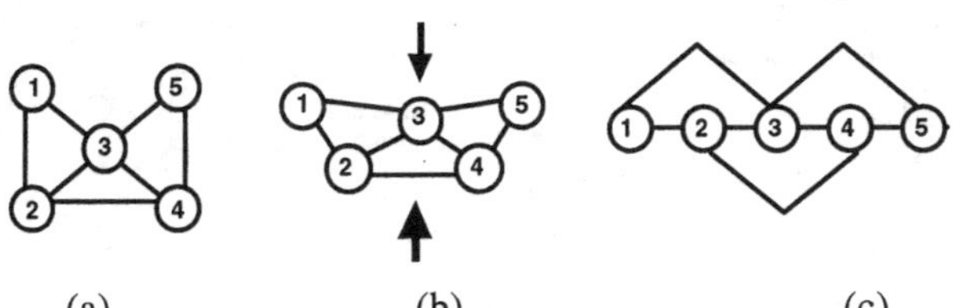

Fig. 4. Subject 1's compression of the board to create a numeric encoding, as given in his protocol. (a) The board itself, (b) the initial pressure , and (c) the collapse into a line

2.3 Subject 2 — A (Relatively) Quick Response

Subject 2, a Ph.D. candidate in cognitive psychology, originally offered no verbal protocol with his analysis, only an elaborate, tree-like diagram on a single sheet of paper, with some small annotations described below. Subject 2 drew his diagram in a few hours, stopping "when I thought it was complete…." He did not use a game board, rather, "At the beginning I tried to imagine the situation on it. Then I just manipulated the letters and numbers. I drew a sketch when I was afraid I was lost, to check it."

It is possible to deduce a good deal of Subject 2's process from his diagram (which is difficult to reproduce and therefore omitted). Figure 5(a) replicates the top portion. In the upper left corner is the start state from Figure 1. Below it are two sketches symbolizing how white could win. (Line segments were indeed omitted, and most sketches, as reproduced here, were also more square than the original in Figure 1.) To the right of the start state is an encoding of it (WW–BB), from which one may deduce the location numbering: 1 = upper left, 2 = upper right, 3 = center, 4 = lower left, 5 = lower right. "12345" lists the labels assigned to the vertices. Immediately below it is a list of the seven pairs of locations between which a move may occur.

There are only two other sketches of the board on the page, both to the right. The first, Figure 5(b), is near the two losing states highest in the diagram, about eight levels down in the tree. It represents the state just before a possible win for either mover. The second, Figure 5(c), appears next to a section of the game tree that is placed far to the right but connected to the diagram (the *subtree-to-the-right*). Figure 5(c) represents a state from which a white mover can win; if black is the mover the contest will merely continue. These are the only sketches Subject 2 drew to check his reasoning.

Most of Subject 2's page is occupied by a diagram in the spirit of Figure 2. Each node is a five-letter string representing the contents of the five locations on the board. Nodes that are the same number of moves from the start state are horizontally aligned in levels on the unlined paper. In most of the tree, a W or a B to the left on each level indicates the mover in the previous state; in the subtree-to-the-right these labels were to the right. The two alternative first moves for black appear beneath the start state. Subject 2 chose to follow only the move from 4 to 3, shown below and to the left of WW-BB in Figure 5(a). At most levels there are only two or three nodes, but occasionally there are as many as six. The diagram reaches all margins of the paper, and includes 80 nodes in all. One node, on the eleventh level, points with a long arrow to

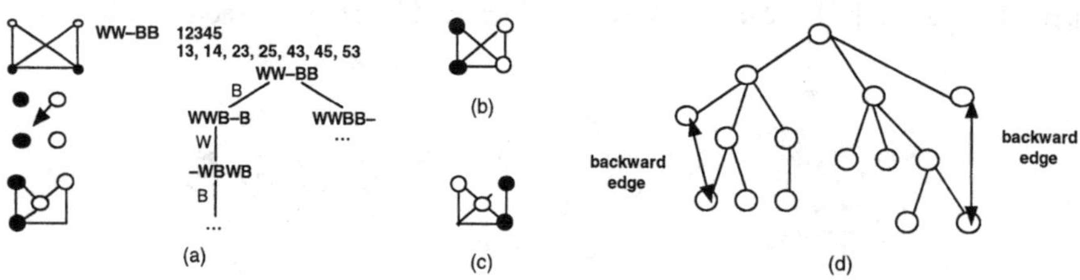

Fig. 5. (a) The top portion of Subject 2's protocol (ellipses are the author's), (b) a board configuration penultimate to a loss; either contestant can win from here, (c) a board configuration from which white, but not black, can win, and (d) orientation of forward and backward edges in Subject 2's tree

the subtree-to-the-right. Otherwise, exploration appears to have had a strong downward tendency, and to have moved from left to right at every level.

Most tree edges are simply lines in his diagram, but at some point Subject 2 noticed that some states repeated. When he found a repeated state (encoding and mover), he drew an unlabeled edge that was *directed* (had an arrow on at least one endpoint), as in Figure 5(d). A *backward edge* begins at one node and runs to another *higher* (closer to the start state) in the tree; there are 11 backward edges. One backward edge in particular, from the leftmost portion of the subtree-to-the-right, returned to the start state and presumably eliminated the need to explore the opening move to WWBB– on the right in Figure 5(a). A *forward edge* runs between two nodes and has arrows on both ends. The diagram includes two forward edges; the only difference apparent between backward edges and forward ones is their length — the repeating pairs joined by a forward edge are those farthest apart on the page. There is also one directed edge between two nodes on the same level, and one dashed forward edge between identical nodes associated with different movers. In the latter case, Subject 1 presumably observed the similarity, noted it with an edge, and judged it not worth pursuing.

When a node has no others beneath it in Subject 2's diagram, it is either the source of a backward edge (indicating repetition in play) or is surrounded by a rectangle (indicating the end of a contest). There are eight such nodes, each with a W or B label indicating the winner in that state. Only four of these are distinct: two wins for white and two for black. There is one significant error: the forward edge from the unexplored first move WWBB– is not to a later copy of itself. There is, however, another copy of WWBB– (one unremarked upon by Subject 2) that is correctly considered.

2.4 Subject 3 — An Obsession

Subject 3, a Ph.D. student in linguistics and cognitive science, has a strong mathematical background. He had been among the students at lunch that day with Subject 1. The day after the lecture, Subject 3 proudly displayed a diagram he had drawn of the search space. He seemed exhausted, and questioning revealed that he had devoted a good deal of time and energy to the task. Although he thought little of them, he had not yet discarded his earlier diagrams, which chronicled his path. He remembered a great deal of the process he had been through in the past 24 hours, so together the author and Subject 3 numbered, discussed, and annotated his early diagrams. From these, Subject 3 began to write what eventually became a 13,000 word protocol, covering about 6 hours of intensive exploration and many more of retrospective reconstruction. The material in this section is based on observation of his diagrams, the debriefing session the day after the assignment, and the original and later versions of the written protocols. Subject 3 identifies 9 stages in his path to expertise at pong hau k'i.

Stage 1. (15 minutes) Subject 3 attempted to "get a feeling for the game and its state space. Initially, he played "very much at random without even knowing how a [losing] state might look...." Indeed, he hypothesized one state as a loss which was not a loss at all. Nonetheless, "After a few more (random) moves my opponent pointed out that his [pieces] were stuck and that he had lost. I looked and realized that indeed, by accident, I had won this first game. Both his [pieces] were on the [right side] of the board and I was blocking him from moving…. From seeing this I figured that *one could lose if (and only if) one ended up having both [pieces] on the same (left/right) side of the game board.*" In a second contest, Subject 3 lost. Then, as the others watched and commented, he began to experiment alone, moving "… according to the rules, without consciously following any strategy. … I got the feeling that the dynamics of the game contained several symmetries.… by a few straightforward moves I could invert the initial configuration (interchange the black and white pieces). And I also got the feeling that I would always end up in these same two [states] (the initial one and its inverse)…. I had the sensation of running into 'attractor states.' Then I tried to … [force either player to lose]. It only occurred occasionally and rather by accident than by controlled force. I could not really find out which moves one had to do/avoid in order to win."

Stage 2. (20 minutes) Subject 3 began to draw his first diagram. He computed 30 possible *configurations*, ways to place the pieces on the board; with the label for the mover, he arrived at 60 possible states. "However, I was not yet sure that all of these states could actually be reached by legal moves from the initial configuration." For compactness, Subject 3 now shifted to a rectangular representation for a node. (He was drawing on graph paper, where a sequence of squares is easy to draw quickly and clearly.) Because black moves first, Subject 3 labeled the 2 lower locations 1 and 2 (left to right, respectively). He labeled the center 3 because he felt it retained "some of the symmetry of the original game board." Originally the top locations were (left to right) 4 and 5, but he interchanged them (5 and then 4) so that 2 to 4 and 1 to 5 were moves. This produced the numbering in Figure 6(a). Using this numbering he wrote sample nodes, such as Figure 6(b), to confirm his calculation of the possible states.

Still at lunch with his friends, Subject 3 now began to draw his second diagram, his first attempt at the full state space. All nodes were in the format of Figure 6(c), where the additional circle to the right denotes the mover. Although he initially checked his diagram against the pong hau k'i board, Subject 3 eventually relied solely on his new representation. He set up the two opening moves at the top, beneath the start state, just as Subject 1 had. "Then I followed only one of the two now opened paths …depicting white's and black's possible moves in alternating order. At a certain point I would interrupt myself and work a bit on the right path hoping for some cycle to close. The whole diagram became very confusing: with almost every new state that I drew I would get the feeling of having already encountered this particular one." As he searched his diagram looking for cycles, Subject 3 became "rather annoyed by the

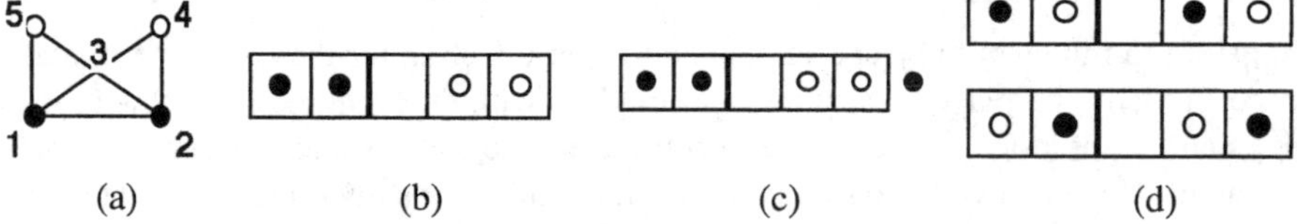

Fig. 6. (a) Subject 3's numbering of the board. (b) The starting board and (c) the start state represented in Subject 3's rectangular coding. (d) The "seeds" for his fourth diagram

fact that I had to distinguish the mover with the state." He decided not to extend "paths starting from those states that occurred for the second time, but with the opposite mover. I believed I could just refer to some sort of symmetry with the paths starting at the corresponding state that I had visited first." At some point he realized that he had forgotten about the edge from 1 to 2: "… indeed, I found two new 'twigs' to be grown in my graph." Returning to it hours later, however, "Everything just became more and more messy and confusing…." He abandoned the task.

Stage 3. (45 minutes) Somewhat later, Subject 3 began alone, with new determination, to draw a third diagram, this time in color. The playing pieces became green and red, edges were colored green or red to denote the mover transforming the state, and the external circle for the mover was eliminated. Edges now had arrows on both ends, indicating that moves were invertible. Subject 3 referred back to his previous diagram regularly, checking for oversights, and frequently found them. Eventually, he extended the space methodically, keeping both sides at about the same depth, and soon reached four winning nodes, two for red and two for green. These were highlighted in orange to denote termination of search. Then he expanded the non-terminal states. "I started to count the states. There were 30 of them. I took this correspondence with my earlier calculation as a strong confirmation for finally having covered the entire state space. … But my representation was just unbearably entangled. (And in fact, I was surprised by the apparent complexity of the state space.) I wanted to shape it up towards a comprehensible new diagram." Subject 3 considered the diagram crowded,

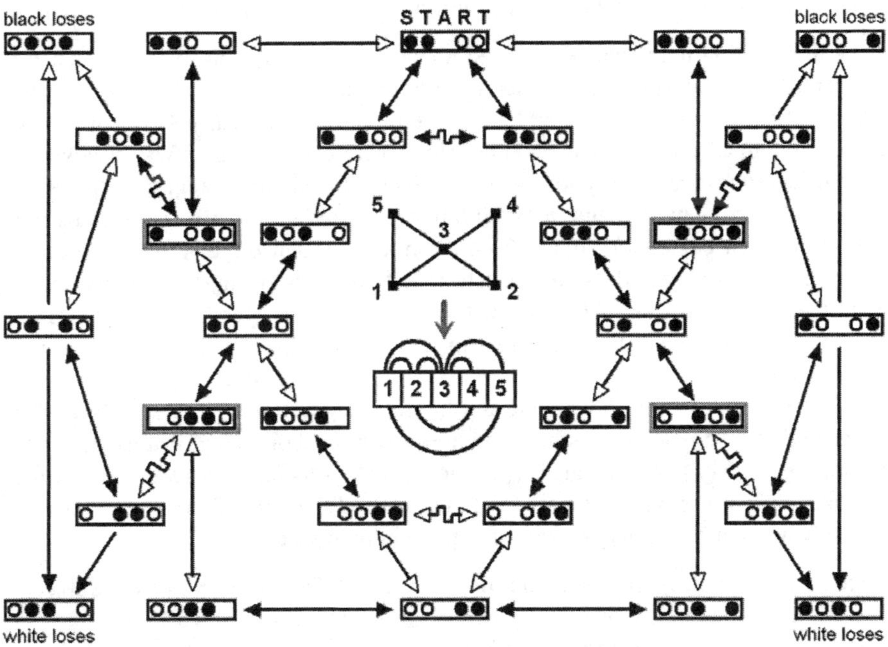

Fig. 7. Subject 3's sixth diagram, the product of 6 hours of intensive work. The four delicate states have gray borders

and it did not support the symmetry he expected; in particular, he "had an expectation that the opposite of the start state would be on the very bottom…naturally."

Stage 4. (50 minutes) Subject 3 drew his fourth diagram, beginning not from the initial state but from "another state of maximal connectivity." He chose the states in Figure 6(d), and positioned them first, horizontally, in the center of his diagram, with the upper one on the left, and the lower one on the right. Subject 3 then traced backward through the previous diagram to position the start state appropriately. He checked off states from his previous diagram as he reached and reproduced them in the left half of the new one, moving outward from the upper state in Figure 6(d). At the same time, he maintained elaborate tests for symmetry and edge color, for states he regarded as "opposites," and for edges that were incorrect but somehow "missing. Although he "was expecting some asymmetries," intersecting edges drove him to redraw it. After the 50 minutes of this stage, he writes that he now considered himself "obsessed."

Stage 5. (50 minutes) His fifth diagram positioned the start state at the top, surrounded by empty rectangular nodes, that is, states without any playing pieces. Subject 3 concentrated on the shape of the space now, not the moves. At some point he filled in the nodes, and then focused on introducing the losing states so that the arrangement was spatially pleasing. To his surprise, Subject 3 found that his desire for symmetry now led to the discovery of edges that he had completely overlooked in earlier diagrams. Moreover, the angle of the edges now became significant to him. Subject 3 initially drew the right side of the diagram with empty nodes and colored edges, and re-derived the nodes before confirming them with his previous diagram.

Stage 6. (duration unknown) Concerned about an asymmetry on the left in his fifth diagram, Subject 3 explained the picture to his roommate. He pointed out four *delicate states* (ones where it is possible to make a fatal error) located between "the secure central area" and the areas on either side where one can lose. "The discovery of the *delicate states* and all my following considerations about strategies were entirely based on the arrow pattern and the state frames – which in my mind appeared almost deprived of their contents." Subject 3 now correctly recognized that pong hau k'i depends entirely on play in the delicate states: "…to prevent losing, I … simply have to avoid the two bad moves… I only need to memorize two specific configurations… whenever my opponent allows me to win, I … simply have to make the one good move…. For this I only need to memorize two further specific configurations…." He was startled, however, by how irrelevant his full diagram was to actual play: "I am surprised that … one would not have to know anything about the global pattern of the state space. … My two strategies … are local rules – local in terms of time (which translates to space on my diagram). … I also noticed that the … strategy …for making the winning move …also serves for …preventing my defeat…."

Stage 7. (20 minutes) The next morning, driven by "global symmetry," Subject 3 drew his sixth diagram, shown in Figure 7. Although he also wanted to renumber the nodes, Subject 3 was pressed for time, and "wanted to focus on the intended rearrangements on the left-hand side of the preceding draft." Therefore he "… drew all 30 empty state frames starting on the right half where I did not have to change much.

Then I added all the arrows until finally I copied the states themselves." He carefully aligned states vertically and horizontally, highlighting terminal states and delicate ones. The result was satisfying — he felt like an expert player and remarked on the symmetry of his diagram. "The stunning thing that was new here is the 'reflection through the center' symmetry and the fact that it operates not only on the state frames and arrow pattern but as well on the state configurations!! I liked this." This was the diagram he initially presented to the author.

Stages 8 and 9. During debriefing that afternoon, Subject 3 recalled his renumbering scheme from Stage 7, which ultimately removed some misunderstandings observed (but not remarked upon) during the conversation. Later that afternoon, in Stage 8, Subject 3 produced his seventh diagram, using that renumbering (Figure 8), so that the patterns in nodes with the same function are symmetric to each other. Over the next few days, in Stage 9, Subject 3 revised and extended his protocol, and produced his final diagram (Figure 9). It classifies states and moves according to their safety and the number of moves they afford; it also maintains vertical, horizontal, and through-the-center symmetry with respect to states and edges. Note that there are no pieces whatsoever in the nodes.

Subject 3 now felt he was an expert, and had been since Stage 5, although he had only competed once (described Section 2.2) since then. "…in a way I had reformulated the game in my own terms and then detached it to some extent from its original shape. My (final) game consisted of boxes … and arrows and I had mastered these …." Why then had he continued to draw? His additional work, he believed, increased

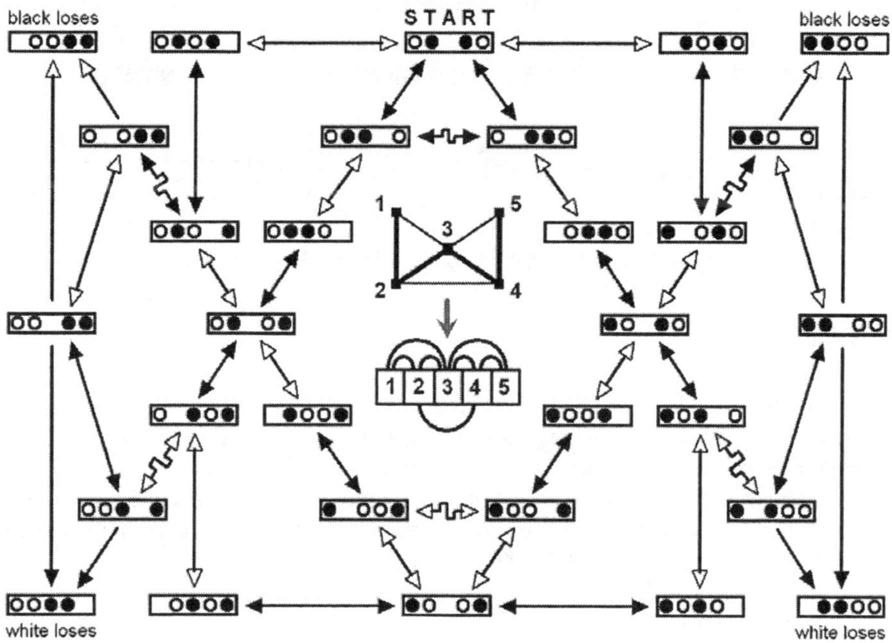

Fig. 8. Subject 3's "W-shaped encoding," seventh diagram, with the board locations renumbered as in the center

his subtle "…knowledge of the structure underlying the game. Knowledge that might not directly influence performance." Even as he dissected his experience, Subject 3 remained puzzled by his own lack of rigor. "… I probably could have proceeded in a much more intelligent/elegant way and in less time. I am not ashamed of this but rather surprised. For before [writing this protocol] … I [thought I had] approached the task in a more controlled and directed way than this documentation has brought to light." Nonetheless, it was a remarkable intellectual journey.

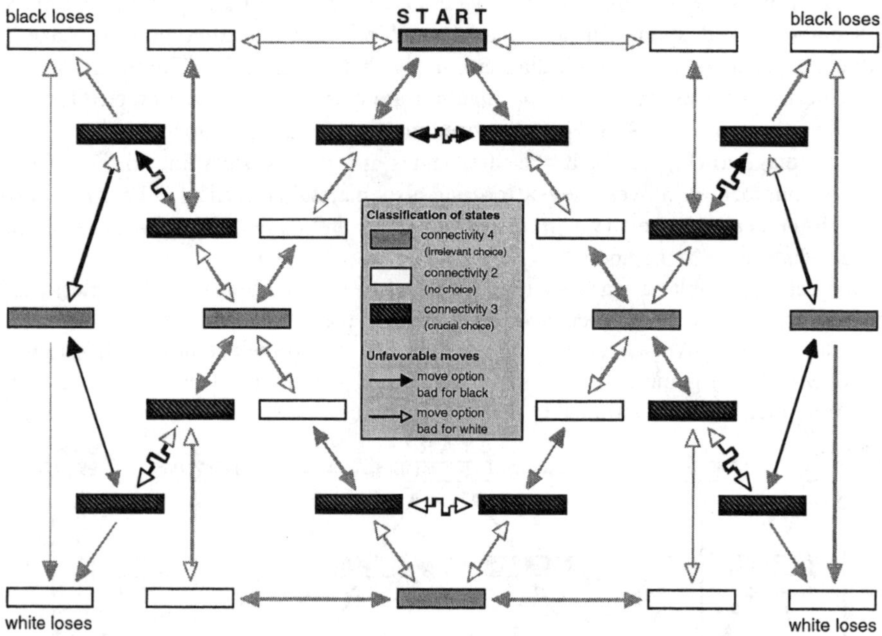

Fig. 9. Subject 3's eighth diagram, showing classes of states

3 Discussion

The instructions to the subjects in this experiment facilitated solution [12]. Memory overload was prevented by asking them to draw a state space diagram. Enough experience to seed planning, learning, and generalization was provided by asking them to compete. Despite identical instructions and similar educational levels, however, their verbal and graphic protocols reveal a surprising diversity of process and of final perspectives, albeit a similar level of skill at the game itself.

3.1 The Right Answers

Pong hau k'i includes 30 possible piece configurations and 60 possible states. Subject 1 wrote code that could draw all possible board states, although he only experienced the ones to which he led it to during competition. Subject 2 found all 60 states (some more than once), but overlooked several edges and in one backward edge paired two different states, asserting an incorrect repetition. Subject 3's first four diagrams included all the configurations, but were missing edges; his others were complete.

Pong hau k'i is a draw game. One cannot win unless one's opponent errs; from most states, any move is safe. Only in the four states Subject 3 called "delicate" is it possible to make a fatal error, and only one of the two moves in each of those states will cause a loss. Even after the wrong move from a delicate state, one's opponent has two possible moves, only one of which wins. Questioned weeks later, Subject 2 wrote that after "a few games with one person" he realized that "it is quite difficult to lose the game, and I had an idea about situations I should avoid." Subject 1 made a similar discovery: "After my first [non-draw] I realised that the only way [to win] is to put the opponent into the side [locations. This describes both states in Figure 5.] ... I tested this idea, and tried to avoid [having my pieces there] ...." Thus, one would expect all three subjects to play expert pong hau k'i; drawing against each other, and winning if an opportunity arises.

The students were asked to construct a labeled diagram of the state space, a rule set, and a good heuristic. Subject 1 constructed a program that "knew" the space, without ever drawing it out. He learned rules and heuristics from experience that moved him through the space as he competed against his program, a kind of virtual representation. Subject 2 only constructed the space, but felt that he had completed the task. Subject 3 developed game graphs that proved the heuristics and the rule set that the other subjects arrived at. His final product is a powerful, rigorous, visual argument.

3.2 Invention of State Representations

Although all these subjects received the same instructions, each of them approached the task differently. To represent an individual state, all three found the physical board itself confining and soon abandoned it. Subject 1's program was in part motivated by the burden of repeatedly drawing the board. Subject 2 switched because he wanted to fit the entire state space on a single page. Each subject began by numbering the board: Subject 1 with a thoughtful, visually-described process which he quickly abandoned; Subject 2 as if he were reading text, left to right and top to bottom; and Subject 3 left to right but from the bottom up. Subjects 1 and 3 considered more than one numbering. In every numbering, however, the central position was always labeled 3, presumably because it is also the median of the numbers from 1 to 5.

Clearly, perception and symmetry are powerful influences on state representation. Each numbering of the locations on the board drives exploration of the search space differently, and there are 120 possible numberings. It is likely more than coincidental that Subject 3's second numbering, in Figure 4, matched Subject 1's. Indeed, Subject 1's original idea was to explore several such numberings as if they were preference strategies, playing one against the other. Remarkably, the other numbering he planned to explore was the numbering used by Subject 2. The preferred numberings either move through the board as if reading English (Subject 2), or trace an "M" (Subject 1's program) or a "W" (Figures 4 and 9) along its lines.

Numbering quickly led to a one-dimensional, horizontal format: Subject 1 encoded a state as an array of numbers that denote color, Subject 2 used a string of B's and W's, and Subject 3 used circles in a row. The colors of the playing pieces were also clearly irrelevant. Subject 1 shifted to red and green, Subject 2 used letters instead, and Subject 1 shifted to red and blue. As observed in Section 1, operators that generate moves quickly and correctly are essential to expertise. Although the horizontal axis is ordinarily a neutral representation [20], it fails to convey key information here: the legal moves as defined by the edges. The edges on the game board serve as a move generator. Once they abandoned the two-dimensional board, the subjects needed a way to produce all possible moves methodically from a given state. Subject 2 extracted moves from an edge list that really contains indices into a string of B's and W's. Subject 1 wrote a simple routine that served the way Subject 2's edge list did. Subject 3 memorized them as the arcs in the centers of Figures 8 and 9.

Each subject established a state syntax with little deliberation. Subject 1 appeared relatively unscathed by his. Subject 2 labored under his arbitrary syntax and, as predicted by [3], struggled to generate moves correctly. Subject 3 began with one syntax and then progressed to another that gave him a more "satisfying" diagram. In every case, the linearity of the representation appeared to have made part of problem solving (move generation) more difficult because the subjects lost an important invariant, connectedness [22]. This accounts for missing nodes, missing edges, and incorrect edges in the diagrams.

There are some indications about these subjects' internal representations of a game state. Subject 1 used the computer as a sketch pad; once his code was complete, he appears to have relied completely on the board from Figure 1 for his own move generation. Subject 2 manipulated his linear state representation comfortably, but periodically converted back to the Figure 1 representation to consider what was really happening at certain points in the space. At crucial moments in its construction, he returned to the board diagram for confirmation, trusting his visual perception more than his mechanical process. Only Subject 3 appears to have made a complete shift to a one-dimensional state representation: "At the beginning I tried to imagine the situation on [the board]. Then I just manipulated the letters and numbers."

3.3 Reliance on a State Space Representation

There is no evidence that any of these subjects was able to retain or represent an entire state space diagram in memory. Two relied upon a diagram for memory, and their state space diagrams are more complex than their state diagrams [10]. This complexity derives from the number of states in the space, the introduction of color, the orientation of states with respect to each other (particularly in the later diagrams of Subject 3), the implicit animation (the arrows in Subject 3's later diagrams), the presence of two competitors, and the uncertainty about the other competitor's response.

Subject 1, although he claimed not to have drawn at all, effectively had his program draw state diagrams for him. While he competed against it, he certainly saw on the screen far many more pictures of the board itself than either of the other two students, but any representation of the state space he had was purely internal. As a result, one can only guess at the operators available to him.

In contrast, Subject 2 was strongly influenced by Figure 2. He explored the game by expanding the tree depth-first, generating moves with his list of edges. He noted repetition in the diagram, but did not attempt to winnow it out — rather, he used it to curtail further search until all nodes had been explored. Rather than imagine the result of a move, Subject 2 represented each result on paper. As a result, he was able to generate all the states correctly, but he also produced duplicates. His challenge then became detecting repeated states. Subject 2's directed edges were an innovation not shown during the lecture; they were in direct response to that need.

Subject 3 began in the style of Figure 2, but soon worked breadth first, that is, level by level. In addition, he became interested in the shape of the space more than its contents. He used repetition, both of states and of piece configuration, as a way to tease out all possibilities. His color-coded operations were player-dependent moves.

A diagram about play must include the moves, the player making them, and the ultimate outcome. Subject 1's program, which could have portrayed movement, simply makes the next state "appear" on the screen, without any transitional sliding. Subject 2 labeled the edges to designate the mover, eschewing the state labels of Figure 2. Subject 3 colored his edges to show how pieces might move. Both Subject 2 and Subject 3 highlighted winning nodes in their diagram, imputing importance to the absence of a cycle. Subject 3 went further, aligning these highlighted nodes. Subject 1 had no tangible record of winning states, only his memory of them.

Subjects 2 and 3 also established a state space syntax. While a game tree represents cycles by endless repetition, a *game graph* links moves backwards to earlier states. Initially, Subject 2 was seriously hindered by his tree representation; it made the detection of repetition difficult, and unexpanded states were not obvious. He soon incorporated elements of a game graph, using predominantly one-headed arrows to indicate repetition. This also allowed him to retain the game tree convention that represents the mover by a node's level. Subject 3, in contrast, soon moved from a game tree to a game graph, but with two-headed arrows. He found, however, that in his new syntax he could no longer "read" the board to identify the mover. This gave rise to incorrect associations among states with the same visual properties but different movers. In response, Subject 3 confirmed correctness and completeness visually, relying on the inherent symmetries of the game. Eventually he enhanced his edge representation with the identity of the mover. Furthermore, Subject 3 repeatedly rearranged the elements of his diagrams to highlight their role in play. By his fourth diagram he had abandoned any pretense at a tree; he then sought to convey not only sequences of states in play, but also the symmetries of their values. His final diagram emphasized the number of moves available from a state, which moves are dangerous, and which states are losses. This variety of perspective is predicted in [3]. Figure 9 includes the perspective of each mover, the perspective of "safe" moves, and the perspective of winning, along with an elegant statement of symmetry.

Within the pong hau k'i state space, some states are particularly salient. Subject 2 sketched (i.e., translated from a string back to a diagram) only two: Figure 5(b) where the mover wins, and Figure 5(c) where white must win or lose. Subject 1 appears to have noticed these in passing. Although Subject 3 eventually focused on four states one ply removed from Figure 5(b), he did not recognize the significance of those delicate states until he explained his diagram verbally to a friend — producing the diagram was not as powerful as interpreting it aloud.

The subjects' state space diagrams served a variety of purposes. Subject 1 used his computer-drawn states to provide him with playing experience; those states were connected temporally but visible only one at a time. Subject 2 used his tree to check whether or not he had considered all possible states, and his occasional sketches to check his reasoning. Subject 3 used his first diagrams the way Subject 2 used his tree, and subsequently used one diagram to check another, and portions of one diagram to generate other portions. He suspects that he began to think about strategy around the same time; his "structural" concerns in his third diagram were replaced by his "semantic" considerations in his fourth.

Kaplan and Simon's result that a good representation produces a smaller space [11] is borne out by Subject 3's switch to mover-less states. There is, however, no evidence here of an effective generator that relies on invariants. Finally, none of these subjects used their diagrams to compete. Each of them believed he had internalized the most important facets of his diagram(s), and could play perfectly without reference to them. The diagrams were thus an exploratory device. Once completed, they were not essential for expert performance.

3.4 Beyond the Assignment

As one becomes immersed in these protocols, it is difficult to remember precisely what the assignment was — these subjects turned it into individual discovery odysseys. The first part of the assignment was to "play it, take protocols on the development of your own expertise, and capture any representational shifts." All three played pong hau k'i, and were able to report on the development of their own experience to varying degrees. Subject 1 animated his numbering decision as Figure 4. Subject 2 drew his representational shift for the state diagrams at the top of the page. Subject 3 detailed the evolution of both his state diagrams and his state space diagrams.

The second part of the assignment was to "draw out the state space for pong hau k'i, label [it] with outcomes, produce a rule set to play it, and produce a good heuristic for it." Subject 1 never drew the state space, but he did realize "that the only way [to win] is to put the opponent into [two locations on the same] side," which is necessary and sufficient for expertise. Subject 2 first played a few contests, during which he realized that "it is quite difficult to lose the game." This gave him "an idea about situations [he] should avoid." His state space diagram contained a few errors, but he too had the right idea. Nonetheless, this experiment indicates that a traditional state space representation obscures the structure that gives rise to heuristics of the kind people readily develop.

Subject 3, in his search for structure, produced an increasingly symmetrical sequence of state space diagrams, eventually *planar* (without crossed edges) ones.

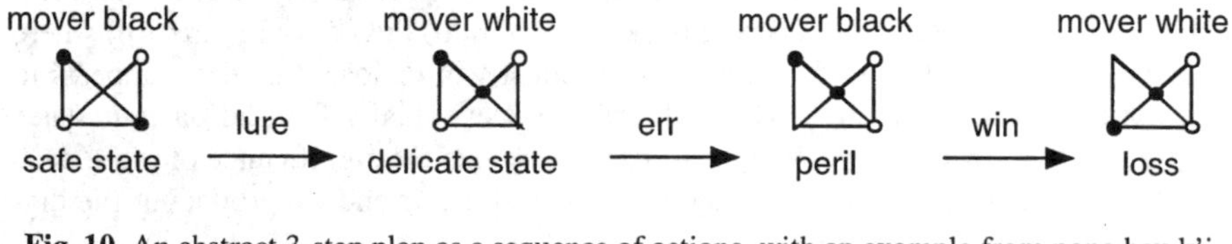

Fig. 10. An abstract 3-step plan as a sequence of actions, with an example from pong hau k'i

These representations were driven by his determination to understand the relationships within the space, long after he could play expertly. His final diagram, Figure 9, classifies states by their strategic importance and the number of choices they offer, rather than by the location of their pieces. It also makes clear where important and irrelevant decisions lie. Thus his final concepts were about vulnerability and freedom, not about states at all. His success is in part attributable to his HUYGENS-like search through diagrams for regularity, and in part to the insight stemming from his search through representation space [11]. Once Subject 3 was able to abandon the mover as part of the state representation, he could attend to the relevant features.

3.5 Discovery and Planning

This task was deliberately posed as a discovery problem. The exploratory behavior predicted by [12] was visible in the subjects' early play: "...we were placing the bread without any strategy, [we] just kept [following] the rules." Discovery in a game is made more difficult by one's own inability to predict the next state where one will be forced to act.

The only way to win at pong hau k'i is to lure one's opponent into a delicate state and hope that she errs. This three-step plan (lure, err, win) relies on the fallibility of the opposition. Figure 10 descriptively labels these actions and the states they transform, each from the perspective of the mover, with an example. The only other move from the delicate state restores the contest to a neutral state. All three subjects were aware of the dangers of the delicate state, but the protocols of Subjects 1 and 2 indicate only defensive use of it. Subject 1 wrote, "I played a few games ... during [which] I realized that it is quite difficult to lose the game, and I had an idea about situations I should avoid." Subject 2 wrote, "...we gave up, because I couldn't lose [either]." Only Subject 3 saw the plan's offensive and defensive aspects, once he discovered delicate states.

Subject 3's planning knowledge is closely linked to his state space diagrams. He writes of "the secure central area" and "the ... terminal columns ... [where] the game would either end in a terminal state or return to the secure central area ... via a delicate state. He also recognized that "it is not even guaranteed that any of the four delicate states will ever be reached." Thus a contest could circle only the innermost loop.

What supported discovery here? For Subject 3, it was the elimination of the mover label on the state, and the way the states oriented themselves on the paper. For Subject 1, the program that made random moves in such a small state space likely led to every state several times. Although this is insufficient to support the development of expertise in a larger space [6], it was adequate for pong hau k'i. For Subject 2, the move generator was essential, as predicted by [12].

There is much empirical evidence that diagrammatic representations support problem solving. Cheng has identified four processes in conceptual learning that diagrams can support: observation of the space, modeling to link and generalize observations, acquisition of new concepts, and integration of those concepts [3]. He argues that a diagram to support learning should encode the laws that govern the relevant phenomena in the problem domain, and highlight differences among concepts. Subject 3's final diagrams meet these criteria.

3.6 Testing the Representation

One hallmark of a good representation is its applicability to other, similar problems [3]. As a test of Subject 3's techniques, the author created a thought experiment: apply these methods to another simple game, called merely "pong." The rules of pong are identical to those of pong hau k'i, and its board also has five locations. Only its layout differs, as shown in Figure 11. At its conception, the author was unaware of the properties of pong's state space, but anticipated that one player could win by trapping the other on the bottom of the triangle.

Subjects in the central experiment and in [12] make clear that it is important to number the locations on the board in a way that supports move generation. Thus the "natural numbering" for pong with 1 at the top was immediately discarded for one that begins at the bottom left, goes up to the top, and down the right side. With this numbering a one-dimensional representation for the start state is BB–WW. Once again there are 30 possible board configurations and 60 possible states. The next step is to lay out the states on paper, and attempt to find a planar representation. Surprisingly, that proved more difficult; checking for repeated states requires careful bookkeeping. A list of all possible configurations proved a helpful check, as did the expectation of symmetry in the state space diagram.

Pong

- **Two-player game**
- **5 locations**
- **First to move plays black**
- **Second to move plays white**
- **Move = slide to the next empty location**
- **Initially, the board is as shown**
- **Mover who cannot slide loses**

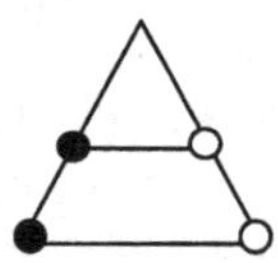

Fig. 11. The (invented) game of pong

When a diagram clarifies things, it becomes a tool. The pong state space diagram of Figure 12 adapts Subject 3's representation of the mover by edge color. That reduces the number of states to 30, represented in Figure 12 only as numbers (the order in which they were generated by hand). Legal play thus involves a sequence of edges that alternate in color. An algorithm to identify all reachable states from such a diagram, using spreading activation, appears in Table 1. It begins from the start state (labeled "1"), and alternately collects states reachable by edges of one color, then states reachable from those just-added states by edges of the opposite color. When the algorithm executes on Figure 12 with 1 as the start state, 10 of the theoretically possible pong states are never reached (those numbered above 20). As a result, the anticipated loss states (two pieces of the same color at the bottom of the triangle) never arise. Instead, the only realizable losses (states 15 and 20) are when a player occupies the top vertex and one of the mid-side locations. (As a matter of fact, only one such loss for each side is accessible, depending upon the first player to move. If white were to go first, 24 and 28 would be accessible instead.)

Could a better start state have been chosen for pong? The diagram in Figure 12 makes clear that no state and designated first mover in pong provide access to the

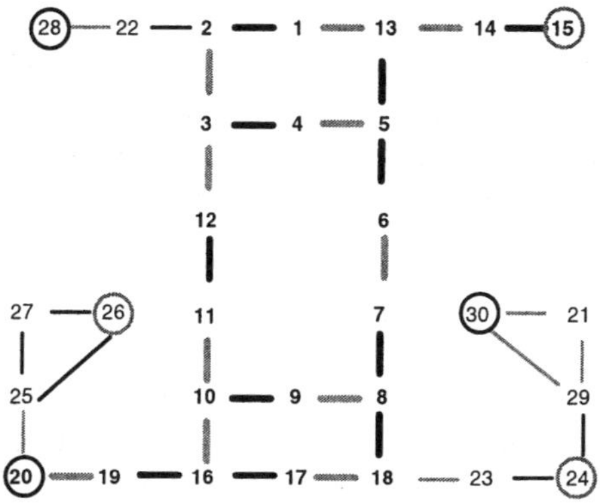

Fig. 12. An abstraction of pong's state space, with states numbered in the order in which they were generated. The start state is labeled 1. States reachable from 1 are in bold. Black lines denote a move for black; gray lines a move for white. Black circles denote a loss for black; gray circles a loss for white

Table 1. A high-level algorithm to identify all reachable states in an edge-colored state space like that in Figure 11. The function `other` toggles between two colors. When the algorithm halts, all nodes accessible from the start state are displayed

```
Reachable ← {start-state}
fringe ← {start-state}
new-fringe ← null
hue ← first mover's color
until fringe is empty
    select node from fringe
        for each edge with color = hue from node to new-node
          unless new-node is in reachable, add new-node to reachable and to new-fringe
    fringe ← new-fringe
    new-fringe ← null
    hue ← other(hue)
display reachable
```

full space. Ideally a diagram teaches the solver about the nature of the problem, that is, the person learns from the diagram she has drawn. The dangerous states in pong are 13 and 16. Since play begins at state 1 and black moves first, state 13 is entered only through a black move, where white has two choices, one of which will lead to its defeat. Similarly, the wrong decision in 16 will cost black the contest. Thus a player can make any legal move, except in the single state of concern to her. Unlike pong hau k'i, once your opponent has made an error, a win in pong requires no skill at all; it is inevitable.

This analysis is a demonstration of the power of a good diagram to highlight the key features of a problem. The author has also applied the same technique to a variety of other 5-location boards, and derived similar insights. The diagram moves one beyond the game itself, to concepts about what is possible within the game's framework.

Figure 12 also facilitates planning. As black, one wants to reach 15 which is only accessible through 13. In turn, 13 is only reachable from 5, which is itself inevitable from 1. So one must plan to move from 5 to 13 (and not to 6), hoping white will move to 14. This is another example of the (lure, err, win) plan structure. If black moves from 5 to 6 instead, white will have an opportunity at 10 to execute a similar plan.

4 Conclusions

Perceptual information from a diagram can provide guidance during problem solving. This experiment has shown how diagrams can be used during problem solving to develop representations, to inventory possible actions, to confirm representational shifts, to identify opportunities and dangers, and to plan. Ideally, the function of a state space diagram is to support display-based problem solving, done entirely within the context of the diagram without additional logical formalisms. Ultimately, display-based problem solving should support the extraction of simple but cogent rules for expertise, as it did here. Thus it is a tool to develop intelligent behavior, one that should eventually be exchanged for the clarity and speed of the heuristics it engenders.

In particular, AI researchers devised the game tree to facilitate their analysis of two-person games. A game tree captures turn-taking and the impact of sequences of decisions, and serves as a repository for states and their relationships. Indeed, this representation has supported the development of many sophisticated search mechanisms that now underlie champion game-playing programs. Nonetheless, in human hands, a game tree does not facilitate either the correct and complete generation of moves, or the detection of duplicates in cyclic games, both of which are crucial to expert play. Furthermore, both the "natural" extensions to game trees for duplicate states make them visually more complex. The mover-free state representation developed here is a step toward more comprehensible diagrams for games where states repeat.

Diagrams can also mislead. Each subject conserved effort by encoding the two-dimensional board in one dimension. Ironically, the linear representations they devised to simplify their task made move generation considerably more difficult. (It is, we suspect, far easier to imagine sliding along visible lines than jumping in a set of apparently disconnected leaps.) Thus they were forced to find some methodical way to generate moves: indexing into a sequence, envisioning with arrows, or writing a routine.

Diagrams are dispensable. In a relatively simple problem, like those addressed here, fortuitous induction may serve equally well. Although the state space diagram is an essential memory device, this experiment demonstrated that it does not necessarily reveal the deeper structure of its associated task, in part because of its size and complexity. To deduce deeper structure, however, some representation of state space is essential, the more compact the better, as a variant on the traditional game tree proved here. A good diagram not only captures the salient portions of a problem, it focuses attention on them. The best of these diagrams did this so well that they themselves became obsolete.

The "more is better" bias previously identified for learning tic-tac-toe and its isomorphs is valid here, but "more of what?" is the issue. There are five (non-disjoint) position categories in pong hau k'i: left, right, top, bottom, and center. The winning invariant is the number of distinct locations adjacent to those you hold. (An obvious error is to prefer more edges through the occupied locations instead. That heuristic prefers a top and a bottom location on the same side to two top locations, and leads to a loss.) Although all three subjects learned to play well, no representation specifically incorporated the winning invariant. Indeed, in some sense, the description of movement in the rules emphasizes the lines on the board, rather than the locations to which they lead. It thereby masks the winning invariant and makes the task more difficult. It would be interesting, therefore, to study isomorphs of this problem, including deformations of the essentially-square board.

One might hypothesize some process for the representational shifts that abound in these protocols, a "think-choose-try-evaluate-revise" cycle. This is supported by all the subjects' selection of a state representation, and by the third subject's intensive development of a search space diagram. In the latter, each new approach is repeatedly criticized and abandoned. This suggests that a meta-level critic ran in parallel as he drew, mediating between the visual representation and some more abstract expectations.

These protocols reveal a wealth of cognitive activity in which diagrams play a central role. The paradigmatic game tree diagram alone cannot, and did not, guarantee completeness, correctness, or well-managed repetition. Subject 1's diagrams were a kind of random travel through the space, but travel adequate to develop expertise. Subject 2's diagram served to satisfy his curiosity, not to organize his thoughts about the problem. Only with Subject 3 might one argue that diagrams led to deeper understanding of the game, an understanding unnecessary to expertise but satisfying in its level of knowledge organization. His final diagrams imposed order and a value system on a set of repetitive chronological sequences. Furthermore, the technique he developed has proved applicable to other simple cyclic games that could be applied for game-analysis and game authoring. It is a model of how diagrams support thought.

Acknowledgements

Thanks to Boicho Kokinov for his invitation to teach at The Cognitive Science Summer School, and for attracting such thoughtful students. Thanks too to the three anonymous, enthusiastic, and cooperative subjects, who transformed a homework assignment into an intriguing demonstration of the role of spatial cognition in human expertise. This work was supported in part by NSF IIS-0328743. This paper develops and expands upon material first presented at the Diagrams 2002 conference.

References

1. Amarel, S.: On Representations of Problems of Reasoning about Actions. In: Michie, D. (ed.): Machine Intelligence 3. Edinburgh University Press, Edinburgh (1968) 131-171
2. Anzai, Y. and Simon, H.: The Theory of Learning by Doing. Psychological Review 36 (1979) 124-140
3. Cheng, P. C.-H.: Unlocking conceptual learning in mathematics and science with effective representational systems. Computers in Education 33 (1999) 109-130
4. Cheng, P. C.-H. and Simon, H. A.: Scientific discovery and creative reasoning with diagrams. In: Smith, S., Ward, T., and Finke, R. (eds.): The Creative Cognition Approach. MIT Press, Cambridge, MA (1995). 205-228

5. Cross, N., Christaans, H., and Dorst, K. (eds.). Analyzing Design Activity. John Wiley and Sons. Chichester (1996)
6. Epstein, S. L.: Toward an Ideal Trainer. Machine Learning 15 (1994) 251-277
7. Falkenhainer, B. C. and Michalski, R. S.: Integrating Quantitative and Qualitative Discovery. Machine Learning 1 (1986) 367-401
8. Gelernter, H.: Realization of a Geometry-Theorem Proving Machine. In: Feigenbaum, E. A. and Feldman, J. (eds.): Computers and Thought. McGraw-Hill, New York (1963) 134-152
9. Hadamard, J.: The Psychology of Invention in the Mathematical Field. Dover Publications. New York (1945)
10. Hegarty, M.: Mental Visualizations and External Visualizations. In: Gray, W. and Schunn, C. (eds.): Proceedings of the 24th Annual Meeting of the Cognitive Science Society, Lawrence Earlbaum, CA (2002) 40-45
11. Kaplan, C. A. and Simon, H. A.: In Search of Insight. Cognitive Psychology 22 (1990) 374-419
12. Kotovsky, K., Hayes, J. R., and Simon, H. A.: Why are some problems hard? Cognitive Psychology 17 (1985) 248-294
13. Langley, P., J. M. Zytkow, Simon, H. A., and Bradshaw, G. L.: The Search for Regularity: Four Aspects of Scientific Discovery. In: Michalski, R. S., Carbonell, J. G. and Mitchell, T. M. (eds.): Machine Learning 2: An Artificial Intelligence Approach. Tioga Publishing, Palo Alto (1986) 425-469
14. Larkin, J. H.: Display-based Problem Solving. In: Klahr, D. and Kotovsky, K. (eds.) Complex Information Processing: The Impact of Herbert A. Simon. Lawrence Erlbaum Associates: Hillsdale, New Jersey (1989) 319-341
15. Pascal, B.: Pensées de Pascal. Éditions Garnier Frères, Paris (1964)
16. Pearson, D. Alexander, G., C., and Webster, R.: Working Memory and Expertise Differences in Design. In: Gero, J. S., Tversky, B. T. and Purcell, T. (eds.) Visual and Spatial Reasoning in Design II. Key Centre of Design Computing and Cognition, University of Sydney, Sydney, Australia, (2001) 237-251
17. Poincaré, H.: La Valeur de la Sciènce. Flammarion, France (1970)
18. Saariluoma, P. and Maarttola, I.: Spatial Mental Content and Visual Design In: Gero, J. S., Tversky, B. T. and Purcell, T. (eds.) Visual and Spatial Reasoning in Design II. Key Centre of Design Computing and Cognition, University of Sydney, Sydney, Australia, (2001) 253-268
19. Suwa, M., B. T. Tversky, Gero, J. S., and Purcell, T.: Seeing into sketches: Regrouping Parts Encourages New Interpretations. In: Gero, J. S., Tversky, B. T. and Purcell, T. (eds.) Visual and Spatial Reasoning in Design II. Key Centre of Design Computing and Cognition, University of Sydney, Sydney, Australia, (2001) 207-219
20. Tversky, B. T.: Multiple Mental Models. In: Gero, J. S., Tversky, B. T. and Purcell, T. (eds.) Visual and Spatial Reasoning in Design II. Key Centre of Design Computing and Cognition, University of Sydney, Sydney, Australia, (2001) 3-13
21. Zaslavsky, C.: Tic Tac Toe and Other Three-in-a-Row Games, from Ancient Egypt to the Modern Computer. Crowell, New York (1982)
22. Zhang, J.: The nature of external representations in problem solving. Cognitive Science 21 (1997) 179-217

The Finest of Its Class: The Natural Point-Based Ternary Calculus $\mathcal{LR}$ for Qualitative Spatial Reasoning

Alexander Scivos and Bernhard Nebel

Institut für Informatik
Albert-Ludwigs-Universität Freiburg,
Georges-Köhler-Allee Geb.52, 79110 Freiburg, Germany
{scivos, nebel}@informatik.uni-freiburg.de

Abstract. In this paper, a ternary qualitative calculus $\mathcal{LR}$ for spatial reasoning is presented that distinguishes between left and right. A theory is outlined for ternary point-based calculi in which all the relations are invariant when all points are mapped by rotations, scalings, or translations (RST relations). For this purpose, we develop methods to determine arbitrary transformations and compositions of RST relations. We pose two criteria which we call practical and natural. 'Practical' means that the relation system should be closed under transformations, compositions and intersections and have a finite base that is jointly exhaustive and pairwise disjoint. This implies that the well-known path consistency algorithm [10] can be used to conclude implicit knowledge. 'Natural' calculi are close to our natural way of thinking because the base relations and their complements are connected. The main result of the paper is the identification of a maximally refined calculus amongst the practical natural RST calculi, which turns out to be very similar to Ligozat's flip-flop calculus. From that it follows, e.g., that there is no finite refinement of the TPCC calculus by Moratz et al that is closed under transformations, composition, and intersection.

1 Introduction

Reasoning about spatial configurations is an important task for many applications such as geographical information systems (GIS), natural language understanding, and automatic geometric proofs. In many cases, no detailed quantitative information of the spatial structures under consideration is available. For example, images show only the relative alignment of objects, and text information often contains rough descriptions such as "coming from point a you have to turn right at point b to reach region c". In such cases, qualitative approaches that define formal representations of everyday descriptions are used. They are an effective way to conclude implications of the given spatial information.

In recent years, a series of qualitative spatial calculi have been proposed and analyzed, such as a calculus for reasoning about topological relations [1, 13], calculi about orderings [5, 9, 14], directions [2, 12], relative position of a point with respect to a line segment [8, 3, 4, 11] and others.

C. Freksa et al. (Eds.): Spatial Cognition IV, LNAI 3343, pp. 283–303, 2005.
© Springer-Verlag Berlin Heidelberg 2005

In this paper, we focus on calculi as the latter ones, whereas we develop a general theory for ternary, point-based relations that are invariant when all points are mapped by rotations, scalings or translations. This means that, e.g. the claim that one point lies on a straight line between two other points remains true when the whole map is rotated, shifted or when the scale is changed. We call such relations *RST relations* according to the first letters of the transformations. We present a new way of describing them that gives each RST relation a standard name or representation. This leads to a calculation of compositions and transformations of qualitative relations.

Starting with the observation that Freksa's double-cross calculus [3, 4] has some unsatisfactory properties [16], we consider the entire class of RST calculi, of which the double-cross calculus is an instance. For practical reasons, we expect a calculus to be finite and closed under certain fundamental operations such as transformations and compositions that are used to conclude implicit knowledge. It is known that Freksa's calculus and its finite refinements do not have this closure property [16]. Furthermore, we require that the relation systems of the calculus has the property that base relations do not denote arbitrary sets of points, but they should be connected regions. The goal of this paper is to identify the most refined RST calculus with the these properties. As it turns out, this is a calculus which is very close to Ligozat's flip-flop calculus, and which we call $\mathcal{LR}$.

The remainder of the paper is structured as follows. In section 1, the calculus $\mathcal{LR}$ is defined as an example for a calculus with ternary RST relations. In Section 3, we define the requirements which are necessary for a practical calculus and prove some basic properties of all ternary calculi not yet dealing with the RST property. In Section 4, the theory of RST Calculi, the way of describing ternary RST relations and some fundamental consequences, are introduced. In Section 5, it is proven that $\mathcal{LR}$ is the finest finite RST calculus with the properties proposed in Section 3. In Section 6, a conclusion is given.

2 The Calculus $\mathcal{LR}$

When human beings or robots proceed along a path (going from point a_1 to point a_2), they always distinguish between things ahead of them or behind their back, and they can distinguish whether objects they pass have been to their right or to their left, or if they have directly met them. Moreover, it is easy to recognize in which order certain points have been reached, hence, which objects are between others on the line. However, without having additional information, it is often not possible to find out at which distance from the path an object is located, or at which angle precisely it can be seen.

All these spatial expressions involve the current standpoint or *starting point* a_1, the walking direction and a focus point a_3. The direction can be easily represented by a goal or *reference point* a_2 along the line. The spatial situation can then be described by sentences like the following ones (refer to Figure 1.a):

1. Looking from a_1 to a_2, a_3 is to the left.
2. Walking from a_1 to a_2, $a_3{}'$ is always at your back.

This idea is formalized to a calculus by introducing relations and operations on the relations. We call it the left-right-distinguishing calculus $\mathcal{LR}$.

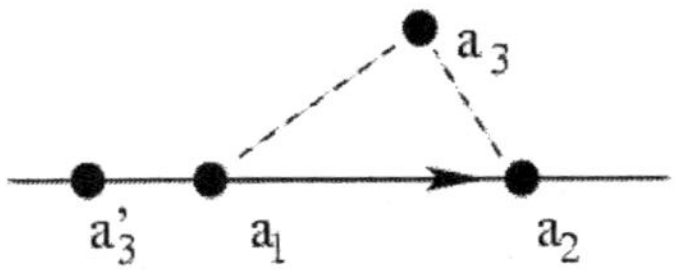

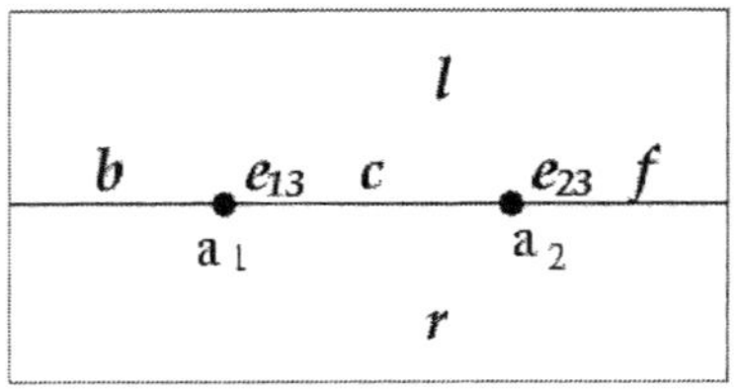

Fig. 1. a. The two examples of relations given in the text

b. The base relations of $\mathcal{LR}$ where $a_1 \neq a_2$ The letters are explained in Table 1

Table 1. Definition of the base relations of $\mathcal{LR}$

$\mathcal{LR}$ Base Relation for triple (a_1, a_2, a_3)		angle at a_1 $\angle(a_2, a_1, a_3)$	angle at a_2 $\angle(a_3, a_2, a_1)$	Meaning
eq	$a_1 = a_2 = a_3$	-	-	a_1, a_2 and a_3 are all equal.
e_{12}	$a_1 = a_2 \neq a_3$	-	-	a_3 is different from $a_1 = a_2$.
e_{13}	$a_1 = a_3 \neq a_2$	-	0°	a_2 is different from $a_1 = a_3$.
e_{23}	$a_1 \neq a_2 = a_3$	0°	-	a_1 is different from $a_2 = a_3$.
	$a_1 \neq a_2 \neq a_3 \neq a_1$:			Looking from a_1 to a_2:
b	back	180°	0°	a_3 is back behind a_1.
c	closer	0°	0°	a_3 is closer to a_1 than a_2.
f	further	0°	180°	a_3 is further ahead.
r	right	$]-180^o; 0^o[$	$]-180^o; 0^o[$	a_3 is to the right.
l	left	$]0^o; 180^o[$	$]0^o; 180^o[$	a_3 is to the left.

2.1 Relations

For each of the different situations we introduce a relation with three arguments a_1, a_2, a_3. Each argument represents a point in the plane $\mathbb{R}^2$. We consider two alternatives in the case that a_1 and a_2 coincide and distinguish between seven situations for a_3 when a_1 and a_2 are different. This leads to a calculus that contains the nine relations as shown in Table 1. Note that for all possible triples of points there is a relation of the calculus (see Figure 1.b), which distinguishes this calculus from Ligozat's flip-flop calculus which does not have the relations eq and e_{12}. In order to express uncertainty, all unions of these nine relations are included in the calculus $\mathcal{LR}$. For example, we write $(a_1, a_2, a_3) \in eq \cup e_{12}$ if we know that $a_1 = a_2$ but do not know anything about a_3. The union of all relations is denoted by $\top$. $(a_1, a_2, a_3) \in \top$ contains no information about (a_1, a_2, a_3).

Definition 1 (Base Relations)
A minimal subset of relations $\mathcal{B} \subset \mathcal{C}$ is called a base *of the calculus $\mathcal{C}$ if any relation R in $\mathcal{C}$ is a set union of some relations in $\mathcal{B}$. The relations in $\mathcal{B}$ are called* base relations *of $\mathcal{C}$.*

2.2 Operations

A calculus provides formal ways to conclude implicit knowledge. There are several ways to derive new claims. Therefore, we define operations on the relations so that the resulting relations represent what we can derive. As standard methods for ternary relations, we use intersection, transformation, and composition, which are generalizations of the corresponding well-known operations for binary relations[7, 17] and of the operations on ternary relations as defined by Isli et al [5].

Intersection: Just as in the case of binary relations, we can combine two claims about the same triple into one claim.

$$(a_1, a_2, a_3) \in \mathsf{R}_1 \quad \wedge \quad (a_1, a_2, a_3) \in \mathsf{R}_2 \Rightarrow (a_1, a_2, a_3) \in (\mathsf{R}_1 \cap \mathsf{R}_2)$$

Transformation: Transformation generalizes the binary concept of the converse (exchanging the arguments of a binary relation). If we have a claim about a triple, we can derive a claim about any permutation of the triple. Therefore, the transformation operations are defined as follows:

Definition 2 (Transformation of a Relation)
Let $\pi \in \mathfrak{S}_3$ be a permutation of the positions of a ternary relation $\mathsf{R} \subset (\mathbb{R}^2)^3 := \mathbb{R}^2 \times \mathbb{R}^2 \times \mathbb{R}^2$, then R^π is the **transformed relation** *with the property*

$$(a_1, a_2, a_3) \in \mathsf{R} :\Longleftrightarrow \bar{\pi}((a_1, a_2, a_3)) := (a_{\pi(1)}, \ldots, a_{\pi(3)}) \in \mathsf{R}^\pi$$

The operation T^π: $\mathsf{R} \mapsto \mathsf{R}^\pi$ is called a transformation.

Examples:
The permutation (231) is a rotation of the indexes that maps 2 to 3, 3 to 1, and 1 to 2. $T^{(231)}$ corresponds with Isli's [5] "rotation" operation.

$$(a_1, a_2, a_3) \in \mathsf{i} \iff (a_2, a_3, a_1) \in \mathsf{b} = \mathsf{i}^{(231)}$$

The permutation (23) is an exchange of the last two indexes that maps 1 to 1, 2 to 3, and 3 to 2. The operation $T^{(23)}$ corresponds with Isli's "converse" operation.

$$(a_1, a_2, a_3) \in \mathsf{l} \cup \mathsf{e}_{12} \iff (a_1, a_3, a_2) \in \mathsf{r} \cup \mathsf{e}_{13} = (\mathsf{l} \cup \mathsf{e}_{12})^{(23)}$$

Table 2 displays all the transformations of $\mathcal{LR}$ (see also [6]).

Composition: Restrictions that originate from a combination of the relations of two overlapping triples are called composition. With ternary relations, one can think of several ways of composing them, depending on the number and order of overlapping points. We proved [15] that the only case in which proper new restrictions can be derived is when the two relations concerning different triples have two common points. Depending on the order of the points in the original relations, we distinguish six different types of composition of which two examples are shown in Figure 2.

Table 2. Transformations of $\mathcal{LR}$

$\mathcal{LR}$ R	$R^{(12)}$	$R^{(13)}$	$R^{(23)}$	$R^{(231)}$	$R^{(321)}$
eq	eq	eq	eq	eq	eq
e_{12}	e_{12}	e_{23}	e_{13}	e_{23}	e_{13}
e_{13}	e_{23}	e_{13}	e_{12}	e_{12}	e_{23}
e_{23}	e_{13}	e_{12}	e_{23}	e_{31}	e_{12}
b	f	c	b	c	f
c	c	b	f	f	b
f	b	f	c	b	c
r	l	l	l	r	r
l	r	r	r	l	l

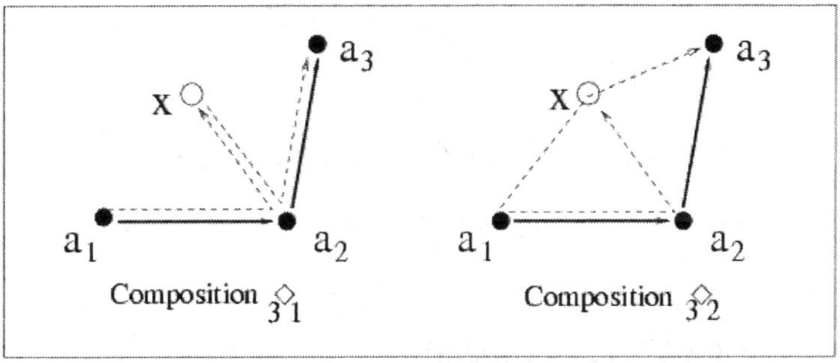

Fig. 2. The idea of compositions. The dotted lines indicate the relations R, S and the solid line the relation $R \diamond S$

Definition 3 (Compositions)
Let $R, S \subset (\mathbb{R}^2)^3$ *be ternary relations. Then for* $\kappa \neq \lambda$ *and* $\kappa, \lambda \in \{1,2,3\}$ *the composition* ${}_\kappa\diamond_\lambda$ *is defined as follows:*

$$(a_1,a_2,a_3) \in R\ {}_3\diamond_2\ S :\Longleftrightarrow \exists x : (a_1,a_2,x) \in R \wedge (a_1,x,a_3) \in S$$

$$(a_1,a_2,a_3) \in R\ {}_3\diamond_1\ S :\Longleftrightarrow \exists x : (a_1,a_2,x) \in R \wedge (x,a_2,a_3) \in S$$

$$(a_1,a_2,a_3) \in R\ {}_2\diamond_3\ S :\Longleftrightarrow \exists x : (a_1,x,a_3) \in R \wedge (a_1,a_2,x) \in S$$

$$(a_1,a_2,a_3) \in R\ {}_2\diamond_1\ S :\Longleftrightarrow \exists x : (a_1,x,a_3) \in R \wedge (x,a_2,a_3) \in S$$

$$(a_1,a_2,a_3) \in R\ {}_1\diamond_3\ S :\Longleftrightarrow \exists x : (x,a_2,a_3) \in R \wedge (a_1,a_2,x) \in S$$

$$(a_1,a_2,a_3) \in R\ {}_1\diamond_2\ S :\Longleftrightarrow \exists x : (x,a_2,a_3) \in R \wedge (a_1,x,a_3) \in S\,.$$

Example:
Suppose there is a house x to the left of the path from a_1 to a_2, and there is a tree a_3 behind the house as seen from a_1. Then the tree a_3 is also to the left of the path since

$$(a_1,a_2,x) \in l,\ (a_1,x,a_3) \in f \Rightarrow (a_1,a_2,a_3) \in l\ {}_3\diamond_2\ f = l.$$

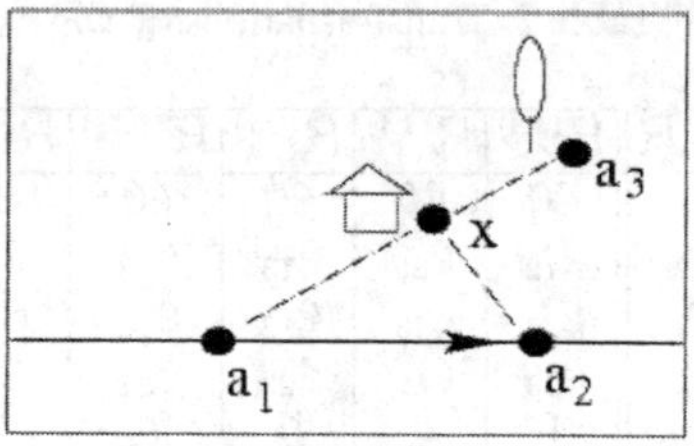

Fig. 3. An example for applying transformation and composition

When looking from a_2 to the house, the tree a_3 will be seen to the right. To derive this, transformation and composition are needed:

$$(a_1, a_2, x) \in l \Rightarrow (a_2, x, a_1) \in l^{(231)} = l.$$

Note that $l \;{}_3\diamond_1\; f = r$, hence

$$(a_2, x, a_1) \in l, (a_1, x, a_3) \in f \Rightarrow (a_2, x, a_3) \in r.$$

Throughout this paper, we will revisit this calculus along with the new concepts we introduce.

3 Properties for Ternary Calculi

3.1 Closure Properties

In $\mathcal{LR}$, any given spatial constellation gives rise to a corresponding set of relations between its triples, and it is possible to describe all conclusions that can be derived using intersection, transformation, and composition of the relations. It is reasonable to generalize these properties as requirements for calculi since this is what calculi are used for.

A calculus $\mathcal{C}$ is called closed under an operation iff for any choice of relations of $\mathcal{C}$ the result of the operation is again a relation of $\mathcal{C}$.

We assume that all calculi are closed under set union to represent uncertainty. If a calculus is closed under intersection, all transformations, and all compositions, then implicit information can be made explicit using the conclusion methods of the calculus mentioned above.

In general, like in the case of $\mathcal{LR}$, we expect that for any triple of points of the plane $\mathbb{R}^2$ there is a unique base relation that describes it. In other words, all base relations jointly cover the set of all possible triples $(\mathbb{R}^2)^3$, and the intersection of different base relations is disjoint. Hence, we have a jointly exhaustive pairwise disjoint basis (JEPD basis). A calculus with a JEPD basis is closed under intersection and set complement. In order to be part of a representation in a computer, this set of base relations has to be finite. Any practical calculus should satisfy these requirements.

Definition 4 (Practical Calculus)
A calculus $\mathcal{C}$ is called practical *if it is closed under transformations, compositions, and intersections and has a finite JEPD basis.*

Practical calculi have advantageous formal properties. A variation of the well-known path-consistency algorithm [10] can be used to find inconsistencies. In this section, some useful algebraic results are presented that are needed for the purpose of this paper: The main result is that there is a tight dependence between composition and transformation: Each composition can be derived from any of the others using transformations. Hence, it is sufficient to store one composition table. Moreover, transformations and compositions distribute over set union. Therefore, in a practical calculus it is sufficient to define these operations on the base relations.

In order to state the dependence property, it is necessary to note that the concatenation of transformations corresponds with concatenation of their underlying permutations [15]. As a consequence we have the following result:

Remark 1 (Inverse Transformation)
For each transformation operation T^π, there exists the inverse transformation operation $(T^\pi)^{-1} := T^{\pi^{-1}}$:

$$(T^\pi)^{-1}(T(\mathsf{R})) = T^{\pi\circ\pi^{-1}}(\mathsf{R}) = T^{id}(\mathsf{R}) = \mathsf{R}.$$

The inverse transformation operation is used to derive one composition from another:

Proposition 1 (Interdependence of Compositions)
For all $\kappa_1, \lambda_1, \kappa_2, \lambda_2 \in \{1,2,3\}$:

$$\mathsf{R}\ {}_{\kappa_2}\diamond_{\lambda_2}\ \mathsf{S} = (\mathsf{R}^\pi\ {}_{\kappa_1}\diamond_{\lambda_1}\ \mathsf{S}^\pi)^{\pi^{-1}},$$

where $\pi \in \mathfrak{S}$ is the permutation for which: $\pi(\kappa_1) = \kappa_2,\ \pi(\lambda_1) = \lambda_2$

Proof:
We introduce the notation $s_\kappa(x)((a_1,a_2,a_3))$ for the substitution of the κ-th element of the triple by x.

First, we prove an equality that holds for any transformation operation π and composition ${}_\kappa\diamond_\lambda$:

$$T^\pi(\mathsf{R}\ {}_\kappa\diamond_\lambda\ \mathsf{S}) = T^\pi(\mathsf{R})\ {}_{\pi^{-1}(\kappa)}\diamond_{\pi^{-1}(\lambda)}\ T^\pi(\mathsf{S}) \qquad (*)$$

By the definition of transformation operation and composition we obtain

$$\begin{aligned}
&(a_1,a_2,a_3) \in T^\pi(\mathsf{R}\ {}_{\kappa_2}\diamond_{\lambda_2}\ \mathsf{S})\\
\Longleftrightarrow\ &(a_{\pi^{-1}(1)},\ldots,a_{\pi^{-1}(3)}) \in (\mathsf{R}\ {}_{\kappa_2}\diamond_{\lambda_2}\ \mathsf{S})\\
\Longleftrightarrow\ &\exists d : (s_{\kappa_2}(d))(a_{\pi^{-1}(1)},\ldots,a_{\pi^{-1}(3)}) \in \mathsf{R} \text{ and } (s_{\lambda_2}(d))(a_{\pi^{-1}(1)},\ldots,a_{\pi^{-1}(3)}) \in \mathsf{S}\\
\Longleftrightarrow\ &\exists d : (s_{\pi^{-1}(\kappa_2)}(d))((a_1,a_2,a_3)) \in \mathsf{R}^\pi \text{ and } (s_{\pi^{-1}(\lambda_2)}(d))((a_1,a_2,a_3)) \in \mathsf{S}^\pi\\
\Longleftrightarrow\ &\exists d : (s_{\kappa_1}(d))((a_1,a_2,a_3)) \in \mathsf{R}^\pi \text{ and } (s_{\lambda_1}(d))((a_1,a_2,a_3)) \in \mathsf{S}^\pi\\
\Longleftrightarrow\ &(a_1,a_2,a_3) \in (\mathsf{R}^\pi\ {}_{\kappa_1}\diamond_{\lambda_1}\ \mathsf{S}^\pi)
\end{aligned}$$

which proves the equality (*).

With the premise on π that $\pi^{-1}(\kappa_1) = \lambda_1$ and $\pi^{-1}(\kappa_2) = \lambda_2$ follows the claim:

$$\mathsf{R} \;{}_{\kappa_2}\!\diamond_{\lambda_2}\; \mathsf{S} = T^{\pi^{-1}}(T^{\pi}(\mathsf{R} \;{}_{\kappa_2}\!\diamond_{\lambda_2}\; \mathsf{S})) \overset{(*)}{=} T^{\pi^{-1}}(T^{\pi}(\mathsf{R}) \;{}_{\kappa_1}\!\diamond_{\lambda_1}\; T^{\pi}(\mathsf{S})).$$

(qed.)

Proposition 2 (Distribution Properties of Practical Calculi)
1. Let $\mathcal{C}$ be closed under transformation and set union. Then for any set of relations $\mathsf{R}_i \in \mathcal{C}$:

$$T^{\pi}(\bigcup_{i\in I} R_i) = \bigcup_{i\in I} T^{\pi}(R_i)$$

2. For each $0 < \kappa, \lambda \leq 3$ and relations R_i, S:

$$(\bigcup_{i\in I} \mathsf{R}_i) \;{}_{\kappa}\!\diamond_{\lambda}\; \mathsf{S} = \bigcup_{i\in I} (\mathsf{R}_i \;{}_{\kappa}\!\diamond_{\lambda}\; \mathsf{S})$$

and

$$\mathsf{S} \;{}_{\kappa}\!\diamond_{\lambda}\; (\bigcup_{i\in I} \mathsf{R}_i) = \bigcup_{i\in I} (\mathsf{S} \;{}_{\kappa}\!\diamond_{\lambda}\; \mathsf{R}_i)$$

Proof:

$$\begin{aligned}
\bar{\pi}((a_1,a_2,a_3)) \in T^{\pi}(\textstyle\bigcup_{i\in I} \mathsf{R}) &\iff (a_1,a_2,a_3) \in (\textstyle\bigcup_{i\in I} R_i) \\
&\iff \exists i \in I: \quad ((a_1,a_2,a_3)) \in R_i \\
&\iff \exists i \in I: \quad \bar{\pi}((a_1,a_2,a_3)) \in T^{\pi}(\mathsf{R}) \\
&\iff \bar{\pi}((a_1,a_2,a_3)) \in (\textstyle\bigcup_{i\in I} (T^{\pi}(\mathsf{R}))
\end{aligned}$$

For all $\kappa, \lambda, \mathsf{R}_i, \mathsf{S}$ holds:

$$\begin{aligned}
& (a_1,a_2,a_3) \in (\textstyle\bigcup_{i\in I} \mathsf{R}_i) \;{}_{\kappa}\!\diamond_{\lambda}\; \mathsf{S} \\
\iff & \exists x: \; s_{\kappa}(x)(a_1,a_2,a_3) \in (\textstyle\bigcup_{i\in I} \mathsf{R}_i) \;\wedge\; s_{\lambda}(x)(a_1,a_2,a_3) \in \mathsf{S} \\
\iff & \exists x: \; (\textstyle\bigvee_{i\in I} (s_{\kappa}(x)(a_1,a_2,a_3) \in \mathsf{R}_i)) \;\wedge\; (s_{\lambda}(x)(a_1,a_2,a_3) \in \mathsf{R}_2) \\
\iff & \exists x: \; \textstyle\bigvee_{i\in I} ((s_{\kappa}(x)(a_1,a_2,a_3) \in \mathsf{R}_i) \;\wedge\; (s_{\lambda}(x)(a_1,a_2,a_3) \in \mathsf{R}_2)) \\
\iff & \textstyle\bigvee_{i\in I} (\exists x: ((s_{\kappa}(x)(a_1,a_2,a_3) \in \mathsf{R}_i) \;\wedge\; (s_{\lambda}(x)(a_1,a_2,a_3) \in \mathsf{R}_2))) \\
\iff & (a_1,a_2,a_3) \in \textstyle\bigcup_{i\in I} (\mathsf{R}_i \;{}_{\kappa}\!\diamond_{\lambda}\; \mathsf{S})
\end{aligned}$$

The second form of this property can be proved analogously. (qed.)

3.2 Refinement

Typically, in a real-world situation, the information available does not match one-to-one the relations of a calculus. We have included the operation set union so that we can describe spatial information for which we have only indefinite knowledge.

Sometimes, conversely, the available knowledge is more detailed than a selected calculus can express. Then, information is lost when building the knowledge base. This loss is smaller if more detailed relations are available. The observation that some expressions are combinations of finer ones, e. g. "on the way from a to b" means either "at a", or "at b", or "in between", motivates the idea of refining some relations. A calculus that comprises another one's relations as unions of its own relations, is called a refinement of the coarser one.

Definition 5 (Refinement)
A calculus $\mathcal{C}_{fine}$ is called finer *than another calculus $\mathcal{C}_{coarse}$ if each relation in $\mathcal{C}_{coarse}$ is a set union of relations of $\mathcal{C}_{fine}$. Then, $\mathcal{C}_{coarse}$ is called* coarser *than $\mathcal{C}_{fine}$.*

At the beginning of Section 5, we introduce another concept that is related to connectivity. A calculus is called natural if its base relations and their complements are connected. Our goal is to find the finest natural practical (hence: finite) calculus. We concentrate on a class of calculi that proved to be of much interest because its relations are independent of scaling, rotation, or shift from fixed reference points: RST calculi.

4 RST Calculi

In this section, the special properties of RST calculi are analyzed and a theory of RST relations is developed.

Definition 6 (RST Calculus)
Rotations, scalings with scaling factor > 0 and translations and concatenations of such mappings are called RST automorphisms. *A ternary relation R over points in the plane $\mathbb{R}^2$ is called* RST relation *if it has the* RST property*:*

For all RST automorphisms ρ holds

$$(a_1, a_2, a_3) \in \mathsf{R} \Rightarrow (\rho(a_1), \rho(a_2), \rho(a_3)) \in \mathsf{R}.$$

If all relations of a calculus are RST relations, it is called an RST calculus.

In the next subsections, we will see that there is a finest RST calculus. Thus, arbitrary RST relations can be understood as union of the base relations of the finest RST calculus. Because of Proposition 2, the operations of transformation and composition of other RST calculi are based on the transformations and compositions of the finest one.

4.1 Standardized Triples

Figure 1b. shows that for fixed points a_1, a_2 as starting and reference point, there are different relations of $\mathcal{LR}$, depending on the location of a_3 with respect to a_1 and a_2. All relations except eq and e_{12} are represented by that region in the plane, which is a possible location for a_3 given the position of a_1, a_2. This scheme is the basis of the following idea: Each triple can be mapped by rotation, scaling, and translation to a standardized triple (b_1, b_2, b_3) such that $b_1 = (0, 0)$ and $b_2 = (0, 0)$ (if a_1 and a_2 coincide), or $b_2 = (1, 0)$ (if $a_1 \neq a_2$).

To make this idea explicit, we present the function η that maps a triple to its standardized triple. This can most easily be done if we identify a point in $\mathbb{R}^2$ with a complex number using the standard isomorphism between $\mathbb{R}^2$ and $\mathbb{C}$ because the RST automorphisms of $\mathbb{R}^2$ are exactly those mappings that correspond with simple arithmetic operations in $\mathbb{C}$: The addition of a complex number corresponds with a translation, multiplication with a scalar value $r \in \mathbb{R}$ corresponds with a scaling, and multiplication with a purely imaginary number $r\imath$ ($r \in \mathbb{R}$) corresponds with a rotation of the complex plane.

To simplify the reading, for $z_i = x_i + \imath y_i \in \mathbb{C}$ let (z_1, z_2, z_3) denote the triple $((x_1, y_1), (x_2, y_2), (x_3, y_3))$. We distinguish three cases:

A. For $z_1 = z_2 = z_3$ (i.e. $(z_1, z_2, z_3) \in \mathsf{eq}$), we define the standardization as follows: $\eta((z_1, z_2, z_3)) := (0, 0, 0)$. This corresponds with a shift of the plane.
B. For $z_1 = z_2 \neq z_3$ (i.e. $(z_1, z_2, z_3) \in \mathsf{e}_{12}$), set $\eta((z_1, z_2, z_3)) := (0, 0, 1)$.
C. For $z_1 \neq z_2$, set

$$\eta((z_1, z_2, z_3)) := (0, 1, \frac{z_3 - z_1}{z_2 - z_1}).$$

These functions are well-defined.

In addition to a shorter notation, this $\mathbb{C}$ based representation is motivated by the following strategy of our proof: We show that any RST relation can be identified with a set of complex numbers. The advantage of such a notation is that we can determine complex functions that correspond with compositions and transformations on RST relations. By investigating consequences of possible results of these functions, we will find constraints on the set of relations in a practical calculus. By combining several of these constraints, we prove that the finest practical calculus cannot be further refined. We will only use arithmetic properties of $\mathbb{C}$.

In order to motivate our first step, we show that triples have the same standardization η only if there is no RST relation that can distinguish between them.

Proposition 3 (Standardization)
In a ternary RST calculus with η defined as above, holds:
$\eta((z_1, z_2, z_3)) = \eta((z_1', z_2', z_3'))$
iff there is an RST automorphism α that maps (z_1, z_2, z_3) to (z_1', z_2', z_3')
iff for all RST relations R holds: $(z_1, z_2, z_3) \in \mathsf{R} \Leftrightarrow (z_1', z_2', z_3') \in \mathsf{R}$.

Proof:
First, we show: $\eta((z_1, z_2, z_3)) = \eta((z_1', z_2', z_3'))$ implies that there is an RST automorphism α with $\alpha(z_1, z_2, z_3)) = (z_1', z_2', z_3')$.

By definition, on (z_1, z_2, z_3), η is equal to a concatenation of the complex subtraction

$$\tau : z \mapsto z - z_1$$

(corresponding with a translation of points in $\mathbb{R}^2$) and the multiplication (depending on the case A., B. or C.)

$$\rho_A : z \mapsto 1 \cdot z \qquad \text{or} \qquad \rho_B : z \mapsto \frac{1}{z_3 - z_1} \cdot z \qquad \text{or} \qquad \rho_C : z \mapsto \frac{1}{z_2 - z_1} \cdot z$$

(corresponding with a rotation and scaling of $\mathbb{R}^2$). Similarly, there are τ', ρ' so that $\eta((z_1', z_2', z_3')) = \rho'(\tau'((z_1', z_2', z_3')))$. Hence $\alpha = (\tau')^{-1}(\rho')^{-1}\rho\tau$ is the desired RST autorphism.

For the reverse direction (i.e. for each RST automorphism α: $\eta((z_1, z_2, z_3)) = \eta(\alpha((z_1, z_2, z_3)))$ holds), it is sufficient to show this property for each rotation-scaling $\alpha : z \mapsto z \cdot c$ ($c \in \mathbb{C}$) and for each translation $\alpha : z \mapsto z+d$ ($d \in \mathbb{C}$). It is easy to verify that $\eta((z_1, z_2, z_3))\eta((z_1 \cdot c, z_2 \cdot c, z_3 \cdot c))$ and $\eta((z_1, z_2, z_3)) = \eta((z_1+d, z_2+d, z_3+d))$.

Still, we have to prove the second equivalence. For triples $a = (a_1, a_2, a_3)$, $b = (b_1, b_2, b_3)$ regard the relation $a \sim_{RST} b$ that holds iff there is an RST automorphism α so that $\alpha(a) = b$. Clearly, $\sim_{RST}$ is an equivalence relation.

By definition of $\sim_{RST}$, the equivalence classes induced by $\sim_{RST}$ and their set unions are RST relations. The RST property means that if an RST relation contains a triple of an equivalence class, it contains the whole class. Hence each RST relation is a set union of equivalence classes. A triple belongs exactly to all RST relations that are supersets of its $\sim_{RST}$-equivalence class. This proves the claim. (qed.)

4.2 The Finest RST Calculus $\mathcal{F}$

The proof shows that RST relations are exactly the supersets of $\sim_{RST}$-equivalence classes. The RST calculus $\mathcal{F}$ whose base relations are the $\sim_{RST}$-equivalence classes hence is the finest RST calculus. Our goal is to describe the transformations and compositions of $\mathcal{F}$. Then, we can derive any RST calculus from $\mathcal{F}$, as unions of $\mathcal{F}$'s transformations and compositions because of the Proposition 2(Distribution Property). As we have seen, each equivalence class is characterized by its standardized triple. Almost all these triples only differ in its third number. To be able to calculate, we identify each such triple $\eta(z_1, z_2, z_3) = (0, 1, z_3')$ with the single complex number $z_3' = \frac{z_3 - z_1}{z_2 - z_1}$. An RST relation R contains e_{12} or eq or consists of triples whose reference and starting point are different. In the latter case, RST relations can be represented and denoted by the region (set of points) in $\mathbb{C}\{z_3 | \ (0, 1, z_3') \in \mathsf{R}\}$.

Definition 7 (Representability, RST Relations as Regions)
Let $\mathsf{R} \subset (\mathsf{T} \setminus \mathsf{e}_{23} \setminus \mathsf{eq})$ *be an RST relation. Then*

$$Reg\mathsf{R} := \{z = x + \imath y \in \mathbb{C} | \ \ ((0,0), (1,0), (x,y)) \in \mathsf{R}\}$$

is called the region *of* R. z *is said to be contained in* R. R *is called* representable, *and* represented *by the Region* $Reg\ \mathsf{R}$.

If $\mathsf{e}_{23} \subset \mathsf{R}$, *and* $\mathsf{R} \setminus \mathsf{e}_{23}$ *is represented by* R', *then* $R' \cup \{\infty\}$ *is called the* (Riemann) representation *of* R, *and* R *is Riemann representable.*

Note that triples that are in the same RST relation can have different representations. But, triples with the same representation are always in the same relation.

A connected representable relation has a connected region.

Example:
As an example, we give the regions of the relations of $\mathcal{LR}$:

eq is not representable. All others are Riemann representable, as follows:

$$\begin{array}{lll} \mathsf{e}_{12} = \{\infty\} & \mathsf{e}_{13} = \{0\} & \mathsf{e}_{23} = \{1\} \\ \mathsf{b} = \,]-\infty, 0[& \mathsf{c} = \,]0,1[& \mathsf{f} = \,]1,\infty[\\ \mathsf{l} = \{z \in \mathbb{C} \mid \Im z > 0\} & \mathsf{r} = \{z \in \mathbb{C} \mid \Im z < 0\} & \end{array}$$

4.3 Mathematical Approach to Derive Transformations and Compositions

The following two propositions allow to derive the operations for any RST relations. They are fundamental tools in our central proof in the next section.

Lemma 4 (Transformation Lemma)
1. The result of a transformation of an RST relation is again an RST relation.
2. The transformations of Riemann representable relations are Riemann representable relations. For their representations holds:

$$\begin{aligned} z \in \mathit{Reg}\ \mathsf{R} &\iff (1-z) \in \mathit{Reg}\ \mathsf{R}^{(12)} \\ &\iff \tfrac{1}{z} \in \mathit{Reg}\ \mathsf{R}^{(23)} \\ &\iff \tfrac{z}{z-1} \in \mathit{Reg}\ \mathsf{R}^{(13)} \\ &\iff \tfrac{1}{1-z} \in \mathit{Reg}\ \mathsf{R}^{(231)} \\ &\iff 1-\tfrac{1}{z} \in \mathit{Reg}\ \mathsf{R}^{(321)} \end{aligned}$$

Proof:
1. The assertion is trivial for the identity.
Let α be an RST automorphism, R an RST relation. Then:

$$\begin{aligned} (a_{\pi(1)}, \ldots, .a_{\pi(3)}) \in \mathsf{R}^{\pi} &\iff ((a_1, a_2, a_3)) \in \mathsf{R} \\ &\iff (\alpha(a)_1, \ldots, \alpha(a)_3) \in \mathsf{R} \\ &\iff (\alpha(a_{\pi(1)}), \ldots, .\alpha(a_{\pi(3)})) \in \mathsf{R}^{\pi} \end{aligned}$$

This proves that the RST property is preserved.
2. First, assume $z_1 \neq z_2$. Then $z_1^{\pi} \neq z_2^{\pi}$, and by applying the definitions, we get

$$\begin{aligned} z = \tfrac{z_3 - z_1}{z_2 - z_1} \in \mathit{Reg}\mathsf{R} &\iff (z_1, z_2, z_3) \in \mathsf{R} \\ &\iff (z_1^{\pi}, z_2^{\pi}, z_3^{\pi}) := (z_{\pi(1)}, z_{\pi(2)}, z_{\pi(3)}) \in \mathsf{R}^{\pi} \\ &\iff z^{\pi} := \tfrac{z_3^{\pi} - z_1^{\pi}}{z_2^{\pi} - z_1^{\pi}} \in \mathit{Reg}\mathsf{R}^{\pi} \end{aligned}$$

After substituting $z_i^{\pi} = z_{\pi(i)}$ and $z_1 = 0,\ z_2 = 1,\ z_3 = z$, we get the five results for the transformed points z^{π}, e. g.:

$$\begin{aligned} z_1^{(12)} &= z_2 = 1, \\ z_2^{(12)} &= z_1 = 0, \\ z_3^{(12)} &= z_3 = z, \implies z^{(12)} = \tfrac{z-1}{0-1} = 1 - z \end{aligned}$$

The other formulas are derived analogously.

In the case $z_1 = z_2$, a case analysis shows that this arithmetic holds for Riemann representations if we set

$$\frac{1}{0} = \infty,\ \ \infty - 1 = \infty,\ \ 1 - \infty = \infty,\ \ \frac{1}{\infty} = 0 \text{ and } \frac{\infty}{\infty} = 1$$

(qed.)

Example:
This property allows to "calculate" with RST relations. For example, it is possible to determine all $\mathcal{F}$ base relations R for which $\mathsf{R} = \mathsf{R}^{(231)}$. Apart from eq (eq is not Riemann representable and must be considered separately), these are exactly the relations { w } for which $w = \frac{1}{1-w}$ holds. There are two solutions:

$$w = \frac{1}{2} + \frac{\sqrt{3}}{2}\imath;\ \bar{w} = \frac{1}{2} + \frac{\sqrt{3}}{2}\imath.$$

Like for transformations, there are arithmetic ways to determine all compositions.

Lemma 5 (Composition Lemma)
1. Compositions of RST relations are RST relations.
2. If R_1, R_2 and their transformations are representable, then the composition is continuous, and for the representing regions holds:

$$\begin{aligned}
\mathsf{R}_1\ {}_3\diamond_2\ \mathsf{R}_2 &= \{(z_1 \cdot z_2) \mid\ z_1 \in Reg(\mathsf{R}_1) \wedge z_2 \in Reg(\mathsf{R}_2)\}\\
\mathsf{R}_1\ {}_3\diamond_1\ \mathsf{R}_2 &= \{(z_1 + z_2 - z_1 z_2) \mid\ z_1 \in Reg(\mathsf{R}_1) \wedge z_2 \in Reg(\mathsf{R}_2)\}\\
\mathsf{R}_1\ {}_2\diamond_1\ \mathsf{R}_2 &= \{\tfrac{(z_1 \cdot z_2)}{z_1 + z_2 - 1} \mid\ z_1 \in Reg(\mathsf{R}_1) \wedge z_2 \in Reg(\mathsf{R}_2)\}
\end{aligned}$$

3. *For all* $0 \leq \kappa, \lambda \leq 3$:

$$\mathsf{R}_1\ {}_\kappa\diamond_\lambda\ \mathsf{R}_2 = \mathsf{R}_2\ {}_\kappa\diamond_\lambda\ \mathsf{R}_1$$

$$(\mathsf{R}_1\ {}_\kappa\diamond_\lambda\ \mathsf{R}_2)\ {}_\kappa\diamond_\lambda\ \mathsf{R}_3 = \mathsf{R}_1\ {}_\kappa\diamond_\lambda\ (\mathsf{R}_2\ {}_\kappa\diamond_\lambda\ \mathsf{R}_3)$$

Proof:
For 1., we show that if R_1 and R_2 are RST relations, then $\mathsf{R}_1\ {}_3\diamond_2\ \mathsf{R}_2$ is an RST relation. Let α be an RST automorphism, then for the RST relations R_i holds:

$$(a_1, a_2, a_3) \in \mathsf{R}_i \iff (\alpha(a_1), \alpha(a_2), \alpha(a_3)) \in \mathsf{R}_i \text{ for } i \in \{1, 2\}.$$

$$\begin{array}{lll}
\text{Hence} & (a_1, a_2, a_3) \in \mathsf{R}_1\ {}_3\diamond_2\ \mathsf{R}_2 & \\
\iff \exists x : & (a_1, a_2, x) \in \mathsf{R}_1 & \text{and } (a_1, x, a_3) \in \mathsf{R}_2\\
\iff \exists \alpha(x) : & (\alpha(a_1), \alpha(a_2), \alpha(x)) \in \mathsf{R}_1 & \text{and } (\alpha(a_1), \alpha(x), \alpha(a_3)) \in \mathsf{R}_2\\
\iff & (\alpha(a_1), \alpha(a_2), \alpha(a_3)) \in (\mathsf{R}_1\ {}_3\diamond_2\ \mathsf{R}_2) &
\end{array}$$

From Lemma 4 (Tranformation Lemma) and Lemma 1 (Lemma of Interdependence of Compositions), the property follows for all compositions.

For 2., first we regard the composition $(z_a, z_b, z_x) \in \mathsf{R}_1\ {}_3\diamond_2\ (z_a, z_x, z_c) \in \mathsf{R}_2$.

If all transformations are representable in $\mathbb{C} \setminus \infty$, then $\frac{1}{z_b - z_a}$, $\frac{1}{z_b - z_c}$ and $\frac{1}{z_d - z_a}$ are defined. Thus

$$\begin{aligned}
z_3 \in (R_1 \ {}_3\diamond_2 \ R_2) &\iff \exists\, z_a, z_b, z_c \in \mathbb{C}: \quad z_3 = \tfrac{z_c - z_a}{z_b - z_c},\ (z_a, z_b, z_c) \in R_1 \ {}_3\diamond_2 \ R_2 \\
&\iff \exists\, z_a, z_b, z_c, z_x \in \mathbb{C}: (z_a, z_b, z_x) \in R_1 \text{ and } (z_a, z_x, z_c) \in R_2 \\
&\iff \exists\, z_a, z_b, z_c, z_x \in \mathbb{C}: \tfrac{z_x - z_a}{z_b - z_a} \in R_1 \text{ and } \tfrac{z_c - z_a}{z_x - z_a} \in R_2 \\
&\iff \exists\, z_1, z_2 \in \mathbb{C}: \quad z_1 \in R_1 \text{ and } z_2 \in R_2 \text{ and } z_3 = z_1 z_2.
\end{aligned}$$

The other compositions are calculated using Proposition 1 (the conversion formula) and Lemma 4 (the transformation formula), e. g.

$$\begin{aligned}
z_3 \in (R_1 \ {}_3\diamond_1 \ R_2) &\iff \quad z_3 \in (R_1^{(12)} \ {}_3\diamond_2 \ R_2^{(12)})^{(12)} \\
&\iff \quad (1 - z_3) \in (R_1^{(12)} \ {}_3\diamond_2 \ R_2^{(12)}) \\
&\iff \exists\, z_1^{(12)} \in R_1^{(12)}, z_2^{(12)} \in R_2^{(12)} : 1 - z_3 = z_1^{(12)} \cdot z_2^{(12)} \\
&\iff \exists\, z_1 \in R_1, z_2 \in R_2 : \quad 1 - z_3 = (1 - z_1)(1 - z_2) \\
&\iff \exists\, z_1 \in R_1, z_2 \in R_2 : \quad z_3 = z_1 + z_2 - z_1 z_2
\end{aligned}$$

As the formulas are symmetric in z_1 and z_2, exchanging z_1 and z_2 does not change the result. Hence, for all $\kappa, \lambda \in \{1, 2, 3\} : R_1 \ {}_\kappa\diamond_\lambda \ R_2 = R_2 \ {}_\kappa\diamond_\lambda \ R_1$.
Further calculations show the last equality, e. g.

$$\begin{aligned}
&z_4 \in R_1 \ {}_3\diamond_2 \ (R_2 \ {}_3\diamond_2 \ R_3); \\
&z_4 \in (R_1 \ {}_3\diamond_1 \ R_2) \ {}_3\diamond_1 \ R_3 \iff \exists z_1, z_2, z_3 : \\
&\qquad z_4 = (z_1 + z_2 - z_1 z_2) + z_3 - (z_1 + z_2 - z_1 z_2) z_3 \\
&\qquad\quad = z_1 z_2 z_3 - z_1 z_2 - z_2 z_3 - z_3 z_1 + z_1 + z_2 + z_3 \\
&\qquad \iff z_4 \in R_1 \ {}_3\diamond_1 \ (R_2 \ {}_3\diamond_1 \ R_3)
\end{aligned}$$

All these mappings are continuous. The other compositions are continuous as concatenations of the composition ${}_3\diamond_2$ and transformation operations. (qed.)

As we see, compositions and transformations of $\mathcal{F}$ are represented by sets of complex numbers, which represent relations in $\mathcal{F}$. This means that $\mathcal{F}$ is closed under composition and transformation. Although being fundamental for the definition of compositions of other RST calculi, $\mathcal{F}$ is not practical because $\mathcal{F}$ has infinitely many relations (For any complex number, there is a base relation of $\mathcal{F}$). We want to find a practical, i. e. finite coarser calculus that is closed under composition and transformation operation.

Is our standard example, $\mathcal{LR}$, closed under composition? From Propositions 1 and 2, we know that it is sufficient to regard one composition, e. g. ${}_3\diamond_2$, on the base relations. With the theory of RST relations, the complete composition table of $\mathcal{LR}$ can be generated. There are two cases: Either one of the base relations is eq, e_{12} or e_{13}. Then the resulting constraints on the equality of points needs to be checked.

All other compositions, mostly involving refinements of $d = b \cup c \cup f \cup l \cup r$, can be calculated using the composition lemma. Table 3 lists the compositions of ${}_3\diamond_2$. This shows that $\mathcal{LR}$ is practical.

5 The Special Role of $\mathcal{LR}$

When studying point-based qualitative representation calculi, one notices that they all share a particular property. For all tuples in one base relation, we obtain connected

Table 3. Compositions of $\mathcal{LR}$. Note that d is an abbreviation for $b \cup c \cup f \cup l \cup r$

$_3\diamond_2$	eq	e_{12}	e_{13}	e_{23}	b	c	f	l	r
eq	eq	e_{12}	$\emptyset$	$\emptyset$	$\emptyset$	$\emptyset$	$\emptyset$	$\emptyset$	$\emptyset$
e_{12}	e_{12}	$\emptyset$	eq	e_{12}	e_{12}	e_{12}	e_{12}	e_{12}	e_{12}
e_{13}	e_{13}	$d \cup e_{23}$	$\emptyset$	$\emptyset$	$\emptyset$	$\emptyset$	$\emptyset$	$\emptyset$	$\emptyset$
e_{23}	$\emptyset$	$\emptyset$	e_{13}	e_{23}	b	c	f	l	r
b	$\emptyset$	$\emptyset$	e_{13}	b	$c \cup e_{23} \cup f$	b	b	r	l
c	$\emptyset$	$\emptyset$	e_{13}	c	b	c	$c \cup e_{23} \cup f$	l	r
f	$\emptyset$	$\emptyset$	e_{13}	f	b	$c \cup e_{23} \cup f$	f	l	r
l	$\emptyset$	$\emptyset$	e_{13}	l	r	l	l	$l \cup r \cup b$	$d \cup e_{23}$
r	$\emptyset$	$\emptyset$	e_{13}	r	l	r	r	$d \cup e_{23}$	$l \cup r \cup b$

regions when we vary one point and leave all other points constant. In fact, it would appear to be very "unnatural" if one would get unconnected regions. Such sets of unconnected regions one would only expect if a relation is truly disjunctive. Since this property turns out to be very important, we give it a name in the next definition.

Definition 8 (Natural Calculus)
An RST calculus is called natural *iff for all its representable base relations* B, $Reg\,\mathsf{B}$ *and* $\mathbb{C} \setminus Reg\,\mathsf{B}$ *are connected.*

Example:
In $\mathcal{LR}$, all base relations and the complements of base relations are connected.

In this section, we will prove that $\mathcal{LR}$ is the finest practical natural RST calculus.

5.1 Some Definitions and Lemmata

More definitions and lemmata are required in order to structure the proof.

Definition 9 (Bounded Relation)
A representable relation R *except* $\{0\}, \{1\}$ *is called bounded iff* $Reg(\mathsf{R})$ *is bounded.*

The exceptions $\{0\}, \{1\}$ are omitted because not all of their transformations are representable.

In the remainder, using some properties of transformations and compositions, we derive limitations for bounded relations in practical calculi. We then prove for all $\mathcal{LR}$ relations that no connected refinement of $\mathcal{LR}$ relations will fulfill these criteria.

Lemma 6 (Bounded Composition and Inverse)
For representable relations holds:

1. If R *is bounded, then* $\mathsf{R}^{(12)}$ *is bounded.*

2. If R_1 *and* R_2 *are bounded, then the composition* $\mathsf{R} = \mathsf{R}_1 \;_3\diamond_2\; \mathsf{R}_2$ *is bounded, or* $\mathsf{R} = \{1\}$

Proof:
1.This follows directly from Lemma 5 (Composition Lemma) and Lemma 4 (Transformation Lemma). Let $z \in Reg\mathsf{R}$, then $1-z \in Reg(\mathsf{R}^{(12)})$. Let S be the upper boundary for $Reg(\mathsf{R})$, hence $|z| < S$ for all z in R, then $S+1$ is an upper boundary for $|1-z|$. The exceptions, $\{0\}, \{1\}$ cannot be the resulting relation because $\{0\}^{(12)} = \{1\}$, and $\{1\}^{(12)} = \{0\}$, but R is neither $\{0\}$ nor $\{1\}$.

2. $\mathsf{R}_i \neq \{0\} \Rightarrow \exists z_i \neq 0 : \; z_i \in R_i \Rightarrow z := z_1 \cdot z_2 \in R, z \neq 0$. Hence $\mathsf{R} \neq \{0\}$. With the Lemma 5 (Composition Lemma) we obtain: If S_i is upper boundary for $Reg(\mathsf{R}_i)$, then $S_1 \cdot S_2$ is upper boundary for the composition because $z \in \mathsf{R}_1 \; {}_3\diamond_2 \; \mathsf{R}_2$ iff $z = z_1 \cdot z_2$ (with $|z_i| \leq S_i$). Then $|z| \leq S_1 \cdot S_2$. The exception $\mathsf{R}_1 \; {}_3\diamond_2 \; \mathsf{R}_2 = \{0\}$ cannot occur because both relations contain values $z_1, z_2 \neq 0$, thus $z = z_1 \cdot z_2 \in \mathsf{R}_1 \; {}_3\diamond_2 \; \mathsf{R}_2, z \neq 0$. (qed.)

Lemma 7 (0-1-Lemma)
Let $\mathcal{C}$ be a practical RST calculus. Then :

1. *For each relation $\mathsf{R} \not\subseteq \{\mathsf{eq}, \mathsf{e}_{12}, \mathsf{b}, \mathsf{e}_{13}\}$ in $\mathcal{C}$: $1 \in \partial(Reg\mathsf{R})$*
2. *If R is bounded, then $\limsup_{z\in\mathsf{R}} |z| = 1$.*
3. *If R is bounded, then $Reg(\mathsf{R}) \subseteq \mathbb{R}^+$*
4. *If R is bounded, then $0 \in \partial(Reg\mathsf{R})$.*

This shows that only some bounded relations occur in practical calculi. As we will see later, refining $\mathcal{LR}$ would require bounded regions that do not fulfill the criteria of the 0-1-Lemma.

Proof:
First, we prove properties 2 - 4, and then derive property 1 for possibly unbounded relations at the end of this proof, as a generalization of property 4 for bounded relations. We start with the proof of property 2.

Let R be bounded and $S = \limsup_{z\in\mathsf{R}} |z|$ be the smallest upper limit. We claim that S=1. Due to the Lemma 5 (Composition Lemma) we know that

$$\mathsf{R}^k := \mathsf{R} \; {}_3\diamond_2 \; \mathsf{R} \ldots \; {}_3\diamond_2 \; \mathsf{R} = \{z \mid \exists z_1, \ldots, z_k \in \mathsf{R} : z = z_1 z_2 \ldots z_k\}.$$

Multiplication is a continuous and monotonous function, hence

$$\limsup_{z\in\mathsf{R}^k} |z| = S^k$$

Suppose $S \neq 1$, then all these infinitely many relations $(\mathsf{R}^k)_{k\in\mathbb{N}}$ were pairwise different, and all of them were contained in $\mathcal{C}$ because $\mathcal{C}$ is closed under compositions. But $\mathcal{C}$ is finite, hence $S = 1$. This proves property 2.

For 3., suppose, $|z| = r(cos\phi + \imath\, sin\phi)$ $(\phi \neq 0)$ is contained in R. Then R^k contains $z^k = r^k(\cos(k\phi) + \imath\,\sin(k\phi))$. For some k, $\Re(z^k) < 0$. Due to the previous Lemma 6 the relation $(\mathsf{R}^k)^{(12)}$ is bounded. By Lemma 4(Transformation Lemma), $(\mathsf{R}^k)^{(12)}$ contains $1 - z^k$ with $|1 - z^k| \geq \Re(1 - z^k) > 1$ in contradiction to 2. This proves property 3. (See Figure 4)

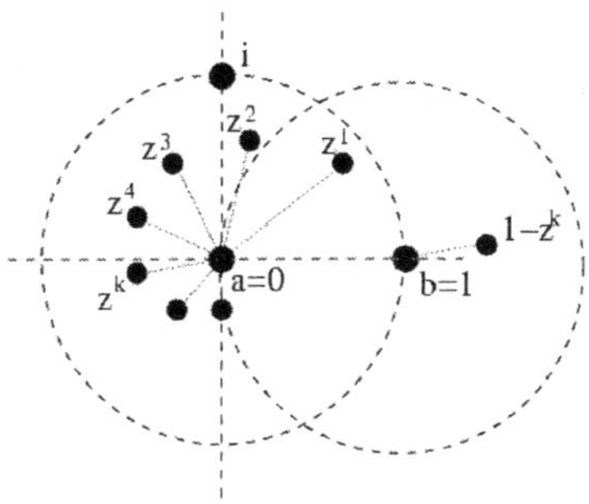

Fig. 4. Sketch for the proof of Lemma 4. For some k, we have $1-z^k \in (\mathsf{R}^k)^{(12)}$ and $|1-z^k|>1$

For the proof of 4., suppose, $0 \notin \partial\ \mathsf{R}$. Then $\mathsf{R}^{(23)}$ is bounded because $\mathsf{R}^{(23)} = \{z \mid \frac{1}{z} \in \mathsf{R}\}$ (Lemma 4, Transformation Lemma). For $\mathsf{R}^{(23)}$ in $\mathcal{C}$, we have

$$\limsup_{z\in \mathsf{R}^{(23)}} |z| = 1, \text{ hence } \liminf_{z\in \mathsf{R}} |z| = 1.$$

Thus, for all $z \in \mathsf{R} : |z| = 1$. From property 3. follows $\mathsf{R} = \{1\}$, but by definition of boundedness, $\{\ 1\ \}$ is not bounded. Hence, the supposition is wrong, which proves 4.

Property 1 for possibly unbounded relations can be derived from property 4. It is sufficient to prove the claim in $1 \in \partial(\mathsf{R} \setminus \{1\})$. Let $\mathsf{R}\setminus\{1\}$ be an arbitrary relation $\mathsf{R} \not\subseteq \{\mathsf{eq}, \mathsf{e}_{12}, \mathsf{b}, \mathsf{e}_{13}\}$. Suppose, $1 \notin \partial\ \mathsf{R}$. Then $0 \notin \partial \mathsf{R}^{(12)}$. By the Lemma 4 (Transformation Lemma) $(\mathsf{R}^{(12)})^{(23)}$ is bounded. From 3. and 4., we know $(\mathsf{R}^{(12)})^{(23)} \subseteq [0,1]$. Hence, $\mathsf{R}^{(12)} \subseteq [1,\infty]$ and $\mathsf{R}^{(12)} \subseteq [-\infty,0] = \mathsf{b} \cup \mathsf{e}_{13}$, in contradiction to the prerequisite. This completes the proof. (qed.)

The restrictions in the 0-1-Lemma mainly apply because a practical calculus has to be finite. As a consequence, the demand for a finite calculus limits the number of refinements of RST calculi.

5.2 The Central Theorem

Theorem 1 (Central Theorem)
Any practical natural refinement of $\mathcal{LR}$ is $\mathcal{LR}$ itself.

The proof is based on the fact that the relations of practical calculi fulfill the properties of the Lemma 7 (0-1-Lemma). A close consequence is the following lemma that shows that the relation C cannot be further refined.

Lemma 8 (Connected Bounded Relations)
If $\mathcal{C}$ is a practical *calculus, then any connected bounded relation in $\mathcal{C}$ is a whole interval from* 0 *to* 1.

Proof:
Let R be a connected bounded relation in $\mathcal{C}$. According to the 0-1-lemma, a bounded relation is a subset of the interval $[0,1]$. Suppose there is some $x \in\ ]0,1[\ , x \notin Reg\mathsf{R}$. Since $Reg\mathsf{R}$ is connected, all points are below or above x:

$$\limsup_{z\in\mathsf{R}} |z| \leq x \quad \text{or} \quad \liminf_{z\in\mathsf{R}} |z| \geq x$$
$$\text{and then } \limsup_{z\in\mathsf{R}^{(12)}} |z| \leq 1-x.$$

In both cases, this contradicts the 0-1-lemma. Hence $]0,1[\ \subset Reg\mathsf{R}$. (qed.)

$Reg(\mathsf{c}) = \ ]0,1[$. Therefore, neither the relation c nor its transformations b and f can be further refined. eq and the point relations e_{12}, e_{13}, e_{23} cannot be refined as they are already relations of $\mathcal{F}$. The task is now to prove that l and r cannot be refined: Unlike with f and b, we cannot refer to bounded transformations because the transformations of l and r are l and r. However, using criteria 1. from the 0-1-lemma, we can show: If there is a refinement and all base relations are connected, then there is a bounded subrelation of l, which contradicts the previous Lemma 8.

In practical calculi, for all base relations B, either

$$\mathsf{B} \cap \mathsf{B}^{(231)} \cap \mathsf{B}^{(321)} = \emptyset$$
$$\text{or } \mathsf{B} \cap \mathsf{B}^{(231)} \cap \mathsf{B}^{(321)} = \mathsf{B},$$

holds because the intersection is a relation of the calculus. We will call base relations for which the second line is true *rotation-symmetric* because they remain unchanged under the "rotating" transformations. (Note that $\mathsf{B}^{(231)}$ and $\mathsf{B}^{(321)}$ are base relations, hence, $\mathsf{B}^{(231)} = \mathsf{B}^{(321)} = \mathsf{B}$).

In order to simplify the proof, we transform the plane of complex numbers so that the "rotating" transformations are actually represented by a rotation of the complex plane with center 0. This allows us to use the inherent symmetry in the proof.

Therefore, regard the Mobius transformation μ, that maps the half plane $Reg(\mathsf{l}) = \{z \mid \Im(z) > 0\}$ to the unit disk :

$$\mu(z) := \frac{z-w}{z-\bar{w}}\ , \text{where } w = \sqrt[3]{-1} = \tfrac{1}{2} + \tfrac{\sqrt{3}}{2}\imath$$

Remark 2 (Rotating Transformations)

$$\mu(0) = w^2,\ \mu(1) = w^4,\ \mu(\infty) = 1,\ \mu(w) = 0$$

$$\mu(\mathsf{R}^{(231)}) = w^2 \cdot \mu(\mathsf{R}), \quad \mu(\mathsf{R}^{(321)}) = w^4 \cdot \mu(\mathsf{R})\ .$$

This shows that the images of rotation-symmetric base relations are rotation-symmetric with angle 120° *around* $\mu(w) = 0$.

Proof:
First we state some equalities: With $w^6 = 1,\ w^5 = \bar{w},\ \bar{w}w = 1 = \bar{w} + w$, we obtain

$$\frac{\frac{1}{1-z}-w}{\frac{1}{1-z}-\bar{w}} = \frac{1-w+wz}{1-\bar{w}+\bar{w}z} = \frac{\bar{w}+wz}{w+\bar{w}z} = \frac{1+w^2z}{w^2+z} = w^2\left(\frac{z-w}{z-\bar{w}}\right)$$
$$\text{and } \frac{1-\frac{1}{z}-w}{1-\frac{1}{z}-\bar{w}} = \frac{z-1-wz}{z-1-\bar{w}z} = \frac{(1-w)z-1}{1-\bar{w}z-1} = \frac{\bar{w}z-1}{wz-1} = \frac{w^5z-w^6}{wz-w\bar{w}} = w^4\left(\frac{z-w}{z-\bar{w}}\right)$$

The Transformation Lemma now completes the proof:

$$\begin{array}{rcccc} z \in Reg\mathsf{R} & \Longleftrightarrow & 1-\frac{1}{z} \in Reg\mathsf{R}^{(321)} & \Longleftrightarrow & \frac{1}{1-z} \in Reg\mathsf{R}^{(231)} \\ \Rightarrow \frac{z-w}{z-\bar{w}} \in \mu(Reg\mathsf{R}) & \Longleftrightarrow & \frac{\frac{1}{1-z}-w}{\frac{1}{1-z}-\bar{w}} \in \mu(Reg\mathsf{R}^{(231)}) & \Longleftrightarrow & w^2(\frac{z-w}{z-\bar{w}}) \in \mu(Reg\mathsf{R}^{(231)}), \\ \frac{z-w}{z-\bar{w}} \in \mu(Reg\mathsf{R}) & \Longleftrightarrow & \frac{1-\frac{1}{z}-w}{1-\frac{1}{z}-\bar{w}} \in \mu(Reg\mathsf{R}^{(321)}) & \Longleftrightarrow & w^4(\frac{z-w}{z-\bar{w}}) \in \mu(Reg\mathsf{R}^{(321)}) \end{array}$$

(qed.)

Note that Mobius transformations preserve connectivity. If $\mu(Reg\mathsf{R})$ is connected, then R is connected.

Lemma 9 (Rotation Symmetry)
For a practical calculus with connected base relations holds: A base relation B is rotation-symmetric iff $w \in \mathsf{B}$ or $\bar{w} = w^5 \in \mathsf{B}$.

Proof:
If $w \in Reg\mathsf{B}$, $\mu(w) = 0$, follows $w \in Reg\mathsf{B}^{(321)}$ by the previous remark (Remark on Rotating Transformations). If $\bar{w} \in Reg\mathsf{B}$, then $1 - \frac{1}{\bar{w}} = \overline{(1 - \frac{1}{w})} = \bar{w} \in Reg\mathsf{B}^{(321)}$, so in both cases $\mathsf{B} \cap \mathsf{B}^{(321)} \neq \emptyset$. Thus $\mathsf{B} = \mathsf{B}^{(321)}$ is rotation-symmetric.

For the reverse direction, we use the following consequence of the Jordan Curve Theorem ([18]): If v is a continuous closed path in $\mathbb{C} \setminus \{\infty\}$ that does not contain w but does go around a point z at least once, then the set of all points that can be reached from z by a continuous path that does not intersect v is bounded. (For the proof refer to [15].)

Suppose that there is a representable rotation-symmetric base relation B that contains neither w nor $\bar{w}$. Then because of connectivity, there is a continuous path v from a point $z \in \mu(Reg\mathsf{B})$ to $w^2 z \in \mu(Reg(\mathsf{B}^{(231)}))$. Then $w^2 \cdot v$ is a continuous path from $w^2 z$ to $w^4 z$, and $w^4 \cdot v$ a continuous path from $w^4 z$ to $w^6 z = z$. Because of the rotation symmetry of $\mathsf{B} = \mathsf{B}^{(231)}$, all these paths remain in $\mu(Reg\mathsf{B})$, and together they form a closed path that goes around 0. By assumption, this path does not hit, but separates $0 = \mu(w)$ and $\infty = \mu(\bar{w})$ because $w, \bar{w} \notin \mathsf{B}$. Either $\mu(w) = 0$ and $\mu(\infty) = 1$ are on different sides of the path, then the base relation B_w , containing w is bounded, or $\mu(\bar{w}) = \infty$ and $\mu(\infty) = 1$ are on different sides of the path, then the base relation $\mathsf{B}_{\bar{w}}$ containing $\bar{w}$ is bounded. Here, we use the connectivity of the base relations.

This contradicts the 0-1-lemma. (qed.)

Now, we complete the proof of Theorem 1 (Central Theorem). We have to show that l and r cannot properly be refined.

Let us assume there is a proper refinement of l in a natural calculus, hence exits $\mathsf{B}_w \subsetneq \mathsf{l}$ with $w \in Reg(\mathsf{B}_w)$. Let $\mathsf{C}_w = \top \setminus \mathsf{B}_w$. In a natural calculus, C_w is connected. In a proper refinement, $\mathsf{C}_w \cap \mathsf{l} \neq \emptyset$. Hence, there is $z \in \mathsf{C}_w \cap \mathsf{l}$. Note that $\Im(z) > 0$, hence $|\mu(z)| < 1$. It is essential to the proof that $\mathsf{C}_w = \mathsf{C}_w^{(231)} = \mathsf{C}_w^{(321)}$ is rotation symmetric being the complement of the rotation-symmetric relation B_w . Then $w^2\mu(z), w^4\mu(z) \in \mu(Reg(\mathsf{C}_w))$, and there is a continuous closed path v from

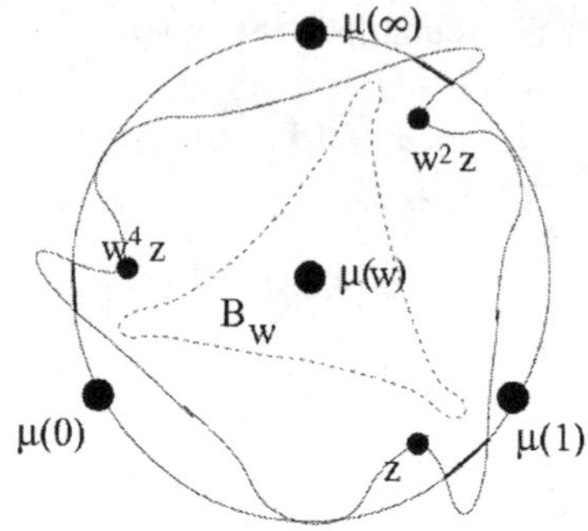

Fig. 5. Sketch for the proof of the Central Theorem

$\mu(z)$ to $w^2\mu(z)$ to $w^4\mu(z)$ to $\mu(z)$ within $\mu(Reg(C_w))$. Without loss of generality, we assume that this path remains inside the closed unit disk $\mu(\{z \mid \Im(z) \geq 0\})$ because $\mathbb{R} \subset Reg(C_w)$ (see Figure 5). Note that $0 = \mu(w)$ is in the inside of the path. As B_w is connected, any point of $\mu(B_w)$ is inside this path, hence B_w is bounded but $Reg B_w \not\subset \mathbb{R}$, in contradiction to the 0-1-Lemma 7. This means, the assumption is wrong - ℓ has no proper refinement in practical natural calculi. By reasons of symmetry, the same holds for r. This proves Theorem 1 (Central Theorem). (qed.)

6 Conclusion

We developed a general theory for ternary point-based calculi such that the relations are invariant when all points are mapped by rotations, scalings or translations. These calculi are called RST calculi. Examples for such RST calculi are Freksa's double cross calculus [3, 4], Ligozat's flip-flop calculus [8], and Moratz et al.'s TPCC calculus [11].

We argued that one requirement on an RST calculus is that it should be "practical," i.e., that it is closed under the usual operations and that it possesses a finite JEPD base. This is a prerequisite for applying Montanari's path-consistency algorithm [10] in a way such that no information loss occurs. Further, we required an RST calculus to be "natural," which means that the region denoted by a relation is internally connected.

The original versions of the double-cross and the flip-flop calculus fail the practicality test since their relations are not jointly exhaustive because some equality relations are missing. After adding these relations one arrives at the following results. As shown elsewhere [16], the double-cross calculus does not have a finite base, while on the other hand, the completion of the much coarser flip-flop calculus, which we named $\mathcal{LR}$, is a practical and natural calculus. As the main result of the paper, we were able to show that $\mathcal{LR}$ cannot be properly refined without loosing this property. From that it follows, e.g., that Moratz et al.'s TPCC [11] does not have a finite base, which was unknown until now.

An interesting direction of further research could be to make use of the method to compute exact compositions and transformations in the infinite finest RST calculus $\mathcal{F}$ presented in this paper. This might provide a new solution to conclude knowledge in some cases, even with infinite calculi.

References

1. M. J. Egenhofer. Reasoning about binary topological relations. In O. Günther and H.-J. Schek, editors, *Proceedings of the Second Symposium on Large Spatial Databases, SSD'91*, volume 525 of *Lecture Notes in Computer Science*, pages 143–160. Springer-Verlag, Berlin, Heidelberg, New York, 1991.
2. A. Frank. Qualitative spatial reasoning with cardinal directions. In *Proceedings of the Seventh Austrian Conference on Artificial Intelligence*, Berlin, Heidelberg, New York, 1991. Springer-Verlag.
3. C. Freksa. Using orientation information for qualitative spatial reasoning. In A. U. Frank, I. Campari, and U. Formentini, editors, *Theories and Methods of Spatio-Temporal Reasoning in Geographic Space*, pages 162–178. Springer-Verlag, Berlin, Heidelberg, New York, 1992.
4. C. Freksa and K. Zimmermann. On the utilization of spatial structures for cognitively plausible and efficient reasoning. In *Proc. of the IEEE International Conference on Systems, Man, and Cybernetics*, pages 261–266, Chicago, IL, 1992. IEEE.
5. A. Isli and A. G. Cohn. A new approach to cyclic ordering of 2D orientations using ternary relation algebras. *Artificial Intelligence*, 122(1-2):137–187, 2000.
6. A. Isli and R. Moratz. Qualitative spatial representation: Algebraic models for relative position. Technical Report FBI-HH-M284/99, Univ. Hamburg, 1999.
7. P. B. Ladkin and R. Maddux. On binary constraint problems. *Journal of the Association for Computing Machinery*, 41(3):435–469, May 1994.
8. G. Ligozat. Qualitative triangulation for spatial reasoning. In *Spatial Information Theory: A Theoretical Basis for GIS (COSIT-93)*, pages 54–68, 1993.
9. N. Megiddo. Partial and complete cyclic orders. *Bull. Am. Math. Soc*, 82:274–276, 1976.
10. U. Montanari. Networks of constraints: fundamental properties and applications to picture processing. *Information Science*, 7:95–132, 1974.
11. R. Moratz, B. Nebel, and C. Freksa. Qualitative spatial reasoning about relative position. In F. et al., editor, *Spatial Cognition III*, pages 385–400. Springer-Verlag, 2003.
12. D. Papadias, M. J. Egenhofer, and J. Sharma. Hierarchical reasoning about direction relations. In *GIS '96, Proceedings of the fourth ACM workshop on Advances on Advances in Geographic Information Systems, November 15-16, 1996, Rockville, Maryland, USA*, pages 105–112, 1996.
13. D. A. Randell, Z. Cui, and A. G. Cohn. A spatial logic based on regions and connection. In B. Nebel, W. Swartout, and C. Rich, editors, *Principles of Knowledge Representation and Reasoning: Proceedings of the 3rd International Conference (KR-92)*, pages 165–176, Cambridge, MA, Oct. 1992. Morgan Kaufmann.
14. C. Schlieder. Reasoning about ordering. In A. U. frank and W. Kuhn, editors, *Spatial Information Theory: A Theoretical Basis for GIS (COSIT-95)*, pages 341–349, 1995.
15. A. Scivos. Einführung in eine Theorie der ternären RST-Kalküle für qualitatives räumliches Schließen. Diplomarbeit, Albert-Ludwigs-Universität Freiburg, Mathematische Fakultät, 2000.
16. A. Scivos and B. Nebel. Double-crossing: Decidability and computational complexity of a qualitative calculus for navigation. In *Spatial Information Theory: Foundations of Geographic Information Science (COSIT-2001)*, Morro Bay, CA, 2001. Springer-Verlag.
17. A. Tarski. On the calculus of relations. *Journal of Symbolic Logic*, 6:73–89, 1941.
18. T. tom Dieck. *Topologie*. de Gruyter, Berlin, 2000.

Exploiting Qualitative Spatial Neighborhoods in the Situation Calculus

Frank Dylla and Reinhard Moratz

SFB/TR 8 Spatial Cognition,
University of Bremen,
Bibliothekstr. 1, 28359 Bremen, Germany
{dylla, moratz}@sfbtr8.uni-bremen.de

Abstract. We present first ideas on how results about qualitative spatial reasoning can be exploited in reasoning about action and change. Current work concentrates on a line segment based calculus, the dipole calculus and necessary extensions for representing navigational concepts like *turn right*. We investigate how its conceptual neighborhood structure can be applied in the situation calculus for reasoning qualitatively about relative positions in dynamic environments.

1 Introduction

Most current reliable robot systems are based on a completely determined geometrical world model. The applied metric methods are tending to fail in frequently changing spatial configurations and when accurate distance and orientation information is not obtainable. Qualitative representation of space abstracts from the physical world and enables computers to make predictions about spatial relations, even when precise quantitative information is not available [4]. Important aspects are topological and positional (orientation and distance) information about in most cases physically extended objects. Calculi dealing with such information have been well investigated over recent years and give general and sound reasoning strategies, e.g. about topological relations between regions as in RCC-8 [37, 38], about the relative position orientation of three points as in Freksa's Double-Cross Calculus [13] or about orientation of two line segments as in the Dipole Calculus [32, 42]. For reasoning with calculi as the above mentioned ones standard constraint-based reasoning techniques can be applied. In [42] Schlieder sketched how a qualitative calculus like the Dipole Calculus might be applied to robot navigation.

In this paper we want to show how to use conceptual neighborhoods [12] for combining Qualitative Spatial Reasoning (QSR) approaches with Reasoning about Action and Change (RAC) approaches for the purpose of robot navigation and path-planning. For the first sketch of our ideas we chose the Dipole Calculus (QSR) and the Situation Calculus [29] respectively Golog [28] (RAC) as agent control language.

Two relations are conceptual neighbors if their spatial configurations can be continuously transformed into each other with only minimal change, e.g. in RCC-8 two disconnected regions (configuration 1) cannot overlap (3) without being externally connected (2) in between. Therefore 1 and 2, as well as 2 and 3 are conceptually neighboring relations but not 1 and 3. Expressing these connections between the relations leads to

C. Freksa et al. (Eds.): Spatial Cognition IV, LNAI 3343, pp. 304–322, 2005.
© Springer-Verlag Berlin Heidelberg 2005

conceptual neighborhood graphs (CNH-graph). For further motivation for qualitative spatial reasoning we refer to [14] .

Frameworks for reasoning about action and change, e.g. the Situation Calculus [29] based programming language Golog [28], also provide facilities for representing and reasoning about sets of spatial locations. Current variants are able to deal with e.g. concurrency [7], continuous change [22] or decision theory [15]. The Golog framework has been applied successfully in real world domains, e.g. in the RHINO museum tour guide project [3] , or for playing robotic soccer in the RoboCup tournaments [10]. Additionally we can integrate navigational and non navigational actions, e.g. $say(.)$, $pick(.)$, or $look(.)$, without any extra effort.

Unfortunately no formal spatial theory, e.g. for relative position terms like left, right or behind, is defined within for dealing with underspecified, coarse knowledge. Therefore every project modeling dynamic domains needs the naive formalization of such a theory by the developer again and again, although such concepts have been formally investigated.

We present first ideas about qualitative navigation on the basis of oriented line segments, which we consider a valuable starting point. In this context we will show one way how the results about conceptual neighborhood can be applied in the Situation Calculus resp. Golog. In the first stage of this work we will only look at simulated scenarios to omit additional complexity caused by real robot control.

The long term goal is a general representation and usage of qualitative spatial concepts about relative position like e.g. $left$, $right$, or $inFrontOf$ within the Situation Calculus or, the programming language Golog, e.g. providing action facilities like $go(leftOf, exhibit_7)$. We do not only expect such formal qualitative concepts being useful in agent programming but also in human machine interaction [44, 30].

In this paper we will present several variants of the Dipole Calculus at different levels of granularity and their corresponding conceptual neighborhoods. We present necessary extensions for expressing robot navigation tasks more adequately. Without doing so we would not be able to formalize navigational behavior with the basic translational and rotational commands intrinsic to every robot system. After a brief introduction to the Situation Calculus and the programming language Golog we will present a first approach combining the Dipole Calculus with the Situation Calculus and Golog. We will clarify our ideas with concrete examples and end with final conclusions.

2 Qualitative Spatial Reasoning

Qualitative Spatial Reasoning (QSR) is an abstraction that summarizes similar quantitative states into one qualitative characterization. From a cognitive perspective the qualitative method *compares* features of the domain rather than *measuring* them in terms of some artificial external scale [13]. The two main directions in QSR are topological reasoning about regions, e.g. the RCC-8, and positional reasoning about point configurations. An overview is given in [5].

Solving navigation tasks involves reasoning about paths as well as reasoning about configurations of objects or landmarks perceived along the way and thus requires the representation of orientation and distance information [41, 25]. Many approaches deal with global allocentric metrical data. For many navigational tasks we do not need absolute allocentric information about the position, instead we need relative egocentric representations and a fast process for updating these relations during navigation [45, 46].

Several calculi dealing with relative positional information have been presented in recent years. Freksa's double cross calculus [13] deals with triples of points and can also be viewed as a calculus dealing with positional binary relations between a dipole and a point. Schlieder proposed a calculus based on line segments with clockwise or counter clockwise orientation of generating start and end points in [42]. He presented a CNH-graph of 14 basic relations. Isli and Cohn [24] presented a ternary algebra for reasoning about orientation. This algebra contains a tractable subset of base relations.

Moratz et al. [32] extend Schlieder's calculus. In a first variant ten additional relations are regarded, where the two dipoles meet in one point, resulting in a relation algebra in the sense of Tarski [26] with 24 basic relations. Also an extended version is introduced such that spatial configurations can be distinguished in a more fine grained fashion. An application oriented calculus dealing with ternary point configurations (TPCC) is presented in [31]. It is suited for both, human robot communication [30] and spatial reasoning in route graphs [31]. Even more fine grained calculi can be used to do path integration for mobile robots [33]. In [47] a line segment approach for shape matching in a robotic context is presented. In this context the line extraction are derived efficiently in $O(n \log n)$ by using the method of Discrete Curve Evolution [47]. In [1] qualitative spatial calculi are linked to ontological engineering.

2.1 Neighborhood-Based Reasoning

Neighborhood-based reasoning describes whether two spatial configurations of objects can be transformed into each other by small changes [12]. The conceptual neighborhood of a qualitative spatial relation which holds for a spatial arrangement is the set of relations into which a relation can be changed with minimal transformations, e.g. by continuous deformation. Such a transformation can be a movement of one object of the configuration in a short period of time. On the discrete level of concepts, neighborhood corresponds to continuity on the geometric or physical level of description: continuous processes map onto identical or neighboring classes of descriptions [14]. Spatial neighborhoods are very natural perceptual and cognitive entities and other neighborhood structures can be derived from spatial neighborhoods, e.g. temporal neighborhoods. The term continuous in the presence of transformation or deformation needs a grounding in spatial change over time. From our point of view the continuous transformation is the continuous motion of a robot r. This can be described by the function $pos(r) : T \rightarrow P$, where T is a set of times and P is a set of possible positions of r. Now assuming T and P being topological spaces, the motion of r is continuous, if the the function $pos(r)$ is continuous [18]. Detailed work on different aspects of continuity were investigated in [2, 6, 16, 17, 23, 35]. Based on different definitions of continuity different neighborhood graphs may arouse. This is also true for different robot kinematics, e.g. comparing differential drive robots versus omnidirectional drive robots.

A movement of an agent can then be modeled qualitatively as a sequence of neighboring spatial relations which hold for adjacent time intervals. Using this qualitative representation of trajectories neighborhood-based spatial reasoning can be used as a sim-

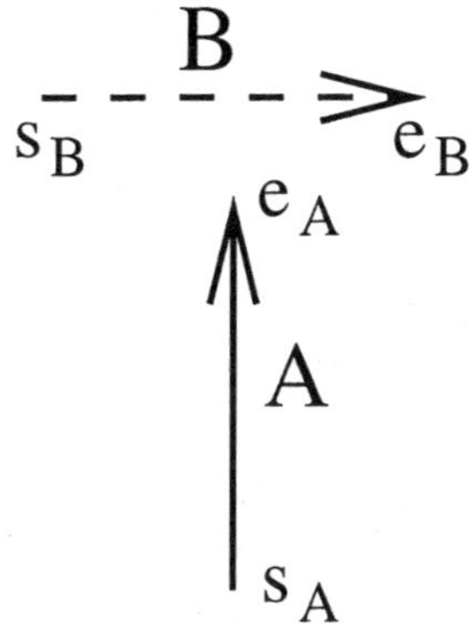

Fig. 1. The lrrr orientation relation between two dipoles

ple, abstract model of robot navigation and exploration. Neighborhoods can be formed recursively and represented by hierarchical tree or lattice structures.

Schlieder [42] mapped orientation onto ordering. He defined the orientation on triangles and for every set with more than three points recursively for every triangle. He extracted 14 basic relations to reason about ordering of line segments[1]. The conceptual neighborhood graph is shown in Fig. 3. The labels are illustrated in Fig. 4.

From the neighborhood graphs of the individual relations, the neighborhood graph of the overall configuration must be derived. In this global neighborhood graph, spatial transformations from a start state to a goal state can be determined. It has been investigated to use the neighborhood graph of two objects for spatial navigation [42]. It has not been investigated yet how a neighborhood graph for a configuration of more complex or even several objects can be constructed using efficient, qualitative methods based on local knowledge.

A problem for the efficient construction of neighborhood graphs for multiple objects is the combinatorial explosion due to the combined neighborhoods of all objects. A potential solution to this problem is to locally restrict the combination of transitions. If we partition the environment of the moving agent into small parts, then only the neighborhood transition graph for these smaller spatial configurations needs to be considered.

2.2 Dipole Relation Algebra

In [32] a qualitative spatial calculus dealing with two directed line segments, in the following also called *dipole*, as basic entities was presented. These dipoles are used for representing spatial objects with intrinsic orientation. A dipole A is defined by two points, the start point s_A and the end point e_A. The presented calculus deals with the orientation of two dipoles. An example of the relation $lrrr$ is shown in Fig. 1. The four letters denote the relative position (e.g. *left* or *right*) of one of the points to the other dipole:

$$A\, lrrr\, B := A\, l\, s_B \wedge A\, r\, e_B \wedge B\, r\, s_A \wedge B\, r\, s_A$$

[1] 16 potential triangle configurations, but two configurations are geometrically impossible.

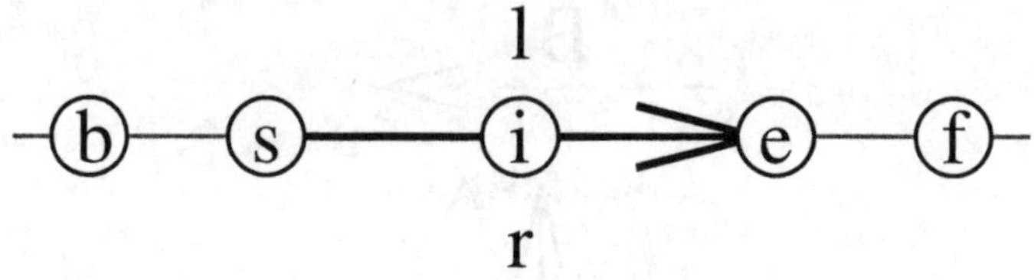

Fig. 2. Extended dipole point relations

Based on a two dimensional continuous space, $\mathbb{R}^2$, the location and orientation of two different dipoles can be distinguished by representing the relative position of start and end points. This means *left* or *right* and the same *start* or *end* point if no more than three points are allowed on a line, and without this restriction *back*, *interior* and *front* additionally (Fig. 2).

The first view leads to 24 *jointly exhaustive and pairwise disjoint* (jepd) basic relations, i.e. between any two dipoles exactly one relation holds at any time. Additionally they build up a relation algebra with 24 basic relations. These relations build up a quite coarse distinction between different orientations, thus we will call this algebra ($\mathcal{DRA}_c$). A visualization is given in Fig. 4. Because of forming a relation algebra standard constraint-based reasoning techniques can be applied. The unrestricted version leads to a relation algebra with 72 basic relations. We will call this fine grained algebra $\mathcal{DRA}_f$. For a detailed description of the calculus' properties we refer to [32].

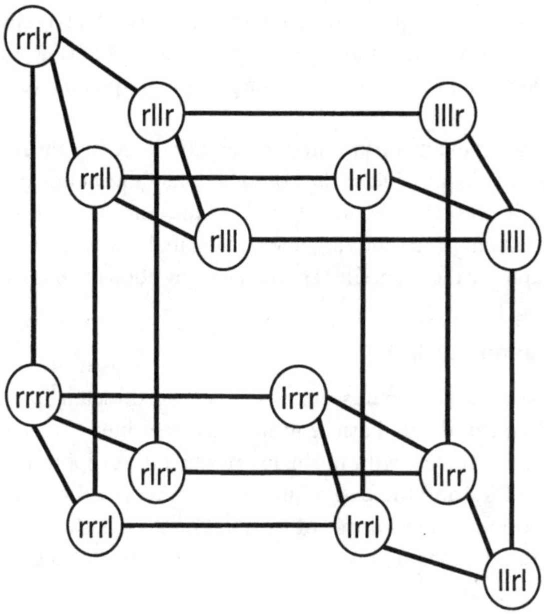

Fig. 3. The conceptual neighborhood graph for the 14 basic relations by Schlieder

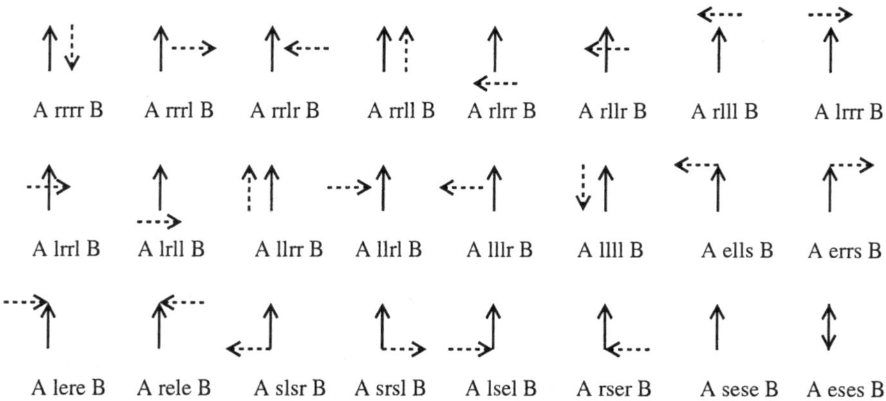

Fig. 4. The 24 atomic relations of the coarse dipole calculus. In the dipole calculus orthogonality is not defined, although the visual presentation might suggest this

2.3 Extended Dipole Relation Algebra

Unfortunately $\mathcal{DRA}_f$ may not be sufficient for robot navigation tasks, because even in this finer grained version many different dipole configurations are pooled in one relation. Thus we extend the representation with additional orientation knowledge and derive a more fine grained relation algebra with additional orientation distinctions. We will call this $\mathcal{DRA}_{fp}$.

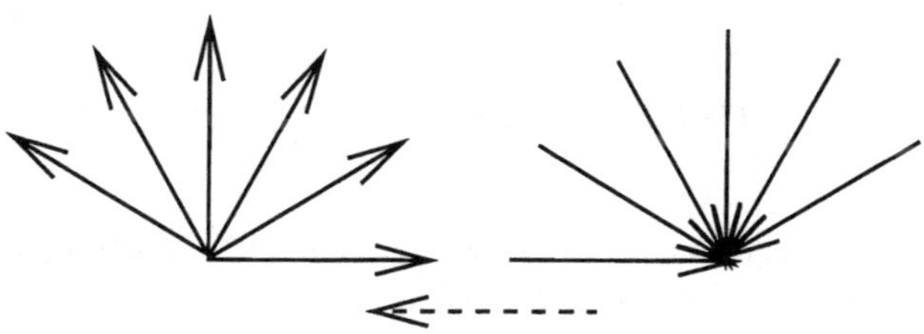

Fig. 5. Pairs of dipoles (A: solid, B: dashed) subsumed by the same relation A(rrrr)B

Fig. 5 for example visualizes the large configuration space for the *rrrr* relation. This might lead to quite squiggly paths if using these concepts for robot navigation. Other relations being extremely coarse are *llrr*, *rrll* and *llll*. We would expect a more goal directed behavior breaking up the relations by regarding the angle spanned by the two dipoles qualitatively. This gives us an important additional distinguishing feature with four distinctive values. These qualitative distinctions are parallelism (P) or

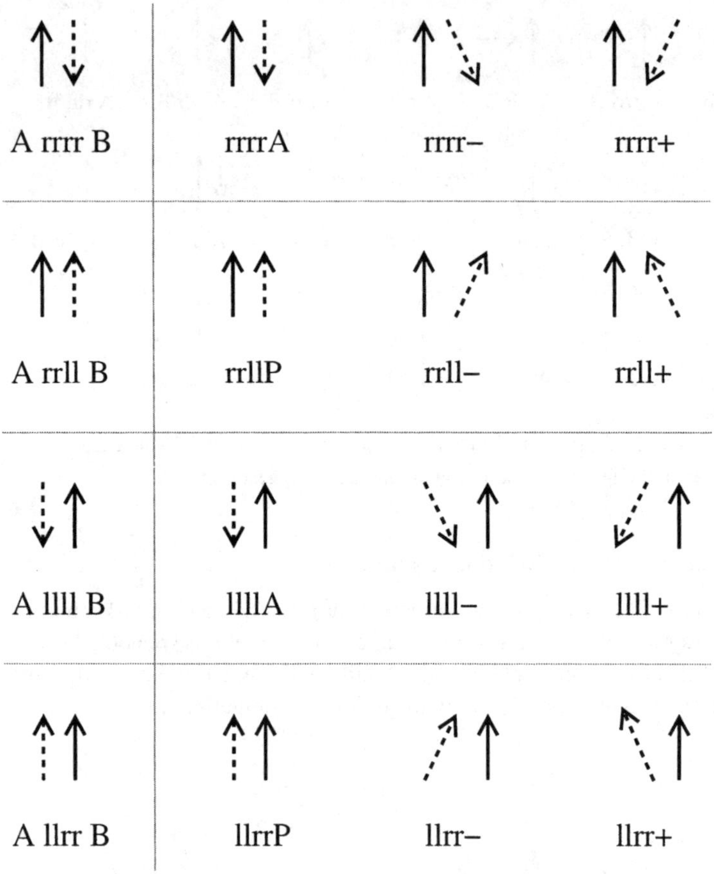

Fig. 6. Refined base relations in $\mathcal{DRA}_{fp}$

anti-parallelism (A) and mathematically positive and negative angles between A and B, leading to three refining relations for each of the four above mentioned relations (Fig. 6). Thus we call this algebra $\mathcal{DRA}_{fp}$ being an extension of the fine grained relation algebra $\mathcal{DRA}_f$ with additional distinctions based on "parallelism".

For the other relations a '+', '−', 'P', or 'A' is already determined by the original base relation. We give a complete list of the resulting $\mathcal{DRA}_{fp}$ algebra:

1. Original relations from $\mathcal{DRA}_c$:
   (a) Extended relations (12):
   $rrrr+$, $rrrrA$, $rrrr-$, $rrll+$, $rrllP$, $rrll-$, $llrr+$, $llrrP$, $llrr-$, $llll+$, $llllA$, $llll-$
   (b) Unmodified relations (20):
   $rrrl-$, $rrlr+$, $rlrr+$, $rllr+$, $rlll+$, $lrrr-$, $lrrl-$, $lrll-$, $llrl-$, $lllr+$ $ells+$, $errs-$, $lere-$, $rele+$, $slsr+$, $srsl-$, $lsel-$, $rser+$, $seseP$, $esesA$

2. Additional cases on one line[2], $seseP$ and $esesA$ are already defined in 1.(b):
   (a) Basic Allen cases (12):
   $ffbbP$, $efbsP$, $ifbiP$, $bfiiP$, $sfsiP$, $beieP$, $bbffP$, $bsefP$, $biifP$, $iibfP$, $sisfP$, $iebeP$
   (b) Converse cases relative to Allen (12):
   $ffffA$, $fefeA$, $fifiA$, $fbiiA$, $fseiA$, $ebisA$, $iifbA$, $eifsA$, $isebA$, $bbbbA$, $sbsbA$, $ibibA$
3. Other additional cases:
   (a) Without converse (12):
   $lllb+$, $llfl-$, $llbr+$, $llrf-$, $lirl+$, $lfrr-$, $lril-$, $lrri+$, $blrr-$, $irrl-$, $frrr+$, $rbrr+$
   (b) The converse (12):
   $lbll-$, $flll+$, $brll-$, $rfll+$, $rlli-$, $rrlf+$, $illr+$, $rilr-$, $rrbl+$, $rlir+$, $rrfr-$, $rrrb-$

For lack of space we refer to our web page[3] for the CNH-graphs and CNH-tables for $\mathcal{DRA}_c$, $\mathcal{DRA}_f$ and $\mathcal{DRA}_{fp}$. Restricting to relations suited for robotic navigational tasks where dipoles represent solid objects[4] we end up with only 40 base relations, thus giving us a condensed CNH-graph.

3 The Situation Calculus

The situation calculus is a second order language for representing and reasoning about dynamic domains. Although many different variants have been developed from the original framework for dealing with, e.g. concurrency [7], continuous change [20, 22] or uncertainty [19], all dialects are based on three sorts: *actions*, *situations* and *fluents*.

All changes in the world are caused by an action a_i in the specific situation s_i resulting in the successor situation s_{i+1}. The special constant S_0 denotes the *initial situation* where no action has been performed before. The binary function $s_{i+1} = do(a_i, s_i)$ starting from S_0 together with a sequence of actions forms a history. Actions are only applicable in the specific situation if preconditions hold which are axiomatized by the predicate $Poss(a, s)$. Fluents are features of the world that might change from situation to situation, e.g. the agents $position$ is changed by a *go*-action. Two fluent types can be distinguished. Relational fluents describe truth values while functional fluents hold general values and both might change over situations. They are denoted by predicate, or function symbols holding the situation as last argument. The action effects on fluents are axiomatized in so called successor state axioms (SSA) [39]. The general form of an SSA for a relational fluent F with its parameters $\boldsymbol{x}$ is

$$\begin{aligned} Poss(a,s) \Rightarrow (F(\boldsymbol{x}, do(a,s)) = true \equiv \\ &\text{the execution of } a \text{ makes } F(\boldsymbol{x}, s) \text{ true} \\ &\vee F(\boldsymbol{x}, s) \text{ already true and } a \text{ makes no change}). \end{aligned}$$

[2] For a relation algebra about this subset of $\mathcal{DRA}_f$ see [40].

[3] WWW.SFBTR8.UNI-BREMEN.DE/PROJECT/R3/CNH/

[4] Other non solid objects like doorways may also be represented by dipoles.

With a basic action theory as presented in [27], namely the action precondition axioms, the successor state axioms, the initial situation and an additional unique names axiom a domain model can be formalized.

Golog [28] is a programming language based on the situation calculus for specifying complex tasks like those typically found in robotic scenarios. Golog offers programming constructs well known from imperative programming languages like *sequence*, *if-then-else,*, *while* and *recursive procedures*. Additionally, a *nondeterministic choice* operator is provided to choose from the given alternatives during runtime. Another important difference compared to most other programming languages is the notion of a *test condition*, which in general can be an arbitrary first order sentence.

We give a list of common programming constructs offered by Golog. The e_i denote legal Golog programs:

1. a: primitive actions (actions in the situation calculus)
2. $[e_1, \ldots, e_n]$: sequence of actions
3. $?(c)$: test whether condition c is true, with c denoting an *arbitrary* first order formula
4. $if(c, e_1, e_2)$: conditional execution of e_1 if c evaluates true and e_2 otherwise
5. $while(c, e)$: while c is true e will be executed
6. $e_1|e_2$: nondeterministic action choice, such that either e_1 or e_2 is executed
7. $star(e)$ or $\star(e)$: nondeterministic iteration, i.e. e is repeated an arbitrary number of times
8. $pi(\vartheta, e)$ or $\Pi(\vartheta, e)$: nondeterministic argument choice, i.e. choose an argument term t and proceed with e substituting all occurrences of ϑ with t
9. P_i: procedures, including recursion

Golog programs, also called procedures, can be viewed as macros for complex actions which are mapped onto primitive actions in the situation calculus. With the above given features Golog serves as integrative framework for programming and planning in dynamic domains. Central for the semantics is the ternary relation $Do(\delta, s, s')$ which is a mapping onto a situation calculus formula. Roughly spoken $Do(\delta, s, s')$ means that given a program δ the situation s' is reachable starting in s. Several extensions e.g. dealing with concurrency [7,36], continuous change [22], sensing [8], probabilistic projections [21] or decision theory [11,43] have been presented. Very important for defining our task are sensing and exogenous actions for on-line robot control. Both actions bind the given results to one or more fluents such that they can be used for controlling the agent. With the help of sensing actions the agent is able to obtain environment information actively. If the robot wants to deliver a letter to a specific person he has to check actively whether the person talking to is the right recipient, e.g. by $get\_name\_of\_person(name)$ with the fluent $name$ holding the result afterwards. In contrast exogenous actions are asynchronous events in the environment, e.g. if someone starts talking to the robot an action with the content might be written in the history by $speech\_input("Could\ you...")$. Modeling reacting to speech by a sensing action would lead to a quite introverted agent only willing to react if he “likes to”.

4 Examples

We have presented on the one hand the situation calculus as framework for reasoning about action and change, which spatial relations are build on an absolute geometrical coordinate system. On the other hand we presented the line segment based dipole calculus together with its conceptual neighborhood (CNH) graph for reasoning about relative position. The CNH-graph describes possible qualitative transitions between adjacent relative configurations by continuous motion.

Regarding only two dipoles (compare to Fig. 1 with the dashed dipole representing an agent and the solid dipole a static object) in $\mathcal{DRA}_c$ the term $behind$ may be defined by relation *rlrr* and *lrll*, and $front$ by *rlll* and *lrrr*. In the following we will restrict dipoles to representing only solid objects.

4.1 General Assumptions and Definitions

Below we will use our newly developed dipole calculus $\mathcal{DRA}_{fp}$, because we consider $\mathcal{DRA}_f$ not being fine grained enough, especially in the context of turning operations. As stated above the CNH-graph is presented on our web page[4]. We define the symmetric binary relation $cnh(p, q)$ holding if two relations p and q are conceptually neighboring. We denote the set of all defined dipoles in the domain with D.

A simple object is a single dipole. A complex object is a polygon, i.e. a sequence of n dipoles $R_i \in D$ where two consecutive dipoles share a common point. For a closed complex object R_0 and R_n must share a common point as well. How such representations can be extracted efficiently and in a compact manner is shown in [47]. The set of all objects is denoted O.

Modeling a robot domain in the situation calculus at least one fluent $pos(s)$ for holding the recent position of the agent A relative to one object resp. dipole is needed: $pos(s) = \langle r_i, o\rangle$ with $r_i \in \mathcal{DRA}_{fp}$ and $o \in O$, i.e. the relation $A\ (r_i)\ o$ holds. In general there will be more than one dipole, or object present in an environment. Therefore more than one positional fluent relative to different dipoles will be necessary for sound navigation, i.e. $pos_j(s) = \langle r_{i_j}, o_j\rangle$ with $j = 1 \ldots n$ and n representing the number of necessary dipoles. In this paper we will show by example, that not all dipoles are important for doing so. In future we have to investigate which dipoles are essential.

In our examples we consider only the basic navigational action $go(r_i, o)$. The precondition that the agent is not blocked holds at any time. Other actions dealing with relative positional information in a domain are e.g. transporting an object R from current position to destination $\langle r_{dest}, o_{dest}\rangle$: $bring(R, r_{dest}, o_{dest})$ or informational questions about spatial configurations.

Because of restricting dipoles to representing only solid objects we can denote subsets (not necessarily disjoint) of relations suitable for intra-object, agent-object and inter-object relations, regarding a dipole and an object. As defined above the subsequent dipoles of the intra-object description need to share a common point. Therefore only relations containing an e or s are suitable for object descriptions. For the sake of simplicity we omit the case of an internal connection of two dipoles. If we assume the agent not being allowed to touch any other object only relations without sharing a start, end or internal point are applicable. Thus we can define a subset of relations $\mathcal{DRA}_{fp}^{object}$ suitable for intra-object definition.

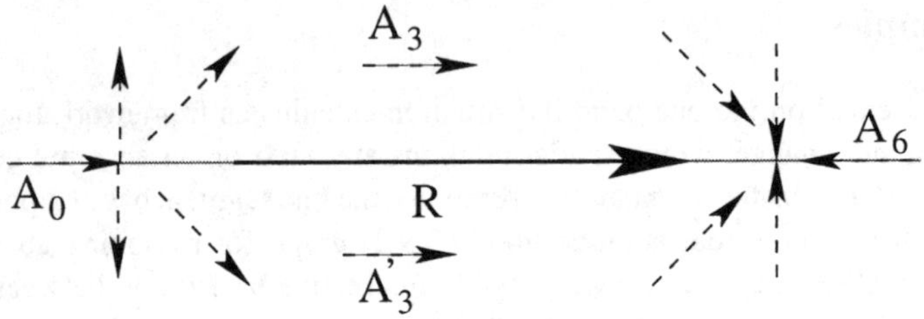

Fig. 7. Simple example with two options for agent A going round object R

$$\mathcal{DRA}_{fp} \supset \mathcal{DRA}_{fp}^{object} = \{\text{ells+,errs-,lere-,rele+,slsr+,srsl-,lsel-,rser+}\}$$

For agent-object relations all other relations except relations containing an internal dipole connection are suitable, for inter-object relations all $\mathcal{DRA}_{fp}$ relations except relations with overlapping dipoles may be used.

4.2 Naive Implementation for Two Dipoles

In a first step we show how a CNH structure might be represented in the situation calculus for two dipoles representing an agent A and an arbitrary object R. The successor state axiom for the *go*-action looks the same as in other domain models without a formal qualitative spatial theory:

$$\begin{aligned} Poss(a,s) \Rightarrow [pos(do(a,s)) = \langle r_j, o\rangle \equiv \\ a = go(r_j, o) \vee \\ [pos(s) = \langle r_j, o\rangle \wedge a \neq go(r_k, o_l)]] \end{aligned}$$

The formula describes the SSA for the *go* action. If action a is possible in situation s, the fluent $pos(s)$ holds for $\langle r_j, o\rangle$ in the successor situation (i.e. after executing a the agent is r_j relative to object o), iff the *go* action just defined to go there, or the agent was already in that position in the originating situation and no *go* action told to go in any other relation to any other object. But the graph structure of the dipole calculus helps us for the definition of the preconditions by exploiting the adjacency of the conceptual neighborhood structure. A movement of the agent to a relative position towards the object is only possible if he is already in a conceptually neighboring configuration. This results in:

$$Poss(go(r_j, o), s) \Leftrightarrow pos(s) = \langle r_i, o\rangle \wedge cnh(r_i, r_j).$$

Assuming an agent A and an object R being in relative position $A(\text{lrrr-})B$ with the goal of being $A(\text{ffffA})B$. The situation calculus and CNH-graph will give the same solution, namely two options to go around R. We sketch the action sequence resp. the transition through neighboring CNH-graph states in Fig. 7.

4.3 Complex Objects (Going Round the Kaaba)

Now we present an example for a complex object. One of the tasks during the hadsch (the great Muslim pilgrimage) is rounding the Kaaba (a cubic building in the main mosque in Mekka) seven times. The knowledge about the Kaaba k (compare Fig. 8) can be represented as follows:

$$R_0(errs-)R_1 \wedge R_1(errs-)R_2 \wedge R_2(errs-)R_3 \wedge R_3(errs-)R_0$$

The agent A may start in position A_0 with $A_0(\mathrm{rrllP})R_0$. At this time the other walls of the Kaaba are of no interest for determining the relative position. Going round the corner of R_0 and R_1 we may get the relations shown in Fig. 9. There are other options traversing the neighborhood graph while turning around, e.g. if the robot starts turning a little earlier. Here we wanted to state the existence of at least one possibility how to turn around the corner.

Looking at all relations for a round trip an analogous situation holds at each corner while the other sides provide no useful knowledge. Thus in this example only two sides are sufficient for describing the relative position of an agent towards the complete object. We expect this being true for more complex, but convex objects, although we have no formal proof so far.

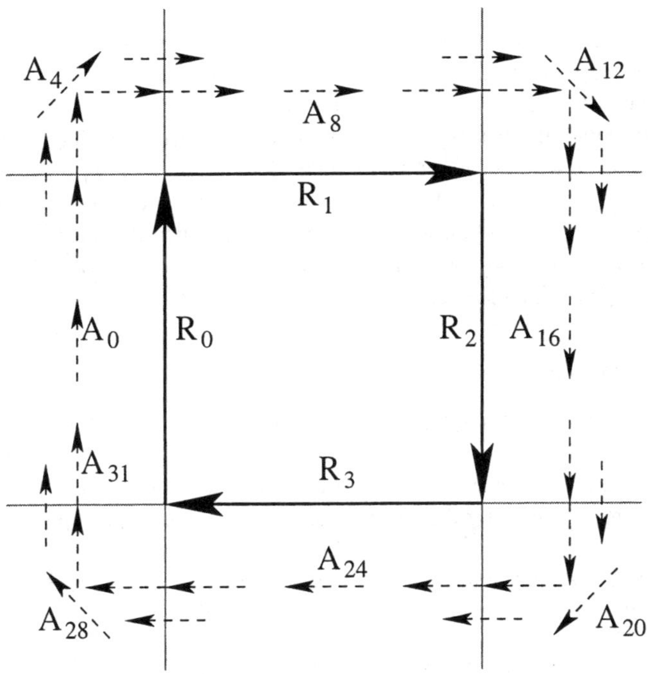

Fig. 8. 32 different qualitative positions an agent A might traverse going round the Kaaba $\{R_0...R_3\}$

4.4 Going Towards Macro Definitions

After extracting the neighborhoods for one complex action like "turn right" we are now heading for some sort of macro definition such that an agent is able to perform a "turn right" on the basis of line segments and imprecise orientation information. We are now looking at the turn problem from a rather communicative perspective.

Imagine being blind, standing at a wall with the task of turning right at the next corner with arbitrary angle and describing it to an external person. The only sensor is one's own right hand extended to the right front which can be seen as some sort of coarse "orientation sensor" transferring the task to a robot. One way describing the process of the first right turn in Fig. 8 might be:

1. Position yourself parallel to the wall and follow the wall until you feel an edge (A_1). If one is moving away or coming closer to the wall it has to be adjusted until the robot is again parallel to the wall.
2. Go a little straight ahead so that the edge is to the right of you (A_2), i.e. the next wall comes just into reach on the right side.
3. Turn (in a bow) right around the corner until you are parallel to the next wall (A_3-A_5).
4. Go a little straight ahead until the first wall just gets out of reach (A_6).
5. Go straight ahead until the corner is right behind you (A_7).
6. Follow the wall (A_8).

All the named actions can be modeled as local behaviors and with the help of the base relations presented in $\mathcal{DRA}_{fp}$. If for example loosing parallelism ($A(rrllP)R_0$) to a wall while following it, we have to look whether we have a mathematically positive or negative orientation towards the relating dipole and turn accordingly. We will take such descriptions as a basis for our macro definitions. At first glance the relations of the ($\mathcal{DRA}_{fp}$) might seem to be too fine grained to represent a simple behavior like turning right adequately. But without the additional relations compared to the ($\mathcal{DRA}_c$) we have not found a way for making the transition from one reference dipole to another (from R_0 to R_1) possible, which is necessary to model going round a corner.

4.5 Macro Definitions

We now have to define an action macro within the framework such that we can model the desired behavior, turning right at a specific corner with the help of neighborhood transitions. Regarding the visualization in Fig. 9 we defined a procedure as shown in Alg. 1. A preliminary to executing this description is being able to percept and distinguish the dipoles.

Turn right at the next opportunity defined in the terms of an initial situation and a goal state might for example lead to the following description:

- $S_0 : pos(R_i, rrllP)$, i.e $A(rrllP)R_i$ with R_i being an aritrary wall the agent stands next to with the heading in the just about correct direction.
- $Goal : A(rrllP)R_j$ if the relation $R_i(errs-)R_j$ holds[5].

[5] We omit the symmetric case of $R_i(rele+)R_j$ here.

R_x / A_y	A_0	A_1	A_2	A_3	A_4	A_5	A_6	A_7	A_8
R_0	$rrllP$	$rrllP$	$rrllP$	$rrllP$	$rrll+$	$rrlf+$	$rrlr+$	$rrfr+$	$rrrr+$
R_1	$rrrr-$	$rrrb-$	$rrrl-$	$rrbl-$	$rrll-$	$rrllP$	$rrllP$	$rrllP$	$rrllP$

Fig. 9. The relations $(A_y(r_{y,x})R_x)$ for an idealized turn right

This means in the initial situation only the wall to the right (R_i) and the dipole which is the next connected one (R_j) in the direction of movement is relevant. We omit the case where a straight wall might be represented with several dipoles being connected via a relation such as *bsefP*. In the goal situation only the relative position towards R_j is important. Thus we have to look how the transitions from $pos(rrllP, R_i)$ to $pos(rrllP, R_j)$ can be defined. Thus in the beginning R_i is the *main anchor* for the relative position while in the end it is R_j. During the period of turning around the corner described by $R_i(errs+)R_j$ the according relation has to be kept as *auxiliary anchor* in mind. Thus we have to split the $pos(.)$ fluent in pos_{main} and pos_{aux}.

Following this result we have to reformulate the precondition and successor state axiom such that only the main anchor is changed by a movement action and the auxiliary anchor is unaffected:

$$Poss(go(r_j, o), s) \Leftrightarrow pos_{main}(s) = \langle r_i, o\rangle \wedge cnh(r_i, r_j)$$

$$Poss(a, s) \Rightarrow [pos_{main}(do(a, s)) = \langle r_j, o\rangle \equiv$$
$$a = go(r_j, o) \vee$$
$$[pos_{main}(s) = \langle r_j, o\rangle \wedge a \neq go(r_x, o_x)]]$$

The question now is where R_i and R_j should flip from serving as main anchor to auxiliary anchor, and vice versa. Analyzing Fig. 9 shows the relations from A_0 to A_3 relative to R_0 being stable, whereas A_5 to A_8 is stable for R_1. Therefore the anchors need to be changed around the relation concerning position A_4. Additionally we have to introduce several new actions which allows us for example to set a specific dipole as auxiliary anchor ($set\_pos_{aux}(R_x)$[6]) and to switch between the two anchors ($switch\_pos_{main}\_pos_{aux}$). We also introduce the special turning actions $rotate\_right(o_i, r_{i_j})$ and $rotate\_left(o_i, r_{i_j})$, although they have almost the same formalization as the $go()$ action except not all conceptual neighbors are reachable, namely the ones needing translation to be reached. In Alg. 1 we use $pos_{main}(R_x, r_x)$ as abbreviation for $pos_{main}(s) = \langle R_x, r_x\rangle$. Additionally a coarse environment description as presented for the Kaaba in section 4.3 is given.

In the initial situation we only know the agent being next to R_0, thus defining pos_{main} and no auxiliary dipole is set. Given the task of turning right at the next possibility (compare $turn\_next\_right$ in Alg. 1) we have to bind a fluent R_i to the current value of the main anchor and look for the next right turn R_j in our environment description. As stated above we omit the case of several dipoles representing a straight wall. Now we have the relevant information available for turning right at the next opportunity.

[6] We are using the term *nil* to reset the fluent.

Algorithm 1 A first approach defining a *turn right* macro

Initial Situation S_0:

$pos_{main}(R_0, rrllP)$, i.e. $A(rrllP)R_0$ and
$pos_{aux}(nil, nil)$

procedure $turn\_next\_right$

1: $(\Pi\ R_i)[\ pos_{main}(R_x, r_x), ?(R_x = R_i),$
2: $\quad (\Pi\ R_j)[\ ?((R_i(errs-)R_j \vee (R_i(rele+)R_j)),$
3: $\quad\quad turn\_right(R_i, R_j)]]$

procedure $turn\_right(R_0, R_1)$

1: $?(R_0(errs-)R_1 \vee R_0(rele+)R_1),$
2: $(\Pi R_i)[pos_{main}(R_x, r_x),$
3: $?(R_i = R_0)],$
4: $set\_pos_{aux}(R_1),$
5: **if** $(r_x = rrll-)$ **then**
6: $\quad rotate\_right(R_0, rrllP)$
7: **else if** $(r_x = rrll+)$ **then**
8: $\quad rotate\_left(R_0, rrllP)$
9: **end if**
10: $switch\_pos_{main}\_pos_{aux}$, // $pos_{main} = R_1$
11: $go(rrrb-, R_1)$, // (A1)
12: $go(rrrl-, R_1)$, // (A2)
13: $go(rrbl-, R_1)$, // (A3)
14: $go(rrll-, R_1)$, // (A4a)
15: $switch\_pos_{main}\_pos_{aux}$, // $pos_{main} = R_0$
16: $go(rrll+, R_0)$, // (A4b)
17: $switch\_pos_{main}\_pos_{aux}$, // $pos_{main} = R_1$
18: $go(rrllP, R_1)$, // (A5a)
19: $switch\_pos_{main}\_pos_{aux}$, // $pos_{main} = R_0$
20: $go(rrlf+, R_0)$, // (A5b)
21: $go(rrlr+, R_0)$, // (A6)
22: $go(rrfr-, R_0)$, // (A7)
23: $go(rrrr+, R_0)$, // (A8)
24: $switch\_pos_{main}\_pos_{aux}$, // $pos_{main} = R_1$
25: $set\_pos_{aux}(nil)$.

The $turn\_right(R_0, R_1)$ procedure makes the robot turn at the specific corner between R_0 and R_1. In the lines one to three we have to check whether R_0 is the main position anchor as well as R_0 and R_1 really form a right corner. Next we set the auxiliary anchor to R_1. So far we have not checked whether we are in the right orientation relative to R_0. So we have to check and turn accordingly. Now we need R_1 as main anchor (line 10). The rest of the procedure is coded straightforward according to Fig. 9. In the end R_1 is the main anchor for the robot's position and we do not need pos_{aux} anymore. After executing this procedure the agent is in position A_8 according to Fig. 8.

5 Conclusion and Outlook

We presented our approach that showed how the concept of conceptual neighborhood can be exploited for reasoning about relative positional information in the situation calculus in the absence of precise quantitative information. We introduced an extended dipole relation algebra $\mathcal{DRA}_{fp}$ better suited for spatial navigation. We expect that every qualitative calculus can be translated in a straightforward manner naively onto preconditions and successor state axioms using its conceptual neighborhood feature. We have shown by example that not all dipoles of a complex object are necessary to determine the relative position towards the object. We expect the results for connected complex objects being applicable for several not connected dipoles. Additionally we extracted several subsets of the base relations for representing a complex object and dynamic agent behavior.

Future work will deal with the question of how to keep the position representation small for more than one dipole respectively object. A naive implementation would lead to a combinatorial explosion, because the relative position of the agent has to be traced for every single dipole. The presented approach with the two anchors will be problematic in general, a set of nearest dipoles as a basis for the positional fluents will be more adequate. A coarse grid partitioning the eight directions *ahead, ahead-left, adjacent-left, behind-left, behind, behind-right, adjacent-right and ahead-right* and the agent in the middle as presented in [9, 34] will serve as a starting point. For each direction the most valuable dipole or object will be held together with the relation between the agent and the dipole. The definition of the term 'most valuable dipole' is one of the major tasks to solve in this context. We will also look on the effects allowing dipoles to represent non-solid entities, e.g. doorways, and potentials to define some sort of general macro definitions for $turnLeft$ and $turnRight$ or $GoAround$ by paths in the conceptual neighborhood graph.

Acknowledgment

The authors like to thank Diedrich Wolter, Christian Freksa, Jochen Renz, and Marco Ragni for fruitful discussions and impulses. And we would like to thank Marc-Björn Seidel for computing the neighborhood graph for different calculi, as well as the anonymous reviewers for their valuable comments. Our work was supported by the DFG Transregional Collaborative Research Center SFB/TR 8 Spatial Cognition.

References

1. John Bateman and Scott Farrar. Modelling models of robot navigation using formal spatial ontology. In *Proceedings of Spatial Cognition Conference 2004 (this volume)*, 2004.
2. Brandon Bennett and Antony P. Galton. A unifying semantics for time and events. *Artificial Intelligence*, 153(1-2):13–48, March 2004.
3. Wolfram Burgard, Armin B. Cremers, Dieter Fox, Dirk Hahnel, Gerhard Lakemeyer, Dirk Schulz, Walter Steiner, and Sebastian Thrun. Experiences with an interactive museum tour-guide robot. *Artificial Intelligence*, 114(1-2):3–55, 1999.

4. Anthony G. Cohn. Qualitative spatial representation and reasoning techniques. In Gerhard Brewka, Christopher Habel, and Bernhard Nebel, editors, *KI-97: Advances in Artificial Intelligence, 21st Annual German Conference on Artificial Intelligence, Freiburg, Germany, September 9-12, 1997, Proceedings*, volume 1303 of *Lecture Notes in Computer Science*, pages 1 – 30, Berlin, 1997. Springer.
5. Anthony G. Cohn and Shyamanta M. Hazarika. Qualitative spatial representation and reasoning: An overview. *Fundamenta Informaticae*, 46(1-2):1–29, 2001.
6. E. Davis. Continuous shape transformation and metrics of shape. In *Fundamenta Informaticae*, volume 46, pages 31–54. May 2001.
7. Giuseppe De Giacomo, Yves Lésperance, and Hector J. Levesque. ConGolog, A concurrent programming language based on situation calculus. *Artificial Intelligence*, 121(1–2):109–169, 2000.
8. Giuseppe De Giacomo and Hector Levesque. An incremental interpreter for high-level programs with sensing. In Hector J. Levesque and Fiora Pirri, editors, *Logical Foundation for Cognitive Agents: Contributions in Honor of Ray Reiter*, pages 86–102. Springer, Berlin, 1999.
9. J. Fernyhough, A.G. Cohn, and D. Hogg. Constructing qualitative event models automatically from video input. *Image and Vision Computing*, 18:81–103, 2000.
10. A. Ferrein, C. Fritz, and G. Lakemeyer. On-line decision-theoretic golog for unpredictable domains. In *Proc. of 4th International Cognitive Robotics Workshop*, 2004.
11. Alexander Ferrein, Christian Fritz, and Gerhard Lakemeyer. Extending DTGolog with Options. In *Proc of the 18th International Joint Conference on Artificial Intelligence*, 2003.
12. Christian Freksa. Conceptual neighborhood and its role in temporal and spatial reasoning. In Madan G. Singh and Luise Travé-Massuyès, editors, *Proceedings of the IMACS Workshop on Decision Support Systems and Qualitative Reasoning*, pages 181 – 187, North-Holland, Amsterdam, 1991. Elsevier.
13. Christian Freksa. Using orientation information for qualitative spatial reasoning. In A. U. Frank, I. Campari, and U. Formentini, editors, *Theories and methods of spatio-temporal reasoning in geographic space*, pages 162 – 178. Springer, Berlin, 1992.
14. Christian Freksa. Spatial Cognition - An AI Prespective. In *Proceedings of 16th European Conference on AI (ECAI 2004)*, 2004.
15. Christian Fritz. Integrating decision-theoretic planning and programming for robot control in highly dynamic domains. Master's thesis, RWTH Aachen (Knowledge-based Systems Group), Aachen, Germany, 2003.
16. Anthony Galton. Space, time, and movement. In Oliviero Stock, editor, *Spatial and Temporal Reasoning*, pages 321–352. Kluwer Academic Publishers, Dordrecht, 1997.
17. Antony Galton. *Qualitative Spatial Change*. Oxford University Press, 2000.
18. A.P. Galton. Continuous motion in discrete space. In A.G. Cohn, F. Giunchiglia, and B. Selman, editors, *Proc. 7th Internat. Conf. on Principles of Knowledge Representation and Reasoning (KR2000)*, pages 26–37. Morgan Kaufmann, San Francisco, CA, 2000.
19. Henrik Grosskreutz. Probabilistic projection and belief update in the pGOLOG framework. In *CogRob-00 at ECAI-00*, 2000.
20. Henrik Grosskreutz and Gerhard Lakemeyer. cc-Golog: Towards more realistic logic-based robot controllers. In *AAAI-00*, 2000.
21. Henrik Grosskreutz and Gerhard Lakemeyer. Turning high-level plans into robot programs in uncertain domains. In *ECAI-00*, 2000.
22. Henrik Grosskreutz and Gerhard Lakemeyer. On-line execution of cc-Golog plans. In *IJCAI-01*, 2001.
23. Shyamanta M. Harzarika and Anthony G. Cohn. Qualitative spatio-temporal continuity. *Lecture Notes in Computer Science*, 2205:92–107, 2001.

24. Amar Isli and Anthony G. Cohn. An algebra for cyclic ordering of 2d orientations. In *AAAI/IAAI*, pages 643–649, 1998.
25. B. J. Kuipers. Representing knowledge of large-scale space. Technical Report 418, 1977.
26. P Ladkin and R Maddux. On binary constraint problems. *Journal of the Association for Computing Machinery*, 41(3):435–469, 1994.
27. H. Levesque, F. Pirri, and R. Reiter. Foundations for the situation calculus. In *Linköping Electronic Articles in Computer and Information Science*, volume 3. 1998.
28. Hector J. Levesque, Raymond Reiter, Yves Lesperance, Fangzhen Lin, and Richard B. Scherl. GOLOG: A logic programming language for dynamic domains. *Journal of Logic Programming*, 31(1-3):59–83, 1997.
29. J. McCarthy. Situations, actions and causal laws. Technical report, Stanford University, 1963.
30. R. Moratz, T. Tenbrink, J. Bateman, and K. Fischer. Spatial knowledge representation for human-robot interaction. In Christian Freksa, W. Brauer, C. Habel, and K. F. Wender, editors, *Spatial Cognition III*. Springer, Berlin, Heidelberg, 2002.
31. Reinhard Moratz, Bernhard Nebel, and Christan Freksa. Qualitative spatial reasoning about relative position: The tradeoff between strong formal properties and successful reasoning about route graphs. In Christian Freksa, Wilfried Brauer, Christopher Habel, and Karl Friedrich Wender, editors, *Spatial Cognition III*, volume 2685 of *Lecture Notes in Artificial Intelligence*, pages 385–400. Springer, Berlin, Heidelberg, 2003.
32. Reinhard Moratz, Jochen Renz, and Diedrich Wolter. Qualitative spatial reasoning about line segments. In *Proceedings of ECAI 2000*, pages 234–238, 2000.
33. Reinhard Moratz and Jan Oliver Wallgrün. Propagation of distance and orientation intervals. In *Proceedings of 2003 IEEE/RSJ International Conference on Intelligent Robots and Systems (IROS)*, pages 3245–3250, 2003.
34. A. Mukerjee and G. Joe. A qualitative model for space. In *Proc. of AAAI-90*, pages 721–727, Boston, MA, 1990.
35. Philippe Muller. A qualitative theory of motion based on spatio-temporal primitives. In Anthony G. Cohn, Lenhart Schubert, and Stuart C. Shapiro, editors, *KR'98: Principles of Knowledge Representation and Reasoning*, pages 131–141. Morgan Kaufmann, San Francisco, California, 1998.
36. Javier Pinto. *Temporal Reasoning in the Situation Calculus*. PhD thesis, Department of Computer Science, University of Toronto, 1994.
37. D. A. Randell and A. G. Cohn. Modelling topological and metrical properties of physical processes. In R. J. Brachman, H. J. Levesque, and R. Reiter, editors, *Proceedings of the 1st International Conference on Principles of Knowledge Representation and Reasoning (KR'89)*, pages 357–368. Morgan Kaufmann, 1989.
38. David A. Randell, Zhan Cui, and Anthony Cohn. A spatial logic based on regions and connection. In Bernhard Nebel, Charles Rich, and William Swartout, editors, *KR'92. Principles of Knowledge Representation and Reasoning: Proceedings of the Third International Conference*, pages 165–176. Morgan Kaufmann, San Mateo, California, 1992.
39. R. Reiter. *Knowledge in Action*. MIT Press, 2001.
40. Jochen Renz. A spatial odyssey of the interval algebra: 1. directed intervals. In *IJCAI*, pages 51–56, 2001.
41. Thomas Röfer. Route navigation using motion analysis. In Christian Freksa and David M. Mark, editors, *Cosit 1999*, pages 21–36. Springer, Berlin, 1999.
42. C. Schlieder. Reasoning about ordering. In A. U. Frank and W. Kuhn, editors, *Spatial Information Theory - A Theoretical Basis for GIS (COSIT'95)*, pages 341–349. Springer, Berlin, Heidelberg, 1995.
43. Mikhail Soutchanski. An on-line decision-theoretic golog interpreter. In *IJCAI*, pages 19–26, 2001.

44. Thora Tenbrink, Kerstin Fischer, and Reinhard Moratz. Spatial strategies in linguistic human-robot communication. In Christian Freksa, editor, *KI-Journal: Special Issue on Spatial Cognition*. arenDTaP Verlag, 2002.
45. R.F. Wang and E.S. Spelke. Updating egocentric representations in human navigation. *Cognition*, 2000.
46. R.F. Wang and E.S. Spelke. Human spatial representation: Insights from animals. *Trends in Cognitive Sciences*, 6(9):376–382, 2002.
47. Diedrich Wolter and Longin Jan Latecki. Shape matching for robot mapping. In Chengqi Zhang, Hans W. Guesgen, and Wai K. Yeap, editors, *Proceedings of 8th Pacific Rim International Conference on Artificial Intelligence*, Auckland, New Zealand, August 2004.

Branching Allen

Reasoning with Intervals in Branching Time

Marco Ragni and Stefan Wölfl

Institut für Informatik, Albert-Ludwigs-Universität Freiburg,
Georges-Köhler-Allee, 79110 Freiburg, Germany
{ragni, woelfl}@informatik.uni-freiburg.de

Abstract. Allen's interval calculus is one of the most prominent formalisms in the domain of qualitative spatial and temporal reasoning. Applications of this calculus, however, are restricted to domains that deal with linear flows of time. But how the fundamental ideas of Allen's calculus can be extended to other, weaker structures than linear orders has gained only little attention in the literature. In this paper we will investigate intervals in branching flows of time, which are of special interest for temporal reasoning, since they allow for representing indeterministic aspects of systems, scenarios, planning tasks, etc. As well, branching time models, i. e., treelike non-linear structures, do have interesting applications in the field of spatial reasoning, for example, for modeling traffic networks. In a first step we discuss interval relations for branching time, thereby comprising various sources from the literature. Then, in a second step, we present some new complexity results concerning constraint satisfaction problems of interval relations in branching time.

1 Introduction

Allen's interval calculus is one of the most prominent formalisms in the domain of qualitative spatial and temporal reasoning. But applications of this calculus are restricted to domains in which intervals in *linear* flows of time are considered. Surprisingly, the question of how the ideas of Allen's calculus can be extended to other, weaker structures than linear orders has gained only little attention in the literature. In this paper we will focus on intervals in branching time. The basic idea of branching time is that at each moment there exists only one possible past, but many possible futures. Hence branching flows of time, which can be modeled by tree-like structures are of special interest for temporal reasoning since they allow for representing indeterministic aspects of systems, scenarios, and planning tasks.

In modal logic, branching time models have been studied intensely in the past two decades. Originating from philosophical logic (cf. [29]), where branching time logics have been investigated for analyses of indeterminism, causality, and action-theoretical concepts, branching time logics such as CTL and CTL* (cf. [10]) and their multi-agent extensions ATL and ATL* (cf. [3]) have been discussed as specification languages, mainly for model checking purposes of closed, reactive systems as well as of systems that interact with their environment.

C. Freksa et al. (Eds.): Spatial Cognition IV, LNAI 3343, pp. 323–343, 2005.
© Springer-Verlag Berlin Heidelberg 2005

Allen's interval algebra is a reasoning formalism in the spirit of Hayes' naïve manifesto [14]. If the instants of a linear flow of time count as primary entities (which is, for example, the point of view of the so-called point algebra of linear time), intervals are *sets* of instants. Thus, the interval algebra can be seen as a shift of perspective from first-order entities (instants) to second-order entities (intervals). This paper will deal with the analogous change of perspective, from moments in branching time towards intervals in branching time.

Algebraic aspects of the point algebra of branching time were first investigated by Düntsch, Wang, and McCloskey [9]. Hirsch [17] showed that local consistency is insufficient for satisfiability testing for the point algebra of branching time. Contrary to the point algebra of linear time, satisfiability testing for branching time is NP-hard. Broxvall [6] discussed tractable subclasses of the point algebra of branching time. Tractable subclasses of the interval algebra of linear time were identified by Nebel and Bürckert [26] and by Ligozat [24].

What is the motivation for considering branching time, tree-like structures? First, tree-like structures are a natural choice for modeling temporal aspects of events. For example, Kutschera [18] defined events as sets of closed intervals in branching time. Tree-like structures are used to model the various courses the world might take. A (complete) branch of a tree represents one specific way in which the world can evolve. The basic idea, then, is to identify an event with the set of its occurrences in time, i. e., with the set of its temporal extensions. An event can occur in many branches — an event is said to occur in a branch if one of its instances is completely contained in that branch. But since events are understood as singular events, an event can occur only once in a branch. The main requirement of Kutschera events is that when an event occurs in two branches that overlap while the event occurs in at least one of them, then the event starts in both branches at the same moment. This little discussion already indicates how reasoning with Allen-style interval relations (adapted for branching flows of time) could be used for reasoning about events.

With respect to more spatial domains, a theory of intervals in tree-like structures may have interesting applications, for example, when routes in traffic networks are represented by qualitative concepts. Of course, most street networks are not tree-like, but many railroad networks are. Modeling street networks by tree-like structures may be applicable, especially when one focuses on "small" traffic scenarios. To illustrate this, let us assume that the spatial configuration of an intersection of highways is to be represented by qualitative means. Then one can distinguish lane segments according to the traffic regulations that hold within these segments. These segments may be related to each other by any of the base relations of Allen's interval algebra. For example, a segment in which passing is forbidden may start a segment in which speed is limited. Thus, lane segments are a natural candidate for intervals. But also cars and accumulations of cars can be represented by intervals. Hence, a congestion in a lane segment can be modeled by two intervals, with one contained in the other. The branching aspect comes into play since, in this qualitative language, we can describe a car driving off the road or a road connecting two highways.

Finally, branching structures can also be applied in planning domains. Planning deals with the question of how a certain goal state can be reached from an initial state by executing a sequence of actions. Usually, planning tasks can be modeled by Kripke style transition graphs, and these graphs can be *unwinded* to tree-like structure. The method of unwinding a transition graph is applied implicitly, when heuristic forward search is used in planning algorithms.

The paper is organized as follows: In section 2 we review some basic concepts of the theory of tree-like structures, and we sketch some results concerning the point algebra of branching time. In section 3 we present the base relations between intervals in branching time. More precisely, we define two algebras of base relations, where one is a refinement of the other one. Section 4 deals with the conceptual neighborhood graph for interval relations in branching time and discusses its relationship to the linear time version. In section 5 we investigate the computational complexity of constraint satisfaction problems of the algebras presented previously. In particular, we show that the satisfiability problem with respect to the coarser algebra of interval relations is NP-complete. In section 6 we work out some of the particularities of the composition table of interval relations in branching time. Finally, section 7 summarizes the results of the paper and gives a short overview of some questions that are left open in this paper.

2 Branching Time Theory

To begin with, let us recall some basic concepts from the theory of tree-like structures.

Definition 1. A *tree* is an ordered pair $\mathcal{B} = \langle T, \prec \rangle$ consisting of a non-void set of *nodes*, T, and a binary relation, $\prec$, satisfying the following properties:

(a) $\prec$ is a partial order on T, i. e., $\prec$ is irreflexive and transitive;

(b) $\prec$ does not allow for *backward branching*, i. e., $\prec$ is linear-to-the-left;[1]

(c) T is *connected* via $\prec$, i. e., for all $t, t' \in T$ with $t \not\prec t'$, $t' \not\prec t$, and $t \neq t'$, there exists a $t'' \in T$ such that $t'' \prec t$ and $t'' \prec t'$.

We read $t \prec t'$ as "t is earlier than t'". Symbols such as $\preceq$, $\succ$, and $\succ$ are used in the natural manner. For sets X and X' of nodes, let $X \preceq X'$ ($X \prec X'$) be defined as: for all $t \in X$ and $t' \in X'$, it holds that $t \preceq t'$ ($t \prec t'$). Finally, $X \preceq t$ is an abbreviation of $X \preceq \{t\}$, etc. Nodes t and t' are said to be *unrelated*, $t \parallel t'$, if neither $t \preceq t'$ nor $t' \preceq t$.

A *chain* of nodes is a set of nodes that is linearly ordered by the relation of earlier-than, i. e., each pair of nodes chosen in the chain are comparable with respect to $\prec$. A chain k is said to be *upper-bounded* if there is a node t with $k \preceq t$. In an analogous manner concepts such as *lower-bounded* or *bounded* are introduced.

Maximal $\prec$-chains in a tree $\mathcal{B}$ are said to be *branches*, and the set of all branches is denoted by B. For a given node t, let $B\langle t \rangle$ be the set of all branches that contain t as an element. Furthermore, we will use the following terminology: Branches b and b' are *undivided* at node t if there exists a node $t' \succ t$ with $t' \in b \cap b'$. Branches b and b' *split* at node t if t is the maximal element of $b \cap b'$. Note that the intersection of a pair of

[1] This means that for all nodes t and t', it holds either $t \prec t'$ or $t = t'$ or $t' \prec t$, provided there is an t'' with $t, t' \prec t''$.

branches need not have a maximal element, even if they intersect. A node t is a *splitting point* if there exist branches that split at t. Branches b and b' are *separated* at node t if either $t \in b \setminus b'$ or $t \in b' \setminus b$.

With these notations we can replace condition 1 (c) by each of the following conditions:

(c_1) If $t \parallel t'$, then there exists a t'' with $t'' \prec t$ and $t'' \prec t'$.
(c_2) There exists a t'' with $t'' \preceq t$ and $t'' \preceq t'$.
(c_3) For each pair of branches b and b', $b \cap b' \neq \emptyset$.

And if $\prec$ is infinite-to-the-left, then condition (c) is equivalent to:

(c_4) There exists a t'' with $t'' \prec t$ and $t'' \prec t'$.

A tree $\mathcal{B} = \langle T, \prec \rangle$ is said to be *dense* if for each pair of nodes $t, t' \in T$ with $t \prec t'$, there exists a node $t'' \in T$ such that $t \prec t'' \prec t'$; $\mathcal{B}$ is said to be *branching dense* if, for each pair of nodes $t, t' \in T$ with $t \prec t'$, there exists a $t'' \in T$ such that $t \prec t''$ and $t' \parallel t''$. Obviously, density does not follow from branching density, and vice versa. Note that there exist finite and branching dense trees, but that no tree is both finite and dense. Nevertheless, branching density is a very strong condition since, in a finite branching dense tree, each node that is not the endpoint of some branch is a splitting point. Finally, it is worth mentioning that trees are not required to have roots.

The intended models for Allen's calculus are dense linear flows of times without endpoints, for example, the linear order of the rationals or that of the reals. A typical example of a dense and infinite tree is any instance of a $\mathbb{Q}$- or an $\mathbb{R}$-tree.

Definition 2. A tree $\mathcal{B} = \langle T, \prec \rangle$ is said to be a $\mathbb{Q}$*-tree* (an $\mathbb{R}$*-tree*, resp.) if there exists a family $(\iota_b)_{b \in B}$ of order isomorphisms $\iota_b \colon b \longrightarrow \mathbb{Q}$ (or $\iota_b \colon b \longrightarrow \mathbb{R}$, resp.) such that for all $b, b' \in B$ and each $x \in b \cap b'$, $\iota_b(x) = \iota_{b'}(x)$.

Hence in a $\mathbb{Q}$-tree, each node of a branch can be labeled by a rational number via an order isomorphism, and the labeling of nodes in one branch respects the labeling of nodes in another branch as long as both branches intersect.

In the class of all dense trees, it is reasonable to distinguish two tree types with regard to the structure of how branches of the tree actually split:

(a) $\mathcal{B}$ is said to be of *type 1* if, for each pair of distinct branches $b, b' \in B$, the intersection of b and b' has a maximum, i. e., there exists a node at which b and b' split.[2]
(b) $\mathcal{B}$ is said to be of *type 2* if, for each pair of distinct branches b and b', the intersection of b and b' has no maximum, i. e., b and b' are undivided at each node $t \in b \cap b'$.

This list is not exhaustive since further splitting types can be defined. As well, there are trees that do not have a uniform splitting type. However, when a scenario is represented in terms of $\mathbb{Q}$-trees, it is reasonable to fix a specific splitting structure according to the typology presented here. Note that finite trees are always of type 1. We will discuss this point in more detail later.

[2] This condition is also known as the *semi-lattice condition.*

Table 1. The composition table of the point algebra for branching flows of time (cf. [6])

	$\prec$	$\succ$	$\parallel$	$=$
$\prec$	$\prec$	$\prec,=,\succ$	$\prec,\parallel$	$\prec$
$\succ$	$\top$	$\succ$	$\parallel$	$\succ$
$\parallel$	$\parallel$	$\succ,\parallel$	$\top$	$\parallel$
$=$	$\prec$	$\succ$	$\parallel$	$=$

In Allen's interval calculus, an interval is represented by a pair of points, namely the start and the endpoint of the interval. Thus in a weak sense, reasoning with intervals can be reduced to reasoning with points of the underlying linear flow of time. On the other hand, reasoning with instants of a linear flow of time can be done by employing the so-called point algebra for linear time, PA_{lin}. The point algebra for branching flows of time, PA_{br}, has been investigated by Broxvall and Jonsson [7,6]. In PA_{br}, the relations $\prec$, $=$, $\succ$, and $\parallel$ count as base relations. More precisely, these relations are the atoms of the relation algebra that is defined on the set of all (set-theoretical) unions of base relations via the composition table shown in Table 1. In qualitative reasoning, unions of base relations are considered to model imprecise knowledge in a given scenario.

Note that given an atomic relation algebra A with finite atom set $B(A)$ (i. e., $B(A)$ is the set of all base relations), each relation $r \in A$ can be written in a unique manner as a union of base relations $b_1, \dots, b_n$. Hence, algebraic functions such as composition, converse, intersection, union, and complement, can be computed from base relations by applying the following equations:

$$\begin{aligned}
(b_1 \cup \cdots \cup b_n) \circ (b'_1 \cup \cdots \cup b'_m) &= \bigcup_{1 \le i \le n, 1 \le j \le m} (b_i \circ b'_j) \\
(b_1 \cup \cdots \cup b_n)^{-1} &= (b_1^{-1} \cup \cdots \cup b_n^{-1}) \\
(r \cap r') &= \bigcup \{b \in B(A) : b \subseteq r \text{ and } b \subseteq r'\} \\
(r \cup r') &= \bigcup \{b \in B(A) : b \subseteq r \text{ or } b \subseteq r'\}
\end{aligned}$$

It is worthwhile to remark that the general constraint satisfaction problem for PA_{br} is NP-hard, while it is in P for the point algebra of linear time. Broxvall [6] identified five maximal tractable subsets of the point algebra for branching time and showed that these are the only maximal tractable subsets.

3 Intervals and Branching Time

As said before, in Allen's theory intervals are identified with pairs of points of a given linear order $\langle T, < \rangle$, which is assumed to be isomorphic to the linear order of the rationals. More precisely, an *Allen interval* is a pair of points $\langle t_1, t_2 \rangle \in T^2$ with $t_1 < t_2$. By considering start and endpoints of intervals, Allen identified thirteen jointly exhaustive and pairwise disjoint relations known in the literature as the Allen 13 relations (cf. Table 2).

Table 2. The 13 base relations of Allen's interval algebra

Relation	Converse	Pictorial Representation
$I\, b\, J$	$J\, bi\, I$	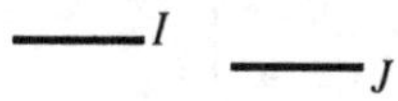
$I\, m\, J$	$J\, mi\, I$	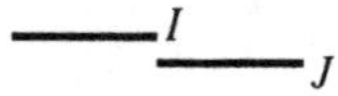
$I\, o\, J$	$J\, oi\, I$	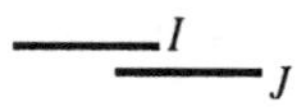
$I\, d\, J$	$J\, di\, I$	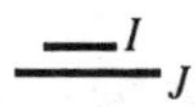
$I\, s\, J$	$J\, si\, I$	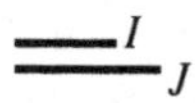
$I\, f\, J$	$J\, fi\, I$	
$I\, e\, J$	$J\, e\, I$	

In contrast to Allen's theory, we will use the notion of *interval* in the following sense:

Definition 3. Let $\mathcal{B} = \langle T, \prec \rangle$ be a tree. An *interval* in $\mathcal{B}$ is a convex and bounded $\prec$-chain in T.

It is worthwhile to recall that in $\mathbb{R}$ each interval (in terms of this definition) can be represented as an Allen interval, i. e., via start and endpoint. With respect to the linear order of the rationals, this situation is different since $\mathbb{Q}$ is not Dedekind-complete. Note that representing intervals as Allen intervals is not unique. For example, the intervals $(0,1)$, $(0,1]$, $[0,1)$, and $[0,1]$ do have the same Allen representation $\langle 0,1 \rangle$. In Allen's approach this in some sense imprecise representation is chosen intentionally to abstract from the exact boundary structure of intervals.[3] But when we are to present models for Allen relations it is reasonable, for the sake of simplicity, to fix the exact structure of intervals because the boundary structure depends on the interpretation of Allen primitives such as "starts", "meets", etc. For if the primitive "meet" is understood in the sense that "I meets J" entails that I and J intersect, then Allen intervals are understood as closed intervals, i. e., as intervals of the form $[t,t']$. If "I meets J" means "there is no interval K with $I \leq K \leq J$", then we are free to fix the boundary structure.

[3] Note that there are two distinct meanings in which abstraction from the boundary structure can be understood. Let us assume that we are given a set of Allen relation constraints, C, a linear flow of time, and an assignment of the variables occurring in C, a, that satisfies C and interpretes intervals as, say, closed intervals. In its first sense, then, abstraction means that we obtain a model of C by replacing a by any assignment a' that interpretes *some* of the intervals in C as open intervals. In the weaker sense, abstraction means that we have to replace *uniformly*, for example, a closed interval interpretation by an open interval interpretation to obtain a satisfying model. In what is to follow we will use abstraction in the first sense.

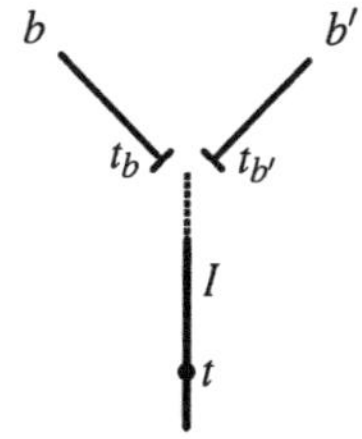

Fig. 1. A branching scenario of type 2. Interval I starts at t and continues until b and b' are split. I does not contain one of its supreme points, t_b and $t_{b'}$. How could I be represented as an Allen interval?

In branching time the boundary structure of intervals is more crucial than it is in linear flows of time. To illustrate this, let us assume that $\mathcal{B} = \langle T, \prec \rangle$ is an $\mathbb{R}$-tree of type 2. Let b and b' be branches that intersect, and choose $t \in b \cap b'$. Consider then the interval $I := \{t' \in b \cap b' : t \preceq t'\}$. Obviously, in each of the branches b and b' there exists a supremum of I, t_b and $t_{b'}$ respectively, but t_b is not contained in b', and, vice versa, $t_{b'}$ is not contained in b. Now if we represented I as the Allen interval $\langle t, t_b \rangle$, then the interval $I = [t, t_b) = [t, t_{b'})$ will have the same Allen representation as the interval $I_b := [t, t_b]$. But from a conceptual point of view, I and I_b are essentially distinct, since I is contained in b', while I_b is not. The difference between I and I_b can even be expressed in terms of Allen relations. For if t' is a node with $t' \succ t_{b'}$, then it makes perfect sense to say that interval I meets interval $J := [t_{b'}, t')$, while one would be inclined to say that I_b does not meet J.

This small discussion shows that the Allen representation of intervals is at least problematic or even inappropriate in the context of branching time. One could argue that the example just discussed crucially depends on the splitting structure of the tree. But analogous examples could be constructed if the tree at hand is of type 1.

Following we will present in an informal manner a set of 24 relations that may hold between two intervals in a tree. To our knowledge, these relations were first discussed by Anger, Ladkin, and Rodriguez [4], and we widely adopt the notations from their paper. To present the set of 24 base relations, we will assume that intervals are closed, in the sense that they contain a minimal and a maximal element (a start and an endpoint): this is advantageous, since in this case we need not take care of the splitting type of the tree. Furthermore, we will use concepts such as "I is connected with J" or "I meets J" in the sense that intervals I and J do intersect. Let now I and J be intervals. We start by presenting those relations where both intervals I and J are contained in some branch b. Obviously, these relations are exactly the 13 base relations known from Allen's interval algebra, now in a branching context (cf. Fig. 2). In what follows these relations will be referred to as *linear relations*.

Let us now assume that I is completely contained in some branch b, while J is not. From this it follows that there is a subinterval of I that is not comparable to J via one of the Allen relations. Then we may distinguish two cases: In the first case another subinterval, more precisely an initial segment, of I is related to J by one of the Allen relations. This enables us to differentiate seven additional relations (cf. Fig. 3).

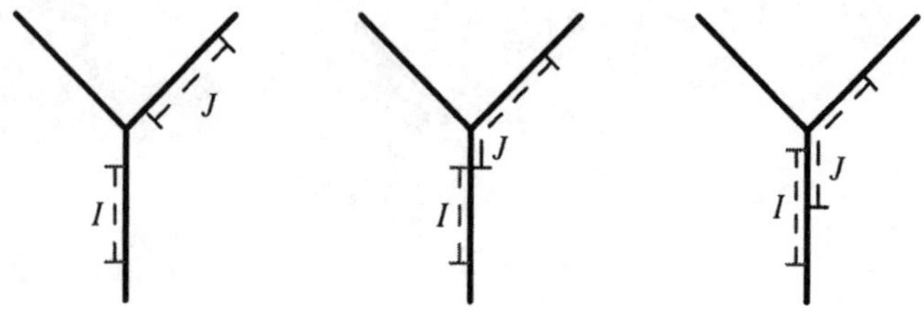

(a) *I* before *J* (symb.: *I b J*, converse: *J bi I*) (b) *I* meets *J* (symb.: *I m J*, conv.: *J mi I*) (c) *I* overlaps *J* (symb.: *I o J*, conv.: *J oi I*)

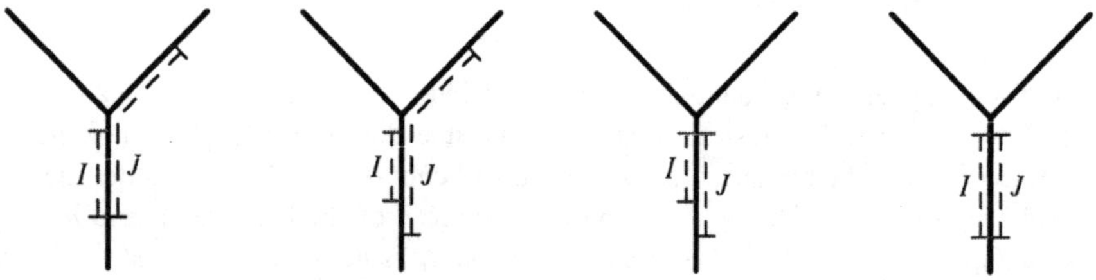

(d) *I* starts *J* (symb.: *I s J*, conv.: *J si I*) (e) *I* during *J* (symb.: *I d J*, conv.: *J di I*) (f) *I* finishes *J* (symb.: *I f J*, conv.: *J fi I*) (g) *I* equals *J* (symb.: *I e J*, symmetric)

Fig. 2. The 13 Allen relations in a branching context

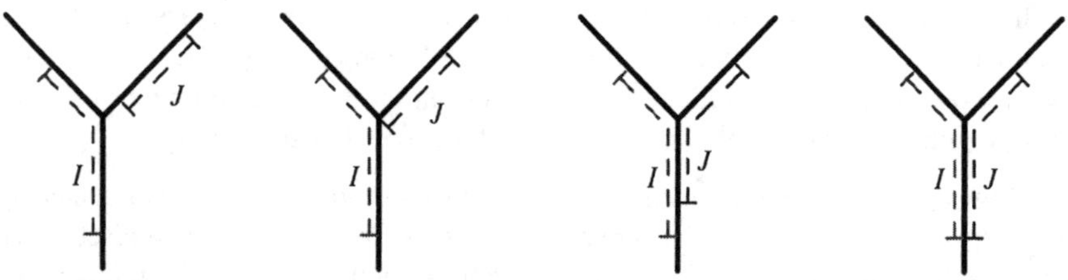

(a) *I* partially before *J* (symb.: *I pb J*, conv.: *J pbi I*) (b) *I* partially meets *J* (symb.: *I pm J*, conv.: *J pmi I*) (c) *I* partially overlaps *J* (symb.: *I po J*, conv.: *J poi I*) (d) *I* partially starts *J* (symb.: *I ps J*, symmetric)

Fig. 3. Splitting point within one of the intervals. Seven relations can be distinguished if there exist two subintervals of *I* (or *J*, resp.), where one is comparable to *J* (or *I*, resp.) via Allen relations, and the other one is not

In the second case, we may assume that no initial segment of *I* is related to *J* via some Allen relation. In this case we can distinguish four further relations as depicted in Fig. 4. Note that these relations can be characterized by quantifying over intervals. For example, if we assume that *I* and *J* are not comparable by one of the 20 relations presented before, then *I* is unrelated to *J* if and only if there are intervals I' and J' such

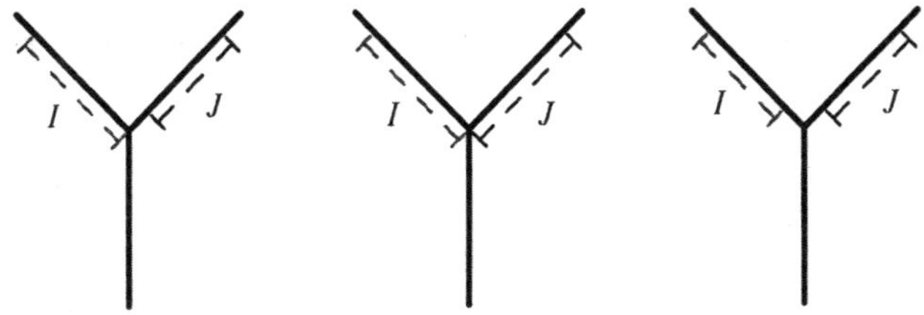

(a) I adjacent to J (symb.: $I\ a\ J$, conv.: $J\ ai\ I$)

(b) I touches J (symb.: $I\ t\ J$, symmetric)

(c) I unrelated to J (symb.: $I\ u\ J$, symmetric)

Fig. 4. Splitting point preceding the intervals. Four relations can be distinguished if there is no subinterval of I (or J, resp.) that is comparable to J (or I, resp.) by an Allen relation

that I' meets I, J' meets J, and I' and J' are not related by one of the 13 Allen relations. Then the relations a and t can be defined as follows: I is adjacent to J if each interval that finishes I is unrelated to J and if no interval that starts I is before J (note that this definition depends on the assumption that the tree at hand is dense). And, I touches J if for each pair of intervals I' and J' with $I'\ f\ I$ and $J'\ f\ J$, I' is unrelated to J'.

It is worthwhile to remark that only 19 base relations are definable if we restrict consideration to the start and endpoints of intervals. "Definable" here means definable in the constraint language of endpoints, which does not allow quantification over splitting points. Table 3 provides an overview (cf. Euzenat [11]). By comparing the defining constraints for the set of 19 base relations (depicted in Table 3) with those constraints for the refined set of 24 base relations (cf. figures 2–4 and Table 4), one can readily check that the following equivalences hold:

$$
\begin{aligned}
I\ ib\ J &\iff I\ pb\ J \vee I\ a\ J \\
I\ ie\ J &\iff I\ ps\ J \vee I\ t\ J \\
I\ im\ J &\iff I\ pm\ J \vee I\ po\ J
\end{aligned}
$$

The following remarks will round off the presentation of interval relations in branching time.

Remark 4. If connectedness of trees (condition 1(c)) is not enforced, a tree may have a forest-like structure consisting of several (genuine) tree components. Then the relation u can be partitioned into two subrelations, namely "unrelated, but contained in the same tree component" and "contained in disjoint tree components". Here the term "tree component" is used to denote a tree in which all branches are connected.

Remark 5. The 19 base relations as well as the 24 base relations defined in this section can be shown to be jointly exhaustive and pairwise disjoint. This follows immediately from the defining constraints presented in tables 3 and 4. But this result depends crucially on the closed-interval interpretation we have chosen in the presentation. A more general result can be obtained by introducing a first order theory of intervals in branching time in the style of Allen and Hayes' axiomatization of the meet relation for linear time (cf. [1, 15]). Yet, a detailed discussion of such a theory would go beyond the scope of this paper.

Table 3. A set of 19 base relations definable by interval endpoints in branching time

Symbol	Relation	Defining Constraints	Pictorial Representation
b (conv.: bi)	I before J	$e_I \prec s_J$	$\cdots\ s_I \rightarrow e_I \rightarrow s_J \rightarrow e_J\ \cdots$
m (mi)	I meets J	$e_I = s_J$	$\cdots\ s_I \rightarrow e_I = s_J \rightarrow e_J\ \cdots$
o (oi)	I overlaps J	$s_I \prec s_J,\ s_J \prec e_I,$ $e_I \prec e_J$	$\cdots\ s_I \rightarrow s_J \rightarrow e_I \rightarrow e_J\ \cdots$
d (di)	I during J	$s_J \prec s_I,\ e_I \prec e_J$	$\cdots\ s_J \rightarrow s_I \rightarrow e_I \rightarrow e_J\ \cdots$
s (si)	I starts J	$s_I = s_J,\ e_I \prec e_J$	$\cdots\ s_I = s_J \rightarrow e_I \rightarrow e_J\ \cdots$
f (fi)	I finishes J	$s_J \prec s_I,\ s_I \prec e_J,$ $e_J = e_I$	$\cdots\ s_J \rightarrow s_I \rightarrow e_J = e_I\ \cdots$
e	I equals J	$s_I = s_J,\ e_I = e_J$	$\cdots\ s_I = s_J \rightarrow e_I = e_J\ \cdots$
ib (ibi)	I initially before J	$s_I \prec s_J,\ s_J \parallel e_I$	$\cdots\ s_I \nearrow e_I\ \cdots$; $s_I \searrow s_J \rightarrow e_J\ \cdots$
im (imi)	I initially meets J	$s_I \prec s_J,\ s_J \prec e_I,$ $e_I \parallel e_J$	$\cdots\ s_I \rightarrow s_J \nearrow e_I\ \cdots$; $s_J \searrow e_J\ \cdots$
ie	I initially equals J	$s_I = s_J,\ e_I \parallel e_J$	$\cdots\ s_I = s_J \nearrow e_I\ \cdots$; $s_I = s_J \searrow e_J\ \cdots$
u	I unrelated to J	$s_I \parallel s_J$	$\cdots\ s_I \rightarrow e_I\ \cdots$; $\cdots\ s_J \rightarrow e_J\ \cdots$

This table shows all possibilities of how two Allen intervals $I = \langle s_I, e_I \rangle$ and $J = \langle s_J, e_J \rangle$ can be related in a tree $\mathcal{B} = \langle T, \prec \rangle$. It is always assumed that $s_I \prec e_I$ and $s_J \prec e_J$.

Remark 6. Are all base relations satisfiable in each model? Of course not, since each linear order is a tree in which all non-linear base relations are not satisfiable. The pictorial representations in table 3 and table 4, however, suggest that there is a tree consisting of seven nodes in which all base relations are satisfiable.

Table 4. Refining the 19 base relations

Symbol	Relation	Defining Constraints	Pictorial Representation
pb (pbi)	I partially before J	$s_J \parallel e_I, \exists t(s_I \prec t \wedge t \prec e_I \wedge t \prec s_J)$	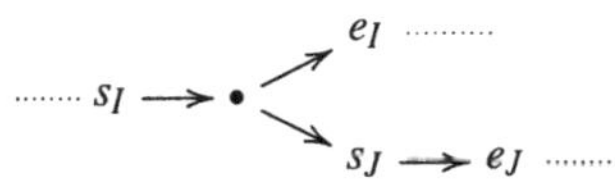
a (ai)	I adjacent to J	$s_I \prec s_J, s_J \parallel e_I, \neg\exists t(s_I \prec t \wedge t \prec e_I \wedge t \prec s_J)$	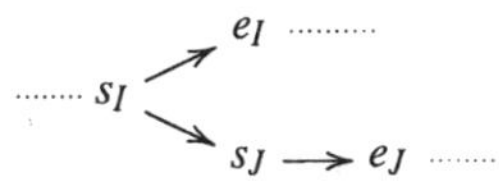
po (poi)	I partially overlaps J	$s_I \prec s_J, e_I \parallel e_J, \exists t(s_J \prec t \wedge t \prec e_I \wedge t \prec e_J)$	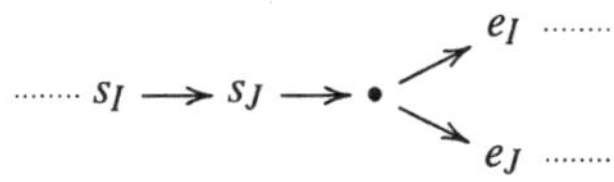
pm (pmi)	I partially meets J	$s_I \prec s_J, s_J \prec e_I, e_I \parallel e_J, \neg\exists t(s_J \prec t \wedge t \prec e_I \wedge t \prec e_J)$	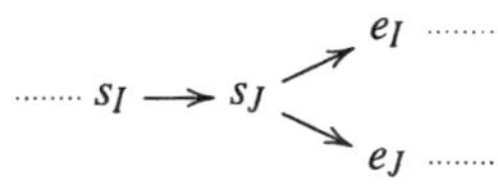
ps	I partially starts J	$s_I = s_J, e_I \parallel e_J, \exists t(s_I \prec t \wedge t \prec e_I \wedge t \prec e_J)$	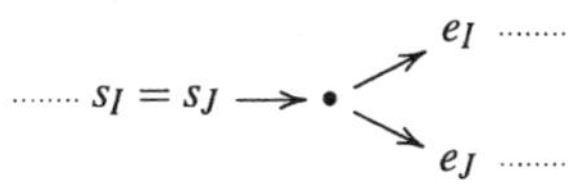
t	I touches J	$s_I = s_J, e_I \parallel e_J, \neg\exists t(s_I \prec t \wedge t \prec e_I \wedge t \prec e_J)$	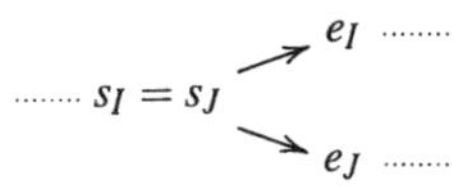

Ten base relations are definable by interval endpoints in branching time if we allow for quantifying over nodes. The table refines some of the entries in Table 3, namely *ib*, *im*, and *ie*.

Remark 7. How does satisfiability of base relations depend on the splitting type of the tree at hand? We will not give a complete and satisfactory answer to this question, but some comments will help to illustrate the problem. Consider first a $\mathbb{Q}$-tree that branches only at Dedekind gaps of $\mathbb{Q}$. Then, for example, the relation *pm* is satisfiable in that tree only if interval terms denote left-open intervals. For $\mathbb{R}$-trees of splitting type 1 the problem is less crucial since one can show that each interpretation of interval variables in such a tree can be transformed into an equivalent closed-interval interpretation. The same does not hold true for $\mathbb{R}$-trees of type 2. For example, let $\mathcal{B}$ be any tree of this kind that has at least two branches. Consider then the sentence "I meets both of the two disjoint intervals J and J'", which can be expressed by $I\, m\, J \wedge I\, m\, J' \wedge \neg J\; ps\; J'$. Although this formula is satisfiable in $\mathcal{B}$, it is not satisfiable by a closed-interval interpretation in that tree.

4 The Neighborhood Graph

Assume that two intervals are related by one of the base relations presented in the previous section. What happens if we move one of the intervals (in very small "steps") along one of the branches in which it is contained? Which relation could hold between the intervals if we increased or decreased one of them? These questions are usually answered by presenting a neighborhood graph. The neighborhood graph is also often understood as a similarity measure for the conceptual neighborhood of relations. This technique was first discussed by Freksa [12].

In what follows we will only investigate the first of these two questions, i. e., we enforce the size persistency constraint. Obviously, the neighborhood graph for interval relations in branching time (cf. Fig. 6) contains the corresponding graph for linear time, which is depicted in Fig. 5.

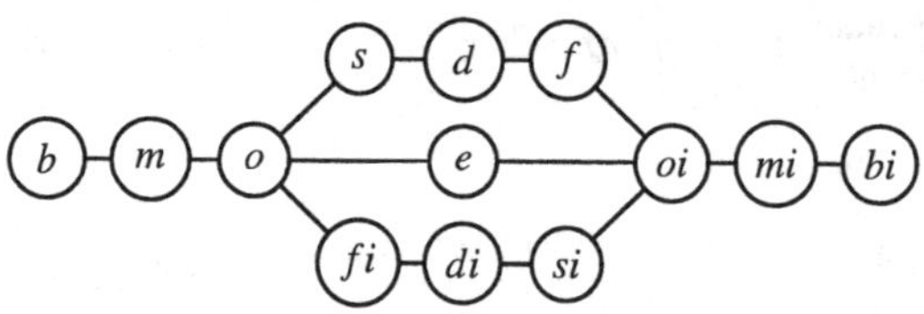

Fig. 5. The neighborhood graph of the interval algebra for linear time

In contrast to the neighborhood graph for linear time, the graph for branching time is not unique, i. e., it makes a significant difference for the neighborhood graph whether we fix one interval (for instance the second relatum) and allow for moving the other interval, or if we admit that at each step one of both intervals can be moved. A neighborhood graph with the first property will be referred to as a neighborhood graph of *type 1*. Neighborhood graphs in the weaker sense are said to be of *type 2*. It is immediately clear that the neighborhood graph of type 1 is a base of the neighborhood graph of type 2. Moreover, neighborhood graphs of type 2 are symmetric in the following sense: If the graph contains a transition between two relations, then it also contains a transition between the converses of these relations. Thus, the neighborhood graph of type 2 can be obtained from the type 1 graph by closing the latter one under converses.

It is remarkable that the neighborhood graph of type 1 is more precise, since it encodes the relation transitions that are admitted if only the interval at the first relation argument is allowed to be moved. To our knowledge, there is no other neighborhood graph in the literature that shows a similar behavior. The different types of neighborhood graphs mirror the underlying tree structure. Furthermore, in the neighborhood graph of type 2 each linear relation except f, fi, and e has at least one non-linear neighbor. Hence, if I and J are related by one of these relations and if the tree at hand is branching

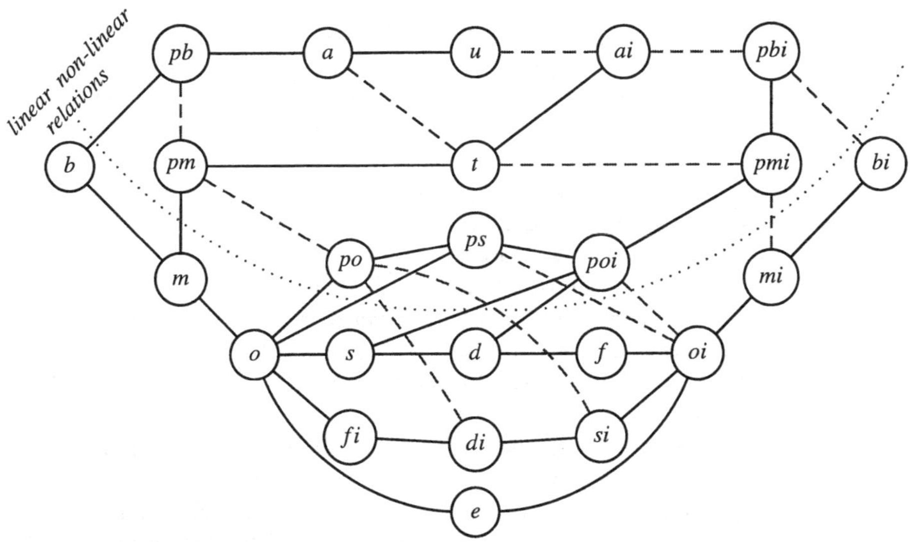

Fig. 6. The neighborhood graph of branching time interval relations. The undashed lines represent the relation transitions that are possible if the first relatum is moved, while the second is fixed. Dashed lines represent those relation transitions that are admitted additionally if one of both relata can be moved

dense, then we can move J in such a way that, afterwards, I and J are related by a non-linear relation. In the cases f, fi, and e, we always have to move J first in a position that is linearly related to I.

A number of non-linear relations have only non-linear relations as neighbors, namely u, t, a, and ai. If two intervals are related by one of these relations, then they have to be contained in distinct branches that split before one of the intervals ends. Thus, in each of these situations an interval cannot be moved in one step such that it is completely contained in the branch of the other interval.

5 Computational Complexity

Assume that a set of relations between some intervals is given. One question might be whether this set of relations is consistent. In other words: Is it possible to construct a tree in which all these relations are satisfiable? What is the computational effort to construct such a tree? Another interesting question might be whether the new, non-linear relations of branching time, especially the "partial relations", provide additional complexity and whether the general satisfiability problem is still NP-complete (cf. [33]).

Complexity analyses for the point algebra of branching time have been carried out by Broxvall [6]. Broxvall shows that the complexity of the satisfiability problem for the point algebra, which is in P for linear time models, is NP-complete for branching time models. Moreover, Broxvall also identified five maximal tractable subclasses. One of these is the class $\Gamma_A = \{\prec, \preceq, \prec\succ, \prec=\succ, \|, =\|, =, \neq, \prec\|, \preceq\|\}$, which will be of interest in what is to follow.[4]

We start by introducing some semantical concepts in a more formal way. The constraint language of the point algebra PA_{br} consists of infinitely many (point) variables $v \in V(PA_{\mathrm{br}})$ and a relation symbol for each of the 16 relations of PA_{br} (we will not sharply distinguish relations and relation symbols, etc., provided the meaning is clear from the context). Formulae of this language (called *PA_{br} constraints*) are expressions of the form $v\,r\,v'$, where v and v' are variables and $r \in PA_{\mathrm{br}}$. A *PA_{br} model* is an ordered pair $\mathcal{M} = \langle \mathcal{B}, a \rangle$ consisting of a tree $\mathcal{B} = \langle T, \prec \rangle$ and a (variable) assignment $a\colon V(PA_{\mathrm{br}}) \longrightarrow T$. The satisfiability relation is defined in the natural way, namely $\mathcal{M} \models v \prec v'$ if and only if $a(v) \prec a(v')$, etc. A constraint set of PA_{br} (i. e., a finite set of PA_{br} constraints), C, is said to be *satisfiable* if there exists a PA_{br} model such that $\mathcal{M} \models \phi$ for each $\phi \in C$.

Analogously, the constraint languages of the interval algebras IA^{19}_{br} (cf. Table 3) and IA^{24}_{br} (cf. Table 4) are introduced. An *IA_{br} model* is an ordered pair $\mathcal{M} = \langle \mathcal{B}, a \rangle$, where $\mathcal{B} = \langle T, \prec \rangle$ is a tree and a is a map that assigns to each interval variable I a closed interval $a(I) = [s_I, e_I]$ of $\mathcal{B}$. The model relation for IA^{19}_{br} and IA^{24}_{br} is introduced as sketched in Table 3 and in Table 4, respectively. For example,

$$\mathcal{M} \models I\; pb\; J \iff s_J \parallel e_I \text{ and there is a } t \in T \text{ with } s_I \prec t,\, t \prec e_I, \text{ and } t \prec s_J.$$

It is now an easy exercise to prove the following proposition:

Proposition 8. *Let C be a constraint set of IA^{19}_{br} or IA^{24}_{br}. Equivalent are:*

(i) *C is satisfiable.*

(ii) *C is satisfiable in a finite tree.*

(iii) *C is satisfiable in a $\mathbb{Q}$-tree (of type 1 or type 2).*

(iv) *C is satisfiable in an $\mathbb{R}$-tree (of type 1 or type 2).* □

It is important to note that these equivalences hold only if we restrict consideration to closed-interval interpretations.

In the remainder of the section we will focus on the algebra IA^{19}_{br}. We aim at showing that reasoning with IA^{19}_{br} relations is NP-complete. To show this we define first a mapping, Φ, that translates constraint sets of IA^{19}_{br} containing base relations only into constraint sets of PA_{br} (it is obvious how this translation could be extended to arbitrary IA^{19}_{br} relations). First we partition the set of PA_{br} variables, $V(PA_{\mathrm{br}})$, into two infinite subsets S and E. Let $V(IA_{\mathrm{br}}) \to S,\; I \mapsto s_I$, and $V(IA_{\mathrm{br}}) \to E,\; I \mapsto e_I$, be injective mappings. According to the defining constraints in Table 3 we translate interval relation constraints as follows:

[4] Here and in what follows we will use symbols such as "$=\|$" to denote the union of the relations $=$ and $\|$ (cf. section 2).

$$\Phi(I\, b\, J) := \{s_I \prec e_I, e_I \prec s_J, s_J \prec e_J\}$$
$$\Phi(I\, m\, J) := \{s_I \prec e_I, e_I = s_J, s_J \prec e_J\}$$
$$\vdots$$
$$\Phi(I\, e\, J) := \{s_I \prec e_I, s_J \prec e_J, s_I = s_J, e_I = e_J\}$$
$$\Phi(I\, ib\, J) := \{s_I \prec e_I, s_I \prec s_J, s_J \prec e_J\}$$
$$\Phi(I\, im\, J) := \{s_I \prec s_J, s_J \prec e_I, s_J \prec e_J\}$$
$$\Phi(I\, ie\, J) := \{s_I = s_J, s_I \prec e_I, s_J \prec e_J\}$$
$$\Phi(I\, u\, J) := \{s_I \prec e_I, s_J \prec e_J, s_I \parallel s_J\}$$

Finally we set

$$\Phi(\{\phi_1, \dots, \phi_n\}) := \bigcup_{i=1}^{n} \Phi(\phi_i).$$

It is now easy to prove the following lemmata:

Lemma 9. *Let $\mathcal{M} = \langle \mathcal{B}, a \rangle$ be a finite IA_{br} model, C be a constraint set of IA^{19}_{br}, and $V(C)$ be the set of interval variables occurring in C. If $\mathcal{M}$ satisfies C, then there exists an assignment $\check{a}\colon V(PA_{br}) \longrightarrow \mathcal{B}$ such that $\langle \mathcal{B}, \check{a} \rangle$ satisfies $\Phi(C)$.* □

Lemma 10. *Let $\mathcal{M} = \langle \mathcal{B}, a \rangle$ be a finite PA_{br} model and let C be a constraint set of IA^{19}_{br} such that $\mathcal{M}$ satisfies $\Phi(C)$. Let $\hat{a}\colon V(IA_{br}) \longrightarrow \mathcal{B}$ be defined by $\hat{a}(I) := [a(s_I), a(e_I))$. Then $\langle \mathcal{B}, \hat{a} \rangle$ satisfies C.* □

Thus, we obtain as a result:

Proposition 11. *Let C be a constraint set of IA^{19}_{br}. Then C is satisfiable if and only if $\Phi(C)$ is satisfiable.* □

We are now ready to present a complexity result for constraint satisfaction problems in IA^{19}_{br}. The general satisfiability problem in IA^{19}_{br} is defined as follows:

Given: A constraint set C of IA^{19}_{br}.
Question: Does there exist any (finite) tree $\mathcal{B}$ in which all constraints of C are satisfiable?

To discuss this problem in more detail, let us first consider the analogous problem for PA_{br}. Broxvall [6] showed that the general constraint satisfaction problem for PA_{br} is NP-hard. But he also presented a polynomial time algorithm *Branch* that is sound and complete for the subset Γ_A of point relations in branching time. Broxvall's algorithm takes as input a problem instance of the form $\Pi = (V, C)$, where V is a set of variables and C is a set of point constraints between these variables. The algorithm then continues as follows: First it partitions the set of variables into distinct components such that only variables related by a linear relation fall into the same component. Hence, on each such component one obtains a (total) linear order in a natural manner. In the next step the algorithm computes the subgraphs, followed by a computation of the transitive and reflexive closure of all these subgraphs. Finally, the root nodes of all components are identified, and the subgraphs are merged to a tree via these root nodes. As the partitioning was defined such that unrelated points are contained in distinct components, for each pair of branches, there exists exactly one node where the branches are merged. In the case of acceptance, the algorithm *Branch* outputs a tree satisfying the input constraint set C.

Lemma 12 (Broxvall [6]). *The algorithm Branch runs in polynomial time and is sound and complete for* Γ_A. □

To discuss the constraint satisfactions problem for IA^{19}_{br}, we consider first *basic constraint sets*. A constraint set C is said to be *basic* if C contains only base relations and if for each pair of interval variables I and J occurring in C, C contains exactly one constraint $I\,r\,J$. A *scenario*, then, is a model of a basic constraint set.

Then the satisfiability of a given basic constraint set of IA^{19}_{br}, C, can be decided by the following algorithm *IntervalBranch*:

1. Translate the interval constraint set C into the point constraint set $\Phi(C)$ of PA_{br}.
2. Compute the transitive closure $\Phi(C)^*$ of $\Phi(C)$.
3. Apply Broxvall's algorithm *Branch* on $\Phi(C)^*$.

In fact, the algorithm *IntervalBranch* is polynomial, and it is sound and complete for basic constraint sets of IA^{19}_{br}. This is an immediate consequence of Proposition 11, Lemma 12, and the following observation:

Lemma 13. *For each basic constraint set C of IA^{19}_{br}, the constraint set $\Phi(C)^*$ is basic too. Moreover, in $\Phi(C)^*$ only relations of the set $\{\prec,\preceq,=,\|\}\subseteq\Gamma_A$ occur.* □

Corollary 14. *Testing satisfiability of basic IA^{19}_{br} constraint sets is tractable.* □

Theorem 15. *The general satisfiability problem of IA^{19}_{br} is* NP*-complete.*

Proof. NP-hardness follows straightforward by a reduction from 3-colorability: Let $G=(V,E)$, $V=\{v_1,\ldots,v_n\}$ be an instance of 3-colorability. Then we use the following interval symbols $\{v_1,\ldots,v_n,c_1,c_2,c_3\}$ with the following constraints:

$$\begin{array}{ccc} c_1 & \{m\} & c_2 \\ c_2 & \{m\} & c_3 \\ v_i & \{m,e,mi\} & c_2 \\ v_i & \{b,m,mi,bi\} & v_j \quad \forall (v_i,v_j)\in E \end{array}$$

It is immediately clear that this reduction is polynomial. Now it is obvious that the constraint system is satisfiable if and only if G can be colored with 3 colors. Therefore the problem is NP-hard.

Membership in NP follows from the following short description of a non-deterministic algorithm: Let C be a constraint set of IA^{19}_{br}. Guess (non-deterministically) a scenario for C. Since each scenario is basic, we only need to check, whether this scenario satisfies C. But this can be done in polynomial time, which is shown in Corollary 14. □

6 The Composition Table

A composition table for the interval algebra of branching time was first presented by Anger, Ladkin, and Rodriguez [4]. In this section we briefly discuss the differences between the composition table for branching time and that for linear time. This means that we focus on the composition of linear relations in the context of branching time.[5]

Table 5. Composing linear relations with linear relations in branching time

	b	*m*	*o*	*d*	*s*
bi	*I*	*ai, poi, bi, mi, oi, f, d, pbi, pmi*	*poi, bi, mi, oi, f, d, pbi, pmi*	*poi, bi, mi, oi, f, d, pbi, pmi*	*poi, bi, mi, oi, f, d, pbi, pmi*
mi	*a, po, b, m, o, fi, di, pb, pm*	*ps, s, si, e*	*pmi, poi, oi, d, f*	*pmi, poi, oi, d, f*	*pmi, poi, oi, d, f*
oi	*pb, pm, po, b, m, o, fi, di*	*pm, po, o, di, fi*	*ps, s, si, e, po, o, di, fi, poi, oi, d, f*	*poi, oi, d, f*	*poi, oi, d, f*
di	*pb, pm, po, b, m, o, fi, di*	*pm, po, o, di, fi*	*po, o, di, fi*	*ps, s, si, e, po, o, di, fi, poi, oi, d, f*	*po, o, di, fi*
si	*pb, pm, po, b, m, o, fi, di*	*pm, po, o, di, fi*	*po, o, di, fi*	*poi, oi, d, f*	*t, ps, s, si, e*

I is defined as the union of all basic *linear* relations.

First note that the composition table presented in Table 5 has to be read in the consistency-based sense (cf. [5]). This means that, in the general case (without further requirement on the class of intended models), the composition table cannot be read extensionally. Composition is understood extensionally if the algebraic function of

[5] The complete composition table is available at `ftp://ftp.informatik.uni-freiburg.de/documents/papers/ki/ragni-woelfl-brallen-comp.pdf`. It is worth noting that Reich [28] corrected one of the entries in the original table.

composition of relations (as used in relation algebras) coincides with its set-theoretical characterization. More precisely, if $\mathcal{K}$ is the class of intended models, then composition can be read extensionally w. r. t. $\mathcal{K}$ if and only for each model $\mathcal{M}$ in $\mathcal{K}$ and for each "composition" r_{ij} of relations r_i and r_j (as indicated in the composition table), it holds:

$$\mathcal{M} \models x r_{ij} y \leftrightarrow \exists z (x r_i z \wedge z r_j y).$$

The consistency-based reading only requires the right-to-left implication in this equivalence. Recall that the composition tables for points or intervals in linear time can be read extensionally for the class of dense linear orders, but not for the class of finite orders.

Composition of interval relations for branching time cannot be read extensionally even for the class of dense flows of time. For example, though $ps \circ u \subseteq u$ is true in that class, $ps \circ u \supseteq u$ is not. The latter subsumption entails that for each pair of intervals with $I\,u\,K$, there exists an interval J such that $I\,ps\,J$ and $J\,u\,K$. But such an interval could only be found if a splitting point can be found in interval I that witnesses the existential quantification in the definition of the relation ps. Note that Anger, Ladkin, and Rodriguez [4] restrict consideration on branching dense trees. Presumably, this (very strong) requirement is owed to the extensional reading of composition. But to our knowledge, it has not yet been proved that, for the class of dense and branching dense trees, composition can always be read extensionally.

Is it possible that the composition of linear relations admits non-linear relations? For answering this question consider the following situation. Let I, J, and K be intervals such that $I\,r\,J$, $J\,r'\,K$, and both relations r and r' are linear Allen relations. It is immediately clear that the start points of I and J, respectively, are comparable with respect to $\prec$ and that the same holds true for the start points of J and K. But the start points of I and K need not be comparable. As an example of this situation, assume that interval J is before I and K and that intervals I and K start after a splitting point of two branches b and b' containing I and J, respectively.

In this situation a non-linear relation can occur only if the endpoint of interval J, e_J, is before the endpoint of I and the endpoint of K. The node e_J has to be before the endpoints e_I and e_K because e_I and e_K are unrelated. But the only linear relations where e_J is before e_I are the relations *bi*, *mi*, *oi*, *di*, and *si*. An analogous statement holds true for intervals J and K. For this reason composition of two linear base relations r and r' can admit a non-linear relation only if $r \in \{bi, mi, oi, di, si\}$ and $r' \in \{b, m, o, d, s\}$. The relations f, fi, and e imply the equality of the endpoints of two intervals and therefore they enforce that all three intervals are related by one of the linear relations.

7 Summary and Outlook

Starting from the question of how qualitative reasoning in non-linear time differs from reasoning in linear time, we first investigated non-linear relations between intervals in branching time. We showed that a naïve Allen representation of intervals is problematic in branching time contexts. To put this in other words, the boundary structure of intervals in branching time is more sophisticated than it is for intervals in linear time. Moreover, we saw that both the boundary structure of intervals and the splitting type of the tree at hand may influence the set of base relations that can be distinguished reasonably. But since our remarks on this topic are still somewhat sketchy, a more detailed analysis needs to be done.

In a second step we indicated that different algebras for intervals in branching time can be defined, depending on whether one allows for quantifying over tree nodes or not. From a cognitive point of view, it could be interesting to see if people actually distinguish between these two conceptual schemata.

This leads to another interesting question: The neighborhood graph for interval relations in branching time shows how interval relations can change in time. But does this graph also mirror a *conceptual* neighborhood of the presented relations? For example, it seems that each linear relation is conceptually more similar to a given linear relation than is each non-linear relation. Moreover, we saw that the underlying tree structure implied two different kinds of neighborhood graphs, which is remarkable since most neighborhood graphs known in the literature are unique. The neighborhood graph presented here did enforce size persistency. It will be interesting to compare this graph with a neighborhood graph figuring transitions of relations between increasing or decreasing intervals.

We investigated the complexity of the general satisfiability problem for the interval algebra IA_{br}^{19} and showed that it is NP-complete. Of course, an analogous result for IA_{br}^{24} would be desirable, but seems harder.

The complexity considerations are strongly related with the question of how to reason with intervals in branching time. Therefore, we analyzed the composition table of branching time interval relations and presented a "heuristics" explaining why the composition of two linear base relations may contain a non-linear base relation. We aim at showing the correctness of this table by a computer-aided proof.

Acknowledgments

This work was partially supported by the Deutsche Forschungsgemeinschaft (DFG) as part of the Transregional Collaborative Research Center SFB/TR 8 Spatial Cognition. We like to thank Tomoko Kitamura for her help in designing figures and in typesetting the composition tables. We also owe thanks to both referees for various helpful comments.

References

1. J. Allen and P. Hayes. A common-sense theory of time. In *Proceedings of the 9th International Joint Conference on Artificial Intelligence (IJCAI-85)*, pages 528–531, Los Angeles, CA, USA, 1985.
2. J. F. Allen. Maintaining knowledge about temporal intervals. *Communications of the ACM*, 26(11):832–843, 1983. Reprinted in D. S. Weld and J. de Kleer, editors, *Readings in Qualitative Reasoning about Physical Systems*, pages 361–372. Kaufmann, San Mateo, CA, 1990.
3. R. Alur, T. A. Henzinger, and O. Kupferman. Alternating-time temporal logic. In *Proceedings of the 38th Symposium on Foundations of Computer Science*, 1997.

4. F. Anger, P. Ladkin, and R. Rodriguez. Atomic temporal interval relations in branching time: Calculation and application. In *Actes 9th SPIE Conference on Applications of Artificial Conference*, Orlando, FL, USA, 1991.
5. B. Bennett, A. Isli, and A. G. Cohn. When does a composition table provide a complete and tractable proof procedure for a relational constraint language? In *Proceedings of the IJCAI97 Workshop on Spatial and Temporal Reasonin*, 1997.
6. M. Broxvall. The point algebra for branching time revisited. In *Proceedings of the Joint German/Austrian Conference on Artificial Intelligence (KI-2001)*, pages 106–121, 2001.
7. M. Broxvall and P. Jonsson. Towards a complete classification of tractability in point algebras for nonlinear time. In *Proceedings of the 5th International Conference on Principles and Practice of Constraint Programming (CP-99)*, pages 129–143, 1999.
8. T. Drakengren and P. Jonsson. A complete classification of tractability in Allen's algebra relative to subsets of basic relations. *Artificial Intelligence*, 106(2):205–219, 1998.
9. I. Düntsch, H. Wang, and S. McCloskey. Relation algebras in qualitative spatial reasoning. *Fundamenta Informatiae*, 39(3):229–249, 1999.
10. E. A. Emerson and J. Srinivasan. Branching time temporal logic. In *Proceedings of REX Workshop 1988*, 1988.
11. J. Euzenat. Algèbres d'intervalles sur des domaines temporels arborescents. In *Actes 11ième RFIA*, Clermont-Ferrand, 1998.
12. C. Freksa. Conceptual neighborhood and its role in temporal and spatial reasoning. In *Decision Support Systems and Qualitative Reasoning*, pages 181 – 187. North-Holland, Amsterdam, 1991.
13. C. Freksa. Temporal reasoning based on semi-intervals. *Artificial Intelligence*, 54:199–227, 1992.
14. P. J. Hayes. The naive physics manifesto. In D. Michie, editor, *Expert Systems in the Micro-Electronic Age*. Edinburgh University Press, 1978.
15. P. J. Hayes and J. F. Allen. Short time periods. In *Proceedings of the 10th International Joint Conference on Artificial Intelligence (IJCAI-87)*, 1987.
16. R. Hirsch. Relation algebras of intervals. *Artificial Intelligence*, 83, 1996.
17. R. Hirsch. Expressive power and complexity in algebraic logic. *Journal of Logic and Computation*, 7(3):309–351, 1997.
18. F. v. Kutschera. Sebastian's strolls. *Grazer Philosophische Studien*, 45:75–88, 1993.
19. P. Ladkin. Models of axioms for time intervals. In *Proceedings of AAAI-87*, 1987.
20. P. Ladkin. Satisfying first-order constraints about time intervals. In *Proceedings of AAAI-88*, pages 512–517, 1988.
21. P. B. Ladkin. Time representation: A taxonomy of interval relations. In *Proceedings of AAAI-86*, pages 360–366, 1986.
22. P. B. Ladkin. The completeness of a natural system for reasoning with time intervals. In *Proceedings of IJCAI-87*, pages 462–467, 1987.
23. P. B. Ladkin and R. D. Maddux. On binary constraint problems. *Journal of the ACM*, 1994.
24. G. Ligozat. A new proof of tractability for ORD-Horn relations. In *Proceedings of the National Conference on Artificial Intelligence (AAAI-96)*, Portland, Oregon.
25. R. Moratz, J. Renz, and D. Wolter. Qualitative spatial reasoning about line segments. In *14th European Conference on Artificial Intelligence (ECAI'00)*, Berlin, Germany, 2000.
26. B. Nebel and H.-J. Bürckert. Reasoning about temporal relations: A maximal tractable subclass of Allen's interval algebra. Technical Report RR-93-11, Deutsches Forschungszentrum für Künstliche Intelligenz GmbH, Kaiserslautern, Germany, 1993.
27. A. Prior. *Past, Present and Future*. Clarendon Press, Oxford, 1967.
28. A. J. Reich. Intervals, points, and branching time. In S. D. Goodwin and H. J. Hamilton, editors, *Proceedings of the TIME-94 International Workshop on Temporal Reasoning*, pages 121–133, Regina, SK, Canada, 1994. University of Regina.

29. R. H. Thomason. Combinations of tense and modality. In D. M. Gabbay and F. Guenthner, editors, *Handbook of Philosophical Logic: Extensions of Classical Logic*, volume II, pages 135–165. D. Reidel, Dordrecht, 1984.
30. G. Valiente. *Algorithms on Trees and Graphs*. Springer-Verlag, Berlin, 2002.
31. J. van Benthem. *The Logic of Time*. Reidel, 1983.
32. J. van Benthem. Time, logic, and computation. In J. de Bakker, W.-P. de Roever, and G. Rozenberg, editors, *Branching Time and Partial Order in Logics and Models for Concurrency*, pages 1–49. Springer Verlag, New York, 1989.
33. M. B. Vilain, H. A. Kautz, and P. G. van Beek. Contraint propagation algorithms for temporal reasoning: A revised report. In D. S. Weld and J. de Kleer, editors, *Readings in Qualitative Reasoning about Physical Systems*, pages 373–381. Morgan Kaufmann, San Mateo, CA, 1989.
34. S. Wölfl. Events in branching time. Accepted for publication by *Studia Logica*, 2004.

SNAPVis and SPANVis: Ontologies for Recognizing Variable Vista Spatial Environments

Tiansi Dong

Cognitive Systems Group,
Department of Math and Informatics, University of Bremen,
28334 Bremen, Germany
tiansi@informatik.uni-bremen.de

Abstract. This paper gives the SNAP and SPAN ontologies relating to recognizing variable vista spatial environments, namely, SNAPVis and SPANVis. It proposes that recognizing spatial environments is a judgment process of whether the perceived environment is compatible with the remembered one. Their compatibility is based on both their spatial changes and the commonsense knowledge of objects' stabilities. The recognition result is determined by whether such changes are due to possible movements of related objects or not. This paper presents six SNAPVis ontologies: fiat boundary, near region, fiat parts (the three are fiat regions), classic topologic relations, qualitative orientations, and qualitative distances (the three are spatial relations) and one SPANVis ontology: the commonsense knowledge of stability of spatial objects. The paper briefly presents a cognitive map of vista spatial environments and the process of recognition.

1 An Introduction

"The structure of space can be described from the point of view of behavior", cf. (Piaget, 1954, pp. 212). Vista spatial environments refer to Montello's vista space, cf. (Montello, 1993), that is "projectively as large as the human body and that can be apprehended from one place without necessary locomotion". Typical vista spaces are single rooms, offices, small valleys, etc.

Vista spatial environments are dynamic. For example, chairs and books are often placed here and there in your office, your home. However, you can recognize your office, your home.

Intuitively, we recognize a place not by checking everything in it. The air in the room, a sheet of paper, an apple and contents in the dustbin do not help to recognize the room. This paper addresses the task of recognizing spatial environments by asking what exists in an environment that makes it be that environment. It proposes formal ontologies for recognizing vista spatial environments by extending the $RCC8$ theory, cf. (Randell et al., 1992).

C. Freksa et al. (Eds.): Spatial Cognition IV, LNAI 3343, pp. 344–365, 2005.
© Springer-Verlag Berlin Heidelberg 2005

2 The State of the Art

Recognizing spatial environments relates to perception, cognition and corresponding linguistic expressions, which are all *fiat*, cf. (Smith, 2001). It has nothing to do with finger print checking, molecule analysis or DNA testing. Therefore this paper strictly distinguishes *fiat* ontologies from *bona fide* ontologies, cf. (Smith and Varzi, 2000).

Rosch, et al. (1976) reported that humans recognize single objects at the basic level category. At this level (of granularity) humans find it easiest to name objects and recognize them the fastest. This introduces the question of the formation of object constancy: If you look at an apple for one minute, how do you know that the apple you saw at the beginning of the minute is the same apple you saw at the end of the minute? Piaget (1954) Marr and Nishihara (1978) and Rock (1983) suggested that information about spatial relations of an object or parts of an object are important in object constancy. This is consistent with the case of a brain impaired patient: Wilson, et al. (1999) reported a woman, LE, who suffered from a rare memory problem. LE cannot retrieve images from her memory, thus, she cannot distinguish two windows whose glasses have different images and she even has difficulty in recognizing her husband's face. What she can retrieve are only contours of objects. On the other hand, she can locate objects. Amazingly, she can recognize spatial environments, such as her home[1]. This shows that recognizing spatial environments does not require much information about single objects but rather the spatial relations among them.

The starting point of this paper is that objects in an environment are recognized at basic level categories and that spatial relations among them are known. This paper presents ontologies for recognizing a vista spatial environment on this basis.

This paper follows the philosophy that dynamic spatial ontology should combine two distinct types of inventory of the entities and relationships in reality: on the one hand, a purely spatial ontology supporting snapshot views of the world at successive instants of time: SNAP; on the other hand, a purely spatiotemporal ontology of change and process: SPAN, cf. (Grenon and Smith, 2004). Recognizing a vista spatial environment relates its two snapshots (one is remembered in the mind, the other is currently perceived) and the spatiotemporal relation between them: can they (or to what extent can they) participate in the same SPAN?

The snapshot of a spatial environment in mind is termed a cognitive map in the literature, cf. (Tolman, 1948; Kuipers, 1978). Structures of cognitive maps concerning environmental spaces and geographical spaces are hierarchical, cf. (Kosslyn et al., 1974; Siegel and White, 1975; Steven and Coupe, 1978; Hirtle and Jonides, 1985; Tversky, 1991). This paper proposes that the structure of the cognitive map of a vista spatial environment is also hierarchical and this hierarchy is revealed in the selection of reference objects in spatial linguistic de-

[1] Personal communication with Allan Baddeley, Sofia, Bulgaria, July, 2003.

scriptions: a location object is at lower or the same level in the hierarchy as its reference object(s). For example, if people say "the cup is on the table", then in their cognitive map the cup is located lower than or at the same position as the table in the hierarchy. This paper proposes that the level of an object in the hierarchy of a person's cognitive map is determined by commonsense knowledge of its stability in this environment, which is a SPAN ontology.

There is some work on representations of spatial relations among extended objects, cf. (de Laguna, 1922; Randell et al., 1992; Clementini and Di Felice, 1997; Goyal, 2000; Schmidtke, 2001; Schmidtke, 2003). This paper briefly introduces topological definitions of subjective distance and orientation relations between regions for the task of recognizing vista spatial environments.

The paper is structured as follows: Section 3 and section 4 present SNAPVis and SPANVis for recognizing vista spatial environments; section 5 presents the structure of a cognitive map; section 6 briefly outlines the recognition issues; section 7 summarizes the paper.

3 SNAPVis for Recognizing Vista Spatial Environments

Following Smith's ontologies of SNAP-SPAN, this article proposes SNAPVis and SPANVis for vista spatial environments. SNAPVis is an ontology for continuants of vista spatial environments; SPANVis is an ontology for occurrents of vista spatial environments.

3.1 Fiat Regions

Fiat Boundaries. When we look around, we do not see clusters of atoms or molecules. Rather, we perceive objects and name them at categories. A correspondence between perception and its language description has been proposed by Tversky and Lee (1999) as follows: *to the extent that space is schematized similarly in language and cognition, language will be successful in conveying space* (the Schematization Similarity Conjecture). Thus, objects recognized through perception are *fiat* in the sense that they relate to both the objects and human cognition. The fiat boundary of an object refers to its conceptual boundary through perception and the perceptual object is named at the basic level category, such as rooms, windows, doors, furniture, etc. A fiat boundary is an instantiation of a basic level category. Fiat boundaries are named by the object names at basic level categories (may be followed by a natural number to distinguish different instances of the same category), such as "$room$", "$room_1$", "$window_2$" and capital lettered names are used as basic level categories, such as "$ROOM$", "$WINDOW$". '$BasicLevel(w)$' is a predicate standing for "w is a basic level category"; '$FiatBoundary(x)$' is a predicate which stands for "x is a kind of fiat boundary at the basic level category"; '$Constituent(x, w)$[2]' is a predicate for "x is an instantiation of the category of w".

[2] This terminology '*Constituent*' is adopted from (Grenon and Smith, 2004).

Formulae in this paper have the form of "$(Q\ x|p \bullet q)$", Q is the quantifier, x is the bound variable, p is the constraint of x, and q is the predicate. It is read as "for all x satisfying p, q holds (or there is x satisfying p such that q)." Any fiat boundary of an extended object belongs to a class.

$$\forall x \exists w \bullet FiatBoundary(x) \rightarrow Constituent(x, w) \wedge BasicLevel(w) \quad (1)$$

A fiat boundary is an instantiation of only one basic level category:

$$\begin{aligned} \forall x, w_1, w_2 | FiatBoundary(x) \wedge Constituent(x, w_1) \wedge BasicLevel(w_1) \\ \wedge Constituent(x, w_2) \wedge BasicLevel(w_2) \\ \bullet w_1 = w_2 \end{aligned} \quad (2)$$

Fiat boundaries of the same basic level category are indistinguishable in isolation. The predicate '$CanBeTheSame(x, y)$' stands for "x and y are indistinguishable in isolation."

$$\begin{aligned} &\forall x, y | FiatBoundary(x) \wedge FiatBoundary(y) \bullet \\ &\quad (\exists w | BasicLevel(w) \bullet Constituent(x, w) \wedge Constituent(y, w)) \\ &\qquad \rightarrow CanBeTheSame(x, y) \end{aligned} \quad (3)$$

Fiat boundaries of different basic level categories must be different fiat boundaries.

$$\begin{aligned} &\forall x, y | FiatBoundary(x) \wedge FiatBoundary(y) \bullet \\ &\quad (\forall w_1, w_2 | Constituent(x, w_1) \wedge Constituent(y, w_2) \\ &\qquad \wedge BasicLevel(w_1) \wedge BasicLevel(w_2) \\ &\qquad \bullet w_1 \neq w_2 \rightarrow \neg CanBeTheSame(x, y)) \end{aligned} \quad (4)$$

Fiat Parts. Humans recognize spatial objects by perceiving their partial images, cf. (Buelthoff and Edelman, 1992; Humphrey and Khan, 1992; Tarr, 1995).

'$RecognizablePart(y, x)$' stands for "y is a recognizable part of fiat boundary x, which can be used to recognize x"; '$recognition(y)$' is a function which returns the basic level category of a recognizable part y.

Recognizable parts of a fiat boundary are used to identify its basic level category:

$$\begin{aligned} &\forall x, y, w | FiatBoundary(x) \wedge RecognizablePart(y, x) \\ &\quad \wedge BasicLevel(w) \bullet w = recognition(y) \rightarrow Constituent(x, w) \end{aligned} \quad (5)$$

Recognizable parts of an object are often different. Your face seen from the front is completely different from that seen from the left side, though both are recognizable parts of the face. '$front(x)$','$left(x)$','$right(x)$' and '$behind(x)$'

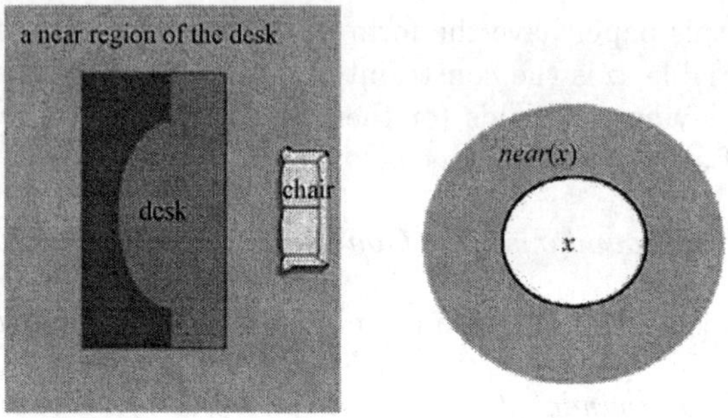

Fig. 1. Fiat extension of spatial region: the near region of the desk

stand for the front, left, right and behind recognizable parts of the fiat boundary x, respectively. They are *linguistic fiat* after (Smith, 2001).

$$\forall x, y | FiatBoundary(x) \bullet y = front(x) \rightarrow RecognizablePart(y, x) \quad (6)$$

$$\forall x, y | FiatBoundary(x) \bullet y = left(x) \rightarrow RecognizablePart(y, x) \quad (7)$$

$$\forall x, y | FiatBoundary(x) \bullet y = right(x) \rightarrow RecognizablePart(y, x) \quad (8)$$

$$\forall x, y | FiatBoundary(x) \bullet y = behind(x) \rightarrow RecognizablePart(y, x) \quad (9)$$

Fiat Extensions of Spatial Regions. For a description like "the chair is near the writing-desk", it is normally explained that the spatial relation between the chair and the writing-desk is *near*. This article takes an alternative view as follows: the fiat boundary of the chair is overlapped with the near region of the desk, shown in Figure 1.

The *Near* Function. If x is a fiat boundary, then $near(x)$ is a fiat region which denotes the nearby region of x. The closure of the $near(x)$ region is a ring which has two boundaries: an inner boundary and a fiat outer boundary. The inner boundary coincides with the fiat boundary x, as we would not say an object is near to a region when it is already connected with the region. The outer boundary is determined by both the objects and human cognition. So the $near(x)$ region is *fiat* and its closure is externally connected with the region[3] x. The *near* function is similar with the 'penumbra region' in (Freksa, 1981) and 'Egg-Yolk' in (Cohn and Gotts, 1996), however, it can be applied recursively. That is, a near region can have its own near region. Formal definitions are given in the appendix, as detailed discussion is beyond the scope of the article.

The *near* function is only a special case of fiat extensions of fiat boundaries. "A is nearer to B than to C" can be further explained as: there is a fiat extension of A which is connected with the fiat boundary B and is disconnected with the fiat boundary C. The function of '$fiat\_extension^n(x)$' is defined as:

[3] The fiat boundary of a 3-dimensional object is a 2-dimensional region.

$$fiat\_extension^n(x) \stackrel{\text{def}}{=} x \cup near(x) \cup near^2(x) \cup \cdots \cup near^n(x) \tag{10}$$

Then:

$$\begin{aligned} &\forall x, y, z | FiatBoundary(x) \wedge FiatBoundary(y) \\ &\quad \wedge FiatBoundary(z) \bullet nearer(x, y, z) \stackrel{\text{def}}{=} \\ &\exists n \bullet C(fiat\_extension^n(x), y) \wedge DC(fiat\_extension^n(x), z) \end{aligned} \tag{11}$$

where C and DC mean "connected with" and "disconnected with" respectively (following the notions in RCC).

3.2 Spatial Relations

Spatial relations include classic topologic relations, qualitative distance relations and qualitative orientation relations, cf. (Stock, 1997).

Classic Topologic Relations. Classic topological relations depict connectedness relations between two regions, as described by $RCC8$.

Qualitative Orientation Relations. The different observed recognizable parts result in the orientation relations between the observer and the object. When you are watching television (i.e., you are looking at the recognizable part of the TV set which provides sequences of images), you know that you are in front of the TV set.

Such orientation relations can be formalized by a comparison of qualitative distances: "the chair is in front of the desk" means that the chair is nearer to the front fiat part of the desk than its other fiat parts, shown in Figure 2.

'$Front(x, y)$' stands for "x is in front of y or in the front of y":

$$\begin{aligned} &\forall x, y | FiatBoundary(x) \wedge FiatBoundary(y) \bullet Front(x, y) \stackrel{\text{def}}{=} \\ &\forall z | RecognizablePart(z, y) \bullet z \neq front(y) \rightarrow nearer(x, front(y), z) \end{aligned} \tag{12}$$

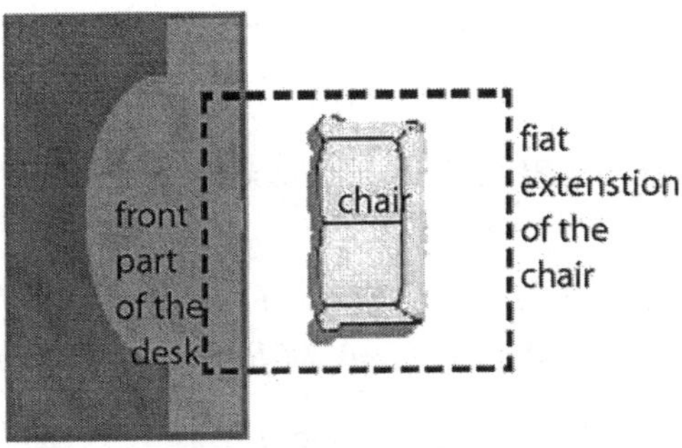

Fig. 2. The qualitative orientation relation is explained by the comparison of qualitative distances among fiat extension region of the location object and fiat parts of the reference object

Similarly, '$Left(x, y)$' stands for "x is located left of y or left to y"; '$Right(x, y)$' for "x is located right of y or right to y"; '$Behind(x, y)$' for "x is located behind y or at behind part of y":

$$\forall x, y | FiatBoundary(x) \wedge FiatBoundary(y) \bullet Left(x, y) \stackrel{\text{def}}{=} \forall z | RecognizablePart(z, y) \bullet z \neq left(y) \rightarrow nearer(x, left(y), z) \quad (13)$$

$$\forall x, y | FiatBoundary(x) \wedge FiatBoundary(y) \bullet Right(x, y) \stackrel{\text{def}}{=} \forall z | RecognizablePart(z, y) \bullet z \neq right(y) \rightarrow nearer(x, right(y), z) \quad (14)$$

$$\forall x, y | FiatBoundary(x) \wedge FiatBoundary(y) \bullet Behind(x, y) \stackrel{\text{def}}{=} \forall z | RecognizablePart(z, y) \bullet z \neq behind(y) \rightarrow nearer(x, behind(y), z) \quad (15)$$

Qualitative Distance Relations. Qualitative distance relations specify the '*DC*' relation in RCC by giving qualitative distance relations such as '*NR*'(near), '*PR*' (penumbra far-or-near), and '*FR*'(far), as we seldom use an expression like "the chair is disconnected with the desk", but instead we give qualitative distance relations, such as "the chair is near the desk".

Fiat boundary x near fiat boundary y is defined as: x is disconnected with y and overlapped with the near region of y:

$$\forall x, y | FiatBoundary(x) \wedge FiatBoundary(y) \bullet NR(x, y) \stackrel{\text{def}}{=} DC(x, y) \wedge O(x, near(y)) \quad (16)$$

Fiat boundary x penumbra far or near y is defined as: x is disconnected with y and externally connected with its near region:

$$\forall x, y | FiatBoundary(x) \wedge FiatBoundary(y) \bullet PR(x, y) \stackrel{\text{def}}{=} DC(x, y) \wedge EC(x, near(y)) \quad (17)$$

Fiat boundary x far away from y is defined as: x is disconnected with y and its near region:

$$\forall x, y | FiatBoundary(x) \wedge FiatBoundary(y) \bullet FR(x, y) \stackrel{\text{def}}{=} DC(x, y) \wedge DC(x, near(y)) \quad (18)$$

It is obvious that *FR, PR, NR* are pairwise disjoint. *FR, PR, NR* also jointly exhausts the *DC* relation, as shown below:

$$\forall x, y | FiatBoundary(x) \wedge FiatBoundary(y) \bullet DC(x, y) \equiv NR(x, y) \vee PR(x, y) \vee FR(x, y) \quad (19)$$

Proof sketch: For all fiat boundaries x and y

$NR(x, y) \vee PR(x, y) \vee FR(x, y)$
[(16), (17), (18)]

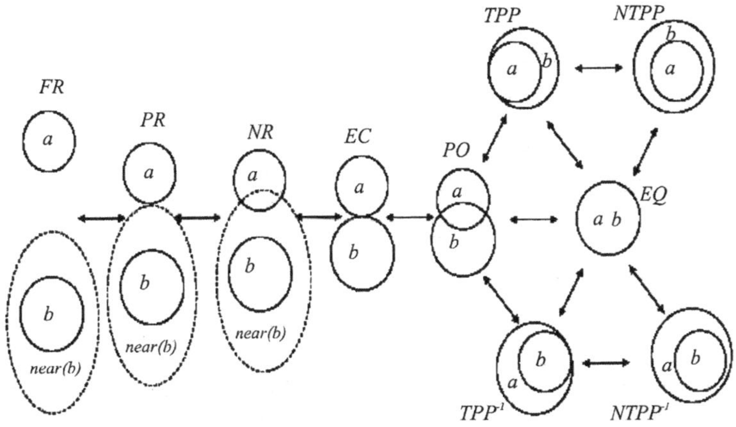

Fig. 3. The ten topological relations $RCC10$ for vista spatial environments. The disconnected (DC) relation in $RCC8$ is specified by three qualitative distance relations: far (FR), penumbra far-or-near (PR) and near (NR)

$$
\begin{aligned}
&\stackrel{\text{def}}{=} DC(x,y) \wedge O(x, near(y)) \vee DC(x,y) \wedge EC(x, near(y)) \\
&\qquad\qquad \vee DC(x,y) \wedge DC(x, near(y)) \\
&\quad [(q \wedge a) \vee (q \wedge b) \vee (q \wedge c) \equiv q \wedge (a \vee b \vee c))] \\
&\equiv DC(x,y) \wedge (O(x, near(y)) \vee EC(x, near(y)) \vee DC(x, near(y))) \\
&\quad [\textit{Definition of EC and DC in RCC}] \\
&\equiv DC(x,y) \wedge (O(x, near(y)) \vee (C(x, near(y)) \wedge \neg O(x, near(y))) \\
&\qquad\qquad \vee \neg C(x, near(y))) \\
&\quad [p \vee (s \wedge t) \equiv (p \vee s) \wedge (p \vee t)] \\
&\equiv DC(x,y) \wedge (O(x, near(y)) \vee \neg C(x, near(y)) \vee C(x, near(y))) \\
&\qquad\qquad \wedge (O(x, near(y)) \vee \neg C(x, near(y)) \vee \neg O(x, near(y))) \\
&\quad [p \vee \neg p \equiv T; T \vee p \equiv T; T \wedge T \equiv T; q \wedge T \equiv q] \\
&\equiv DC(x,y)
\end{aligned}
$$

This paper therefore proposes a new **RCC10** by splitting the DC of $RCC8$ into three qualitative distance relations for vista spatial recognition. The *conceptual neighborhoods network*, after the notion in (Freksa, 1991), of $RCC10$ is shown in Figure 3.

4 SPANVis for Recognizing Vista Spatial Environments

4.1 The Stability

The stability of a fiat boundary is commonsense knowledge about the spatiotemporal property of the fiat boundary, which affects the selection of reference objects in spatial linguistic descriptions. For example, we neither say "the table is

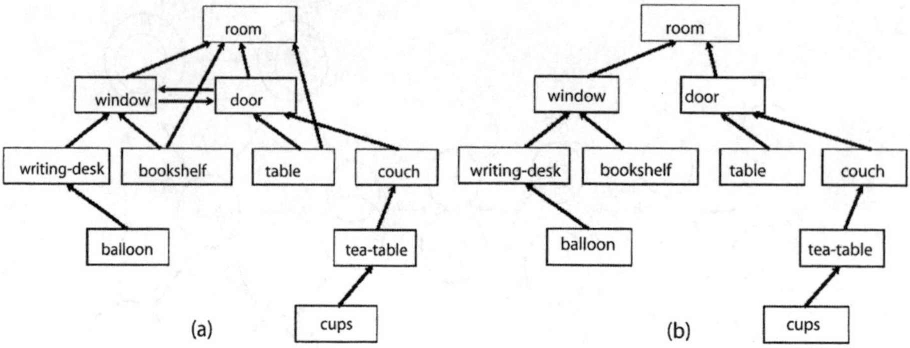

Fig. 4. (a) shows the linguistic reference relations; (b) shows the partial order lattice of direct reference pairs that is generated from (a)

under the book" nor "the wall is behind the picture". Because normally books and pictures are less stable than tables and walls, locating an object by a less stable object provides little information about its location. If a pilot in a plane has lost his current location information, he expects something like "*you are above the South Pole*" not "*you are in your plane*". '$stability(x)$' stands for "the stability of fiat boundary x."

4.2 The Partial Order Lattice of References

Given a set of spatial linguistic descriptions about a snapshot layout of an environment, the direct reference pair (A, B) is defined as such:

(1) A is used as the reference to B and there is no third object C such that (A, C) and (C, B);

(2) B is not used as the reference to A.

All of the direct reference pairs form a partial order lattice. '$POL(pl)$' standing for "*pl* is a partial order lattice". I define here for convenience: '$top\_level(pl)$', '$second\_level(pl)$', '$third\_level(pl)$' and '$fourth\_level(pl)$' stands for "top level, second level, third level and forth level spatial objects in *pl*, respectively". Objects that are lower than the fourth level cannot help to recognize the room, therefore, they are neglected.

4.3 An Example

Given the spatial linguistic descriptions of a remembered scenario, RS, as follows: *the door is in the wall and opposite to the window; the window is in the wall and opposite to the door; the writing-desk is next to the window; the bookshelf is close to the wall and near the window; the couch is on the left of the door; the balloon is in front of the writing-desk; the tea-table is before the couch, tea-cups are on the tea-table; the table is near the wall and on the right, if you come through the door*, linguistic reference relations are shown in Figure 4(a). Then, its partial order lattice of stability, pl_{RS}, can be generated as shown in

Figure 4(b). $top\_level(pl_{RS}) = \{room\}$, $second\_level(pl_{RS}) = \{window, door\}$, $third\_level(pl_{RS}) = \{writingdesk, bookshelf, table, couch\}$, $fourth\_level(pl_{RS}) = \{balloon, teatable\}$.

5 The Cognitive Map of Vista Spatial Environments

To recognize your office, you do not pay attention to the location of apples, books, pens, etc., because they are too un-stable and their locations cannot help you to decide whether it is your target room. Cognitive maps of a vista space only include some of the objects in it.

Foos (1980) investigated constructing cognitive maps from language descriptions. Talmy (1983) discussed how language is effective for conveying spatial information. He proposed that language schematizes space by selecting certain aspects of a referent scene to represent the whole, while disregarding others. The schematization of living spaces in (Ullmer-Ehrich, 1982) discarded all these small objects, like apples, books, pens, etc. and only selected big ones. This suggests that cognitive maps of vista spatial environments only include some objects in the higher levels of the partial order lattice.

A cognitive map of a vista spatial environment, written as '*CogM*', is a representation of the subjective knowledge a person has about the environment. It has a hierarchical structure of spatial relations between fiat regions. The hierarchy is formed according to the commonsense knowledge of stability of spatial objects; and objects are anchored to reference objects at the same level or at the neighboring level in the hierarchy. So, it is a partial-hierarchical structure after (McNamara, 1986). A cognitive map represents a set of SNAPVis ontologies partial-hierarchically structured by an implicit SPANVis ontology: the stability. Different persons may have different spatial knowledge; even the same person will use different spatial knowledge for different tasks. When people recognize a vista spatial environment, normally, the first thing they notice is the shape of the environment, then these non-moveable objects, such as doors and windows, if any, and then big furniture[4]. This paper assumes the following structure of a cognitive map: its root is the room; its second level contains windows and doors; its third level contains big furniture like desks, shelves; its fourth level contains chairs, tea-tables, etc. Normally, objects in the top two levels are non-movable; objects in the third level are seldom moved; objects in the fourth level are often moved.

5.1 The Fiat Container

The fiat container of an object is a fiat region delineated by a spatial relation and this object. For example, "in front of the desk" "behind the desk" "left of the desk" and "right of the desk" delineate four fiat containers by orientation rela-

[4] Personal communication with Jack Loomis, Bad Zwischenahn, Germany, August, 2003.

tions and the desk. From the perspective of the fiat container, the spatial relation between the location object and the reference object is interpreted as which fiat container the location object is located in. An object is always located in fiat containers created by upper level or same level objects. '$fiat\_container(refO, rel)$' or for short '$\mathfrak{S}_{refO}(rel)$' stands for the fiat container delineated by the reference object *refO* and the spatial relation *rel*, which can be one of the $RCC10$ relations or an orientation relation. "`.objects`" is an operator of a fiat container which returns the set of objects that are located in this fiat container. For example, $fiat\_container(desk_1, NR)$.`objects` $= \{chair_1\}(\mathfrak{S}_{desk_1}(NR)$.`objects` $= \{chair_1\})$ stands that $chair_1$ is located in the near region of $desk_1$; similarly, $fiat\_container(desk_1, Front)$.`objects` $= \{chair_1\}$ $(\mathfrak{S}_{desk_1}(Front)$.`objects` $= \{chair_1\})$ stands that $chair_1$ is located in front of or the front part of $desk_1$.

5.2 Fiat Containers for Locations

The location of an object refers to the fiat container where the object is located. '$location\_distance(x, refO) = rel_{RCC10}$' is a function that returns a RCC10 relation, which means that x is located in the fiat container created by *refO* and rel_{RCC10}. E.g., '$location\_distance(chair_1, desk_1) = near$' means "$chair_1$ is located in the *near* fiat container of $desk_1$". '$location\_orientation(x, refO) = rel_{ORI}$' is a function that returns an orientation relation, which means that x is located in the fiat container created by *refO* and rel_{ORI}. E.g., '$location\_orientation$ $(chair_1, desk_1) = front$' means "$chair_1$ is located in the *front* fiat container of $desk_1$".

5.3 Postures as Orientations

The posture of an object refers to how it is located, e.g., it faces to the reference object or backs to it. The posture of an object to its reference object can be interpreted as the orientation of the reference object to the object. For example, "the chair faces towards the desk" is interpreted as "the desk is in front of the chair"; if now the chair turns back to the desk, then it is interpreted as "the desk is behind the chair".

5.4 To Continue the Example

Suppose the spatial relations in the linguistic descriptions in §4.3 are formalized in Table 1, then, the fiat containers are:

$\mathfrak{S}_{room}(PO)$.`objects` $= \{window, door\}$, $\mathfrak{S}_{room}(TPP)$.`objects` $= \{bookshelf, table\}$
$\mathfrak{S}_{window}(FR)$.`objects` $= \{door\}$, $\mathfrak{S}_{window}(NR)$.`objects` $= \{writingdesk, bookshelf\}$
$\mathfrak{S}_{door}(FR)$.`objects` $= \{window\}$, $\mathfrak{S}_{door}(Left)$.`objects` $= \{couch\}$
$\mathfrak{S}_{door}(Right)$.`objects` $= \{table\}$, $\mathfrak{S}_{writingdesk}(Front)$.`objects` $= \{balloon\}$
$\mathfrak{S}_{couch}(Front)$.`objects` $= \{teatable\}$

Table 1. Formal representations of spatial relations in the spatial linguistic descriptions

descriptions	relation	type	formal relation
The door is in the wall	in	RCC10	*PO*
and opposite to the window	opposite to	RCC10	*FR*
the window is in the wall	in	RCC10	*PO*
and opposite to the door	opposite to	RCC10	*FR*
the writing-desk is next to the window	next to	RCC10	*NR*
the bookshelf is close to the wall	close to	RCC10	*TPP*
and near the window	near	RCC10	*NR*
the couch is on the left of the door	on the left	ORIENTATION	*Left*
the balloon is in front of the writing-desk	in front of	ORIENTATION	*Front*
the table is near the wall	near	RCC10	*TPP*
and on the right, if you come through the door	on the right	ORIENTATION	*Right*
the tea-table is before the couch	before	ORIENTATION	*Front*

6 The Spatial Oriented Recognition Process

Generally speaking, the SPAN structure of a spatial environment is indeterminate in the sense that the movements of spatial objects inside are unpredictable. The exact traces of cars and people on the streets are unpredictable at a given time; the exact locations of books, flowers are unpredictable on the next day. Recognizing spatial environments is not a problem of whether the perceived SNAP participates in the expected SPAN of the environment, rather a problem of whether two SNAPs can participate in the same SPAN of the environment. One relates to the spatial environment before the eyes; the other is the one in the memory.

Considering the variability of vista spatial environments, the recognition problem is refined as follows: recognizing a vista spatial environment means that the perceived vista spatial environment is *compatible* to the one in mind; which means that the layout of the perceived environment can be easily changed into the layout of the one in mind.

Relating to different stabilities (correspond to different levels in the hierarchy) of fiat boundaries, their easiness of spatial changes differs accordingly. For example, spatial changes of a room are extremely difficult or impossible; location changes of windows or doors are also extremely difficult and in most cases impossible; location changes of desks or shelves are difficult, though possible; location changes of chairs or sitting balls are easy; location changes of books are so easy that they do not help to recognize the spatial environment.

Two *CogMs* are *not compatible*, when there are spatial differences between objects in the top two levels in hierarchy; two *CogMs might be compatible*, when there are no spatial differences between objects in the top two levels and there are spatial differences between objects in the third level; two $CogM$s are *compatible*, when there are no spatial differences between objects in the top three levels and there are spatial differences between objects in the fourth level; two $CogM$s are *very compatible*, when there are no spatial differences in the top four levels.

6.1 The Qualitative Creation or Destruction

Following (Grenon and Smith, 2004) there are fiat boundaries in the perceived $CogM$ that are not in the memorized $CogM$, namely *qualitative creation*, such as a new chair, new flowers, etc. There are fiat boundaries in the memorized $CogM$ but not in the perceived $CogM$, namely *qualitative destruction*, such as chairs are moved out, flowers withered and thrown away. Comparison of fiat boundaries is at the basic category level, which means that fiat boundaries of the same basic level category are indistinguishable in isolation, though they are made of different atoms and that entities of different basic level category are distinguishable.

A set of fiat boundaries F determines a set of basic level categories B_F such that for each element $b \in B_F$ there is $f \in F$ such that $Constituent(f, b)$ and for each element $f \in F$ its basic level category is in B_F. The sets F and B_F further determine the characteristic category set BC_F by replacing each element $b \in B_F$ with the pair (b, n_b), where n_b is the number of elements in F which are of the basic level category b. For example, $F = \{chair_1, chair_2, chair_3, desk_1, desk_2\}$, then $B_F = \{CHAIR, DESK\}$, $BC_F = \{(CHAIR, 3), (DESK, 2)\}$.

'$=_{blc}$' stands for the *basic level equal* between two sets of fiat boundaries. Let F_1, F_2 be two sets of fiat boundaries, they are at basic level equal, if and only if their characteristic category sets are equal. That is, $F_1 =_{blc} F_2$ iff $BC_{F_1} = BC_{F_2}$.

'$\ominus$' stands for the *basic level minus* between one set and one element. Let F be a set of fiat boundaries: $F = \{f_1, f_2, f_3, \cdots, f_n\}$ and $f' \notin F$.

$$\begin{aligned} & F \ominus f' = \qquad (20) \\ & \iota F_{result}(\exists f, w | f \in F \wedge BasicLevel(w) \wedge Constituent(f, w) \\ & \quad \wedge Constituent(f', w) \bullet F_{result} = F - \{f\}) \end{aligned}$$

'$\iota y(p)$': 'ι' is the descriptor, '$\iota y(p)$' means such y that satisfies p. Though $F \ominus f'$ may result in different set of fiat boundaries, they have a unique characteristic category set.

The *basic level minus* between two sets F and F' is defined as:

$$F \ominus \emptyset = F \qquad (21)$$

$$F \ominus F' = \iota F_{result}(\exists f \in F' \bullet F_{result} = (F \ominus f) \ominus (F' - \{f\})) \qquad (22)$$

'$AppearOrDisappear(F_1, F_2)$' stands for "qualitative creations and destructions between F_1 and F_2".

$$AppearOrDisappear(F_1, F_2) = F_1 \ominus F_2 \cup F_2 \ominus F_1$$

6.2 Qualitative Spatial Changes

There are fiat boundaries in the perceived $CogM$ that are located differently in the memorized $CogM$, such as chairs are moved from near the desk to near the shelf. As fiat boundaries are located in fiat containers created by reference objects in the upper level or the same level, finding location change is pursued by checking whether there are qualitative creations or destructions in fiat containers created by peer objects with the same spatial relation.

Two fiat containers are *peer fiat containers*, if and only if the two reference objects are of the same basic level category and the two relations are the same.

Suppose that fiat containers shown in §5.4 are of the remembered environment and that fiat containers of the current perceived environment are as follows:

$\mathfrak{S}_{room_2}(PO).\mathtt{objects} = \{window_2, door_2\}$

$\mathfrak{S}_{room_2}(TPP).\mathtt{objects} = \{bookshelf_2, table_2\}$, $\mathfrak{S}_{window_2}(FR).\mathtt{objects} = \{door_2\}$

$\mathfrak{S}_{window_2}(NR).\mathtt{objects} = \{writingdesk_2, bookshelf_2\}$

$\mathfrak{S}_{door_2}(FR).\mathtt{objects} = \{window_2\}$, $\mathfrak{S}_{door_2}(Right).\mathtt{objects} = \{table_2, couch_2\}$

$\mathfrak{S}_{door_2}(Left).\mathtt{objects} = \{couch_2\}$, $\mathfrak{S}_{couch_2}(Front).\mathtt{objects} = \{teatable_2, balloon_2\}$

Then, peer fiat containers are:
$\mathfrak{S}_{room}(PO)$ and $\mathfrak{S}_{room_2}(PO)$, $\mathfrak{S}_{room}(TPP)$ and $\mathfrak{S}_{room_2}(TPP)$,
$\mathfrak{S}_{window}(FR)$ and $\mathfrak{S}_{window_2}(FR)$, $\mathfrak{S}_{window}(NR)$ and $\mathfrak{S}_{window_2}(NR)$,
$\mathfrak{S}_{door}(FR)$ and $\mathfrak{S}_{door_2}(FR)$, $\mathfrak{S}_{door}(Right)$ and $\mathfrak{S}_{door_2}(Right)$,
$\mathfrak{S}_{door}(Left)$ and $\mathfrak{S}_{door_2}(Left)$, $\mathfrak{S}_{couch}(Front)$ and $\mathfrak{S}_{couch_2}(Front)$

'$ExistingSpatialChange(CogM_1, CogM_2)$' stands for "there are qualitative creations or destructions between the two cognitive maps $CogM_1$ and $CogM_2$. It returns the union of the object sets of all the qualitative creations or destructions of their peer fiat containers and non-peered fiat containers in $CogM_1$ and $CogM_2$.

For example, let $CogM_{RS}$ be the cognitive map of the remembered scenario whose partial order lattice is shown in §4.3 and fiat containers are shown in §5.4, $CogM_{PS}$ be the cognitive map of the perceived scenario whose fiat containers are shown above, then their spatial differences are observed as follows:

$$
\begin{aligned}
&ExistingSpatialChange(CogM_{RS}, CogM_{PS})\\
&= AppearOrDisappear(\mathfrak{S}_{room}(PO).\mathtt{objects}, \mathfrak{S}_{room_2}(PO).\mathtt{objects})\\
&\quad \cup AppearOrDisappear(\mathfrak{S}_{room}(TPP).\mathtt{objects}, \mathfrak{S}_{room_2}(TPP).\mathtt{objects})\\
&\quad \cup AppearOrDisappear(\mathfrak{S}_{window}(FR).\mathtt{objects}, \mathfrak{S}_{window_2}(FR).\mathtt{objects})\\
&\quad \cup AppearOrDisappear(\mathfrak{S}_{window}(NR).\mathtt{objects}, \mathfrak{S}_{window_2}(NR).\mathtt{objects})\\
&\quad \cup AppearOrDisappear(\mathfrak{S}_{door}(FR).\mathtt{objects}, \mathfrak{S}_{door_2}(FR).\mathtt{objects})\\
&\quad \cup AppearOrDisappear(\mathfrak{S}_{door}(Right).\mathtt{objects}, \mathfrak{S}_{door_2}(Right).\mathtt{objects})\\
&\quad \cup AppearOrDisappear(\mathfrak{S}_{couch}(Front).\mathtt{objects}, \mathfrak{S}_{couch_2}(Front).\mathtt{objects})\\
&\quad \cup AppearOrDisappear(\mathfrak{S}_{door}(Left).\mathtt{objects}, \mathfrak{S}_{door_2}(Left).\mathtt{objects})\\
&\quad \cup \mathfrak{S}_{writingdesk}(Front).\mathtt{objects}\\
&= \{window, door\} \ominus \{window_2, door_2\} \cup \{window_2, door_2\} \ominus \{window, door\}
\end{aligned}
$$

$$\begin{aligned}
&\cup\{bookshelf, table\} \ominus \{bookshelf_2, table_2\}\\
&\cup\{bookshelf_2, table_2\}\ominus\{bookshelf, table\}\cup\{door\}\ominus\{door_2\}\cup\{door_2\}\ominus\{door\}\\
&\cup\{writingdesk, bookshelf\} \ominus \{writingdesk_2, bookshelf_2\}\\
&\cup\{writingdesk_2, bookshelf_2\} \ominus \{writingdesk, bookshelf\}\\
&\cup\{window\} \ominus \{window_2\} \cup \{window_2\} \ominus \{window\}\\
&\cup\{table\} \ominus \{table_2\} \cup \{table_2\} \ominus \{table\}\\
&\cup\{teatable\} \ominus \{teatable_2, balloon_2\} \cup \{teatable_2, balloon_2\} \ominus \{teatable\}\\
&\cup\{couch\} \ominus \{couch_2\} \cup \{couch_2\} \ominus \{couch\} \cup \{balloon\}\\
= &\{balloon_2\} \cup \{balloon\}\\
= &\{balloon_2, balloon\}
\end{aligned}$$

6.3 The Process of Recognizing Spatial Environments

The recognition process is *active* and *top-down*. *Active* means that when we go into a vista spatial environment, we have an anticipation of which vista spatial environment it should be. Thus, we have the SNAPVis and the SPANVis of the expected environment in mind, represented by $CogM_{rem}$. *Top-down* means that spatial objects in the perceived SNAPVis are checked from the top level downwards in the hierarchy.

Let $CogM_{rem}$ and $CogM_{per}$ be the cognitive map remembered and the cognitive map perceived, pl_{rem} and pl_{per} be the partial order lattices of the two cognitive maps, respectively.

The perceived $CogM$ cannot be compatible with the remembered, if there are any spatial differences of non-movable objects. These objects are located at the top level or the second level in the stability hierarchies.

$$\begin{aligned}
&NotCompatible(CogM_{rem},\ CogM_{per}) \stackrel{\text{def}}{=}\\
&(ExistingSpatialChanges(CogM_{rem},\ CogM_{per})\\
&\quad\cap((top_level\ pl_{rem}) \cup (second_level\ pl_{rem}))) \neq \emptyset\\
&\vee(ExistingSpatialChanges(CogM_{rem},\ CogM_{per})\\
&\quad\cap((top_level\ pl_{per}) \cup (second_level\ pl_{per}))) \neq \emptyset
\end{aligned} \tag{23}$$

The perceived $CogM$ might be compatible with the remembered, if there are no spatial differences of non-movable objects and there are spatial differences among big furniture, which are at the third level in the hierarchies.

$$\begin{aligned}
&MightCompatible(CogM_{rem},\ CogM_{per}) \stackrel{\text{def}}{=}\\
&\neg NotCompatible(CogM_{rem}, CogM_{per})\\
&\wedge(ExistingSpatialChanges(CogM_{rem},\ CogM_{per})\\
&\quad\cap((third_level\ pl_{rem}) \cup (third_level\ pl_{per}))) \neq \emptyset
\end{aligned} \tag{24}$$

The perceived $CogM$ are compatible with the remembered, if there are neither spatial differences of non-movable objects nor spatial differences of big furniture and there are spatial differences of small pieces of furniture which are at the fourth level in the hierarchies.

$$\begin{aligned}&Compatible(CogM_{rem},\ CogM_{per})\stackrel{\text{def}}{=}\\&\quad\neg NotCompatible(CogM_{rem},\ CogM_{rem})\\&\quad\wedge\neg MightCompatible(CogM_{rem},\ CogM_{per})\\&\quad\wedge(ExistingSpatialChanges(CogM_{rem},\ CogM_{per})\\&\qquad\cap((fourth\_level\ pl_{rem})\cup(fourth\_level\ pl_{per})))\neq\emptyset\end{aligned}\tag{25}$$

The perceived SNAPVis are very compatible with the remembered, if there is no spatial differences in the top four levels.

$$\begin{aligned}&VeryCompatible(CogM_{rem},\ CogM_{per})\stackrel{\text{def}}{=}\\&\quad\neg NotCompatible(CogM_{rem},\ CogM_{per})\\&\quad\wedge\neg MightCompatible(CogM_{rem},\ CogM_{per})\\&\quad\wedge\neg Compatible(CogM_{rem},\ CogM_{per})\\&\quad\wedge(ExistingSpatialChanges(CogM_{rem},\ CogM_{per})\\&\qquad\cap((fourth\_level\ pl_{rem})\cup(fourth\_level\ pl_{per})))=\emptyset\end{aligned}\tag{26}$$

Suppose the top four levels of partial order lattice of $CogM_{PS}$ in §6.2 are as follows: $top\_level(pl_{PS}) = \{room_2\}$, $second\_level(pl_{PS}) = \{window_2, door_2\}$, $third\_level(pl_{PS}) = \{writingdesk_2, bookshelf_2, table_2, couch_2\}$, $fourth\_level(pl_{PS}) = \{balloon_2, teatable_2\}$, then the perceived scenario is *compatible* with the remembered one:

$$\begin{aligned}&NotCompatible(CogM_{RS},\ CogM_{PS})\\&\stackrel{\text{def}}{=}(ExistingSpatialChanges(CogM_{RS},\ CogM_{PS})\\&\qquad\cap((top\_level\ pl_{RS})\cup(second\_level\ pl_{RS})))\neq\emptyset\\&\quad\vee(ExistingSpatialChanges(CogM_{RS},\ CogM_{PS})\\&\qquad\cap((top\_level\ pl_{PS})\cup(second\_level\ pl_{PS})))\neq\emptyset\\&\equiv\{balloon_2, balloon\}\cap(\{room\}\cup\{window, door\})\neq\emptyset\\&\quad\vee\{balloon_2, balloon\}\cap(\{room_2\}\cup\{window_2, door_2\})\neq\emptyset\\&\equiv F\\&MightCompatible(CogM_{RS},\ CogM_{PS})\\&\stackrel{\text{def}}{=}\neg NotCompatible(CogM_{RS}, CogM_{PS})\\&\quad\wedge(ExistingSpatialChanges(CogM_{RS},\ CogM_{PS})\\&\qquad\cap((third\_level\ pl_{RS})\cup(third\_level\ pl_{PS})))\neq\emptyset\\&\equiv\neg F\wedge\{balloon_2, balloon\}\cap(\{writingdesk, bookshelf, table, couch\}\\&\qquad\cup\{writingdesk_2, bookshelf_2, table_2, couch_2\})\neq\emptyset\\&\equiv F\end{aligned}$$

$$
\begin{aligned}
&Compatible(CogM_{RS},\ CogM_{PS})\\
&\stackrel{\text{def}}{=} \neg NotCompatible(CogM_{RS},\ CogM_{PS})\\
&\quad \wedge \neg MightCompatible(CogM_{RS},\ CogM_{PS})\\
&\quad \wedge (ExistingSpatialChanges(CogM_{RS},\ CogM_{PS})\\
&\qquad \cap ((fourth\_level\ pl_{RS}) \cup (fourth\_level\ pl_{PS}))) \neq \emptyset\\
&\equiv \neg F \wedge \neg F \wedge \{balloon_2, balloon\} \cap (\{balloon, teatable\}\\
&\qquad\qquad\qquad\qquad\qquad\qquad \cup \{balloon_2, teatable_2\}) \neq \emptyset\\
&\equiv T
\end{aligned}
$$

A symbolic simulation system, the LIVE model, has been successfully implemented in Lisp. One example is shown in Figure 5.

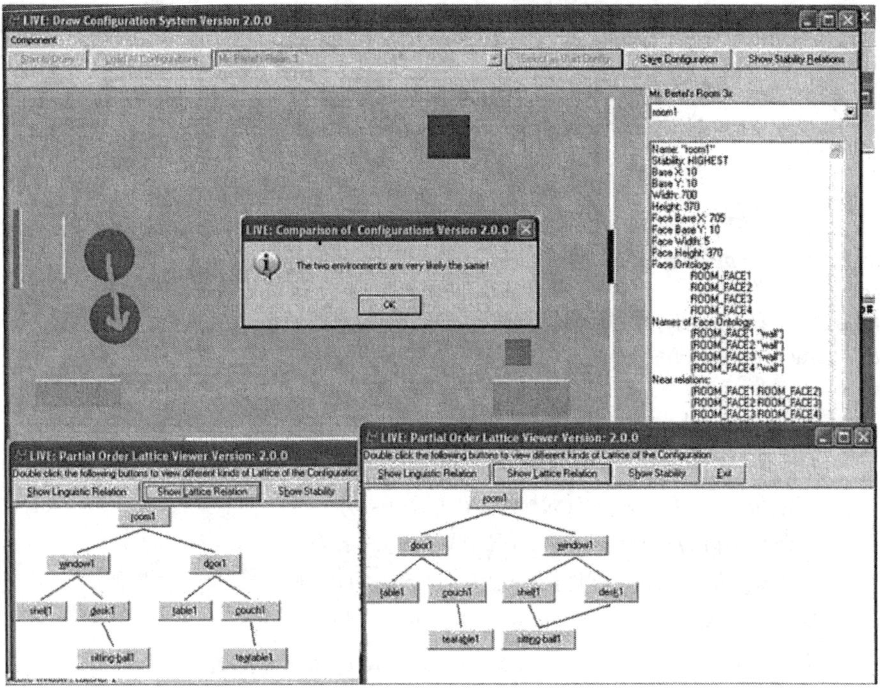

Fig. 5. The sitting-ball in Mr. Bertel's room has been changed a little; its partial order lattice changes accordingly. The recognition process shows that it is Mr. Bertel's room notwithstanding. "to be very likely the same" means: to be the same in most cases

7 Conclusions

This article presents basic formal ontologies for recognizing vista spatial environments. The basic topic in vista spatial cognition is how to define spatial relations between extended objects. Qualitative distance relations are defined as the "connectedness" relations between the location object and the near region of the reference object; qualitative orientation relations are defined as the "connectedness" relations between the location object and the fiat parts of the reference object. The SNAPVis ontologies are, therefore, constructed by regions occupied by extended objects and the connectedness relations between fiat regions determined by extended objects and human cognition. The stability (a SPANVis ontology) is introduced through the observation of the asymmetry between the location object and its reference object in spatial linguistic descriptions. The basic task of vista spatial cognition is how to recognize variable vista environments. This article proposes that recognizing vista spatial environments is the problem of the compatibility between the remembered environment and the perceived environment, which is determined by spatial differences and stabilities of related objects.

Acknowledgements

This work has been carried out in the framework of the SFB/TR 8 Spatial Cognition. Financial supports by DAAD IQN (International Quality Network) scholarship and DAAD research scholarship are greatly acknowledged. I am indebted to Christian Freksa for his supervision, to Diedrich Wolter for his early comments that helped me to shape the concept of the near extension region presented in the appendix, to John Bateman for critiques, comments and language improvements on the earlier drafts of this paper, to Barry Smith, Hardwig Hochmair, Jan Oliver Wallgrün, Till Mossakowski, Sven Bertel, Shou-Ren Lu and Christian Freksa for comments on the later drafts, and to three anonymous reviewers for their critiques, comments and supports.

References

Buelthoff, H. H. and Edelman, S. (1992). Psychophysical support for a two-dimensional view interpolation theory of object recognition. In *Proceedings of National Academy of Science*, pages 60–64. USA.

Clementini, E. and Di Felice, P. (1997). Approximate topological relations. *International Journal of Approximate Reasoning*, 16:173–204.

Cohn, A. G. and Gotts, N. M. (1996). The 'egg-yolk' representation of regions with indeterminate boundaries. In Burrough, P. and Frank, A. U., editors, *Proceedings, GISDATA Specialist Meeting on Geographical Objects with Undetermined Boundaries*, pages 171–187. Francis & Taylor.

de Laguna, T. (1922). Point, line and surface as sets of solids. *The Journal of Philosophy*, 19:449–461.

Foos, P. W. (1980). Constructing Cognitive Maps From Sentences. *Journal of Experimental Psychology: Human Learning and Memory*, 6(1):25–38.

Freksa, C. (1981). *Linguistic Pattern Characterization and Analysis*. PhD thesis, Department of Electrical Engineering and Computer Science, University of California, Berkeley.

Freksa, C. (1991). Conceptual Neighborhood and its role in temporal and spatial reasoning. In Singh, M. and Trav-Massuys, L., editors, *Decision Support Systems and Qualitative Reasoning*, pages 181–187. Elsevier Science Publishers, North-Holland.

Goyal, R. (2000). *Similarity Assessment for Cardinal Directions between Extended Spatial Objects.* PhD thesis, Spatial Information Science and Engineering, University of Maine.

Grenon, P. and Smith, B. (2004). SNAP and SPAN: Towards Dynamic Spatial Ontology. *Spatial Cognition and Computation*, 4(1):69–103. Lawrence Erlbaum Associates, Inc.

Hirtle, S. C. and Jonides, J. (1985). Evidence of hierarchies in cognitive maps. *Memory & Cognition*, 13(3):208–217.

Humphrey, G. K. and Khan, S. C. (1992). Recognizing novel views of three-dimensional objects. *Canadian Journal of Psychology*, 46:170–190.

Kosslyn, S. M., Pick, H. L., and Fariello, G. R. (1974). Cognitive maps in children and men. *Child Development*, 45:707–716.

Kuipers, B. (1978). Modeling spatial knowledge. *Cognitive Science*, pages 129–153.

Marr, D. and Nishihara, H. K. (1978). Representation and recognition of the spatial organization of three dimensional shapes. In *Proceedings of the Royal Society of London*, pages 269–294. B 200.

McNamara, T. P. (1986). Mental Representation of Spatial Relations. *Cognitive Psychology*, 18:87–121.

Montello, D. (1993). Scale and Multiple Psychologies of Space. In Frank, A. and I.Campari, editors, *Spatial information theory: A theoretical basis for GIS*, pages 312–321. Springer, Berlin.

Piaget, J. (1954). *The Construction of Reality in the Child.* Routledge & Kegan Paul Ltd.

Randell, D., Cui, Z., and Cohn, A. (1992). A spatial logic based on regions and connection. In Nebel, B., Swartout, W., and Rich, C., editors, *Proc. 3rd Int. Conf. on Knowledge Representation and Reasoning*, pages 165–176, San Mateo. Morgan Kaufmann.

Rock, I. (1983). *The logic of perception.* MIT Press, Cambridge, MA.

Rosch, E., Mervis, C. B., Gray, W., Johnson, D., and Boyes-Braem, P. (1976). Basic objects in natural categories. *Cognitive Psychology*, 8:382–439.

Schmidtke, H. (2001). The House Is North of the River: Relative Localization of Extended Objects. In Montello, D. R., editor, *COSIT 2001*, pages 415–430. Springer-Verlag.

Schmidtke, H. (2003). A Geometry for Places: Representing Extension and Extended Objects. In Kuhn, W., Worboys, M. F., and Timpf, S., editors, *COSIT 2003*, pages 221–238. Springer-Verlag.

Siegel, A. W. and White, S. H. (1975). The development of spatial representation of large-scale environments. In Reese, H., editor, *Advances in child development and behaviour*, pages 9–55. Academic Press, San Diego, CA.

Smith, B. (2001). *Fiat* objects. *Topoi*, 20(2):131–148.

Smith, B. and Varzi, A. C. (2000). *Fiat* and *bona fide* boundaries. *Philosophy and Phenomenological Research*, 60(2):401–420.

Steven, A. and Coupe, P. (1978). Distance estimation from cognitive maps. *Cognitive Psychology*, 13:526–550.

Stock, O., editor (1997). *Spatial and Temporal Reasoning.* Kluwer Academic Publishers.

Talmy, L. (1983). How Language Structures Space. In Pick, H. and Acredolo, L., editors, *Spatial Orientation: Theory, Research and Application*, pages 225–281. Plenum Press.

Tarr, M. J. (1995). Rotating objects to recognize them: A case study of the role of mental transformations in the recognition of three-dimensional objects. *Psychonomic Bulletin and Review*, 2:55–82.

Tolman, E. C. (1948). Cognitive Maps in Rats and Men. *The Psychological Review*, 55(4):189–208.

Tversky, B. (1991). Spatial Mental Models. *The Psychology of Learning and Motivation*, 27:109–145.

Tversky, B. and Lee, P. (1999). How space structures language. In Freksa, C., Habel, C., and Wender, K. F., editors, *Spatial Cognition*, volume 1404 of *LNAI*, pages 157–176. Springer-Verlag.

Ullmer-Ehrich, V. (1982). The Structure of Living Space Descriptions. In Jarvella, R. J. and Klein, W., editors, *Speech, Place, and Action*, pages 219–249. John Wiley & Sons Ltd.

Wilson, B., Baddeley, A., and Young, A. (1999). LE, A Person Who Lost Her 'Mind's Eye'. *Neurocase*, 5:119–127.

A Formal Definitions

This appendix presents simple descriptions of three basic terminologies: the *near* extension, the *near* region, and the extension region.

Regions are denoted by `mathcal` capital letters, such as $\mathcal{A}$, $\mathcal{B}$, $\mathcal{C}$, ... Regions belong to classes and the classes of regions are denoted by `typewriter` capital letters, such as $\mathtt{A}$, $\mathtt{B}$, $\mathtt{C}$, ...

A.1 Primitives and Postulates

The 'Region' is Primitive. A "region" refers to *the space that is occupied by an extended object*[5]. This is not a definition of regions, rather how it is understood. A "region" has the following properties, which are used as postulates.

The Postulate of Categories. Humans are preferable to recognize single objects at the basic level category. Accordingly, a "region" belongs to a class. "CL" stands for the set of all the classes of regions that a person has.

Axiom-CL 1. CL *is non-empty.*

$$\mathsf{CL} \neq \emptyset$$

For any recognized object, it belongs to one and only one element in CL.

Axiom-CL 2. *Any region belongs to a class which is an element of* CL.

$$\forall \mathcal{A} \exists \mathtt{A} | \mathtt{A} \in \mathsf{CL} \bullet \mathcal{A} \in \mathtt{A}$$

Axiom-CL 3. *Any region belongs to only one element in* CL.

$$\forall \mathcal{A}, \mathtt{A}, \mathtt{B} | \mathtt{A} \in \mathsf{CL} \wedge \mathtt{B} \in \mathsf{CL} \bullet \mathcal{A} \in \mathtt{A} \wedge \mathcal{A} \in \mathtt{B} \rightarrow \mathtt{A} = \mathtt{B}$$

[5] By "an object", I do not exclude objects, such as *a hole*, or *a niche*. Thus, the "region" here refers to the space occupied by a conceptual object, rather than substances.

'Connectedness' is Primitive. The only primitive relation between regions is "connected with": **C**.

A.2 The *Near* Extension, the *Near* Region

Let $\mathcal{A}$ (of the class $\mathtt{A} \in \mathsf{CL}$) and $\mathcal{X}$ (of the class $\mathtt{X} \in \mathsf{CL}$) be two regions. The *near* extension of $\mathcal{A}$ by $\mathcal{X}$, written as $\mathcal{A}^{\mathcal{X}}$, refers to the sum of all regions of the class $\mathtt{X}$ that are connected with the region $\mathcal{A}$. $\mathcal{X}$ is called the extension region.

Formally, the *near* extension of $\mathcal{A}$ by $\mathcal{X}$ can be defined as the region $\mathcal{Y}$ such that, given any region $\mathcal{W}$, $\mathcal{W}$ connects with $\mathcal{Y}$ if and only if $\mathcal{W}$ connects with a certain region $\mathcal{V}$ of the class $\mathtt{X}$ such that $\mathcal{V}$ connects $\mathcal{A}$.

$$\mathcal{A}^{\mathcal{X}} \stackrel{\text{def}}{=} \tag{27}$$
$$\iota\mathcal{Y}(\forall\mathcal{W} \bullet (\mathbf{C}(\mathcal{W},\mathcal{Y}) \equiv \exists\mathcal{V}|\mathcal{V} \in \mathtt{X} \wedge \mathcal{X} \in \mathtt{X} \wedge \mathtt{X} \in \mathsf{CL} \bullet \mathbf{C}(\mathcal{A},\mathcal{V}) \wedge \mathbf{C}(\mathcal{W},\mathcal{V})))$$

The *near* region of $\mathcal{A}$ by $\mathcal{X}$, written as $\mathcal{N}^{\mathcal{X}}_{\mathcal{A}}$, can be defined as the difference of $\mathcal{A}^{\mathcal{X}}$ with $\mathcal{A}$: $\mathcal{N}^{\mathcal{X}}_{\mathcal{A}} = \mathbf{diff}(\mathcal{A}^{\mathcal{X}}, \mathcal{A})$, **diff** following (Randell et al., 1992).

$\mathcal{A}^{\mathcal{X}}$ and $\mathcal{N}^{\mathcal{X}}_{\mathcal{A}}$ are also regions, therefore, according to **Axiom-CL 2** and **Axiom-CL 3** each belongs to one and only one class, which is an element of CL. The class of $\mathcal{A}^{\mathcal{X}}$ is written as $\mathtt{A}^{\mathtt{X}}$, the class of $\mathcal{N}^{\mathcal{X}}_{\mathcal{A}}$ is written as $\mathtt{N}^{\mathtt{X}}_{\mathtt{A}}$.

The set CL is, therefore, expanded as follows: if $\mathtt{X} \in \mathsf{CL}$ and $\mathtt{Y} \in \mathsf{CL}$, then $\mathtt{X}^{\mathtt{X}}$, $\mathtt{X}^{\mathtt{Y}}$ and $\mathtt{N}^{\mathtt{X}}_{\mathtt{Y}}$ are also elements of CL. For example, suppose that $\mathtt{BIF} \in \mathsf{CL}$ represents the British imperial foot, then $\mathtt{BIF}^{\mathtt{BIF}^{\mathtt{BIF}}} \in \mathsf{CL}$, which represents the class of the yard, see §**A.3**.

A.3 The Extension Regions

The *near* extension of $\mathcal{A}$ by $\mathcal{X}$ depends on the class of $\mathcal{X}$, not on the particular $\mathcal{X}$.

For example, British people select their imperial foot as the extension region (*the British imperial foot*) to measure distance. The British unit of distance were the *yard* (a *yard* equals to three *feet*), and the *fathom* (a *fathom* equals to six *feet*). The British imperial foot is used as the class whose elements have the same length as the British imperial foot. The old French unit of distance was the *Paris feet* and the *toise* (a *toise* equals to six *Paris feet*).

Ancient Egyptians used the "Royal Egyptian Cubit" as the extension region, which was equal to "the length of the forearm from the bent elbow to the tip of the extended middle finger plus the width of the palm of the hand of the Pharaoh or King ruling at that time"[6]. When ancient Egyptians said that the width of the door was less than five Cubits, it meant that five connected objects (each has the size of "Royal Egyptian Cubit") can connect the two sides of the door. Therefore, the "Royal Egyptian Cubit" was used as the class whose elements have the same length as the Pharaoh's forearm, rather than the concrete Pharaoh's forearm with extended middle finger.

[6] http://www.ncsli.org/misc/cubit.cfm

The old German unit of distance[7] were *Elle*, the *double feet* series, *a day's journey*, etc. The *Elle* was similar to the "Royal Egyptian Cubit" which was defined as the segment between the bent elbow and the point of extended middle finger; the *double feet* series used two connected feet as the length-unit, such as *Frankfurt double feet*, *Oldenburg double feet*, *Bavaria double feet*, *Vienna double feet*, even *Hamburg short double feet*, *Hamburg long double feet*, etc.; *a day's journey* was defined as the distance that can be covered, especially by a horse with cart, in one day[8]. If the horse with cart is replaced with the light and one day is replaced with one year, then we have an extension region in the modern physics—the lightyear that is the distance that light travels in one year.

Ancient Chinese used *Du*, *Cun*, *Chi*, etc. as the extension regions. A *Du* was "two consecutive steps by different legs of a person"[9]. When ancient Chinese measured the length of a road by *Du*, they walked along the road and some of their steps would be inevitably a little bit different from others. *Du* was, therefore, used as the class of all two consecutive steps of a person. *Cun* is the body segment between the wrist striation behind the thumb and the pulsing point of the radial artery[10]; *Chi* is the body segment between the wrist striation and the striation at the acupuncture point called "Qu-Chi" in Chinese medicine[11]. Therefore, *Cun* and *Chi* are used as classes whose members differ person from person.

---

[7] http://matheboard.de/lexikon/Hauptseite,definition.htm

[8] *Tagereise: einen Tag dauernde Reise (bes. mit Pferd u. Wagen).*

[9] Personal communication with Shou-Ren Lu.

[10] The definitions of 'Chi' and 'Cun' in **Shuo Wen Jie Zi** (Origin of Chinese Characters), by *XU Shen* (58-147 AD) of the Eastern Han Dynasty.

[11] Personal communication with Shou-Ren Lu.

Modelling Models of Robot Navigation Using Formal Spatial Ontology*

John Bateman and Scott Farrar

University of Bremen,
FB10, Sprach- und Literaturwissenschaften,
Bibliothekstr. 1, Bremen 28359, Germany
{bateman, farrar}@uni-bremen.de
http://www.sfbtr8.uni-bremen.de/i1

Abstract. In this paper we apply a formal ontological framework in order to deconstruct two prominent approaches to navigation from cognitive robotics, the Spatial Semantic Hierarchy of Kuipers and the Route Graph of Krieg-Brückner, Werner and others. The ontological framework is based on our current work on ontology specification, where we are investigating Masolo *et al.*'s Descriptive Ontology for Linguistic and Cognitive Engineering (DOLCE) extended particularly for space and navigation by incorporating aspects of Smith *et al.*'s Basic Formal Ontology (BFO). Our conclusion is that ontology should necessarily play an important role in the design and modelling of cognitive robotic systems: comparability between approaches is improved, modelling gaps and weaknesses are highlighted, re-use of existing formalisations is facilitated, and extensions for interaction with other components, such as natural language systems, are directly supported.

1 Introduction

The use of formal ontology in the field of cognitive robotics has until recently been quite limited. We argue in this paper, however, that the sophistication required of current cognitive models, the functionalities required of cognitive robots, and the state of the art in formal and computational ontology all combine to suggest that a closer interaction between ontology and cognitive robotic modelling is now appropriate. The explicit adoption of computational ontology brings a stronger set of modelling constraints to bear on the necessary issues, and also provides a much richer set of re-usable building blocks for modelling.

In general one can envision at least three scenarios for incorporating ontologies into cognitive robotics. First, ontology can be used to enhance the design

* The work reported in this paper was supported by the Deutsche Forschungsgemeinschaft (DFG) as project I1-[OntoSpace] of the Transregional Collaborative Research Center SFB/TR 8 Spatial Cognition. The authors also thank particularly the anonymous reviewers for their extensive corrections and comments, as well as the other members of the SFB/TR 8.

C. Freksa et al. (Eds.): Spatial Cognition IV, LNAI 3343, pp. 366–389, 2005.
© Springer-Verlag Berlin Heidelberg 2005

of robot knowledge representations; that is, ontology can help to clarify the relations among various representational levels and to provide a semantically coherent account of the entities used in symbolic reasoning. Second, ontology can be used to develop and constrain a sharable conceptualization of the environment in the interaction of intelligent agents, including both robot-robot or human-robot interaction. Third, ontology can contribute to solutions for the problem of partial information at the sensory-symbol interface; that is, partial sensory input can be augmented with knowledge from an ontology to build a more accurate symbolic representation. For example, the laws of mereotopology (see below) can be leveraged to verify the existence of necessary parts from poor or incomplete sensory data.

In this paper we focus primarily on the first and second scenarios by investigating the use of spatial ontologies in the modelling of navigational capabilities for cognitive robotics. We do this concretely with respect to two navigational models currently being developed and used in cognitive robotics: the Spatial Semantic Hierarchy of Kuipers [1] and the Route Graph of Werner, Krieg-Brückner and others [2, 3]. These were chosen on the one hand due to their importance for robot navigation and, on the other, because they have already made considerable moves towards compatibility with ontology-based design. We will show how such models can be placed beneficially against a broader ontological background, adopting for this purpose the Descriptive Ontology for Linguistic and Cognitive Engineering (DOLCE) [4] together with some proposals that we are currently developing for extensions in the area of spatial ontology.

We structure the paper as follows. In Section 2, we set out the two selected models of robot navigation. In Section 3, we briefly introduce the account of ontology that we wish to draw upon, concentrating specifically on issues of space. In Section 4, we show how the two individual navigation models can be inter-related, placing them against the ontological background of our framework. Finally, in Section 5, we conclude with an explicit discussion of the benefit of incorporating ontology in the modelling of cognitive systems.

2 Two Models of Robot Navigation

Despite the limited application of ontologies within cognitive robots and navigation, there are now approaches that explicitly draw on ontology in their formalization. Both the Spatial Semantic Hierarchy and the Route Graph are of this kind. Our discussion of each will follow a similar pattern. We first identify the major ontological domains adopted or assumed by each model and then set out briefly the place and nature of relationships *between* those domains that the model specifies.

2.1 The Spatial Semantic Hierarchy

Kuipers' Spatial Semantic Hierarchy (SSH: [5, 6, 1]) is an approach to robot navigation which decomposes a robot's knowledge of its environment across several distinct layers in an *ontology hierarchy.* These layers allow distinct kinds of representations to co-describe the robot's experience and plans. This co-description serves to abstract an agent's spatial knowledge away from the details of its environment and its sensorimotor apparatus [6, p2]. Spatial knowledge can then be used that is derived, or abducted, from the sensory level for purposes of navigation, rather than simply having the agent rely on a reactive/sensory level. This approach finds its motivation in cognitive robotics and draws on research from cognitive science and psychology in an attempt to solve problems via high-level reasoning.

The SSH layers involve the following levels of abstraction; the symbolic levels are given in bold-face type: the sensory level, control level, **causal level, topological level**, and **metrical level**. Each level has, as Kuipers describes it, "its own *ontology* (the set of objects and relations it uses for describing the world) and its own set of inference and problem-solving methods" [6, p2].

The first two levels, *sensory* and *control*, concern the continuous output of sensors such as vision, laser or sonar range-sensing. This numerical data is represented by a sensory input vector $s(t) = [s_1(t), ..., s_n(t)]$ where $s(t)$ is the state of the agent at time t and $s_1(t), ..., s_n(t)$ are individual sensor outputs. These data are then abstracted via a set of *control laws* at the control level to discrete states encompassing position and orientation. This is the primary mechanism by which continuous descriptions and discrete, symbolic descriptions of behavior are related: the continuous numerical entities at the sensory and control levels are abstracted to a discrete symbolic representation for use at the causal and topological levels. We will have little more to say about the sensory and control levels in this paper, however. We will also not address the *metrical* level particularly. This consists simply of a global 2-D geometric map of the environment in a single frame of reference, a so-called "Map in the Head" [1, p195]. Our central concern will be on the central two levels, the causal and the topological.

The ontology of the *causal level* defines views, actions, events and the causal relations among them. Intuitively, a view is some state of affairs as perceived by the robot at some given moment while actions and events describe the motions that a robot may initiate. Views are thus defined as symbolic abstractions over the sensory input vector obtained at a locally distinctive state [1, p205]. A 'locally distinctive state' is another vector $s = (x, y, \theta)$ indicating the agent's position (x, y) and orientation θ within the environment. Views may be both complete and partial, i.e., attending to all or only to some subset of the available sensory inputs.[1] Changes in view are brought about by actions and the SSH defines two types: turns and travels. Turns leave the agent in the same place, while travels change the agent's location [1, p206]. All actions are caused by the agent applying one or more control laws in some distinctive state. The purpose of this abstraction in terms of actions is to lose "the details of how views are defined or how actions are implemented in particular circumstances" [1, p195]. Events are

[1] The sensor values are also considered to be functions of the agent's state [1, p199] but we will not consider this complication further here.

then used to describe the complex of a change in view via an action. They are represented by schemas of the form $\langle V, A, V' \rangle$, where an action A causes view V to change to view V'. Finally, a routine is defined as a set of such schemas indexed by the initial view [1, p207].

Reasoning on entities at the causal level is performed within the SSH with the aid of McCarthy's Situation Calculus [7, 8]. Within the causal layer the fact that some view V holds and that some action A is carried out at the current moment, *now*, is additionally associated with a particular situation, s_o. The state of affairs can then be represented by the Situation Calculus statements: $holds(V, s_o)$ and $do(A, now)$. Accordingly, the state of the world, some situation, is said to change when an action is applied, thus producing a new situation.

The next ontological layer is the *topological* level. This includes the categories of places, paths and regions, with their associated connectivity and containment relations. Places are defined simply as zero-dimensional entities which may lie on a path. A path imposes an order on the places lying on it by virtue of its direction: a path is thus a one-dimensional *subspace* leading from one place to another and having one of two possible directions (dir=±1). A topological, or place, graph can then be constructed as a map of the environment consisting of sets of places and their connecting paths. Paths also serve as the boundaries for regions. A region is defined as a two-dimensional subset of the environment, i.e., a set of places. Path directedness also allows a reference system to be determined. Each directed path divides the world into two regions: one on the right and one on the left. A bounded region is then defined by a directed path with the region on this path's right or on its left.

The SSH also uses regions to define a hierarchical view of space, as now amply motivated psychologically. This allows maps to be pitched at various levels of granularity with sets of places within a map of greater detail being represented within a map of lesser detail by single 'abstraction regions'. There are therefore two *levels of abstraction* within the topological layer, one for place and one for region. A place is still a zero-dimensional point but may also function as an abstraction of a region. To relate these levels Kuipers describes both upward and downward mappings. An *upward mapping* holds when "a place at a lower level is mapped to the place corresponding to the abstraction region that contains it" [1, p212]; a *downward mapping* holds when "a $\langle place, path, dir \rangle$ tuple at the higher level is mapped to a corresponding $\langle place, path, dir \rangle$ tuple at the lower level" [1, p212]. These levels of abstraction are not to be confused with the particular ontological levels defined by the ontology hierarchy. Places and regions occupy two different levels of abstraction within the same ontological level, while the causal, topological, and metric levels are considered ontologically distinct.

We can summarize those relations defined within the topological level by drawing on Kuipers' listing [1, p210] as follows:

- $on(place, path)$: *place* is on *path*;
- $order(path, place_1, place_2, dir)$: the order on *path* from $place_1$ to $place_2$ is *dir*;
- *right-of*$(path, dir, region)$: *path*, facing direction *dir*, has *region* on its right;

- *left-of*(*path*, *dir*, *region*): *path*, facing direction *dir*, has *region* on its left;
- *in*(*place*, *region*): *place* is in *region*.

In general, transforming from consecutive ontological layers in the SSH is carried out via a process of *abduction*. Thus, the places, paths and regions of the topological level are created by deducing some minimal description that is sufficient to explain the regularities found among the observed views and actions of the causal level [1, p209]. An example of an abduction rule from the causal level to the topological level is given as:

$$\forall view \exists place\ (at(view, place))$$

which means that an association will be established between views and particular locations. The following relations defined by the model combine topological and causal level constructs and so can be considered to be 'inter-ontology' relationships. We will bring out the special nature of these kinds of relationships further below.

- *at*(*view*, *place*): *view* is seen at *place*;
- *along*(*view*, *place*, *dir*): *view* is seen along *path* in direction *dir*;

Entities from the control and topological levels may also be mixed in what the SSH terms 'axioms of commonsense'. Consider the gloss of one such axiom: "If the agent travels along a certain directed path, turns right, then travels again to reach a certain place, then that place lies within the region right-of that directed path" [1, p211]. This relates a routine consisting of three events to a region via several places; it also clearly requires several further assumptions to be made concerning the 'shape' of paths and the non-identity of the places mentioned in order to be accurate. Another example of an axiom is the following: $\langle V, (turn\ \alpha), V' \rangle \rightarrow \exists place[at(V, place) \wedge at(V', place)]$, glossed as "A turn action leaves the traveler at the same place" [1, p210]. This combines levels similarly, relating an event to a place. Note that in order to support sophisticated behavior, a considerable number of these kinds of axioms are necessary; providing such characterizations of the 'world' is precisely one of the tasks of ontology.

2.2 The Route Graph

The second area of research we discuss is the Route Graph (RG: [9, 2, 10, 3]). The RG was originally developed for practical robot navigation in real application contexts and so is also faced with the problems of mediating direct sensory input with abstract path control. The essence of the RG was that information concerning different routes can be integrated within a single network-like structure, combining a variety of data sources [2, p297]. Route graphs have accordingly been characterized in a number of different ways and several alternative descriptions of the RG that are closer to actual robot control structures have also been given in the literature [10, 11]. In our discussion here, however, we will draw particularly on the characterization of Krieg-Brückner *et al.* in this volume [3]. This brings a number of advantages for us—in particular that the RG is already defined there in terms of several distinct ontological areas, similar to the approach seen for the SSH. We will assume for present purposes, therefore, that the varied RG descriptions are all broady compatible with this latter ontologically-inspired account.

The starting point for the definition is basic graph theory, yielding nodes and edges. The edges are also directed, and so each has a source and a target. Edges may be combined into a sequence of edges, which may in turn be specialized as a path (possibly containing cycles), which may be specialized further as a route (containing no repeated edges). The Route Graph then refines the generic graph notions in a number of ways suited for concrete robot navigation and motion control involving a real physical robot, with dimensions and sensor capabilities.

First, nodes are refined to places. Places are anywhere that a robot can 'be'. They include in their specification a width (provided by a Voronoi representation of the free space constituting a place: cf. [3, 11]), and an origin that is used for defining the relative (angular) positions, or 'bearings' of all the nodes' incoming and outgoing edges. The origin constitutes a *reference system* for a place [2, p307]; orientation is therefore strictly local. Second, edges are refined to route segments leading from one place to another. Each route segment includes the angular displacements of its edge with respect to both the origins of its source and its target. This means that it is possible to specify, and also physically to rotate, a robot positioned with respect to the origin of the starting node, so that it is ready to follow a given route segment and, after having followed the segment, again to rotate the robot so that it faces in the direction of the origin of the target. Each segment specifies in addition its own length (so that the robotic agent knows how far to travel before expecting to be at its target) and its width (so that the robotic agent can know whether or not it will fit through the segment).

The RG also includes a notion of abstraction similar to that of the SSH. For example, at a particular level of abstraction an entry and an exit ramp of a highway can be considered as two different places (nodes in the RG), whereas, at a higher level of abstraction, the two nodes could be considered one place corresponding to the complex notion of a 'road junction'. Similarly, the complex possibilities for navigation within a train station might quite appropriately, at a higher level of abstraction, be collapsed to a single place: 'the station'. This is modelled formally in terms of an AbstractsTo relation, defined in terms of graph morphisms between RGs of differing granularities.

This abstract definition of a route graph is intended to be neutral across a wide variety of possible route planning tasks. Some intuitive examples of routes discussed include: "CommuterTrainLine, ShipRoute, FootPassage, CityRoad, Highway or Labyrinth" [2, p307]. To accomodate this, the RG incorporates a notion of *layers* where "each Layer represents a Route Graph of a particular Kind" [2, p310]. *Kind* refers to the nature of the places found within the RG. That is, the places in a railway system, e.g., the stations and various transfer points, are of the same kind, but are different from the places in a RG for of-

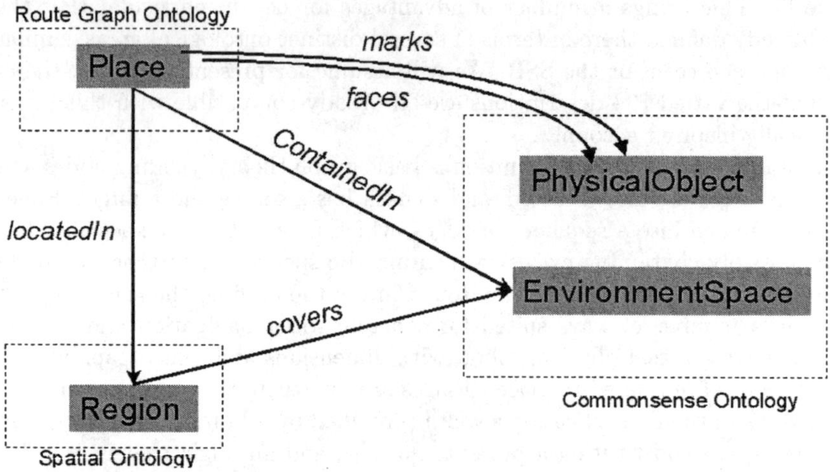

Fig. 1. Relations between RG, spatial, and commonsense ontologies: compiled from Krieg-Brückner *et al.* (2004)

fice navigation, e.g., doorways, corridors, and rooms. When the source and target nodes of a route segment are of the same kind, such RGs are called *homogeneous*. Heterogeneous route graphs do not allow direct transitions between them. Instead, a special type of route segment, whose source and target nodes can be of different kinds is introduced, called a *Transfer* [2, p311]. Krieg-Brückner *et al.* suggest that each of the RG-kind specializations may bring to bear additional kinds of information relevant for that specific application domain. For example, route segments may bring with them certain conditions that must be met before proceeding along the course of the segment.

This latest definition of route graphs also sets out some of the relations that are required between the world of route graphs (consisting of an ontology of places, route segments, paths, etc.) and other ontological domains more clearly than has hitherto been the case. Two such domains are explicitly identified: a spatial ontology, that is expected to provide *spatial regions*, and a 'commonsense ontology', providing everyday objects such as rooms, offices, corridors, and so on. The relationships provided are intended to allow inferences back and forth between places in a route graph, the spatial regions that such places occupy, and the everyday objects that those regions 'cover'. Thus, a place in a route graph can be defined as being locatedIn some given spatial region, and that spatial region might itself cover some physical 'environment space', such as a room; alternatively, the route graph place might be specified directly as being containedIn that environment space. The containedIn relationship is thus a composition of locatedIn and covers.

Two further inter-ontology relations are defined: marks and faces. Both relate a place in a route graph to a commonsense object. For example, a particular place in the route graph can be defined as a *routemark* which marks a commonsense object such as a window. With this in place, a route instruction such as "Go to the window" can be directly related to a route graph description which could drive a robot's motion to the required place. Similarly, a place can be said to face a particular object, so that the robot can be told to "face the elevator" with an appropriate realization of this at the level of the route graph description. A *landmark* is similar but marks points that lie not on a route but in the distance; this is simply modelled by incorporating special kind of edge (a vector) that leads to the landmark in question. Such marks are provided to support localization by triangulation.

We shall argue below that these inter-relationships are in fact crucial for combining the navigation and control-oriented conception of a route graph with further components, such as dedicated spatial reasoning systems or natural language interaction modules. Figure 1 collects together the current set of inter-ontology relationships proposed by Krieg-Brückner *et al.*, summarizing the ontological domains to which they refer.

3 Formal Ontology and Space

In this section we present the necessary background to orient the reader concerning formal ontology, and in particular the place of space within formal ontology. The reader is also referred to Bateman and Farrar [12] for an almost exhaustive account of the state-of-the-art concerning current approaches to space within ontology.

3.1 Formal Ontology: Background

The kind of formal and computational modelling that is our concern here has been described by Nicola Guarino as belonging to the *ontological level* [13]. Such descriptions are intended to provide a 'meaningful' structuring of some domain of concern that goes beyond purely logical adequacy, conforming with the necessary regularities of the world of our experience. As an example of such a description, Guarino considers the following logical expression:

$$\exists x \ (ball(x) \wedge red(x))$$

According to such a representation there is no ontological (or any other) difference between 'ball' and 'red'. Logically they are both unary predicates but, in terms of our experience of the world, this misses several important distinctions. For an adequate knowledge representation, we would rather state that actually a 'ball' is a concept and 'redness' is some property that such concepts can carry—e.g., 'red' is the value of a 'hasColor' relation. This begins to impose far more *structure* on the information being represented than is evident in the initial logical expression alone.

The ontological questions that arise then revolve around just what kind of categories can enter into 'hasColor' relationships, what kind of relation is 'hasColor', and what is the full set of such categories and relationships. An ontology is a way of making explicit those commitments and structural necessities that follow from the fact that we are modelling not knowledge in the abstract, but concrete objects, qualities, relations and events of the known world subject to a rich web of non-arbitrary constraints. As a corollary, an ontology is also a way of specifying explicitly just what follows from particular kinds of modelling decisions: were we to state that 'ballness' is a quality that inheres in certain colors, this would be a strong ontological statement and many consequences would follow from it. This means that the choice of logical representation is no longer arbitrary. The ontology therefore establishes a methodology and a set of principles for deciding in what way entities, relations, activities and so on are to be captured in a formal representation. Since the resulting ontologies are motivated by their anchoring in the world, it is generally hoped that representations that respect those ontologies will provide a more robust basis for inter-operability and knowledge sharing.

The specification of an ontology starts with a modelling language, which is used to represent the elements in the intended domain of discourse. Depending on the approach, there may be another language called an *ontology meta-language* used to describe the modelling language itself. The constants of the meta-language are used as predicates in the modelling language. The relationship between an ontology modelling language and a meta-language can be made to do some useful work by setting out clear methodological criteria for how ontology construction may proceed in terms of properties that need to hold at the meta-level. A successful example of this can be found in the *OntoClean* methodology [14, 15] which uses the notion of a meta-language extensively in its definition of *meta-properties.* Meta-properties are properties of properties, not of objects in the world and are used to constrain ontology development and to evaluate particular proposed ontological organizations. The meta-properties particularly important for OntoClean are: *rigidity*, *identity*, *unity*, and *dependence.* Rigidity refers to essential properties, i.e., properties that an entity cannot loose without ceasing to be itself; identity refers to criteria for discriminating entities from each other or for recognizing when one has a particular kind of entity; unity refers to the 'wholeness' of an entity, whether it has parts, boundaries and so on; and dependence reflects whether an entity can exist independently or whether it needs to be 'carried' by another (e.g., the color of an object is dependent for its existence on that object, the hole of a doughnut is dependent for its existence on that doughnut).

In the construction of formal ontologies, several issues immediately confront the knowledge (or, rather, ontological) engineer. The first is to consider the basic categories and their interrelationships so as to build up an organizational backbone for further specialization. We argue that representing such very basic, foundational features of the world in general, and of spatial objects in particular, is a prerequisite for constructing intelligent spatially-aware agents. Such systems can then operate in terms of situations or settings that are very much more like the kinds of settings that humans take for granted without the need for more *ad hoc* axiomatizations: this is the traditional link that is made to naive physics and modelling situations for intelligent behavior [16]. The foundational ontological properties are anchored into the representation in rather more fundamental ways than is the case with contingent knowledge that may vary or be effected by events in the world. No matter what occurs, basic ontological relationships between objects, their constituting matter, and the locations of that matter will not be effected.

3.2 Generic Upper Ontologies

We are currently constructing a particular view of ontology which builds upon several state of the art formal ontology specifications. We assume a collection of abstract generic ontological modules that are used for the definition of more specific subontologies. Although the relation between ontological modules can be complex and requires more extensive discussion, we will not foreground this aspect here. We will simply assume for present purposes that relations between ontological modules take the form of structured mappings between the classes and relations of the modules involved. In the simplest case, an ontology submodule may straightforwardly extend a more generic ontology by the subsumption relation.

For reasons we have set out at length elsewhere [17, 12], we select the Descriptive Ontology for Linguistic and Cognitive Engineering (DOLCE) [4] as our main organizing framework. DOLCE was originally part of the WonderWeb project[2] whose aim was the development of foundational ontology libraries for the Semantic Web. DOLCE's upper level, a portion of which relevant for space is shown in Figure 2, provides a generically re-usable high-level characterization of the entities of the world that is particularly appropriate as a basis for further development. It was constructed applying the principles set out in the OntoClean methodology [15, 14] mentioned above and is supported by detailed axiomatization. In the following, we pick out particularly those aspects of DOLCE that are necessary and useful for our consideration of space in the context of robot navigation.

The most fundamental divisions made in DOLCE assert that the world can be divided into four classes of entities: first, there is a fundamental division between entities that unfold in time, called PERDURANTS, and entities which are present 'all-at-once' in time, called ENDURANTS, and second, there are QUALITIES, which inhere in other entities, and ABSTRACT entities. The physical objects generally of most concern to robots are a particular subclass of endurant (physical endurant: PED). Physical endurants are distinguished from non-physical endurants primarily by their relation to space: they are necessarily located spatially.

3.3 Spatial Ontology Within DOLCE

A very basic question of traditional philosophical importance is whether space exists independently of any objects that happen to have locations within space or, alternatively, whether space is mainly a matter of inter-relationships between

[2] IST Project 2001-33052 WonderWeb: Ontology Infrastructure for the Semantic Web.

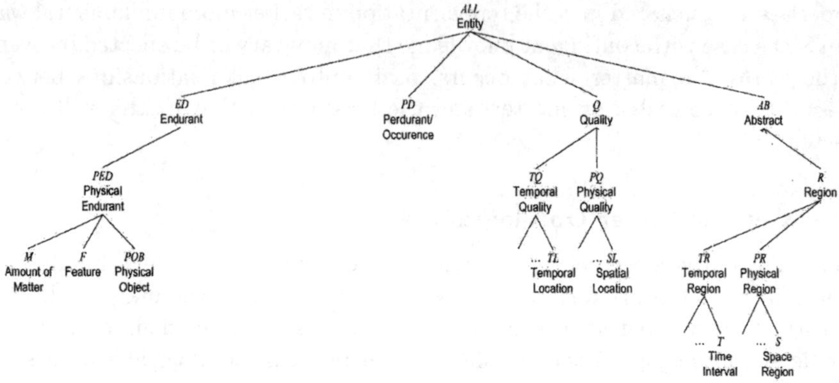

Fig. 2. DOLCE taxonomy relevant for space: extracted from Masolo *et al.* (2003)

objects [18, p2]. The first view is termed the *Newtonian*, *Galilean* or *absolutist* view of space, and the second the *Leibnizian* or *relationist* view. This distinction has important implications for how to explicitly model space in a representation, how space might be used for inference, and how it may be talked about during communication. When building an ontology under the Newtonian approach, for example, space may be modelled directly as a category in that ontology. It then enters into a range of relationships with other entities and should be axiomatized accordingly. In contrast, this need not be the case in a Leibnizian ontology, where space is only present indirectly as relations between objects themselves.

The treatment of space and location in DOLCE is to consider location analogously to other 'qualities' that a physical endurant may 'possess', such as color or weight. Qualities are bound very closely to their bearing entities: thus, the color of some particular rose is uniquely *that* rose's color and no other's. Exactly parallel, a rose's location is uniquely the location of that rose. The color and the location of the rose cannot be separated from that rose. Comparison between entities in terms of their qualities is only possible by a further step of relating the quality to ABSTRACT regions that give them values. These regions are 'quality spaces' in the sense of Gärdenfors' *conceptual spaces* [19] and are thus not to be confused with the spatial notion of 'region' used in formalizations of space. Instances of QUALITY are said to "inhere" in their associated hosts, but their values are defined as points (elements) in a corresponding abstract quality space. This supports an ontologically sound understanding of situations described, for example, as "the color of the rose changed from red to brown" where it is clearly not some color, e.g., 'red', that changes but only the intrinsic color that inheres in the rose. That is to say that the position of that intrinsic color with respect to the abstract physical region, or quality space, of color changes. When applied to location, therefore, DOLCE maintains that each physical endurant necessarily has a quality location and this in turn receives a 'value' in terms of the SPACE REGION.

More formally, using the DOLCE axiomatization, particular spatial locations (SL: cf. Figure 2) are themselves unstructured and are kinds of PHYSICAL

QUALITY (PQ). These are related by a QUALITY relation ('qt') to entities organized within the SPACE REGION (S), a subcategory of ABSTRACT (AB) PHYSICAL REGIONS (PR). Significantly, the specification of particular possibilities for structuring the SPACE REGION is not refined further. Entities that are spatially present, e.g., physical endurants, are then bound into space by virtue of what DOLCE terms a *spatial mutual specific dependency* relationship (MSD_S). Specific dependency is defined in terms of mutual disjointness and the necessary existence of a spatial dependency between particulars such that it is necessarily the case that those particulars are present in the same setting and at the same time. Being present simply requires that there be some spatial location but does not restrict further how that might be specified. The generic quality axiom $MSD_S(PQ, PED)$ of DOLCE then states that physical qualities and physical endurants are mutually dependent and the qualities and their objects will necessarily co-locate.

One further spatially relevant subcategory of physical endurant within DOLCE is FEATURE. A FEATURE refers to those tangible, physical characteristics of an object that are 'parasitic' in that they cannot exist without the existence of their hosts (a DOLCE *one-sided generic dependence*: OGD). They are not then distinguishable as 'parts' of an object in the sense that they could be isolated (even potentially) from their wholes and include traditional problem cases such as 'holes' (cf. [20])—as in the holes of donuts—gulfs, openings, boundaries and so on. DOLCE distinguishes two kinds of features:

- *Relevant parts* of entities, such as a bump or an edge;
- *Places* such as 'underneath the table', 'in front of the house', etc.

We consider 'relevant parts' to be related to spatial qualities proper and will pick out 'places' for separate treatment below.

3.4 Extensions to the DOLCE View of Spatial Ontology

We now focus in on the characterization of space and what precisely it may mean for physical objects (or events) to be *located at* particular locations in space. This raises questions about both how the objects concerned and the locations are to be identified. Our goal is to reach a specification that is adequate for an ontological account of robot navigation models and which will support inter-operability with other components.

First, we consider critically along the preliminary lines that we began setting out in [21] the treatment of locations within DOLCE as spatial qualities. It appears to us beneficial to separate clearly between notions of spatial extent and a broader notion of location. Whereas the former is indisputably an inherent quality of a physical object, any physical object necessarily takes up the space that it does and in the particular morphological form that it does, the latter is more complex. A reasonable specification of location, in the sense of where a physical object is located, requires reference to a broader scheme of possible positions. Only when such a scheme has been specified can we talk of the position of an entity at all. In this sense, then, there appears to be a subtle difference between the quality and qualia of color and the notions of location/position. Whereas it is not necessary to fix on a color space in order to acknowledge the existence of a rose's color: the rose would have the color quality regardless, it *is* necessary to fix on some decomposition of space in order to see that rose as 'positioned' at all. We will term such a spatial decomposition a LOCATION SCHEME and it is only through such location schemes that spatial access to entities is provided. Thus, although analogous to a quality space, a location scheme is more like a quale supporting qualities of an entire physical setting rather than for any particular entity.

Whereas particular colors (i.e., color qualities) that may exist (such as the particular red of this particular rose) depend on their bearing objects for their existence, particular positions depend on their locational scheme. A color quality is given a 'value' by a selected color quale, whereas a location/position *only exists* given a selected location scheme. We take this to be the case for both Newtonian/Galilean and Leibnitzian views of space. Any physical object will always be placed within an entire framework of spatial relationships. These relationships may make reference only to other objects (the Leibnizian view) or may rely on pre-structured space as such (the Newtonian/Gallilean view). The precise characterization of the placement depends on the location scheme adopted. Moreover, the position may exist regardless of any object that happens to occupy that position: the existence of the location scheme is sufficient.

Candidates for location schemes within a foundational ontology then include all of the formal accounts of space, of topology, of regions and so on that have been developed in the qualitative spatial reasoning and representation community [22, 12]. When such an account is also axiomatized in a manner compatible with the axiomatization provided for the rest of the foundational ontology, it can then directly constitute a parameterized ontology module in the sense of a submodule mentioned above. The provision of such ontology modules constitutes one of the ongoing research tasks of the Collaborative Research Center on Spatial Cognition within which our own research is situated [23].

To distinguish locations, positions and location schemes from the quality-quale axiomatization in DOLCE, we propose here to adopt the spatial primitives set out by Grenon and Smith in their Basic Formal Ontology (BFO: [24]). This also requires an adjustment to the categories found under physical endurants in DOLCE. Grenon and Smith posit both a purely spatial ontological category, SPATIAL REGION, including points, lines, surfaces, volumes, etc. as necessary (and is therefore Newtonian and absolutist), and a class of endurants that are intrinsically spatial, called SITES. Sites are just like other physical endurants as far as necessarily having a location in space, but also function themselves further as 'spaces' within which other physical objects (BFO: substances, DOLCE: physical endurants) can be. Grenon and Smith list holes, cavities and both real and fiat enclosed spaces as possible sites drawing on the formalization of such entities worked out by Casati and Varzi [20].

Sites are important for robot navigation because most of the objects with which we are concerned there, rooms, corridors and the like, are of this kind rather than simple physical objects. It should be clear that to talk of a 'room' ontologically as a physical object in the same way as a 'chair or a 'table' would be problematic. A room is a 'hole' of a particular kind that is defined by the physical object of its surrounding walls. In our conceptualizations of rooms, however, it is not the walls that are usually prominent—even though it is probably the walls that are more prominent to a robot's sensors. Bridging between these levels of reality is one of the tasks taken up by the modelling we have seen in the SSH and the RG, but it is also fundamentally involved in the specification of an adequate formal ontology. The most recent formalization of the RG then talks quite properly of a relationship between places and ENVIRONMENT SPACES: this maps quite directly to the notion of site as employed here.

The ontological relationships for location posited within the BFO are:

- SpatialLocation (or located at in Smith and Varzi [25]): a functional notion that associates an entity with a unique spatial region (a part of pure space);
- OccupiedBy: a relationship between a site and (one of) its occupants that ensures that the associated spatial regions align properly;
- place-of: a further functional notion associating occupants with their minimal places (functionally identified sites).

Although there are several further subtleties that also belong in a full account, we simply adopt these for present purposes without further discussion.

The notion of pure space we assume is taken to come with the structuring of a location scheme as suggested above. In addition to standard qualitative treatments of space, we also adopt here the notions of variable granularity and *qualitative coordinates* introduced by Bittner and Smith [26, 27]. This gives us the power necessary to talk of varying levels of abstraction such as a 'room' being a point in a network of routes, which can be used to position other rooms, or itself as a place in which movement can be pursued. The 'hole' or 'cavity' that is the room has itself a (parasitic) spatial extent and so can of necessity also serve as a component of a scheme for structuring location that can be utilized for qualitative coordinates. All such objects are then further subject to granularity selections in that a description can pick out positions 'in the room', 'in the corner of the room', 'in the drawer of the desk in the corner of the room', etc. Just as with the case with quality regions and color, the *labels* for the qualities are drawn from the quality region (e.g., 'red' for color and 'in', 'in the corner', etc. for position). The particular location descriptions are similarly drawn from the make-up of a space region as defined by its location scheme. We can also at this point align the DOLCE dependent category of PLACES (a type of FEATURE: cf. Figure 2) with the BFO notion of site. As Grenon and Smith [24] write:

> "A room in a house is a site, as also is a landing strip, a meadow, the interior of your car or of your airline cockpit. The corner of a room is a site, as also is your alimentary tract or the interior of an oil pipeline."

We discuss this further elsewhere [21], where we also offer an illustrative informal example of the combination of location schemes with qualitative coordinates.

4 Towards an Ontological Account of Robot Navigation

In this section we take the final step of relating elements of the two models of robotic navigation described in Section 2 to the ontology framework sketched in the previous section. Our goal is to show that once this is done, we will have placed the individual formalizations within an ontologically broader context that can be used to shape subsequent design decisions more effectively. The deconstruction will also allow us to clarify some of the differences between the models and to consider how they may best be related to other components of a complete system.

4.1 Ontology and the SSH

As we saw above, the design of the SSH already makes reference to various levels whose elements are described, at least superficially, as originating from an ontology. 'Superficially' here means that the ontological levels described make little contact with generic ontological frameworks of the kind introduced above and so have remained isolated.

The first level of representation in the SSH that can be sensibly related to the terms of a spatial ontology is the causal level. First of all in the SSH implementation presented in [1], the entities at the causal level are tied to the ontology assumed by the Situation Calculus. As we saw above, this ontology is rather neutral, consisting of only first-order entities (*view*, *situation*, and *action*), plus second-order predicates for operating over them (*holds*, *result*, *do*). Here, we consider only the first order terms and not the second-order predicates of the Situation Calculus, since the SSH is not strictly dependent on them.

According to the model, views are *observed* in particular situations. Views, then, are meant to correspond to configurations of real-world states, such as, 'the chair being next to the desk' or 'the cat being on the mat'. These are intended as representational abstractions over the direct sensory input. The precise terms in which such richly structured representations can be formed is a difficult issue. Here a specification in terms of the everyday objects and relations of a 'commonsense' ontology would provide a sensible target for such abstractions however they are constructed. This also fills in the possible definitions of situations to include more than just the agent's state captured by position and orientation. Actions would need to be assimilated under the DOLCE class of perdurants, as they unfold in time. Clearly, any specific ontological characterization of a domain in which the robotic agent is to act, such as an office environment, or a car navigation scenario, should be able to feed directly into the specification of views, actions, and situations. This needs to be managed as a case of importing generic ontology modules and not as a specific piece of additional axiomatization specific to the SSH.

Perhaps more revealing is the consideration of the topological ontological level of the SSH. Here we can see that the definitions provided by the SSH are directly spatial: places are zero-dimensional points, sets of points define regions, paths are boundaries of regions and so on. In other words, the SSH topological level already commits to a particular location scheme for decomposing space. Moreover, it takes on the task of providing reasoning about space within that location scheme. This location scheme gives rise to a SPACE REGION which has points, (directed) lines, and regions, plus a granularity operation by which regions can be collapsed to places. There are, however, other alternatives explored within the qualitative spatial reasoning community[3] and it might be considered beneficial to have a more parameterizable approach whereby these alternatives can also be selected, under appropriate conditions, for driving reasoning. Such modularity is also an intended goal of ontological specification.

This is also significant for communication with such robotic agents. It is now known that mismatching conceptualizations between users and robots can seriously impede effective communication [29, 30]. Fixing a particular location scheme means that space is conceptualized only in the way that that location scheme provides. Concretely for the SSH, this appears to involve just the relations: *on*, *order*, *in*, *right-of* and *left-of*. As Tenbrink in this volume shows [31], the necessary variation in natural communication is considerably higher and would be difficult to support by this scheme alone. When a user is conceptualizing the navigational task in other terms, a re-organizational overhead will be inevitable.

4.2 Ontology and the Route Graph

Whereas the original Route Graph showed more of its origins in a concrete structure for guiding mobile robot navigation by maintaining several rather distinct kinds of information, ranging from perceptual inputs to linguistic node-labels, the latest specification described above (Section 2.2 and [3]), presents a far more ontologically rigorous account. Here, we see distinct information types being separated out, each allocated to its own appropriate ontological subdomain. This is precisely what an adherence to ontological engineering principles requires and allows reuse of as much formal organizational structure in those domains as is available, leveraging off more detailed and established accounts of various aspects of spatial (and other) knowledge.

Much of the specific spatial information that we see incorporated within the SSH model is therefore in the RG case 'contracted out'. The RG specification commits to certain ontological classes being present in order to function and these serve additionally to position the RG account within a more general lattice of ontological modules. We illustrate this briefly with respect to the basic connectivity of the RG and then turn to more interesting aspects concerned with the explicit inter-ontology relations that are defined.

As we saw above, the basic organizational structure of Route Graphs is defined as a refinement of generic graph theory with graph morphisms for supporting abstraction. Several current ontologies include such graph theory modules and Krieg-Brückner *et al.* [3] also provide such a formalization drawing on the

[3] Some of these are also applied within the situation calculus approach: cf. Dylla and Moratz in this volume [28].

algebraic specification language CASL (cf. [23]). Nodes and edges of generic graphs are accordingly mapped to places and route segments within the ontology of route graphs. This kind of inter-module import is crucial to ontology design and several schemes have been proposed (cf. [32, 3]).

The connectivity of a RG is therefore achieved in a very different way to that of the SSH. In the latter, we see a direct modelling of places and connections in terms of spatial relationships; in the former, we see an abstract model of connectivity that is quite distinct from a concrete spatial instantiation. This allows for the possibility of providing more detail concerning the route graph places and edges than would be coherent with a strictly spatial interpretation. We can also see that 'places' as such need to be seen as very different in the RG and SSH, despite their superficially similar functional roles. Places in the RG need to define reference systems: that is, they have an internal spatial structure that is not compatible with the basic zero-dimensional spatial notion of a place within the SSH.

Proceeding further, we can now state that the most generic way of allocating a spatial structure to places is to relate reference systems directly to location schemes in the sense introduced above. At present, the only location scheme that appears to be envisaged in the RG specification is that of angular displacement, or bearing—although even here there are a number of possibilities; for example, schemes can vary according to their granularity (e.g., 360 degrees *vs.* first-quarter, second-quarter, ...) or according to their orientation (absolute: north, south, east, west *vs.* intrinsic: forwards-backwards-left-right). Since the route graph is also intended to drive robot motion, the more refined and exact reference systems will probably offer more effective choices here. This is quite different when one considers communication, however, where again it is useful to explore the appropriateness of a variety of location schemes. The RG class place must also therefore import properties of an 'intrinsically oriented region'; a kind of entity that has internal spatial parts that lie in some specifiable spatial arrangement that can be determined appealing to the spatial relations provided by some location scheme, i.e., a SPACE REGION.

This means that there is actually little ontological difference between a place in a route graph and the route graph as a whole: both may be related to some spatial region. And this should not come as a surprise since this is precisely what the RG abstraction relation abstractsTo enforces. A place may stand for an entire route graph: the selection is one of granularity.

We see an analogous situation with the ontological class of SITE that we adopted above from BFO. Sites have an occupant, and that is the agent that is situated within the route graph. At one degree of resolution, then, a route graph might most naturally be related to a site. The route graph defines those places that the occupant of the site can be. The site thus defines a certain functional potential for action (movement/navigation), the precise possibilities for which are set out by the connectivity of the route graph and the properties of the route segments. Sites, within our generic ontology, are physical endurants and are therefore positioned in space (via a location scheme). But the occupants of sites are also physical endurants and so the possibility of recursion is built in.

This alignment may be carried further by considering more closely the relations defined between the Route Graph ontology and other ontological domains that Krieg-Brückner *et al.* introduce and which we summarized in Figure 1 above. These 'glue' relations serve to anchor route graphs both to spatial representations and to everyday commonsense representations of the world. We now align these briefly with the resources provided by our generic upper ontology.

The locatedIn relation assumes that some spatial ontology will at least provide REGIONS. This provides a channel for importing any logical specification of the properties and behavior of regions that an adopted location scheme provides. Regions will group route graph places into spatial neighborhoods via their participation in locatedIn relationships. There is so far no guarantee, however, that the connectivity of the route graph is 'well-behaved' with respect to the hierarchy and connection relations over spatial regions. For example, nodes that are immediately connected in the route graph may be spatially positioned in disjoint spatial regions and more distant route graph nodes may be spatially positioned in the same region. Explicitly imposing 'good-behavior' constraints on the route graph during its construction via the properties of spatial regions may be a way of improving the reliability of the construction process—at least for naturalistic route graphs. Route graph neighborhoods would then align with spatial inclusion relationships among regions.

The ContainedIn relationship assumes that some commonsense ontology will provide ENVIRONMENTSPACES. Such a class is already provided by the generic ontology class SITE. A RG node may then be containedIn a commonsense SITE. As we suggested above, a route graph node is then picking out the possible (from the route graph perspective) positions that a site's occupant may occupy. Note, however, that this is precisely what a location scheme would offer for a site in any case: a way of picking out positions within the spatial region that is the site's location. Under this interpretation, the route graph as such might even be accomodated as a structuring of space alongside other such possible structurings—e.g., traditional spatial calculi such as the Region Connection Calculus (RCC: [33]) or, more directly related to current work with route graphs, dipoles [28]. We might then hypothesize that a route graph is a location scheme that decomposes the space of a site according to the possibilities for movement within that site rather than, for example, according to connection and overlap of regions and subregions. A move that may well also be more in line with embedded cognition approaches to space.

This then raises a precisely analogous situation to the 'good-behavior' constraints for spatial inclusion mentioned for the locatedIn relation. When building up a hierarchical nesting of route graphs, there is no *a priori* guarantee that that nesting will reflect commonsense categories. The fact that some nodes are to be containedIn in a room and some others in a connecting corridor is not a necessary consequence of the connectivity of the route graph.[4] Moreover, as noted in Krieg-Brückner *et al.*, there may even be *differences* between the 'commonsense' decompositions provided by robot and user and, even, between different users. For a cognitive agent that is following a route graph to be able to communicate with its users, therefore, we need both to impose sensible ContainedIn relationships and to negotiate these with the particular commonsense SITE categories that some particular user is adopting. The potential for confusion that this raises for communication within a RG-based model is discussed at greater length in Ross *et al.* in this volume [34].

The covers relationship assumed by the RG specification between the spatial and commonsense ontologies stands in for an entire complex of issues, several of which we have discussed above. We have already seen both the spatial region and environment space entities that this relation relies upon. The relationship itself is then simply that of SPATIALLOCATION between a site and its position. How that positioning is achieved is defined by an adopted location scheme.

We can now state some simple correspondences between the RG specification and the classes and relations available in the generic ontology. We write the SPATIALLOCATION relation as l_{LS}, indicating that we consider location to be always relative to a specified location scheme (LS), and the spatial 'part of' relation as $\subset$. The parameters of the relations are also informally typed, drawing on RG or generic ontology constructs as appropriate.

- $covers(Region : R,\ Site : S) \longleftrightarrow l_{LS}(S) = R$
- $containedIn(Place : P,\ Site : S) \longleftrightarrow l_{LS}(P) \subset l_{LS}(S)$
- $locatedIn(Place : P,\ Region : R) \longleftrightarrow l_{LS}(P) \subset R$
- $marks(Place : P,\ PED : x) \longrightarrow l_{LS}(P) = l_{LS}(x)$
- $faces(Place : P,\ PED : x) \longrightarrow l_{LS}(\textit{in-front-of}\ P) = l_{LS}(x)$[5]

In the case that the adopted location scheme is itself that of a route graph, the left-hand location operators can then be ommitted from the generic ontology statements. This is because the route graph specification is then naturally itself already the 'location'. The correspondence for the marks relation, for example, becomes:

$$marks(Place : P, PED : x) \longrightarrow P = l_{RG}(x)$$

That is, the location of the physical endurant x, with respect to a location scheme that is a route graph, is the place in that route graph that marks x. In general, however, the location scheme will not be a route graph and other schemes will mediate attributions of location.

---

[4] Although the free-space geometry naturally generates hypotheses.

[5] Whereas marks is essentially a simple 'naming' relationship based on identity of position, the faces relation includes instrinsic relative-orientation information, in particular, the 'functional relationship' [35] of being 'in front of'. We include this here without further discussion, although its relationship to the possibilities offered by the location scheme needs to be spelled out in more detail.

Finally, we note that these inter-ontology relations themselves provide a backbone for relationships that may be specialized to express more or less detailed correlations between the two levels of ontological abstraction.

4.3 An Illustration of the Benefits of an Ontological Foundation

We have suggested that it is beneficial to embed robotic navigation models within a more generic framework grounded in a broad area of commonsense and spatial information. One area where this is particularly evident is in the relation between navigation models and higher cognitive functionalities, such as communication. We believe that the ontological placement of the various navigational components provides a much improved foundation for building in sophisticated communication functionality. We illustrate this by example, showing how a simple linguistic utterance concerning navigation requires activation of all of the distinct components of the model. Any reduction in the range of ontological modules employed brings with it an automatic restriction in the range of communicative functionality that can be supported.

The following utterance is a realistic directive given to a robotic system in an office environment such as we are working with:

> "Go to the window and follow the corridor until the last room on the right."

We will assume that speech recognition and grammatical and semantic analysis have been carried out so that what remains is a shallow semantic representation unresolved against context (cf. [36, 34]). There still remain significant problems for a navigation system, however; particularly we will consider the process of mapping between such a shallow semantic representation and concrete actions that can be carried out by a robotic system on the basis of a navigational representation such as the route graph. Note that we envision a situation here where there is an ongoing interaction between user and robot—the environment may not have been completely mapped out and labelled. And even there, it is still possible, perhaps likely, that a user may deviate from that labelling.

First "go to the window": the shallow semantics makes it clear that a movement action is being called for and that the destination of that movement is being labelled *by the user* as belonging appropriately to a semantic type window. The use of the definite article also sets a reference resolution problem, the robotic system with dialogue component must be able to locate a real-world entity (PED) that is considered describable by the semantic type. The particular entity will be revealed either by the discourse context (i.e., the window we have been discussing) or by perception. For the latter, we make use of the additional fact that any such utterance must in general be seen as selecting a certain 'ontological granularity'. In this case, the use of the semantic type 'window' selects a subdomain of a commonsense ontology concerning SITES with bona fide boundaries, such as walls, doors, windows. The task of the navigation system using a RG is then to locate a suitable node marked as corresponding to a PED 'window' or at least facing such an entity.

If the nodes of the RG have already been partitioned according to containedIn relationships, then the search for a node can be restricted to a neighborhood defined by an appropriate site. Note that no assumption of the kind that the node corresponding to the window (or facing or marking the window) is an immediate neighbor of the current position holds. The site may itself be a complex arrangement with its own internal route graph structure. If the nodes of the RG have not been so partitioned, then the RG-reasoner will need to follow edges until candidate nodes have been found.

There is also, however, the very real possibility that the user and the robot *disagree* about how exactly a specified physical object is to be described. What for the user may be a window could for the robot be labelled as a 'glass door'. For the robot to resolve this problem gracefully, it is necessary to invoke the commonsense ontological information that both windows and glass doors can form (parts of) boundaries of certain sites and that there is a certain confusion likelihood because of the similarity in the material of the entities. This kind of 'flexible' reference is only possible when substantial real-world knowledge is available; and a link to a commonsense ontology provides just this. Note that this problem can arise *whenever* a semantic type is used: the assumption that such types can be unproblematically resolved by appropriate linguistic labelling of a navigation graph is unfounded. Given sufficient uncertainty, the robot can also engage in precisely focused clarificatory dialogue, for example "do you mean the glass door over there?"

Assuming that the navigation system has located a node in the route graph window that both the robot and the user agree is the 'window', we come to the next component of the directive: "follow the corridor". As before, this may involve confusions concerning 'what is a corridor' while the utterance itself selects a certain ontological granularity. Here we are dealing with SITES such as rooms, corridors, lifts, and so on. The RG nodes containedIn the corridor define the search space for the subsequence search for the 'last room on the right'. In general, however, the precise set of nodes belonging to this corridor might not yet be clear; this can then lead to targetted clarification dialogues such as "are we still in the corridor?"

We can also employ the connection described above between, on the one hand, the potentially recursive structure of SITES and, on the other, route graphs and nodes related by the RG abstractsTo relationship. If we define a relation $\text{containedIn}_{collective}$ that relates the collection of all the RG nodes containedIn a given SITE to that SITE, then this is equivalent to a composition of abstractsTo and marks. Thus, when the linguistic utterance selects a granularity of SITE appropriate for corridors, this also corresponds within the RG to stating that there is some set of nodes that stand in an abstractsTo relation to a more 'abstract' node that may be marked by the site 'corridor'. A corridor will also bring with it from the commonsense ontology attributes that can be used to constrain appropriate collections of RG nodes that are to be grouped: for example, that it is essentially a path with exits.

Finally, assuming that the navigation system has found a collection of RG nodes that user and robot agree can be called a 'corridor' (which automatically allows a set of containedIn relations, an abstractsTo relation, as well as a marks relation to be recorded if not already present), the RG reasoner can be given the task of locating a sequence of nodes of some type that are all on the 'right' hand side of the path through the corridor. Here there needs to be explicit communication with the spatial ontology and some particular location scheme that supports interpretations of 'left'/'right' and such *intrinsic* references. Moreover, the problem of potential confusion occurs here as always: the semantic type 'room' used in the last component of the directive may refer for the user to a very different set of potential places than it does for the robot; more discussion of this particular problem is given in the description of our dialogue system presented by Ross *et al.* [34].

Without the provision of the commonsense ontology, suggesting possible candidates for confusion and fixing appropriate granularities, the spatial ontology, for determining spatial relationships, and the route graph ontology itself for handling navigation, flexible communication and, above all, natural resolutions of communicative problems during that communication, would not be conceivable.

5 Conclusion

We have seen that most of the 'knowledge-level' issues that are involved in robotic modelling and which have been considered within the SSH and RG models also have their correlates in a thorough formal ontological modelling of space and the possibilities for movement that space entails. Explicitly building into a model possibilities for importing and exporting the information necessary should provide for significantly improved development. The graph-like nature of a navigation graph, for example, can be modelled directly by importing a generic ontology of graphs and graph morphisms. Also, the explicit link between navigation graphs and spatial regions provides access to calculi for reasoning about the various connection relations, etc.

A further benefit is that with such ontologies in place, the relation of further components, for example, those of natural language, to the various ontological levels of robot navigation is clarified. Kuipers suggests that verbal route directions correspond naturally to the causal level of the SSH, i.e., as sequences of imperative corresponding to actions [1, p228]; our discussion of the previous section should have made clear how far this is away from the flexibility required for genuinely natural interaction and its possibilities for misunderstanding and self-correction. The RG account, while not itself providing a model of natural language interaction, comes closer in that the interfaces between the ontological modules are identified and defined.

We can therefore place approaches to robot navigation along a continuum ranging across: (i) no use of an ontological foundation, where an account has to provide its own knowledge modelling and reasoning capabilities from scratch; through (ii) the adoption of ontological modules as a design principle, although the contents of these modules are also provided from scratch (cf. the SSH); to (iii) an ontologically aware design that decomposes the problem across distinct modules, only some of which need to be developed for navigation alone (cf. the RG). We have motivated the utility of adopting the third option and are currently developing our own account of human-robot interaction on this basis [30, 34].

References

1. Kuipers, B.: The spatial semantic hierarchy. Artificial Intelligence **19** (2000) 191–233
2. Werner, S., Krieg-Brückner, B., Herrmann, T.: Modelling navigational knowledge by route graphs. In Freksa, C., Brauer, W., Habel, C., Wender, K., eds.: Spatial Cognition II - Integrating Abstract Theories, Empirical Studies, Formal Methods, and Practical Applications. Springer, Berlin (2000) 295–316
3. Krieg-Brückner, B., Frese, U., Lüttich, K., Mandel, C., Mossakowski, T., Ross, R.: Specification of an Ontology for Route Graphs. In: this volume. (2004)
4. Masolo, C., Borgo, S., Gangemi, A., Guarino, N., Oltramari, A.: Ontologies library (final). WonderWeb Deliverable D18, ISTC-CNR, Padova, Italy (2003)
5. Kuipers, B., Levitt, T.: Navigation and mapping in a large-scale space. AI Magazine **9** (1988) 25–43 Reprinted in *Advances in Spatial Reasoning*, Volume 2, Sushing Chen (Ed.), Norwood NJ: Ablex Publishing, 1990.
6. Kuipers, B.: A hierarchy of qualitative representations for space. In: Proceedings of the 10th International Workshop on Qualitative Reasoning about Physical Systems, Menlo Park, CA, AAAI Press (1996) 113–120 AAAI Technical Report WS-96-01.
7. McCarthy, J.: Situations, actions and causal laws. In Minsky, M., ed.: Semantic Information Processing. MIT Press (1968) 410–417
8. Reiter, R.: Knowledge in Action: Logical Foundations for Describing and Implementing Dynamical Systems. MIT Press, Bradford Books, Cambridge, MA (2001)
9. Krieg-Brückner, B., Röfer, T., Carmesin, H.O., Müller, R.: A taxonomy of spatial knowledge for navigation and its application to the bremen autonomous wheelchair. In Freksa, C., Habel, C., Wender, K., eds.: Spatial Cognition I - An interdisciplinary approach to representing and processing spatial knowledge. Springer, Berlin (1998) 373–397
10. Röfer, T., Lankenau, A.: Route-based robot navigation. Knstliche Intelligenz (2002) Themenheft Spatial Cognition.
11. Wallgrün, J.O.: Autonomous construction of hierarchical Voronoi-based route graph representations. In: this volume. (2004)
12. Bateman, J., Farrar, S.: Spatial ontology baseline. SFB/TR8 internal report I1-[OntoSpace]: D2, Collaborative Research Center for Spatial Cognition, University of Bremen, Germany (2004)
13. Guarino, N.: Formal ontology, conceptual analysis and knowledge representation. International Journal of Human-Computer Studies **43** (1995) 625–640
14. Guarino, N., Welty, C.: An overview of OntoClean. In Staab, S., Studer, R., eds.: Handbook on Ontologies. Springer-Verlag, Heidelberg and Berlin (2004)
15. Guarino, N., Welty, C.: Evaluating ontological decisions with OntoClean. Communications of the ACM **45** (2002) 61–65
16. Hayes, P.J.: The naive physics manifesto. In Michie, D., ed.: Expert systems in the microelectronic age. Edinburgh University Press, Edinburgh, Scotland (1979)

17. Farrar, S., Bateman, J.: General ontology baseline. SFB/TR8 internal report I1-[OntoSpace]: D2, Collaborative Research Center for Spatial Cognition, University of Bremen, Germany (2004)
18. Borgo, S., Guarino, N., Masolo, C.: A pointless theory of space beased on strong congruence and connection. In Aiello, L., Doyle, J., eds.: Principles of knowledge representation and reasoning, Proceedings of the 5th International Conference KR96, San Mateo, CA, Morgan Kaufmann (1996)
19. Gärdenfors, P.: Conceptual spaces: the geometry of thought. MIT Press, Cambridge, MA. (2000)
20. Casati, R., Varzi, A.C.: Holes and other superficialities. MIT Press (Bradford Books), Cambridge, MA and London (1994)
21. Bateman, J.A., Farrar, S.: Towards a generic foundation for spatial ontology. In: Proceedings of the International Conference on Formal Ontology in Information Systems (FOIS-2004), Trento, Italy (2004)
22. Cohn, A., Hazariki, S.: Qualitative spatial representation and reasoning: an overview. Fundamenta Informaticae **43** (2001) 2–32
23. Lüttich, K., Mossakowski, T.: Specification of ontologies in CASL. In: Proceedings of the International Conference on Formal Ontology in Information Systems (FOIS-2004), Trento, Italy (2004)
24. Grenon, P., Smith, B.: SNAP and SPAN: Prolegomenon to geodynamic ontology. In: Spatial Cognition and Computation. (forthcoming)
25. Smith, B., Varzi, A.C.: The Niche. Noûs **33** (1999) 214–238
26. Bittner, T., Smith, B.: A taxonomy of granular partitions. In: Proceedings of the Conference on Spatial Information Theory - COSIT 2001. Lecture Notes in Computer Science. Springer-Verlag, Berlin-Heidelberg (2001) 28–43
27. Bittner, T.: Granularity in reference to spatio-temporal location and relations. In: Proceedings of the 15th Florida Artificial Intelligence Research Symposium (FLAIRS). (2002)
28. Dylla, F., Moratz, R.: Exploiting qualitative spatial neighborhoods in the situation calculus. In: this volume. (2004)
29. Moratz, R., Fischer, K., Tenbrink, T.: Cognitive modelling of spatial reference for human-robot interaction. International Journal On Artificial Intelligence Tools **10** (2001)
30. Moratz, R., Tenbrink, T., Bateman, J., Fischer, K.: Spatial knowledge representation for human-robot interaction. In Freksa, C., Brauer, W., Habel, C., Wender, K., eds.: Spatial Cognition III. Springer, Berlin, Heidelberg (2002)
31. Tenbrink, T.: Identifying objects on the basis of spatial contrast: an empirical study. In: this volume. (2004)
32. Farquhar, A., Fikes, R., Rice, J.: The Ontolingua server: a tool for collaborative ontology construction. In: Proceedings of the 10th Banff Knowledge Acquisition for Knowledge Based Systems Workshop (KAW95), Banff, Canada (1996)
33. Randell, D., Cui, Z., Cohn, A.: A spatial logic based on regions and connection. In: Proceedings of the 3rd. International Conference on Knowledge Representation and Reasoning, San Mateo, Morgan Kaufmann (1992) 165–176
34. Ross, R., Shi, H., Vierhuff, T., Krieg-Brückner, B., Bateman, J.: Towards dialogue based shared control of navigating robots. In: this volume. (2004)
35. Coventry, K.R., Garrod, S.C.: Saying, seeing and acting. The psychological semantics of spatial prepositions. Essays in Cognitive Psychology series. Psychology Press (2004)
36. Alshawi, H., ed.: The Core Language Engine. MIT Press, Cambridge, Massachusetts (1992)

Specification of an Ontology for Route Graphs*

Bernd Krieg-Brückner, Udo Frese, Klaus Lüttich, Christian Mandel,
Till Mossakowski, and Robert J. Ross

FB3 Mathematik und Informatik, Universität Bremen,
Postfach 330 440, D-28334 Bremen
bkb@informatik.uni-bremen.de

Abstract. This paper describes the general concept of Route Graphs, to be used for navigation by various agents in a variety of scenarios. We introduce the concept of an ontology and describe the modelling of general graphs as an example. This approach is then applied to define a "light-weight" ontology of Route Graphs in an indoors environment, giving at first just a taxonomy of (sub)classes and relations between them, as well as to other (spatial) ontologies. Finally, we show how to formalise ontologies using a First Order Logic approach, and give an outline of how to develop actual data structures and algorithms for Route Graphs.

1 Introduction

Route Graphs have been introduced as a general concept ([1]), to be used for navigation by various agents in a variety of scenarios such as humans using railway, underground, road or street networks, as travellers, car drivers or pedestrians, resp. Each application scenario or kind of Route Graph will introduce special attributes on the general structure in a hierarchy of refinements.

While the concept was originally introduced to mediate terminology between artificial intelligence and psychology in spatial cognition, scenarios are not restricted to human users. Route Graphs are also intended for interaction between service robots, such as the Bremen autonomous wheelchair Rolland (cf. Fig. 1), and their users, as well as between robots, for example as a compact data structure for on-line communication in an exploration scenario. Moreover, a bare-bones Route Graph representation is easily constructed from pre-existing map-like representations (at least for floor plans of buildings) or by robot exploration, and we hope that cognitively (more) adequate maps can be constructed from it when the semantic interrelation can be taken into account.

As we are dealing with abstract concepts and interrelations between them and want to standardise or mediate between different uses of terminology, we are using an ontology as *the* central definitional approach and data structure for Route Graphs. We intend to show that an ontological approach is suitable

* This paper has been supported by the Deutsche Forschungsgemeinschaft (DFG) in the SFB/TR 8 Spatial Cognition.

C. Freksa et al. (Eds.): Spatial Cognition IV, LNAI 3343, pp. 390–412, 2005.
© Springer-Verlag Berlin Heidelberg 2005

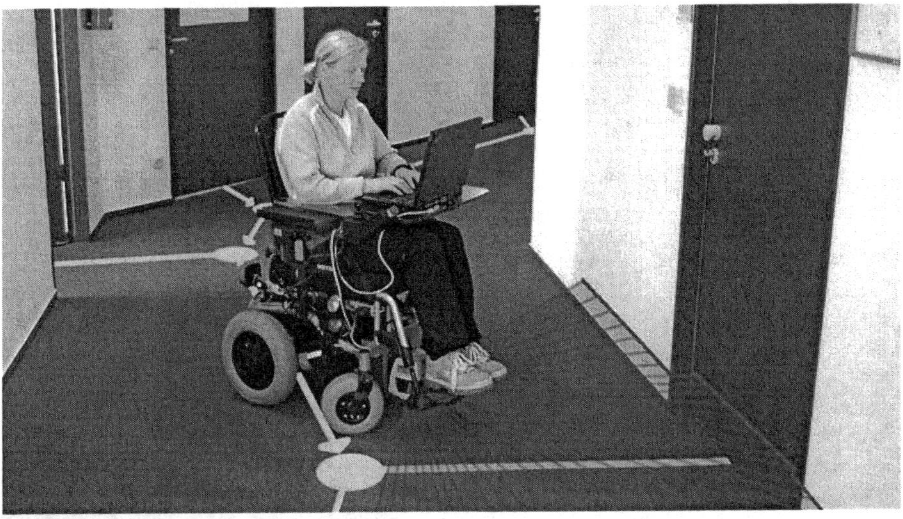

Fig. 1. Rolland following a Route Graph

for both: to define the generic concept of Route Graphs and to instantiate it for a particular scenario in detail, leading eventually to a concrete data structure. Moreover, the example of Route Graphs is suitable for demonstrating that adequate formalisation of ontologies can be introduced step by step in this process.

In this paper, we will show the application to an indoors scenario for Rolland. Thus we have to model detailed information for navigation at the "robot level" as well as more abstract concepts of space at the "user level", and the relationship between these and "common sense" concepts in the environment, such as rooms, doors or windows as route marks (cf. [2]). For a dialogue between user and wheelchair, we want to relate the concepts of Route Graphs to linguistic ontologies; this is described e.g. in [3, 4].

The paper is structured as follows: We first sketch the general concept of Route Graphs in Sect. 2. In Sect. 3 we introduce the concept of an ontology and describe the modelling of general graphs as an example. This approach is then applied to define a "light-weight" ontology of Route Graphs in an indoors environment, giving at first just a taxonomy of (sub)classes and relations between them, as well as to other (spatial) ontologies, in Sect.4. Finally, we show in Sect.5 how to formalise ontologies using a First Order Logic approach, and give an outline of how to develop actual data structures for Route Graphs.

2 Route Graphs

2.1 Sample Scenarios

Figure 2 shows some sample navigation scenarios. We may distinguish between navigation in systems of passages (e.g. road networks or corridors) or in areas of open space (e.g. on a lake; on a market place, surrounded by buildings; in a hall);

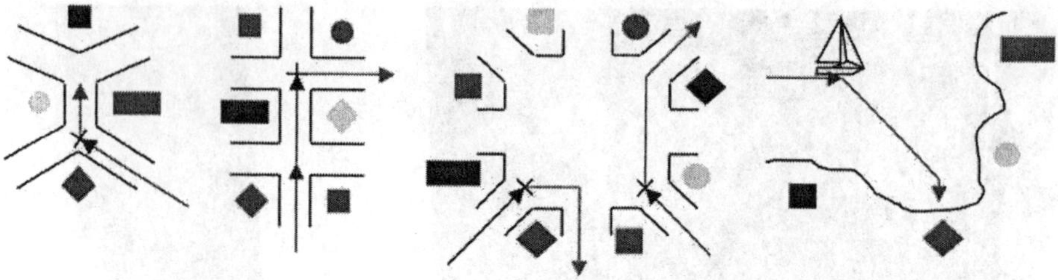

Fig. 2. Examples of routes

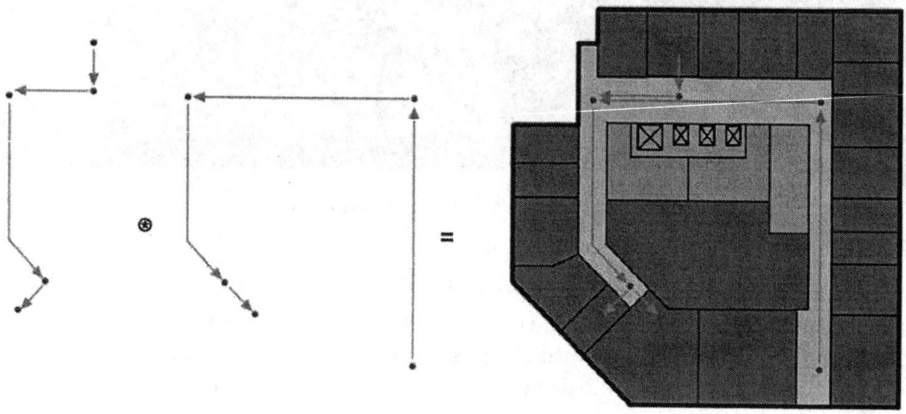

Fig. 3. Two routes and their union

in the former, our course is more or less centred within enclosing borders (e.g. curbstones, walls) and guided by routemarks along the way, in the latter the course is given by a vector to the target and we are guided by global landmarks (e.g. a lighthouse, the sun or a church's spire), cf. [5].

While *Route Graphs* should apply to all such scenarios (cf. [1]), we will concentrate on the indoors scenario here, for Rolland and its user as in Fig. 1, as an example for service robot applications. We briefly introduce the basic concepts of Route Graphs here; more detail will follow in Sect. 3.2 and Sect. 4.

Sample Route: to the Secretary's Office. Consider Fig. 3: two separate routes are united into a simple Route Graph. Let us take the first route as an example. It can be described by directions in natural language as follows:

- Leave Room MZH 8210 into the corridor
- Turn right, go straight ahead to the window
- Turn left, follow the corridor until the last door on the right-hand-side
- Face it, enter Room MZH 8080

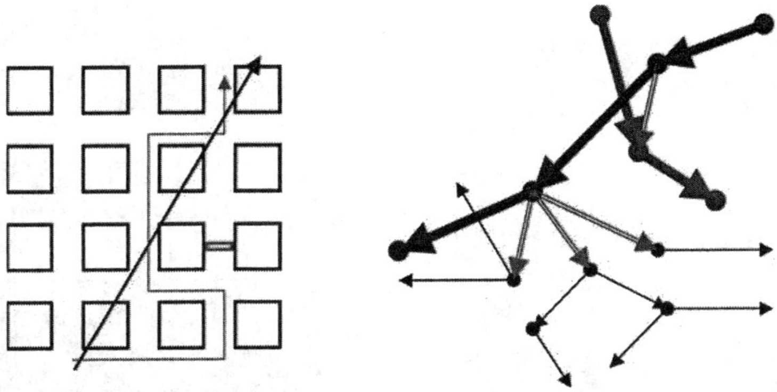

Fig. 4. Layers and transfers

Sample Route Segments. The first two lines can then be translated into the following two route segments (taking additional information about doors, lifts, windows into account, see Sect. 4.4):

Source: room MZH 8210
Entry: turn towards the door
Course: go through the door
Exit: [turn to face the lift]
Target: corridor, facing the lift

Source: corridor, facing the lift
Entry: turn right
Course: follow corridor to the window
Exit: [turn to face the window]
Target: T-crossing, facing the window

Segments. An edge of a Route Graph is directed from a source node to a target node. We call an edge a *(route) segment*; it always has three additional attributes: an entry, a course and an exit. Exactly what information is associated with these attributes is specially defined for each Route Graph instantiation of a particular kind. For example, an entry to a highway may denote a particular ramp, an exit another, while the course is just characterised by a certain length. Additional information may be added, e.g. that the course has three lanes. As another example, the entry and course for a boat route segment may be given as a vector in geo-coordinates, while the exit into a harbour may specify a particular orientation along a quay.

Places. A node of a Route Graph, called a *place*, has a particular position and orientation, its *origin*. Thus each node has its own "reference system"; it may, but need not, be rooted in a (more) global reference system, such as a 3D geo-system. The origin becomes particularly important in a union of routes or Route Graphs, when place integration is required (see Sect. 4.3).

2.2 Homogeneous Route Graphs, Transfers

Fig. 4 shows, on the right hand side, an example where various Route Graphs have been united to one heterogeneous Route Graph. We say a Route Graph of a particular kind *is homogeneous*, if all its segments are of the same kind. In the

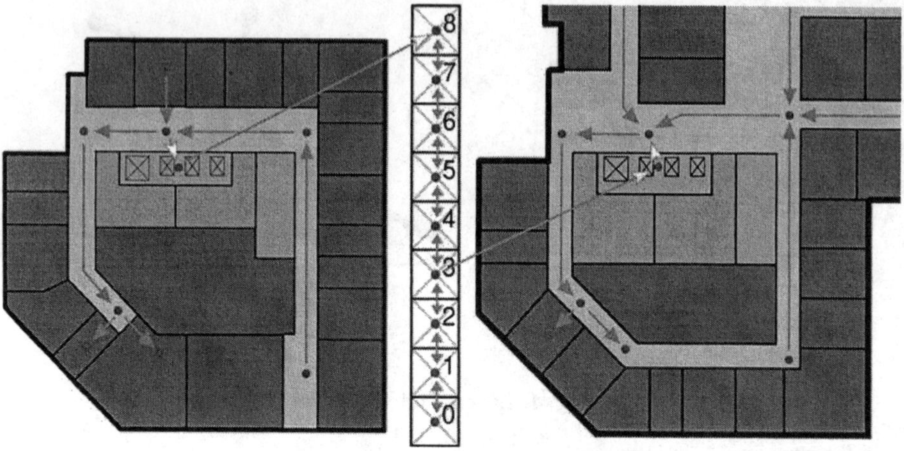

Fig. 5. Two levels and lift

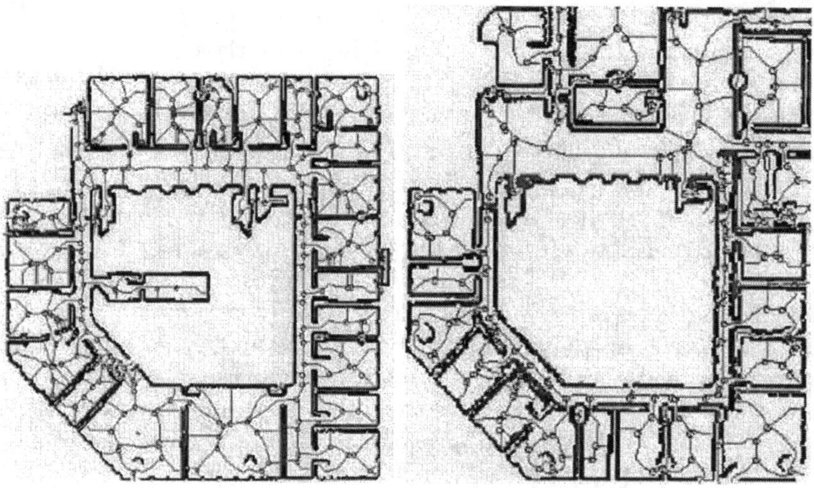

Fig. 6. Voronoi representation for levels 8 and 3

example, the routes depicted by fat arrows might correspond to underground train lines, those with fine arrows to a network of pedestrian passages; both are homogeneous. There is then a need to introduce *transfer segments* for connection, or indeed *transfer Route Graphs* that have transfer segments at their fringes. In the example they would correspond to a pedestrian transfer passage between two underground stations, or several exits from an underground station to the pedestrian network above (note that these might be connected to other exit or entry routes underground). Fig. 5 shows a transfer between a Route Graph at level 8 of our office building, the lift system, and level 3.

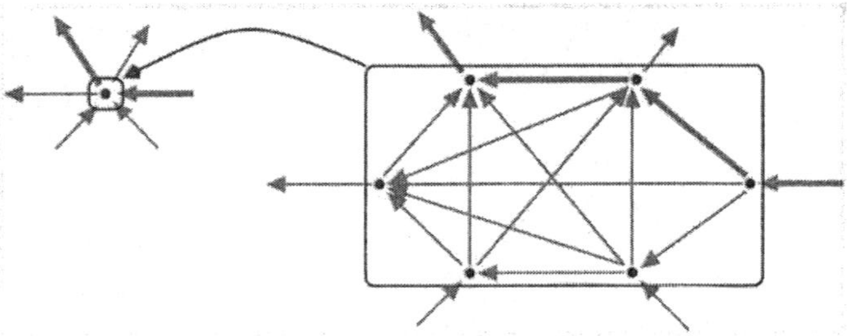

Fig. 7. Abstraction to place

2.3 Layers and Abstraction

We may wish to separate Route Graphs into *layer*s at different levels for conceptual reasons, possibly unconnected. One reason might be that we want to represent route and overview knowledge (cf. [1]). The left hand side of Fig. 4 sketches a route in an urban block scenario, and the target designated by an arrow. When circumnavigating a road blockage, for example, we are likely as humans to use such an arrow as a "sense of direction" while negotiating our way in the block maze – we are using and combining two layers of knowledge at the same time; it is well known that other animals are much better in this respect.

Abstraction. Another reason is abstraction from a more concrete and detailed lower layer to a higher one, as in the abstraction from the robot level to the user level (see also Sect. 4.1). Fig. 5 refers to the user level, while Fig. 6 shows a Voronoi diagram as a representation of space at the robot level (cf. also Sect. 4.2) which corresponds directly to a detailed Route Graph, if we replace the trajectories between places by straight directed edges, in both directions.

What is required here is a relation `abstractsTo` between graphs (in fact a graph morphism), more specifically from a route to a segment, or a graph to a node as in Fig. 7 (cf. also Fig. 16). When a set of nodes SN on the fringe of the graph G at the lower layer is abstracted to a single node N, some obvious conditions must hold, e.g. all incoming/outgoing edges to a node in SN must correspond to incoming/outgoing edges for N; for each pair of incoming and outgoing edges for N there must be a corresponding connecting path in G; etc.

3 Modelling Via Ontologies

Ontologies provide the means for establishing a semantic structure. An ontology is a formal explicit description of concepts in a domain of discourse [6]. Ontologies are becoming increasingly important because they also provide the critical semantic foundations for many rapidly expanding technologies such as software agents, e-commerce, or the "Semantic Web" [7]. In artificial intelligence, ontologies have been used for knowledge representation ("knowledge engineer-

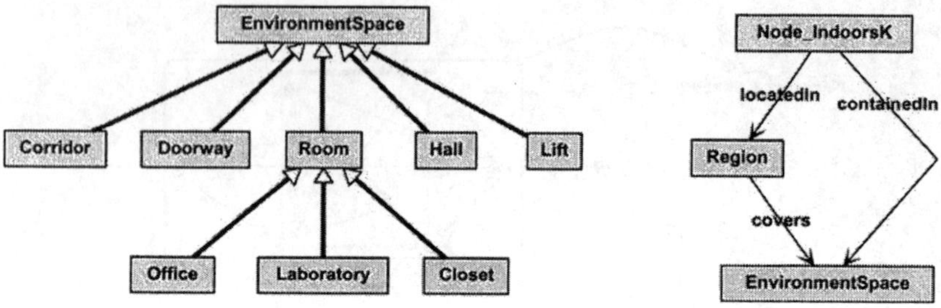

Fig. 8. Subclasses of `EnvironmentSpace` and relations

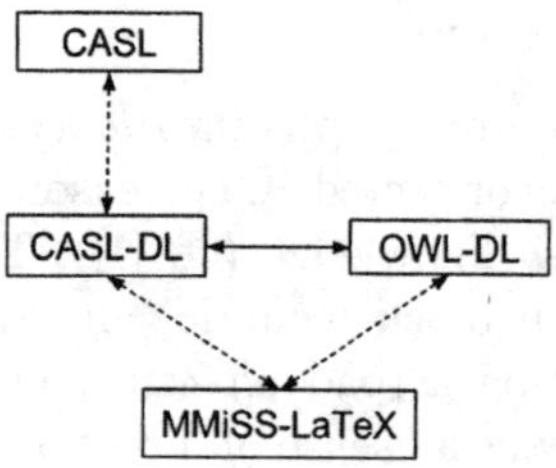

Fig. 9. Various ontology specification formats

ing"). The general idea is to make knowledge explicit by expressing concepts and their relationships formally with the help of mathematical logics.

An *ontology* consists of (a hierarchy of) concepts and relations between these concepts, describing their various features and attributes. Classes and relations are used for defining the abstract semantical level; they provide a vocabulary and properties to characterise the concrete entities corresponding to these concepts. Once the abstract notions are declared in terms of classes, objects can be used to denote entities of semantic concepts. As a simple example for an ontology consider Fig.8. On the left, it depicts a hierarchy of subclasses of the *class* `EnvironmentSpace` – an extract of a taxonomy as it might appear in a "Common Sense Ontology" of space[1], cf. [2]. The relation *is a subclass of* (or "*is a*") is depicted by a fat arrow with a hollow tip. For example, an `Office` is a `Room` which is in turn an `EnvironmentSpace`.

A class represents a set of *objects*; for example `Office8210` is an object (depicted with leading and trailing underscores in the diagrams below) of class `Office`, or `Office8210: Office`.

On the right, we see declarations of *relations* depicted by pointed arrows, with classes as domains and co-domains, to be formalised (see Sect.5) and used eventually for the definition of relations between objects. *containedIn*, for example, relates the class `Node_IndoorsK` of our RouteGraph ontology (to be de-

---

[1] We refer to different ontologies separately here although they are in fact parts of one combined ontology; for the structuring of ontologies see Sect.4.8.

fined below) to `EnvironmentSpace` in the "Common Sense Ontology"; moreover, `Node_IndoorsKs` in the RouteGraph ontology are related by *locatedIn* to (a separate ontology for) a calculus of `Regions`. Thus we should eventually be able to deduce, given
`Place8210: Node_IndoorsK, Region8210: Region,` and `Office8210: Office,` then
`Place8210 locatedIn Region8210` and `Region8210 covers Office8210` implies `Place8210 containedIn Office8210`

3.1 Various Ontology Specification Languages

Description Logic in OWL. Ontologies may come in various specification formats, cf. Fig.9 (or [8]). We adhere to the OWL standard [9]. Its description logic DL has been primarily defined to be decidable for the Semantic Web.

Lightweight Ontologies in LaTeX. For lightweight ontologies, just (sub)classes, objects and relations as in the example above, we use a special LaTeX format that can be translated to OWL and vice-versa (a tool [10], based on the generic graph visualisation tool daVinci [11], supports incremental presentation of and navigation in such graphs, as in the examples shown in this paper). It allows the specification of ontologies for documents to enable their semantic interrelation, enabling much more advanced document management facilities ([12, 13]).

First Order Logic in CASL. In contrast, Casl, the Common Algebraic Specification Language approved by IFIP WG1.3 ([14, 15]), covers full First Order Logic, plus predicates, partial and total functions, and subsorting; more-

MMiSSLaTeX

```
\Class{Graph}{graph}{}
      \RelType{hasNode}{Graph}{Node}
      \RelType{hasEdge}{Graph}{Edge}
\Class{Node}{node}{}
\Class{Edge}{edge}{}
      \RelType{hasSource}{Edge}{Node}
      \RelType{hasTarget}{Edge}{Node}

      \Relation{*-*}{hasNode}{has a node}{}
      \Relation{*-*}{hasEdge}{has an edge}{}
      \Relation{->}{hasSource}{has the source}{}
      \Relation{->}{hasTarget}{has the target}{}

\Class{Sequence_Edge}{\Ref{Edge} list}{}
 \Class{Path}{path}{Sequence_Edge}
  \Class{Route}{route}{Path}
```

Fig. 10. Path and route

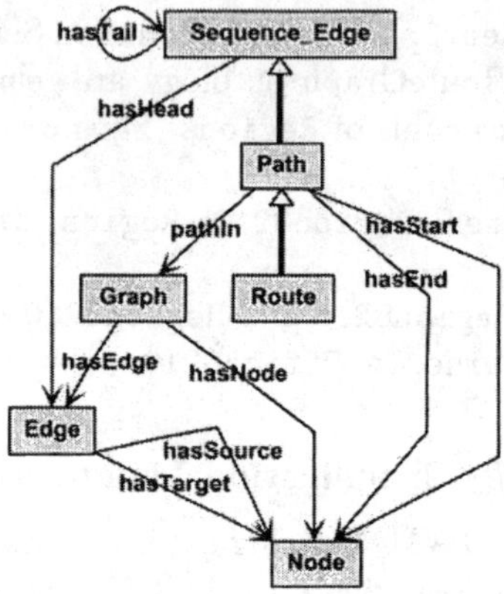

Fig. 11. Simple graph ontology (extract)

over, features for structuring, or "specification in the large", have been included (cf. Sect.4.8). Thus full formalisation of ontologies becomes possible, (as in the DOLCE framework [16, 17]), needed eventually here, see Sect.5. A sublanguage, CASL-DL, has been defined to correspond to OWL DL [18] such that the mappings/embeddings of Fig.9 between the various representations can be realized.

3.2 Ontology of Graphs

As an example, consider a simple ontology of graphs. Fig.10 shows the core in MMiSSLaTeX: A `Graph` has (one ore more) `Nodes` and `Edges`. The MMiSSLaTeX operation `Class` declares these classes: the first parameter denotes the semantics term, the second a default textual phrase (for use in LaTeX documents), the third the superclass.The operation `Relation` is analogous, but contains as an additional first parameter an indication of the kind of relation, e.g. "*-*" to denote "many to many", "->" to denote "onto" (a function), or "<" to denote "strict partial order". With the operation `RelType` relations may be "typed" by source and target classes to allow (static) checking of conformance when objects are related. Thus each `Edge` has exactly one source and one target `Node`.

Note that, at the level of an ontology specification, we do formalise differently than in mathematics or a classical modeling in a specification or programming language: in the latter case, we would introduce a graph as a pair for a set of nodes and edges; here, a `Graph` contains both sets – the abstraction from the pair is perhaps even more natural and definitely adequate as a first go. We show in Sect.5.1 how a more "data structure" oriented design may come back in.

3.3 Paths and Routes

The last lines of Fig.10 show an extension of the basic graph specification. A `Path` is (a subclass of) a `Sequence` of `Edges` (`Sequence` is a basic concept with additional relations, instantiated here), where the target of the first edge is the source of the second, and so on. Note that a Path corresponds to some graph traversal, possibly with cycles. In contrast, a Route contains no repetition of Edges. Such properties will have to be formally specified by additional axioms on the subclasses, successively refined from Sequence_Edge over Path to Route, cf. Sect.5.1. The visualisation in Fig.11 shows a few more relations, sometimes with multiple relation arrows, e.g. hasSource and hasTarget from Edge to Node.

4 A Route Graph Ontology for an Indoors Scenario

4.1 Instantiation to Particular Route Graphs

We assume that the general properties of (labelled) directed graphs are specified along the outline above; similarly, abstract route planning algorithms could be specified at this level without knowing (much) more about Nodes or Edges.

Indoors Instantiation and Kinds. We instantiate this general graph ontology here for an indoors application of Route Graphs, see Fig.13 (cf. also Sect.4.8 for the instantiation aspect). Thus Graph becomes Graph_IndoorsK, Node becomes Node_IndoorsK, and so on. KindRG denotes the kind of Route Graph, where IndoorsK may be refined later on as shown in Fig.12; other kinds might denote RailwayK, UndergroundK or RoadK Route Graphs, for example.

User Level and Robot Level. Our IndoorsK instantiation refers to the (indoors) operating environment of a robot. Note that we do a particular modelling for Route Graphs here which is separate from, but related to, the "Common Sense Ontology" for EnvironmentSpace in Fig.8; we have to expect that they are structured differently. As we shall see below, a Graph_IndoorsK may refer to a rather detailed description at the level of a robot, e.g. the wheelchair Rolland, or its abstraction at the level of an operator, e.g. a Rolland user. User and robot

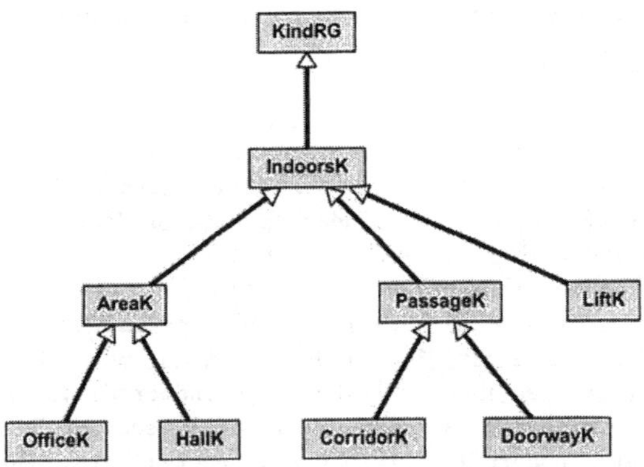

Fig. 12. Ontology of KindRG

interact at this abstract level, while robots may interact among themselves at the detailed level, for example during a multi-robot exploration phase.

4.2 Edges and Route Segments

With the transition from graphs to Route Graphs, an `Edge_IndoorsK` (with source and target) is refined to a `Segment_IndoorsK` (see Fig.14), with an additional entry, course and exit, see Fig.15. Each `KindRG` of Route Graph will introduce its special aspects, in particular (relations to) components.

Vectors and Voronoi Diagrams. In the specialisation of `IndoorsK` Route Graphs, a segment is essentially modelled as a vector in polar coordinates: the entry gives the angle from (the origin of) the source place, the course the distance to the target , and the exit the angle to the target, to assume the orientation of (the origin of) the target place (cf. Sect.4.3). In addition, we attach information derived from a Voronoi diagram representation (see also [19]): each point has a maximal circle of available free space around it, represented by its diameter or `Width`. Thus each `Place_IndoorsK` has a `Width`, and the `Width` of a `Course_IndoorsK` denotes the minimal width of the points along its course.

Abstraction of Segments. Fig.16 shows a Voronoi diagram at the bottom corresponding to a detailed Route Graph for the route in the corridor which has been abstracted to a single segment at the top, showing the `CorridorWidth`.

Consider also the example in Fig.17, corresponding to that in Fig.3 and given verbally in Sect.2.1: `Segment1` represents a passage through a doorway, of `DoorWidth`; `SegmentCorridor2` represents (part of) the corridor, of `CorridorWidth`. The entry and exit angles are depicted by fat little (angular) arrows. Fig.15 and Fig.18 show the ontology of segments and the objects related to the particular segment `SegmentCorridor2`. Note that the entry angle and the distance of the course denote the *direct* vector to the target place thus defining a spatial relation between the two nodes; the actual Voronoi trajectory (cf. e.g. `Segment1`) and its length (which may be larger than the distance) are not represented here; they could be added to the course as additional information, but are

MMiSSLaTeX

```
\Class{Graph_IndoorsK}{\Ref{IndoorsK} graph}{}
      \RelType{hasNode}{Graph_IndoorsK}{Node_IndoorsK}
      \RelType{hasEdge}{Graph_IndoorsK}{Edge_IndoorsK}
\Class{Node_IndoorsK}{\Ref{IndoorsK} node}{}
\Class{Edge_IndoorsK}{\Ref{IndoorsK} edge}{}
      \RelType{hasSource}{Edge_IndoorsK}{Node_IndoorsK}
      \RelType{hasTarget}{Edge_IndoorsK}{Node_IndoorsK}
\Class{Sequence_Edge_IndoorsK}{\Ref{Edge_IndoorsK} list}{}
 \Class{Path_IndoorsK}{\Ref{IndoorsK} path}{Sequence_Edge_IndoorsK}
  \Class{Route_IndoorsK}{\Ref{IndoorsK} route}{Path_IndoorsK}
```

Fig. 13. Indoors Graph ontology

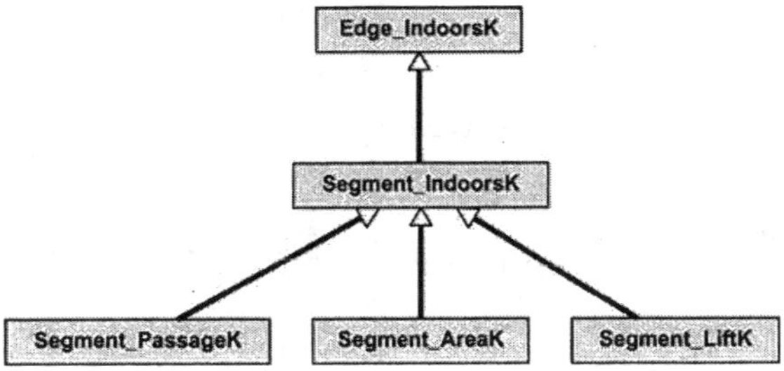

Fig. 14. Ontology of indoors edge subclasses

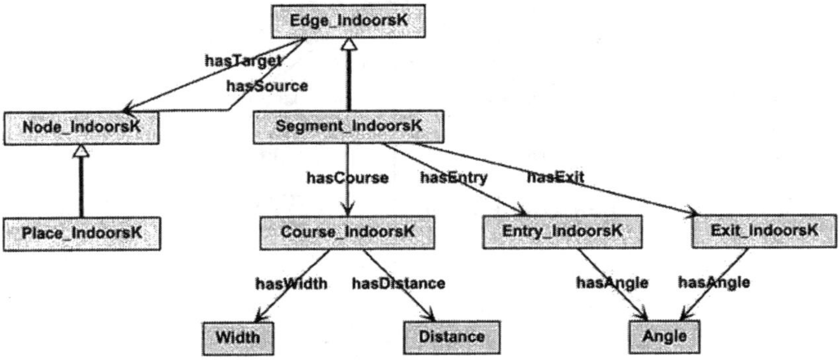

Fig. 15. Ontology of indoors segment subclasses

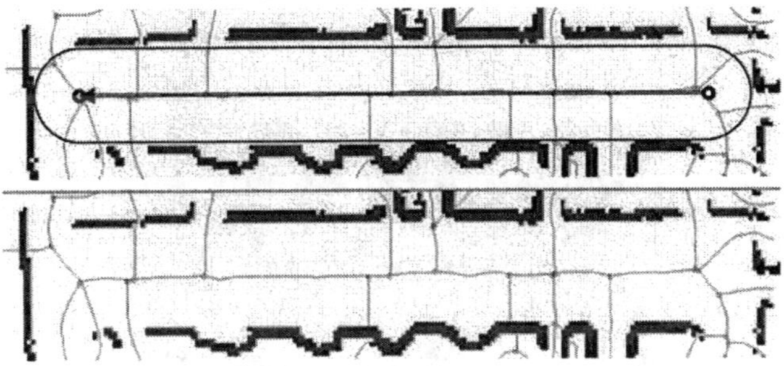

Fig. 16. Abstraction of route in corridor to route segment

probably not necessary since the construction of the actual navigation trajectory from the Route Graph has to take e.g. dynamic obstacles into account.

Modelling Precision. All measurements above (distances, angles, etc.) are intended to be qualitative; in fact no units have been specified so far. Precision can be introduced by an additional attribute, for example an interval of tolerance.

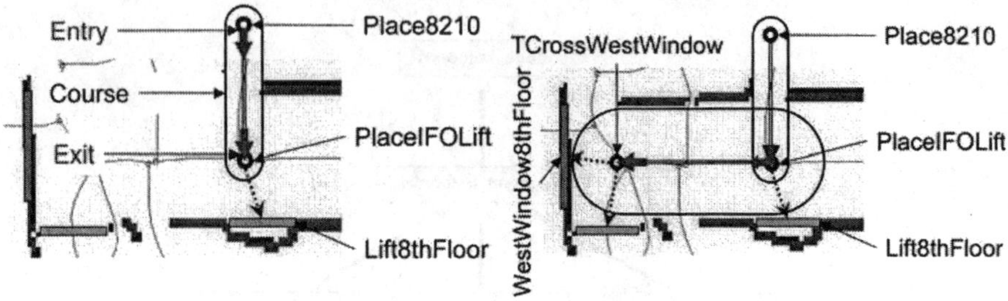

Fig. 17. `Segment1` and `SegmentCorridor2`

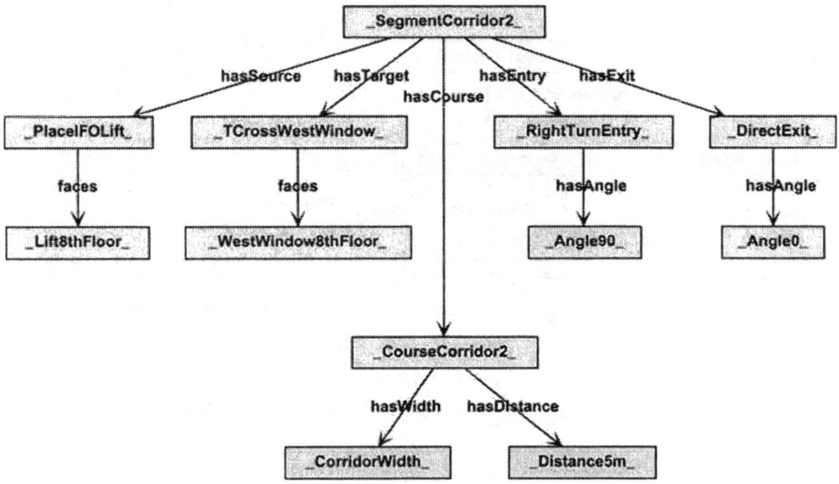

Fig. 18. Ontology of object instantiations representing `SegmentCorridor2`

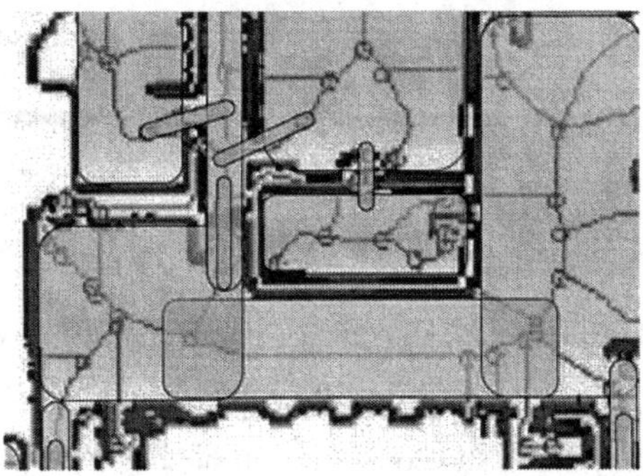

Fig. 19. Overlapping regions

4.3 Nodes and Places

A `Node_IndoorsK` is refined to a `Place_IndoorsK`; each place has its own *origin*, i.e. a well-defined position and orientation with a zero angle. For example (cf.

Fig.18), `Place8210 faces` the office door, `PlaceIFOLift faces` the lift, and `TCrossWestWindow faces` the west window. We might wish to define an origin to coincide with such a `faces` relation, but in fact it does not matter much as it is only important to have a well-defined origin at all.

Place Integration. How is this origin preset? When a node has just one segment attached, its entry angle is zero for an outgoing segment, or its exit angle is zero for an incoming segment.[2] When adding an additional segment, this origin will have to be adhered to. When integrating two routes or in a union of Route Graphs, *place integration* has to be performed for all segments with common places: for any two places to be integrated, a common origin will have to be chosen and the entries/exits of the segments re-computed, if necessary. This is the price to pay for the fact that a place has now become a position of choice for outgoing segments, with different entries in general.

Regions. Places are contained in regions. Recall Fig.8: If a `Place` is `locatedIn` a `Region` and this `Region covers` an `EnvironmentSpace`, say an `Office`, then we may conclude that this `Place` is `containedIn` this `Office`.

Fig.19 shows overlapping and nested regions, where the regions can be classified according to the `KindRGs` in Fig.12.

4.4 Relation to "Common Sense Ontology"

The relations `locatedIn`, `covers` and `containedIn` are examples of relations between different ontologies (or parts of one big joint ontology); this issue is taken up further in [2]. For example, there are various calculi for regions or other spatial configurations (e.g. [20, 21]); the ontology of such calculi, formally specified as suggested in Sect.5, could be associated here.

Facing Windows. is another example. Consider Fig.20 (also the objects in Fig.18): a place can be classified (as a subclass of `Place_IndoorsK`) as a `Place_InFrontOfWindow`, a `Place_InFrontOfLift`, etc.; it may then be related by `faces` to an object in the "Common Sense Ontology", e.g. a `Window` or `Lift`.

Routemarks. In such a situation, we may wish to attach an actual pointer to such an object, i.e. a `Vector_Mark_IndoorsK` having a `Node_IndoorsK`, e.g. a `Place_InFrontOfWindow`, as source and a `Routemark` as target; this `Routemark` is then a `Point` that `marks` a `Window` (e.g. in its centre). As examples, consider the dotted arrows in Fig.17. Routemarks help self-localisation during navigation using such vectors for triangulation.

4.5 Reference Systems

Analogously, a place may be rooted in a global reference system by a vector from its origin; this could also be a 3D Cartesian vector, of course. In fact, it is quite likely that such reference systems correspond to nested regions (Sect. 4.3).

[2] A origin has to be set before a pointer to a routemark is specified, cf. Sect.4.4.

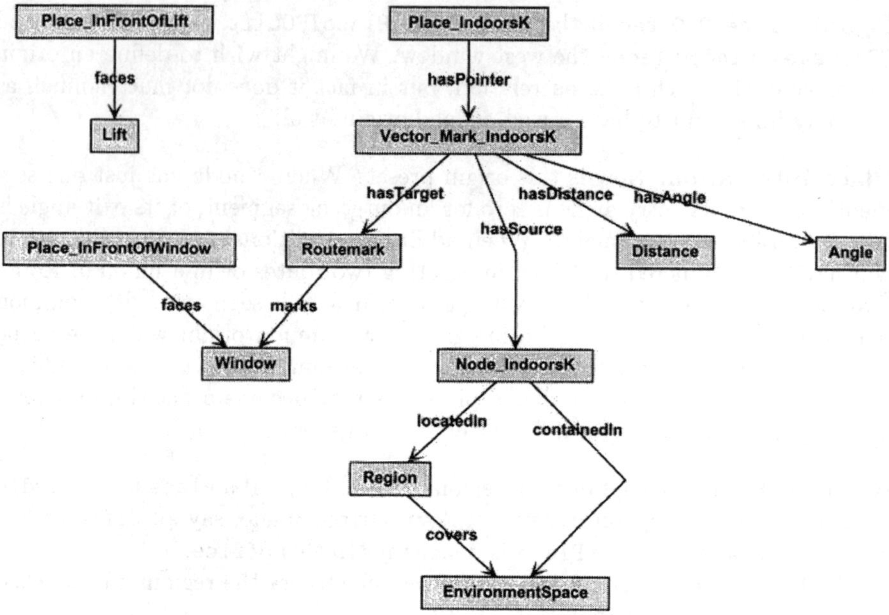

Fig. 20. Places facing a window, and routemarks

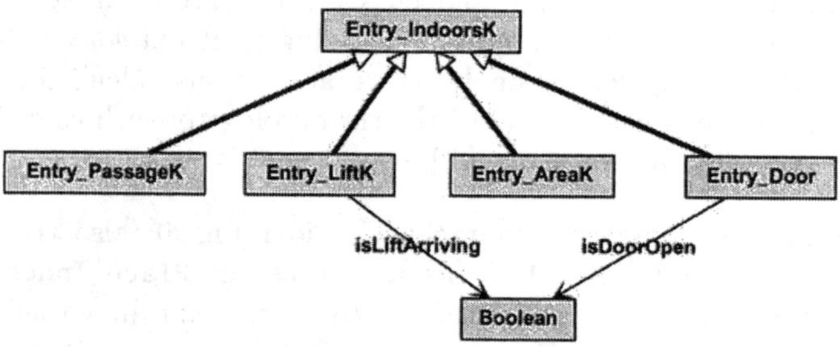

Fig. 21. Entries with dynamic predicates

4.6 Multi-robot Exploration

Multi-Robot Exploration may require additional information to be attached to places, for example a tag marking it as a fringe node for an explored region, to be extended. When various robots co-operate during exploration, place integration becomes of utmost importance, i.e. a criterion for matching two places, to become the same when "closing the loop" (cf. [22, 23, 21]). In such as case, different co-existing Route Graphs denote different possible worlds, with a certain probability attached; identical sub-graphs may of course be shared. We hope that the Route Graph representation will prove to be sufficiently compact for interaction between robots in such situations.

MMiSSLaTeX

```
\NewDecl{GenGraph}[1]  %[KindRG]
{\Class{Graph_#1}{\Ref{#1} graph}{}
      \RelType{hasNode}{Graph_#1}{Node_#1}
      \RelType{hasEdge}{Graph_#1}{Edge_#1}
\Class{Node_#1}{\Ref{#1} node}{}
\Class{Edge_#1}{\Ref{#1} edge}{}
      \RelType{hasSource}{Edge_#1}{Node_#1}
      \RelType{hasTarget}{Edge_#1}{Node_#1}
\GenSequence{Edge_#1}
 \Class{Path_#1}{\Ref{#1} path}{Sequence_Edge_#1}
  \Class{Route_#1}{\Ref{#1} route}{Path_#1}
```

Fig. 22. MMiSSLaTeX source of generic Graph ontology

4.7 Modelling Dynamic Information

So far, all the information associated with Route Graphs was static. For use in navigation, it would be very convenient to be able to model dynamic information as well. Consider Fig.21: entries are guarded by dynamic predicates,[3] to check whether a door is open or a lift is arriving. This way, Rolland may for example choose among segments to several lifts, depending on which is about to arrive. Thus status information is modelled; our graph can be regarded as a kind of state transition graph.

From a puristic point of view, this modelling is adequate for a data structure, but violates the idea of an ontology. However, the ontology (and the "data base" of its associated objects) should be the central source of information, the basic knowledge representation. It will yet have to be seen how these two views can be reconciled better, perhaps by representing dynamic information in a completely separate part of the ontology.

4.8 Structuring Ontologies

Finally, a word about structuring ontologies. Ontologies, as all descriptions or formal specifications, may become quite large and unwieldy; structuring becomes a necessity. We have learned a lot from structuring formal specifications or theories in the context of CASL, cf. [14, 15]: composite specifications may e.g. be constructed by (conservative) extension or union, items may be re-named, morphisms ("views") may be applied, (remote) libraries may be organised as nested folders.

We suggest to apply similar structuring mechanisms to ontologies. One way to do this is the approach of the MMiSS project, where general documents are structured along these lines, and change management supports sustainable development (cf. [12, 13]).

[3] Boolean valued binary relations to emphasise the dynamic query rather than unary predicates represented e.g. as subclasses.

Generic Ontologies. A particularly important structuring mechanism is parameterisation, or abstraction to generic modules, generic (sub)ontologies here. Fig.22 shows a generic graph ontology, as a generalisation of Fig.10: the (only) parameter "#1" may then be instantiated to different specialisations of `KindRG`, e.g. `IndoorsK`, `RailwayK` or `RoadK`. Note that, in its body, `GenSequence` is instantiated, a similarly generalised version of sequences.

5 Formalisation in CASL

This section contains the formalisation of the Route Graph ontology and the RouteGraph data type specification. There are several advantages for using CASL for the entire development of a data structure, together with its functions, from an ontology (see [14, 15, 16, 17], cf. Sect. 3.1, Fig. 9):

CASL covers full First Order Logic, predicates, partial and total functions, and subsorts. Tools in HETS (see below) are available for strong (sub-)type checking, overloading resolution, and interfaces to verification tools.

CASL-DL is a sublanguage of CASL, restricted to the Description Logic underlying OWL DL. It offers precise concepts and definitions together with a translation from and to OWL DL, cf. Fig. 9 and [18]. Thus a variety of tools developed for OWL DL become available for CASL-DL. The LaTeX ontology representation of ??FirstOrderLogic can be translated to CASL-DL (or CASL) to provide a bare-bones specification.

```
spec GenGraph [sort Kind] =
     sorts Graph[Kind], Node[Kind], Edge[Kind]
     preds hasNode : Graph[Kind] × Node[Kind];
           hasEdge : Graph[Kind] × Edge[Kind]
     ops   hasSource, hasTarget : Edge[Kind] → Node[Kind]
     ∀ e: Edge[Kind]; s, t: Node[Kind]; g: Graph[Kind]
     • hasEdge(g, e) ∧ hasSource(e) = s ∧ hasTarget(e) = t ⇒
       hasNode(g, s) ∧ hasNode(g, t)
then
     GenSequence [sort Edge[Kind]] with hasHead, hasTail, EmptySeq, freq
then ...
     op    sources : Sequence[Edge[Kind]] → Sequence[Node[Kind]]
     ...
     pred  connected(l: Sequence[Edge[Kind]]) ⇔
           (∀ e1, e2: Edge[Kind]
             • hasHead(l) = e1 ∧ hasHead(hasTail(l)) = e2 ⇔
               hasTarget(e1) = hasSource(e2)) ∧ connected(hasTail(l))
     sorts Path[Kind] = {l: Sequence[Edge[Kind]] • connected(l)};
           Route[Kind] = {p:Path[Kind]
                           • (∀ n: Node[Kind] • freq(sources(p), n) ≤ 1) ∧
                             (∀ e: Edge[Kind] • freq(p, e) ≤ 1)}
end
```

Fig. 23. Generic Graph ontology in CASL

from BASIC/STRUCTUREDDATATYPES **get** LIST
from BASIC/NUMBERS **get** NAT
from BASIC/GRAPHS **get** NONUNIQUEEDGESGRAPH

spec GRAPH2 [**sort** *Kind*] **given** NAT =
  sorts *Node*, *Edge*
  ops *source*, *target* : *Edge* → *Node*
then
  NONUNIQUEEDGESGRAPH [**sort** *Node*] [**sort** *Edge*] **and**
  LIST2 [**sort** *Edge*] **with** [], __::__, *atMostOnce*
then
  pred *connected* : *List*[*Edge*]
  ∀ *rs1*, *rs2*: *Edge*; *rsl*: *List*[*Edge*]
  • *connected*([]) %(connected_empty)%
  • *connected*([*rs1*]) %(connected_singleton)%
  • *connected*(*rs1* :: *rs2* :: *rsl*) ⇔
    target(*rs1*) = *source*(*rs2*) ∧ *connected*(*rs2* :: *rsl*) %(connected_rec)%
then ...
  op *sources* : *List*[*Edge*] → *List*[*Node*]
  ...
  sorts *Path* = {*l*: *List*[*Edge*] • *connected*(*l*)}; %(def_path)%
    Route = {*l*: *Path* • *atMostOnce*(*sources*(*l*)) ∧ *atMostOnce*(*l*)}
    %(def_route)%
end

Fig. 24. Design specification in CASL

HETS – the Heterogeneous Tool Set – provides support for heterogeneous specification in various different logics (e.g. ModalCASL) and associated tools, most notably the interactive verification system Isabelle [24] and an increasing number of (semi-)automatic theorem provers, e.g. SPASS [25]. Translators to some target programming languages such as HASKELL or JAVA are under development. Moreover, HETS will support re-use of proofs with the Development Graph Manager MAYA [26].

We illustrate the development process proposed here with the example of generic (Route) Graphs.[4] It starts with an ontology stating only classes, relations and their axioms in a very loose fashion. This gives a top-level overview of the intended concepts and their interrelation.

5.1 Generic Graph Ontology in CASL

Fig. 23 shows the ontology of Route Graphs of a specific kind, given as parameter; cf. also Sect. 3.2 and Fig. 22. It starts with the declarations of the basic sorts and relations, and an axiom stating that no dangling edges are allowed.

[4] The formal specification of other parts of the ontology such as parts of the "Common Sense Ontology" or various spatial calculi (cf. Sect. 4.4) would also be quite interesting from the point of view of spatial cognition; we defer this to another paper.

Haskell

```
class Edge a where
    source :: a -> Node
    target :: a -> Node

data Path = forall a . (Eq a,Edge a) => ConPath [a]
data Route = forall a . (Eq a,Edge a) => ConRoute [a]

connected :: Edge a => [a] -> Bool
connected [] = True
connected ([_]) = True
connected (s1:s2:l) = target (s1) == source (s2) && connected (s2:l)

noBranches :: Edge a => [a] -> Bool
noBranches (l) = let sources = map source in atMostOnce (sources l)

isPath :: Edge a => [a] -> Bool
isPath = connected

isRoute :: (Eq a,Edge a) => [a] -> Bool
isRoute l = isPath l && noBranches(l) && atMostOnce(l)
```

Fig. 25. Haskell implementation of Edge

Sequences of Edges. For the relation of a graph to sequences of edges, the parameterized specification GenSequence is instantiated. It provides some basic relations (omitted here) such as *hasHead* and *hasTail* and some auxiliary functions for specification purposes, such as *sources*, yielding the sources of all edges in a graph as a sequence, or *freq* for counting the number of elements in a sequence. Note that parameterized specifications often use compound names which are to be instantiated, e.g. *Edge*[*Kind*] or *Sequence*[*Edge*[*Kind*]].

Paths and Routes. *Path*[*Kind*] is defined as a subsort of *Sequence*[*Edge*[*Kind*]] with the auxiliary predicate *connected*, then *Route*[*Kind*] as a subsort of *Path*[*Kind*] without duplicate branches or edges (this implies no cycles).

5.2 Design Specification in Casl

A Casl specification of Graph2 based on predefined basic data types such as List and NonUniqueEdgesGraph is presented in Fig. 24. It povides the same sorts as the ontological definition of GenGraph in Fig. 23 and is the first "design specification" in a software engineering sense. We can prove rather straightforwardly that this specification satisfies the requirements set out in Fig. 23.

5.3 Implementation of Inheritance in Haskell

We now turn to an operational implementation in the functional programming language Haskell and demonstrate some of the subtleties in the translation.

Fig. 25 shows the polymorphic implementation of the predicates `connected` and `noBranches` based on the type class `Edge`. The implementation of `connected` uses the same recursive definition as in Fig. 24. The existential data types `Path` and `Route` require type classes `Eq` and `Edge` as minimal context. All this, including constructor functions for `Path` and `Route` (not shown), can be implemented without knowledge about the actual structure of later instantiations.

The Route Graph specialisation of class `Edge` to class `Segment` is shown in Fig. 26. Note that providing special information for `entry`, `course` and `exit` is optional and yields a safe `Nothing` when e.g. an instance does not need an exit.

Haskell

```
class Edge a => Segment a where
    entry    :: a -> Maybe Angle
    entry _ = Nothing::Maybe Angle
    course    :: a -> Maybe Course
    course _ = Nothing::Maybe Course
    exit    :: a -> Maybe Angle
    exit _ = Nothing::Maybe Angle

data SegmentIndoorD = SegInd {segI_source :: Node,
                              segI_entry  :: Angle,
                              segI_course :: Course,
                              segI_exit   :: Angle,
                              segI_target :: Node}

instance Edge SegmentIndoorD where
    source = segI_source
    target = segI_target

instance Segment SegmentIndoorD where
    entry  = Just . segI_entry
    course = Just . segI_course
    exit   = Just . segI_exit
```

Fig. 26. Haskell implementation of `SegmentIndoor`

For the instantiation to indoors navigation, the data type `SegmentIndoorsD` is defined with the constructor `SegInd` (with special components/selector functions); it instantiates the classes `Edge` and `Segment`. The subsorting relation of SEGMENT and EDGE in CASL is translated to a class hierarchy in Haskell, where the data type includes all the information needed for all classes of the hierarchy.

Functions in class `Segment` are only accessible for data stored as `Path` or `Route` if we add the context `Segment` to the existential data types. Thus, we have to use the lowest class in the hierarchy that provides access to all information we need when we introduce existential types to hold heterogenous data.

6 Conclusion

We have presented the generic concept of Route Graphs, modelled by an ontology, and its step by step formalisation. We hope that the introduction of a "light-weight" ontology first, restricted to (sub)classes and relations only, will appeal to those who are not so interested in formalisation, and provides a general overview to those who are. Such an ontology specification can be done rather quickly in the special LaTeX format, and ontology visualisation tools (e.g. [10]) can already be applied. Moreover, such an ontology is already suitable for the semantic interrelation of documents (cf. [13]), as e.g. in the repository for a joint interdisciplinary research project such as the SFB/TR8 "Spatial Cognition".

The subsequent formalisation of the Route Graph ontology in CASL will appeal to those who emphasise the need for complete formalisation of ontologies, as in the DOLCE approach [16, 17]; here the structuring concepts of CASL and the availability of verification tools are perhaps the primary advantage.

The development of an actual data structure from a top-level ontology in CASL is a new experiment in Software Engineering since this first (loose) requirement specification is tailored to ontologies and not to software. Nevertheless, we believe that the approach works quite well, and the process will hopefully be further automated as much as possible. In the context of cooperation among and integration between a large interdisciplinary diversity of research and application areas (such as AI, Cognitive Science, Linguistics, and Psychology) we see it as a definite advantage to have a common "language", viz. semantic description formalism, to start from, *the* central (joint) ontology. We hope that those who are only interested in the result, e.g. a data structure for interaction in the robotics domain, will be patient with this "interdisciplinary definitional overhead", and happy with the outcome. We expect tools for the (formalised) ontology and interfaces to the data structure, with appropriate operations to simulate access "as an ontology", to coexist peacefully in the running system.

Finally, we hope to have made a contribution to clarifying concepts and providing a data structure for human–robot and robot–robot interaction for the indoors scenario. In particular, the (indoors) Route Graph ontology shall serve as a basis for linguistic augmentation to enable a dialogue between user and wheelchair, as outlined in [3, 4].

The approach presented in this paper will have to be extended, at the "user level", by a relation to spatial calculi, e.g. for treating "right" and "left", and nested overlapping regions. Formalisation of these calculi in CASL, at the ontology level, should be rather straightforward. The challenge will be the choice and combination of suitable calculi, and the transition between them.

We are looking forward to other instantiations of the generic Route Graph ontology (and data structures) in application domains such as geo-information systems or location-based service scenarios.

Acknowledgement. We thank the other members of the Route Graph and Ontology Working Groups of the SFB/TR 8, particularly John Bateman, Scott Farrar, Achim Mahnke, Reinhard Moratz, Hui Shi, and Tilman Vierhuff.

References

1. Werner, S., Krieg-Brückner, B., Herrmann, T.: Modelling navigational knowledge by route graphs. In Freksa, C., Habel, C., Wender, K., eds.: Spatial Cognition II. Number 1849 in LNAI. Springer (2000) 295–317
2. Bateman, J., Farrar, S.: Modelling models of robot navigation using formal spatial ontology. In Freksa, C., Knauff, M., Krieg-Brückner, B., Nebel, B., Barkowsky, T., eds.: Spatial Cognition IV – Reasoning, Action, Interaction. Number 3343 in LNAI. Springer (2004) 366–389
3. Krieg-Brückner, B., Shi, H., Ross, R.J.: A safe and robust approach to shared control via dialogue. Journal of Software **15(12)** (2004) 1744–1755
4. Ross, R., Shi, H., Vierhuff, T., Krieg-Brückner, B., Bateman, J.: Towards dialogue-based shared control of navigating robots. In Freksa, C., Knauff, M., Krieg-Brückner, B., Nebel, B., Barkowsky, T., eds.: Spatial Cognition IV – Reasoning, Action, Interaction. Number 3343 in LNAI. Springer (2004) 478–499
5. Krieg-Brückner, B., Röfer, T., Carmesin, H.O., Müller, R.: A taxonomy of spatial knowledge for navigation and its application to the bremen autonomous wheelchair. In Freksa, C., Habel, C., Wender, K., eds.: Spatial Cognition. Number 1404 in LNAI. Springer (1998) 373–397
6. Uschold, M., Grüninger, M.: Ontologies: Principles, methods and applications. Knowledge Engineering Review **11** (1996) 93–155
7. Gómez-Pérez, A., Benjamins, V.R., eds.: Knowledge Engineering and Knowledge Management. Ontologies and the Semantic Web, 13th International Conference, EKAW 2002, Siguenza, Spain, October 1-4, 2002, Proceedings. In Gómez-Pérez, A., Benjamins, V.R., eds.: EKAW. Volume 2473 of Lecture Notes in Computer Science., Springer (2002)
8. Corcho, O., Gómez-Pérez, A.: A roadmap to ontology specification languages. In: Proceedings of the 12th European Workshop on Knowledge Acquisition, Modeling and Management, Springer (2000) 80–96
9. Bechhofer, S., van Hermelen, F., Hendler, J., Horrocks, I., McGuinness, D.L., Patel-Schneider, P.F., Stein, L.A.: W3C: OWL Web Ontology Language – Reference. W3C Recommendation `http://www.w3.org/TR/owl-ref/` (2004)
10. (`http://www.informatik.uni-bremen.de/agbkb/publikationen/softlibrary_e.htm`)
11. (`http://www.informatik.uni-bremen.de/~davinci/`)
12. Krieg-Brückner, B., Hutter, D., Lindow, A., Lüth, C., Mahnke, A., Melis, E., Meier, P., Poetzsch-Heffter, A., Roggenbach, M., Russell, G., Smaus, J.G., Wirsing, M.: Multimedia instruction in safe and secure systems. In: Recent Trends in Algebraic Development Techniques. Volume 2755 of Lecture Notes in Computer Science., Springer (2003) 82–117
13. Krieg-Brückner, B., Lindow, A., Lüth, C., Mahnke, A., Russell, G.: Semantic interrelation of documents via an ontology. In Engels, G., Seehusen, S., eds.: DeLFI 2004, Tagungsband der 2. e-Learning Fachtagung Informatik. Volume P-52 of Lecture Notes in Informatics., Springer (2004) 271–282
14. Astesiano, E., Bidoit, M., Krieg-Brückner, B., Kirchner, H., Mosses, P.D., Sannella, D., Tarlecki, A.: CASL – the common algebraic specification language. Theoretical Computer Science **286** (2002) 153–196 `www.cofi.info`.
15. Mosses, P.D., ed.: CASL Reference Manual. Volume 2960 of Lecture Notes in Computer Science. Springer (2004)
16. Gangemi, A., Guarino, N., Masolo, C., Oltramari, A., Schneider, L.: Sweetening ontologies with DOLCE. [7] 166–181

17. Lüttich, K., Mossakowski, T.: Specification of ontologies in CASL. In Varzi, A.C., Vieu, L., eds.: Formal Ontology in Information Systems, Proceedings of the Third International Conference (FOIS 2004). Volume 114 of Frontiers in Artificial Intelligence and Applications., IOS-Press (2004) 140–150
18. Lüttich, K., Krieg-Brückner, B., Mossakowski, T.: Ontologies for the semantic web in CASL. 17th Int. Workshop on Algebraic Development Techniques (2004) (to appear).
19. Wallgrün, J.O.: Autonomous construction of hierarchical voronoi-based route graph representations. In Freksa, C., Knauff, M., Krieg-Brückner, B., Nebel, B., Barkowsky, T., eds.: Spatial Cognition IV – Reasoning, Action, Interaction. Number 3343 in LNAI. Springer (2004) 413–433
20. Randell, D., Cui, Z., Cohn, A.: A spatial logic based on regions and connection. In: Proceedings KR-92, San Mateo, Morgan Kaufmann (1992) 165–176
21. Moratz, R., Wallgrün, J.O.: Spatial reasoning about relative orientation and distance for robot exploration. In Kuhn, W., Worboys, M., Timpf, S., eds.: Spatial Information Theory: Foundations of Geographic Information Science. Conference on Spatial Information Theory (COSIT). Lecture Notes in Computer Science, Springer (2003) 61–74
22. Frese, U., Larsson, P., Duckett, T.: A multigrid algorithm for simultaneous localization and mapping. IEEE Transactions on Robotics (2004) (to appear).
23. Frese, U.: A discussion of simultaneous localization and mapping. Autonomous Robots (2004) (submitted).
24. Nipkow, T., Paulson, L.C., Wenzel, M.: Isabelle/HOL — A Proof Assistant for Higher-Order Logic. Springer Verlag (2002)
25. Weidenbach, C., Brahm, U., Hillenbrand, T., Keen, E., Theobalt, C., Topic, D.: SPASS version 2.0. In Voronkov, A., ed.: Automated Deduction – CADE-18. Volume 2392 of Lecture Notes in Computer Science., Springer (2002) 275–279
26. Autexier, S., Hutter, D., Mossakowski, T., Schairer, A.: The development graph manager MAYA (system description). In Kirchner, H., Reingeissen, C., eds.: Algebraic Methodology and Software Technology, 2002. Volume 2422 of Lecture Notes in Computer Science. Springer (2002) 495–502

Autonomous Construction of Hierarchical Voronoi-Based Route Graph Representations

Jan Oliver Wallgrün

Universität Bremen, Department of Mathematics and Informatics,
Bibliothekstr. 1, 28359 Bremen, Germany
wallgruen@informatik.uni-bremen.de

Abstract. A route graph as proposed in Werner et al. (2000) is a spatial representation of the environment that focuses on integrating qualitatively different routes an agent can use for navigation. In this paper we describe how a route graph based on the generalized Voronoi diagram (GVD) of the environment can be used for mobile robot mapping and navigation tasks in an office-like indoor environment. We propose a hierarchical organization of the graph structure resulting in more abstract layers that represent the environment at coarser levels of granularity. For this purpose, we define relevance measures to weight the meet points in the GVD based on how significant they are for navigation and present an algorithm that utilizes these weights to generate the coarser route graph layers. Computation of the relevance values from either complete or incomplete information about the environment is considered. Besides robot navigation, the techniques developed can be employed for other tasks in which abstract route graph representations are advantageous, e.g. automatically generating route descriptions from floor plans.

1 Introduction

One important design decision in mobile robotics is what kind of spatial representation scheme should be applied to model the robot's knowledge about the environment. Besides facilitating navigation related tasks like localization, path planning, and high-level spatial reasoning, it is often a requirement that the robot is able to construct and maintain this spatial representation autonomously which leads to the simultaneous localization and mapping (SLAM) problem, the problem of incrementally constructing a spatial model of the environment while simultaneously using the same model to localize the robot [17].

In the literature two main classes of representations are frequently distinguished: metric representations, including grid maps [23, 8] and geometric representations [6, 18, 32], on the one hand and topological maps [14, 11, 3] on the other. While metric representations represent the environment by specifying position and orientation of spatial entities with respect to a global coordinate system, topological maps are based on a graph structure that explicitly stores spatial relations (e.g. adjacency or connectivity) between the entities represented by the vertices. In addition, a number of authors have proposed different ways

C. Freksa et al. (Eds.): Spatial Cognition IV, LNAI 3343, pp. 413–433, 2005.
© Springer-Verlag Berlin Heidelberg 2005

(a) (b)

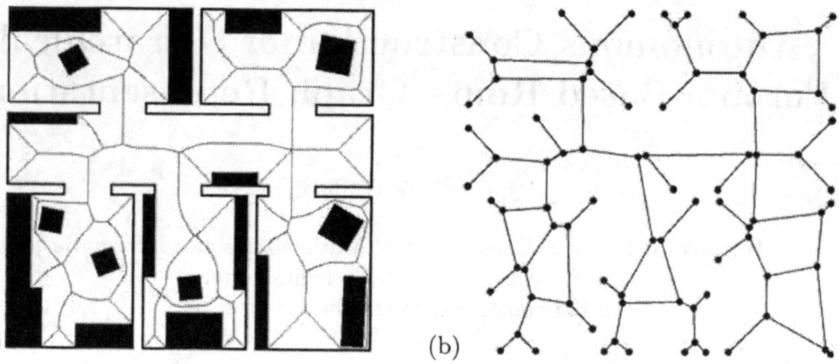

Fig. 1. (a) The generalized Voronoi diagram (GVD) (thin lines) of a 2D environment, (b) the corresponding generalized Voronoi graph (GVG) with vertices placed at the positions of the corresponding meet points for visualization

to combine metric and topological representations to exploit their orthogonal strengths [28, 13, 33].

In the special case in which the graph structure of a topological representation directly reflects the network of qualitatively different routes through the environment with edges corresponding to route segments and vertices corresponding to decision points we will call such a representation a route graph, a concept which has been introduced in [31] as a general model for spatial knowledge obtained during route-based navigation independent of the particular kind of spatial agent – human being, animal, or robot (see [12] in this volume for more details on the route graph concept).

While most current approaches to the SLAM problem are metric ones that employ probability distributions to represent multiple hypotheses about the state of the environment [29, 9, 20], a particular class of route graph approaches is becoming increasingly popular, namely those based on the generalized Voronoi graph (GVG) which is the graph corresponding to the generalized Voronoi diagram (GVD)[3, 34]. The GVD in turn is a generalization of standard Voronoi diagrams [1] that handles other geometric primitives, e.g. line segments, instead of only point sites [16, 10]. It is a retraction of free space onto a network of one-dimensional curves and has been first used in robotics as an intermediate representations to solve motion planning tasks given complete information about the working space of the robot (usually by providing a geometric description of the boundaries of the obstacles) [24, 15]. The GVD is also related to the idea of a shape's skeleton introduced in [2]. The GVG corresponding to a GVD has a vertex for every point at which multiple Voronoi curves meet and for every point at which a Voronoi curve ends in a corner of the environment. The edges of the GVG connect vertices that correspond to adjacent meet points in the GVD (cf. figure 1).

GVG-based route graphs combine advantages from applying a topological representation and from applying the Voronoi diagram for navigation: They are compact representations which allow efficient path planning and a systematic exploration of an unknown environment, provide an interface for high-level reasoning, and do not require special efforts to keep metric information globally consistent. The latter is the case because the robot does not have to localize itself in terms of precise coordinates in a global coordinate system but by tracking its position within the graph structure. In addition, driving along the Voronoi curves keeps the robot away from the obstacle boundaries and can be accomplished by a simple motion behavior.

However, in practice the instability of the underlying Voronoi diagram when computed from noisy and discrete sensor data causes a lot of complications for the autonomous construction of the representation and for localizing within the graph structure. In addition, the GVG still contains vertices and edges that are irrelevant for navigation and thus means to further abstract from the detailed GVG representation are desirable.

In this paper, we therefore propose a hierarchization of the GVG-based route graph to deal with these problems, as well as to make path planning, reasoning, and localization more efficient by allowing it to be done in a hierarchical manner. The hierarchical representation consists of route graph layers that represent the environment at different levels of granularity and that are linked by an abstraction relation that allows to switch to a finer or coarser level.

A prerequisite for a meaningful abstraction from the GVG is a way to distinguish between vertices that represent relevant decision points for navigation and less important ones. For this purpose we introduce relevance measures in section 4 that assess the significance of a vertex and can be used to order vertices with respect to the role they play for navigation. We then describe how the relevance values can be computed either from the complete GVG of an environment or from a partial GVG as required in the context of incremental mapping for a mobile robot (which will be further explained in the next section). In the second case the relevance values computed will often be lower bound estimates of the actual values and have to be updated when new information becomes available. Similar approaches to pruning or hierarchically organizing Voronoi diagrams or skeletons of shapes have been developed in the computer vision community (e.g. [25, 27, 19]). However, these approaches employ the boundary information of the shape to simplify the Voronoi diagram, while for our incremental mapping approach it is crucial that the relevance values can be computed from the information stored in the GVG alone.

In section 5, we present an algorithm that employs the relevance values to produce a coarser route graph layer from the original GVG. Besides navigation and mapping for a mobile robot, the proposed way to abstract from the GVG gives rise to other applications in which the automatic generation of an abstract route graph of the environment will be beneficial, like the automatic production of verbal or pictorial route descriptions based on floor plans or other geometric 2D models of an environment. We will briefly address these applications as well in section 6 in which we present results of the experimental evaluation of our approach.

2 GVG-Based Route Graphs for Robot Mapping and Navigation

In earlier work [30] we proposed a GVG-based representation scheme together with localization and construction procedures to enable a mobile robot equipped with a laser range finder to navigate in an unknown office-like indoor environment while incrementally building up the representation autonomously. The goal of this work was to make localization more robust and exploration more efficient than in the GVG-based mapping approach described by Choset et al. [4, 3][1].

Besides the graph structure of the GVG the following additional information is contained in the proposed route graph representation:

1. Vertices are labeled with a *signature* (see figure 2a) that contains the distance to the generating points (those points at which the maximal inscribed circle centered on the Voronoi vertex touches the obstacle boundaries) and the angles between the lines connecting the vertex with its generating points.
2. The approximate relative position of vertices is represented by annotating the edges with the approximate distances between the connected vertices and by annotating the vertices with the approximate angles between their leaving edges (see figure 2b).
3. Every edge is annotated with a description of the Voronoi curve corresponding to this edge since this curve may deviate from the direct connection between the two vertices.
4. Additional information about which edges are traversable (not too close to obstacles) for the robot and which edges lead to still unexplored areas is also annotated to the graph structure.

While path planning becomes the simple task of using standard graph search techniques to compute a path through the GVG (possibly employing the annotated metric information), the incremental construction and driving along a planned path both require the ability to localize within the graph representation. For this purpose, a matching scheme has been developed that compares a *local* GVG computed from a single 360° scan taken from the robot's current position with the (partially constructed) *global* GVG to identify corresponding vertices and edges while taking into account the vertex signatures, the relative position information, and the odometry information about the last movement. The local GVG describes how the GVG looks in the neighborhood of the robot's current position as far as this can be derived from the sensor data available from this point. Figure 3 illustrates this localization scheme showing on the left a simple environment with the partially constructed global GVG, and in the middle the

[1] We also want to point out that the Hierarchical GVG presented in the work of Choset et al. is an extension of the GVG to higher dimensions that ensures that the graph will always be connected, but does not provide descriptions of the environment at higher levels of abstraction as the hierarchical route graphs we will introduce in section 3.

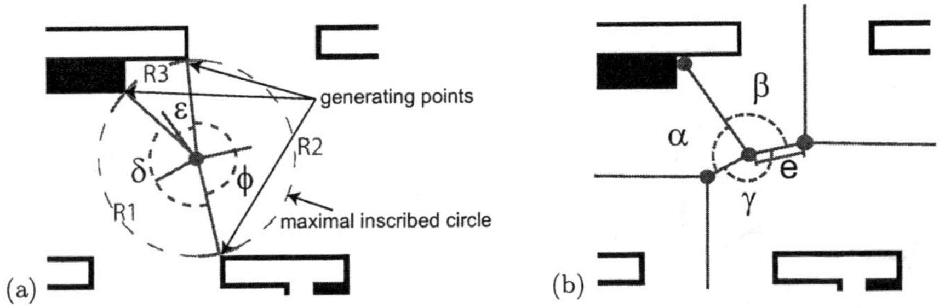

Fig. 2. Different kinds of annotations to the GVG-based route graph: (a) Vertex signature containing the distance of the generating points (radius of the maximally inscribed circle) and angles between the connections to the generating points, (b) relative position information given by the angles between leaving edges and the length of the edges

segmented current scan data together with the local GVG. The dashed arrows indicate vertices that have been identified by the matching algorithm. Based on these associations the position of the robot relative to the vertices of the local GVG can be transferred back to the global GVG and thus localization within the global GVG can be achieved.

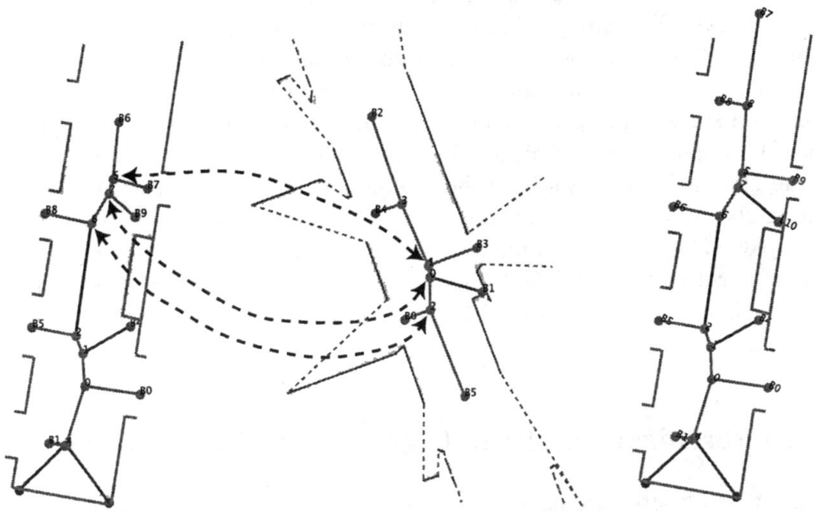

Fig. 3. Localization by matching a local GVG (middle) against the current global GVG (left): The vertices connected by the arrows are identified with each other allowing to transfer the position of the robot within the local GVG to the global GVG. Parts contained in the local but not the global GVG are appended resulting in a new global GVG (right)

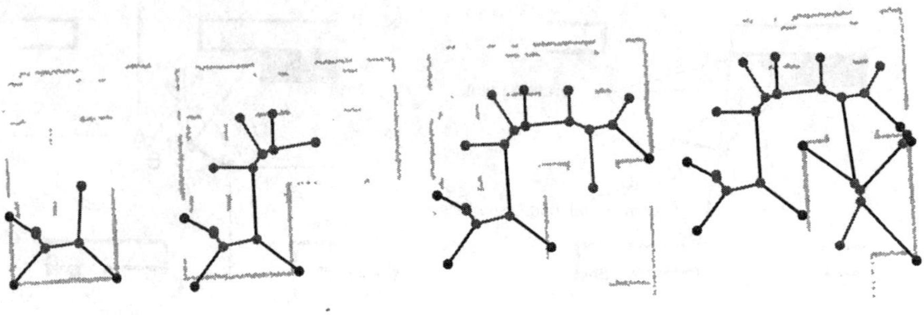

Fig. 4. A sequence of growing GVG-based route graphs (starting with the local GVG of the robot's start position) constructed during the exploration of an unknown environment

The matching scheme has to take into account that the GVD computed from noisy and discrete sensor data is generally not stable. For instance vertices contained in one GVG may be missing in another GVG since they were induced by small dents in the segmented scan caused by noise in the range data. Therefore, the matching algorithm searches for similar subgraphs allowing for certain variations and not for exact isomorphisms (cf. [30] for details on the localization scheme).

The matching scheme as described above allows the robot to track its position relative to the global GVG while driving along a planned path. Furthermore, it enables the robot to build up the global GVG incrementally by sequentially merging local GVGs computed for different positions starting with the local GVG computed for the start position of the robot. Thus, the approach makes maximal use of the information available from each observation avoiding unnecessary exploration steps, an improvement over other construction procedures proposed earlier [4]. Whenever the identification of vertices and edges in the current local and the global GVG allows the robot to replace an edge marked as unexplored in the global GVG by a subgraph from the local GVG, the global GVG will be complemented. The right image of figure 3 shows the new global GVG after the new parts of the local GVG have been appended. Figure 4 demonstrates how the global GVG grows during the first movements through an unknown environment.

3 Hierarchization of the GVG-Based Route Graph

Voronoi-based route graphs as described in the previous section contain the information required for successful navigation. However, they also contain details not required for many tasks since not all meet points of the underlying GVD really correspond to decision points relevant for navigation. While some of them are caused by minor features of the environment like small dents or niches, others are merely the result of noise in the sensor data. For high-level reasoning, planning, and for communication issues a more abstract level of representation

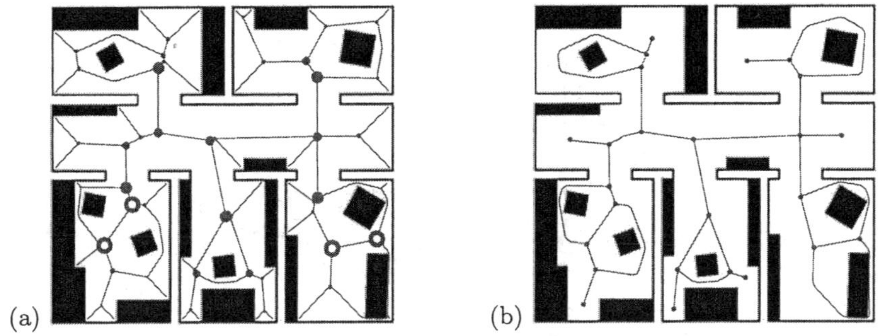

Fig. 5. A coarser route graph (b) generated from the original GVG (a) by the simplification algorithm described in section 5. The radii of the vertices in the original GVG depict their relevance values which will be introduced in section 4. Non-filled circles represent vertices that are maximally relevant because three of their leaving edges are part of cycles in the graph

is preferable if it is still linked to the detailed level required for actually acting within the environment. In addition, the relevant vertices are also those that are very stable and thus less likely to be missing in one of the local GVGs. Hence, localization can also benefit from a more abstract level of representation only containing the relevant vertices.

Therefore, our goal is to construct a hierarchically structured multi-layer route graph representation that bridges from detailed navigational information to abstract high-level route information about the environment and allows to efficiently reason in a hierarchical manner. Every layer of this representation consists of a route graph that models the environment at a certain level of granularity and its features are linked to those of the next higher and next lower layer in a way that allows to switch to a finer or coarser level. To derive more abstract layers from the original GVG measures are required that assess the relevance of individual Voronoi vertices and edges for navigation and we will define such measures in section 4. Figure 5b shows an example of such a coarse route graph generated from the GVG of the environment shown in 5a with the simplification algorithm that will be described later. Figure 6 illustrates the idea of a layered route graph representation with a coarser route graph layer on top of the original GVG. Corresponding features of the layers are linked as indicated by the arrows. Thus, two kinds of edges exist in the hierarchical representation: route graph edges horizontally connecting vertices within the same route graph layer and abstraction edges vertically connecting vertices and edges in one layer with subsets in the layer below.

In the following, we will briefly address how planning, reasoning, and communication can benefit from such a hierarchical organization of the route graph representation.

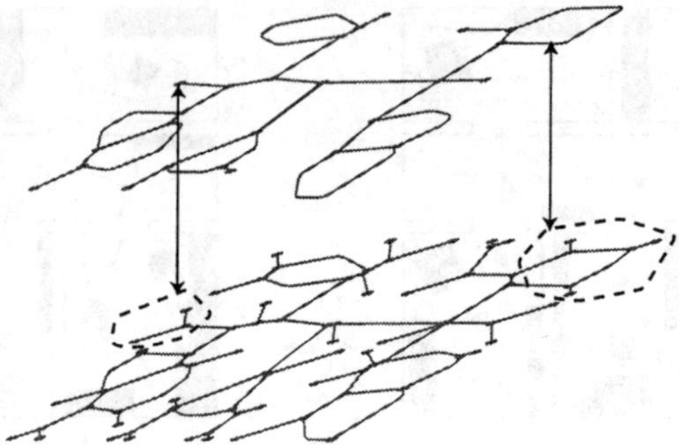

Fig. 6. A two-layer hierarchical route graph representation with the original GVG at the bottom and a coarser route graph layer on top of it. Two examples of how parts of the detailed layer are represented by a vertex or an edge of the coarser layer are shown by the arrows

3.1 Hierarchical Path Planning

A hierarchical route graph representation like the two-layer example in figure 6 can be employed for hierarchical path planning. The edges in the coarser layer correspond to macro operations like driving from one door to the next along a corridor or passing an object on one side. Thus, planning on the higher level (e.g. by using graph search techniques) results in a plan that is not directly executable with the low-level navigation procedures of the robot. However, the abstraction relation allows to recursively break down more abstract operations into finer operations until a plan at the detailed level of the original GVG is reached.

We will define the relevance measures and the simplification algorithm in a way that assures that cycles in the original GVG are retained at a coarser level and no cycle will split up when changing to a higher level of abstraction. This guarantees that a shortest path planned on a higher level will always result in the shortest path at the bottom level as well, when it is recursively transformed into an executable plan.

3.2 Hierarchical Reasoning

In [21] and [22] we described an approach to reason about the relative positions of the decision points within the low-level GVG-based route graph by propagating intervals for the distances and angles (called distance-orientation intervals (DOIs)) annotated to the graph structure along the sequence of edges connecting two vertices. This approach is similar to the composition of spatial relations in qualitative spatial reasoning [5]. The DOIs represent the uncertainty in the metric relative position information assuming certain maximal error boundaries. Reasoning about relative positions of the decision points in the route graph can for instance be applied to determine potential candidates for loops in the environment that need to be closed while constructing the representation during an exploration. Another application is judging if an unexplored junction in a partially constructed route graph might be a good shortcut to a place visited earlier.

This reasoning about routes can also benefit from the hierarchical organization of the route graph representation. Intermediate results from the low-level propagation can be stored as relative position information at the higher levels. This would allow to employ a hierarchical propagation scheme that uses the distance-orientation intervals at the highest level if they are available or switches to a lower level whenever this is not the case, adding the result to the higher level after it has been computed. On the long run the higher level will be completely annotated speeding up the relative position computation significantly.

3.3 Communication

The abstract route graph layers in our representation provide a compact description of the environment that is rather independent of the particular properties of the range sensor of the robot that constructed the representation. Therefore, this information is much better suited to be communicated to another spatial agent than the detailed description given by the original GVG. Scenarios that come to mind here are multi-robot exploration scenarios in which the individual robots exchange knowledge about parts of the environment they have explored so far.

Another application scenario in which the abstract route graph level can be employed beneficially is human-robot communication about routes. Augmenting the route graph with semantic information, for instance stemming from door recognition modules, will allow the robot to generate route instructions to guide a person to a certain goal. In addition, such a representation will make it easier to match route directions given by a human instructor against the robot's model of the environment and translate them into a detailed sequence of actions, since the abstract route graph with all irrelevant vertices and edges removed will be much closer to the route graph the instructor had in mind when generating the route directions[2].

4 Computing the Relevance of Voronoi Vertices

To be able to create a coarser route graph layer from a GVG we will define two relevance measures in this section that will then be used to remove the irrelevant parts of the GVG by the simplification algorithm described in the next section. The first measure assesses how relevant a vertex in the GVG is as a decision point and it is computed from the values a second measure assigns to each leaving edge of the vertex based on how significant the region accessible via this edge is for navigation. Besides defining both measures, we will discuss how to compute the

[2] See [26] in this volume for a general architecture for mapping linguistic expressions to internal representations based on linguistic ontologies and domain ontologies.

relevance values from either complete information given by a fully constructed GVG or from incomplete information, e.g. a partially constructed GVG.

4.1 The Relevance Measures

The GVG as described in section 2 is an undirected Graph RG $= (V, E)$ (with additional annotations) containing only vertices of degree one (the corner vertices) or of degree three or higher (the inner vertices). As figure 2a illustrates, the lines connecting a Voronoi vertex v with its generating points on the obstacle boundaries separate different parts of free space that are accessible via one of v's leaving edges. We will call each such area, that can be reached from v without crossing one of the connecting lines again, a region R_i^v of v. For each Voronoi vertex of degree n there exist n such regions. If v is part of a cycle in the route graph, the regions corresponding to the two edges of v that also belong to this cycle will be identical since the edges provide access to the same part of the environment, just from different directions.

How relevant a Voronoi vertex v is for navigation depends directly on its regions. To be regarded as a decision point, at least three of v's regions need to be significant enough to be judged as different continuations after arriving at this point. Otherwise, no real decision is to be made at this location. We further assume that having two very significant regions can not make up for the third region being insignificant, e.g. a small niche in a corridor will not create a decision point in front of it, irrespective of how long the corridor continues in both direction. Furthermore, having many insignificant regions will not make up for the third most significant one being still insignificant, e.g. two small niches on opposing sides of the corridor will not cause a decision point either.

Therefore, assuming we already have a measure RSM (for *Region Significance Measure*) that assesses the significance of each region of v, it seems reasonable to take the RSM value of v's third most significant region as the relevance value of v. Using maxRSM_k^v to denote the k-highest RSM value over all regions of v, we thus define our *Voronoi Vertex Relevance Measure* (VVRM) for all $v \in V$ with degree$(v) \geq 3$ as:

$$\text{VVRM}(v) = \text{maxRSM}_3^v.$$

We now need to define the RSM measure in a way that captures the notion of a significant region in the context of navigation in an indoor environment. The two major factors that we wanted to account for in our RSM measure are the following: First, the distance from v to the remotest goals belonging to the region should influence the significance of the region, since a region is clearly more significant if one can reach goals within it that are far from the current position. Second, we wanted to include the aspect of visibility to ensure that a region is assessed as less significant if most of it can be perceived from a large area around v.

A very important additional constraint on our RSM measure is that the significance values should be computable from the information contained in the

GVG alone without referring to a geometric description of the boundaries of obstacles, since this is the only available information in our mapping approach. Furthermore, cyclic regions should be treated as maximally significant so that cycles in the graph will never be split up when deriving a coarser route graph from the GVG.

Hence, we define the RSM measure as follows:

1. RSM(R_i^v) $= \infty$, if v and the leaving edge corresponding to R_i^v belong to a cycle in the route graph.
2. Otherwise, the shortest paths from v to the corner vertices belonging to R_i^v in terms of the distance along the GVD are considered. As illustrated in figure 7, it is determined for which corner vertex the length of this path minus the length of the part of this path lying within the maximal inscribed circle of v is maximal, and this is returned as the value RSM(R_i^v).

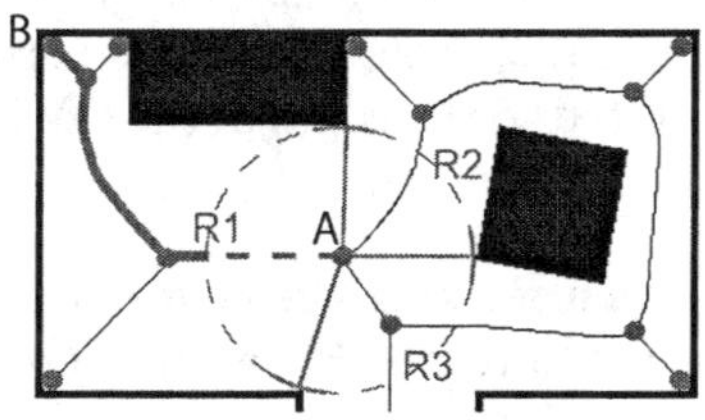

Fig. 7. Computation of the RSM value of region R1 to the left of vertex A: The length of the path to B lying within the maximal inscribed circle (dashed line) is subtracted from the length of the complete path to B yielding the length of the heavy solid drawn part

In the non-cyclic case the distance of the furthest corner vertex contained in the region is used to measure how far the robot could travel into this region. The subtraction of the length of the part that lies in the maximal inscribed circle of v introduces the notion of visibility as mentioned above. For instance in figure 8, the small niche that causes Voronoi vertex A in a wide hallway will be assessed as insignificant since most parts of the paths to B or C are contained within the maximal inscribed circle centered on A, meaning that most parts of the corresponding region are visible from every point within this circle. In figure 5a the individual VVRM values are depicted by the radii of the corresponding circles. Vertices that have a VVRM value of ∞ because at least three of their leaving edges belong to cycles in the GVG are displayed by the non-filled circles.

4.2 Computation from Complete Information

Given the complete GVG describing the environment of the robot, VVRM(v) for a vertex $v \in V$ with degree(v) $= n$ and $n \geq 3$ can be determined by computing RSM(R_i^v) for all $R_i^v \in \{R_1^v, ..., R_n^v\}$ first and then choosing the third highest

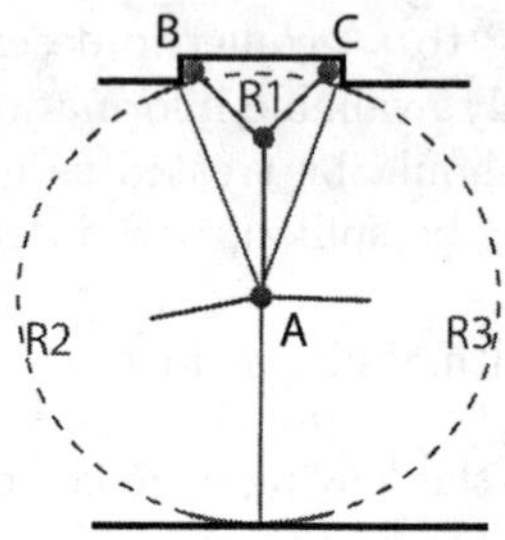

Fig. 8. Region R1 of vertex A is caused by a small niche in a wall. Since A lies in a large open area the distance along the GVD from A to the corner vertices B and C is rather big. However, most part of the connection to these vertices lies within the maximal inscribed circle and will thus be subtracted during the RSM computation resulting in a small RSM value for region R1

value for VVRM(v). RSM(R_i^v) can be computed by employing Dijkstra's single source shortest path algorithm [7] with v as source vertex and using the length of the Voronoi curves annotated to the edges of the GVG as weights. In addition, the following modifications are required:

1. During the first step when relaxing the edges of the start vertex, only the edge $e_{R_i^v}$ leading from v into region R_i^v is considered as connected to v, not those leading into other regions of v.
2. Whenever an edge $e = (u, w)$ is examined, check if either $u = v$ or $w = v$ and $e \neq e_{R_i^v}$. If this is the case, a cycle leading back to v has been found and the computation is terminated returning RSM(R_i^v) $= \infty$.

If the shortest distances from v to the other vertices have been computed by the modified Dijkstra algorithm and no cycle has been detected, these distances have to be corrected for all degree one vertices by subtracting the length of the path lying within the inscribed circle of v. Finally, the corrected distances of the degree one vertices have to be compared to determine the 3rd highest value which will be returned as VVRM(v).

With the time complexity of the dominating Dijkstra algorithm being $O(|E| + |V| \log |V|)$, we end up with a total time complexity for computing VVRM(v) with degree(v) $= n$ of $O(n(|E| + |V| \log |V|))$. The computation can be made more efficient by using a detected cycle to return a RSM value of ∞ for both regions connected by the cycle making the significance computation for the other region superfluous. Furthermore, in cases in which a vertex has no cyclic regions, the Dijkstra computations for the set of regions will only cover disjoint subgraphs of the GVG.

To compute the relevance values for all vertices in the graph, changes the problem into an all-pair shortest path problem in which the shortest paths from every vertex to every other vertex have to be determined. Unfortunately, no algorithm for this problem is known that has a better worst-case complexity than the one we get when we apply Dijkstra's single source shortest path algorithm for every vertex independently which results in a $O(|V|^2 \log |V| + |E||V|)$ time complexity. However, this allows us to employ the modifications as described above and we can restrict the computation to vertices with a degree of three and higher since the relevance measure is only defined for these.

4.3 Computation from Incomplete Information

In the context of mobile robot mapping most of the time no complete GVG is available as assumed in the previous section. However, we still want to compute coarser route graph layers from the partially constructed global GVG that the robot has built up so far. In addition, hierarchical localization based on a matching scheme that utilizes the relevance values of the vertices in both, the global and the local GVG, requires to compute the relevance values for the inevitably incomplete local GVGs.

In both cases we still employ the same algorithm we applied in the case of complete information treating all vertices of degree one that mark the end of an edge that is still unexplored in the same way as the corner vertices. However, without knowing the complete GVG the values computed from an incomplete or local GVG will often be lower bound estimates on the actual relevance values. Every RSM value computed for a non-cyclic region in which the local GVG has unexplored edges will be a lower bound of the real RSM value and will be marked as such. Therefore, VVRM(v) of any vertex v only yields the correct relevance value if all RSM values for this vertex from the 3rd highest on to the smallest are exact values and not just lower bounds. Otherwise, the fact that one of these RSM values could actually be higher could result in a higher 3rd highest RSM value for this vertex. In this case, the computed VVRM value would also be marked as a lower bound estimate of the actual relevance value of v.

Figure 9 illustrates the computation for an incomplete GVG. It shows a local GVG computed for a position in a room with an open door. Vertex B marks a point at which the course of the GVD can not be determined further, since what is outside the door is unknown. As a result, the RSM value computed for region R1 by treating B as a corner vertex is only a lower bound estimate. On the other hand, the RSM values for regions R2 and R3 of A can be computed exactly since they do not contain unexplored edges. Due to the fact that these values are higher than the estimate for R1, VVRM(A) = $\text{maxRSM}_3^{\text{A}}$ = RSM(R1) is also only a lower bound estimate. If either RSM(R2) or RSM(R3) would have been smaller than RSM(R1), VVRM(A) would have been the exact value because the third highest RSM value would have been known exactly.

For our mapping approach this means that lower bound estimates of vertices in the partially constructed global GVG need to be updated whenever new parts are added to a partially constructed global GVG. This will often allow to replace the old estimates by better lower bound estimates closer to the actual relevance values.

5 The Simplification Algorithm

In this section we will describe a basic version of the simplification algorithm that reduces the original GVG to a coarser route graph based on the VVRM values of the vertices and especially the RSM values of the accessible regions. We will leave out the details that are required to construct the hierarchical multi-layer representation and to update the annotations of the graph whenever it is modified. Input of the algorithm are the original GVG and a threshold value θ that determines which vertices will remain in the simplified graph.

The coarser route graph computed by the algorithm satisfies the following properties:

1. The coarser graph will be connected (assuming that the original GVG is connected).
2. All inner vertices of the GVG with VVRM value $\geq \theta$ will remain in the coarser graph.
3. All cycles in the original GVG will remain in the coarser graph.
4. For every inner vertex v remaining in the coarser graph, a leaving edge exists for every region R of v with $\text{RSM}(R) \geq \theta$.
5. The coarser route graph will not contain any vertices with degree two.

The algorithm consists of three subroutines. The first two prune the graph by removing different cases of degree one vertices and are iterated repeatedly until no further pruning is possible. After that, the third subroutine replaces vertices of degree two that have been left over by the first two steps (see algorithm 1 for a pseudocode version).

The operations performed in the three subroutines are illustrated in figure 10. In 10a, a degree one vertex A is connected to an inner vertex B and the RSM value of the region of B to which A belongs is smaller than θ. All such vertices will be removed by the first subroutine (*pruneInsignificantRegions*). Vertex A in 10b on the other hand belongs to a significant region of B (the RSM value of this region is greater than θ) and thus A has to remain if B remains in the coarser

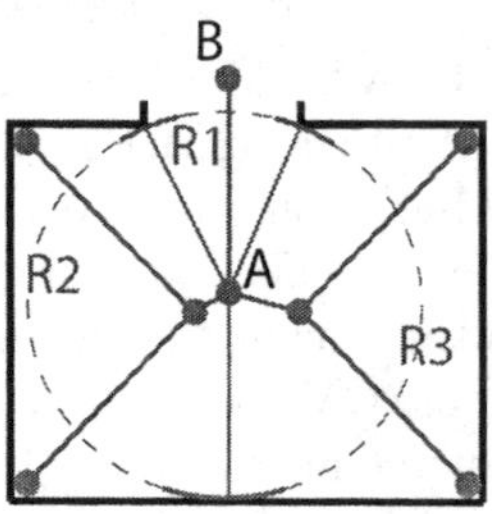

Fig. 9. Computation of the relevance values for a local GVG: While the RSM values for regions R2 and R3 of vertex A can be computed exactly, the value computed for R1 only yields a lower bound on the actual RSM value since it is not known how the GVD continues behind B

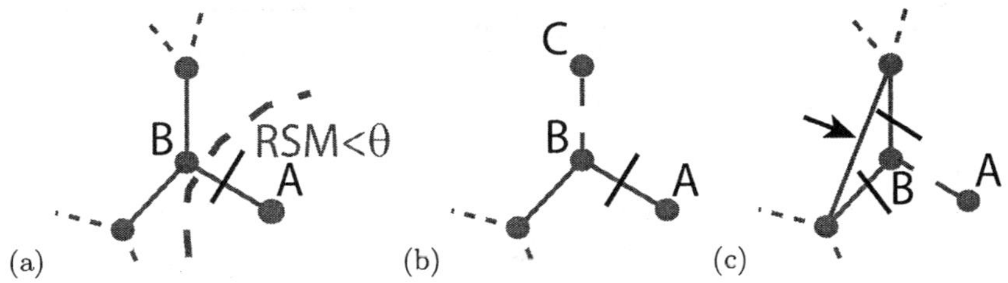

Fig. 10. Operations performed by the different subroutines of the simplification algorithm: (a) Removing a degree one vertex (A) that belongs to an insignificant region of its neighbor (B), (b) removing a degree one vertex (A) from a degree two vertex (B) after C has been removed in an earlier step, and (c) replacing a remaining degree two vertex by an edge directly connecting its neighbors

graph (property 4). However, if the first subroutine ever reduces B to a degree two vertex by removing C, A needs to be removed since it is now clear that B will not be an inner vertex in the coarser graph. Thus, the second subroutine (*pruneSignificantRegions*) removes all degree one vertices that are connected to a degree two vertex. Since both subroutines can create new cases that have to be dealt with by the other subroutine, they are called alternatingly until both fail to remove further vertices.

Finally, the third subroutine (*removeDeg2Vertices*) removes all degree two vertices created by the other subroutines as depicted in figure 10c. Here, vertex A has been removed by the first subroutine leaving a degree two vertex B. B will be removed together with the edges connecting it with its two neighbors and those will be directly connected by a new edge.

As long as the threshold value θ is chosen so that at least one vertex v with VVRM(v) $\geq \theta$ exists in the given GVG, the result of the algorithm is uniquely determined independently of the order in which the vertices are processed. However, when this is not case the GVG will be reduced to one of two degenerated cases: The coarse route graph will either contain a single vertex with an edge leading back to itself (cyclic case) or it will consist of two vertices connected by a single edge (linear case). Which vertices will remain then depends on the processing order. The example shown in figure 5b was constructed with θ set to 1000mm, a value that already produces very abstract representations. The size of the complete environment is approximately 9x10 meter.

6 Experimental Results

In a first experiment we tested how well the relevance measures and the simplification algorithm are able to extract a stable route graph from noisy sensor data. We therefore used a simulation to compute GVGs from range data with varying noise ratios and applied the simplification algorithm with a fixed threshold value of 1000mm. Figures 11a and 11b show the GVGs for the noise ratio of the real laser scanner we use on our robot and for a unrealistically high noise

Algorithm 1 The simplification algorithm.

```
procedure simplify (GVG g, double θ)
  bool changed ← false
  repeat
    pruneInsignificantRegions(g,θ)
    changed ← pruneSignificantRegions(g,θ)
  until not(changed)
  removeDeg2Vertices(g)

procedure pruneInsignificantRegions(GVG g, double θ)
  L ← set of all vertices v of g with degree(v) = 1
  while |L| > 0 do
    v ← arbitrary element of L
    w ← only vertex adjacent to v in g
    r ← RSM value of the region of w containing v
    if r < θ and degree(w) > 1 then
      remove v and its edge from g
      if degree(w) = 1 then
        L ← L ∪ {w}
      end if
    end if
    L ← L \ {v}
  end while

function bool pruneSignificantRegions(GVG g, double θ)
  bool changed ← false
  L ← set of all vertices v of g with degree(v) = 1
  while |L| > 0 do
    v ← arbitrary element of L
    w ← only vertex adjacent to v in g
    if degree(w) = 2 then
      remove v and its edge from g
      changed ← true
      L ← L ∪ {w}
    end if
    L ← L \ {v}
  end while
  return changed

procedure removeDeg2Vertices(GVG g)
  L ← set of all vertices v of g with degree(v) = 2
  while |L| > 0 do
    v ← arbitrary element of L
    w ← one vertex adjacent to v in g
    x ← the other vertex adjacent to v in g
    remove v and both its edges from g
    add edge connecting w and x to g
    L ← L \ {v}
  end while
```

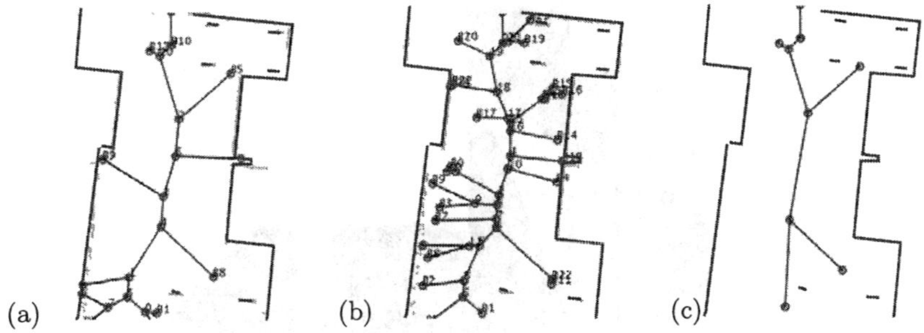

Fig. 11. Simulation of different sensor properties: (a) shows the GVG constructed with a range sensor with low and (b) with high sensor noise. Identical coarse route graphs are computed from both GVGs (c)

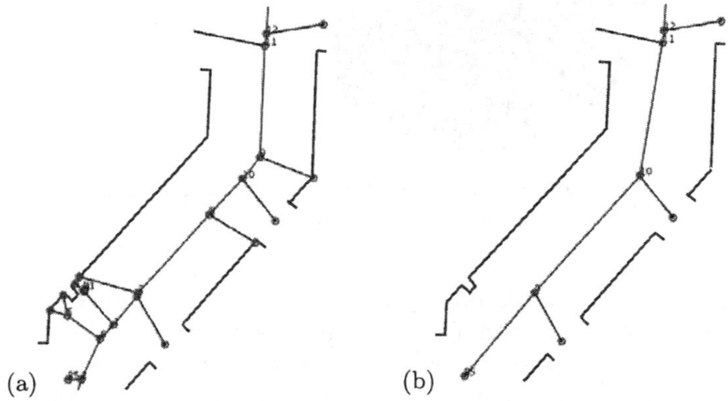

Fig. 12. Example of a coarse route graph (b) computed from a GVG (a) constructed with a real robot that drove down a corridor with flanking offices

ratio, respectively. As is clearly visible, the high noise ratio together with the segmentation of the range data results in a high number of additional vertices and edges in the GVG. In all cases, the simplification resulted in the route graph shown in 11c with only slight variations in the exact positions of the vertices. Further testing by reducing the threshold value revealed that spurious vertices and edges caused by the noise can already be filtered out using a very small threshold value ($\theta < 100mm$), though of course vertices and edges caused by existing concavities of the same order of magnitude will be filtered out, too.

These experiments also demonstrate that our coarser route graph representation is better suited to allow multiple robots equipped with different range sensors to exchange spatial knowledge than the original GVGs.

In a second experiment we tested the approach on real data collected with our Pioneer 2 robot while it was driving along a corridor in our office building. Figure 12a shows a section of the GVG constructed during this exploration run together with a schematization of the environment. Figure 12b shows the route

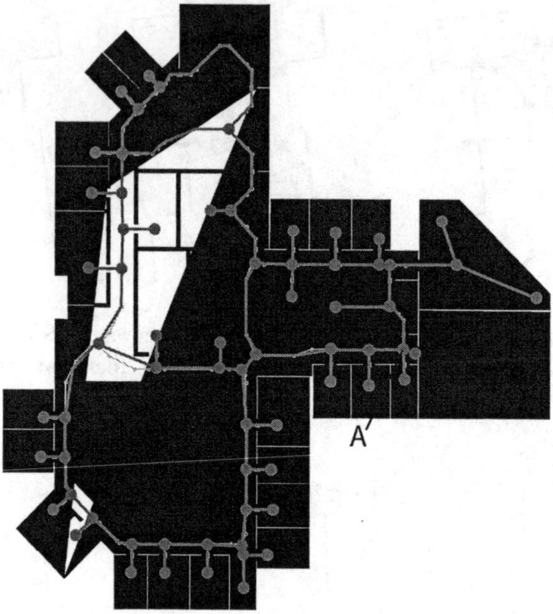

Fig. 13. A coarse route graph computed by the simplification algorithm from a ground plan of a complete floor in the MZH building of the University of Bremen

graph computed from this GVG (again with a threshold value of 1000mm). It demonstrates how the algorithm successfully removes vertices and edges caused by small concavities or noise, resulting in a route graph that only contains edges for traveling along the corridor and for entering the rooms on both sides.

To illustrate that our approach can also be used to automatically generate useful route graphs from 2D floor plans or similar geometric data, we used the 2D model of a floor in the MZH building of the University of Bremen shown in figure 13. Computing the GVG and applying the simplification algorithm afterwards resulted in the route graph depicted in the figure which is well-suited to serve as a starting point for e.g. generating route descriptions leading from a given position to a specific goal location. Extracting or manually annotating additional semantic information (doors, hallways, etc.) should facilitate the generation of route instructions at least in a simple navigation system-like language. The number of vertices has been reduced from 300 in the original GVG to 70 in the simplified route graph and the number of edges from 303 to 73. The computation of the relevance values took about 23 seconds and the simplifcation less than 15 milliseconds on a standard Pentium 4 2-GHz computer.

In this example, a few rooms do not contain a vertex of the simplified route graph. For instance, no edge leads into the big room on the right. The reason for this is that the decision to enter this room is made at vertex A represented by the edge leading towards the room. From there on, no further relevant decision points are encountered and thus entering the room is just a matter of following the taken route until it does not continue any further. That the end point of this edge lies outside the door is a result of the way the simplification algorithm chooses the vertex representing such a region in order to ensure that the result is independent of the order in which the vertices are processed.

7 Conclusions

Route graphs based on the generalized Voronoi diagram are well-suited for a mobile robot equipped with range sensors operating in an indoor environment. As we have argued, a further improvement can be achieved by abstracting from the original GVG and organizing the representation in a hierarchical manner. The two measures we have proposed allow to assess how relevant individual vertices of the GVG are for navigation and together with the described simplification algorithm we have the tools to generate the coarser route graph layers automatically. As the performed experiments demonstrate, the measures perform well in filtering out what we would intuitively judge as relevant decision points. We have discussed the algorithms for computing the relevance values either from complete or incomplete information opening a broad range of applications. In the context of mobile robot mapping and navigation based on such a route graph representation, localization, path-planning, and spatial reasoning benefit from the hierarchical organization of the representation. We also argued that the abstract route graph representations are well-suited to be employed for robot-robot or human-robot communication. We hope to further explore theses applications in the future. In addition, we plan to address other issues involved in generating suitable abstract route graphs like the fact that multiple Voronoi vertices located close to each other may be treated more adequately as a single decision point, an aspect that will be very important for applications involving human-robot communication.

Acknowledgements

The author wishes to thank Christian Freksa, Reinhard Moratz, Frank Dylla, and Diedrich Wolter as well as two anonymous reviewers for valuable comments and suggestions. Thanks also to Martin Held whose software VRONI was used to compute the underlying Voronoi diagrams and to Thomas Arndt who implemented the software used for the experiments. This research was supported by the Deutsche Forschungsgemeinschaft (DFG) grant SFB/TR 8 Spatial Cognition.

References

1. F. Aurenhammer. Voronoi diagrams - A survey of a fundamental geometric data structure. *ACM Computing Surveys*, 23(3):345–405, 1991.
2. H. Blum. A transformation for extracting new descriptors of shape. In W. Wathen-Dunn, editor, *Models for the Perception of Speech and Visual Form*, pages 362–380, Cambridge, MA, 1967. M.I.T. Press.

3. H. Choset and J. Burdick. Sensor based exploration: The Hierarchical Generalized Voronoi Graph. *The International Journal of Robotics Research*, 19(2):96–125, 2000.
4. H. Choset, I. Konukseven, and J. Burdick. Mobile robot navigation: Issues in implementing the generalized Voronoi graph in the plane. In *1996 IEEE/SICE/RSJ International Conference on Multisensor Fusion and Integration for Intelligent Systems, New York, NY, USA*, pages 241–248. IEEE, 1996.
5. A. G. Cohn. Qualitative spatial representation and reasoning techniques. In G. Brewka, C. Habel, and B. Nebel, editors, *KI97: Advances in Artificial Intelligence. Proceedings of the Twenty-first Annual German Conference on Artificial Intelligence*, pages 1 – 30. Springer; Berlin, 1997.
6. J. Crowley. World modeling and position estimation for a mobile robot using ultrasonic ranging. In *Proceedings of IEEE International Conference on Robotics and Automation*, pages 674–680, 1989.
7. E. W. Dijkstra. A note on two problems in connection with graphs. *Numerische Mathematik*, 1:269–271, 1959.
8. Alberto Elfes. Using occupancy grids for mobile robot perception and navigation. *Computer*, 22(6):46–57, 1989.
9. D. Hähnel, D. Fox, W. Burgard, and S. Thrun. A highly efficient FastSLAM algorithm for generating cyclic maps of large-scale environments from raw laser range measurements. In *Proceedings of the Conference on Intelligent Robots and Systems (IROS)*, 2003.
10. D. G. Kirkpatrick. Efficient computation of continuous skeletons. In *20th Annual Symposium on Foundations of Computer Science*, pages 18–27. IEEE, 1979.
11. D. Kortenkamp, T. Weymouth, E. Chown, and S. Kaplan. A scene-based, multilevel representation for mobile robot spatial mapping and navigation. Technical Report CSE-TR-119-92, Computer Science and Engineering Division, University of Michigan, 1992.
12. B. Krieg-Brückner, U. Frese, K. Lüttich, C. Mandel, T. Mossakowski, and R. Ross. Specification of an ontology for route graphs. In C. Freksa, M. Knauff, B. Krieg-Brückner, B. Nebel, and T. Barkowsky, editors, *Spatial Cognition IV - Reasoning, Action, Interaction.* Berlin: Springer, this volume.
13. B. Kuipers. The spatial semantic hierarchy. *Artificial Intelligence*, (119):191–233, 2000.
14. B. J. Kuipers and Y.-T. Byun. A robust, qualitative method for robot spatial learning. In *AAAI 88. Seventh National Conference on Artificial Intelligence*, pages 774–779, 1988.
15. J.-C. Latombe. *Robot Motion Planning.* Kluwer Academic Publishers, 1991.
16. D. T. Lee and R. L. Drysdale III. Generalization of Voronoi diagrams in the plane. *SIAM Journal on Computing*, 10(1):269–271, 1981.
17. J. J. Leonard and H. F. Durrant-Whyte. Simultaneous map building and localization for an autonomous mobile robot. In *Proceedings of IEEE/RSJ International Workshop on Intelligent Robots and Systems*, pages 1442–1447, 1991.
18. F. Lu and E. Milios. Robot pose estimation in unknown environments by matching 2D scans. In *IEEE Computer Vision and Pattern Recognition Conference (CVPR)*, 1994.
19. N. Mayya and V. T. Rajan. Voronoi diagrams of polygons: A framework for shape representation. In *IEEE Conference on Computer Vision and Pattern Recognition*, pages 638–643, 1994.

20. M. Montemerlo, S. Thrun, D. Koller, and B. Wegbreit. FastSLAM: A factored solution to the simultaneous localization and mapping problem. In *Proceedings of the AAAI National Conference on Artificial Intelligence*, Edmonton, Canada, 2002. AAAI.
21. R. Moratz and J. O. Wallgrün. Spatial reasoning about relative orientation and distance for robot exploration. In W. Kuhn, M.F. Worboys, and S. Timpf, editors, *Spatial Information Theory: Foundations of Geographic Information Science. Conference on Spatial Information Theory (COSIT)*, Lecture Notes in Computer Science, pages 61–74. Springer Heidelberg, 2003.
22. Reinhard Moratz and Jan Oliver Wallgrün. Propagation of distance and orientation intervals. In *Proceedings of 2003 IEEE/RSJ International Conference on Intelligent Robots and Systems (IROS)*, pages 3245–3250, 2003.
23. H. P. Moravec. Certainty grids for sensor fusion in mobile robots. *AI Magazine*, 9(2):61–77, 1988.
24. C. Ó'Dúnlaing and C. K. Yap. A retraction method for planning the motion of a disc. *Journal of Algorithms*, 6:104–111, 1982.
25. R. L. Ogniewicz and O. Kübler. Hierarchic Voronoi skeletons. *Pattern Recognition*, 28(3):343–359, 1995.
26. R. Ross, H. Shi, T. Vierhuff, B. Krieg-Brückner, and J. Bateman. Towards dialogue based shared control of navigating robots. In C. Freksa, M. Knauff, B. Krieg-Brückner, B. Nebel, and T. Barkowsky, editors, *Spatial Cognition IV - Reasoning, Action, Interaction.* Berlin: Springer, this volume.
27. K. Siddiqi and B. B. Kimia. Toward a shock grammar for recognition. In *IEEE Conference on Computer Vision and Pattern Recognition*, 1996.
28. S. Thrun. Learning metric-topological maps for indoor mobile robot navigation. *Artificial Intelligence*, 99(1):21–71, 1998.
29. S. Thrun, D. Fox, and W. Burgard. A probabilistic approach to concurrent mapping and localization for mobile robots. *Machine Learning*, 31:29–53, 1998. also appeared in Autonomous Robots 5, 253–271 (joint issue).
30. J. O. Wallgrün. Exploration und Pfadplanung für mobile Roboter basierend auf Generalisierten Voronoi-Graphen. Diploma thesis, Fachbereich Informatik, Universität Hamburg, 2002.
31. S. Werner, B. Krieg-Brückner, and T. Herrmann. Modelling navigational knowledge by route graphs. In *Spatial Cognition II*, volume 1849 of *LNCS,LNAI*, pages 295–316. Springer, 2000.
32. D. Wolter and L. J. Latecki. Shape matching for robot mapping. In Chengqi Zhang, Hans W. Guesgen, and Wai K. Yeap, editors, *Proceedings of 8th Pacific Rim International Conference on Artificial Intelligence*, Auckland, New Zealand, August 2004.
33. W. K. Yeap and M. E. Jefferies. Computing a representation of the local environment. *Aritificial Intelligence*, 107:265–301, 1999.
34. D. van Zwynsvoorde, T. Simeon, and R. Alami. Building topological models for navigation in large scale environments. In *2001 IEEE International Conference on Robotics and Automation (ICRA'2001)*, pages 4256–4261, 2001.

Using 2D and 3D Landmarks to Solve the Correspondence Problem in Cognitive Robot Mapping

Margaret E. Jefferies [1], Michael Cree [2], Michael Mayo [1], and Jesse T. Baker [1]

[1] Department of Computer Science,
University of Waikato, New Zealand
{mjeff, mcc2, mmayo, jtb5}@cs.waikato.ac
[2] Department of Physics and Electrical Engineering,
University of Waikato, New Zealand
cree@phys.waikato.ac.nz

Abstract. We present an approach which uses 2D and 3D landmarks for solving the correspondence problem in Simultaneous Localisation and Mapping (SLAM) in cognitive robot mapping. The nodes in the topological map are a representation for each local space the robot visits. The 2D approach is feature based – a neural network algorithm is used to learn a landmark signature from a set of features extracted from each local space representation. Newly encountered local spaces are classified by the neural network as to how well they match the signatures of the nodes in the topological network. The 3D landmarks are computed from camera views of the local space. Using multiple 2D views, identified landmarks are projected, with their correct location and orientation into 3D world space by scene reconstruction. As the robot moves around the local space, extracted landmarks are integrated into the ASR's scene representation which comprises the 3D landmarks. The landmarks for an ASR scene are compared against the landmark scenes for previously constructed ASRs to determine when the robot is revisiting a place it has been to before.

1 Introduction

In this paper we describe the landmark based approaches we are using to solve the correspondence problem and the associated perceptual aliasing problem in Cognitive Robot Mapping. We use the term Cognitive Robot Mapping because our robot mapping approach is derived from Yeap and Jefferies [1] cognitive mapping system. Yeap and Jefferies' Computational Theory of Cognitive Maps [1, 2] is based on empirical evidence of how the similar mapping problems that robots encounter are solved by animals and humans [3-6]. The challenge is to recognise that parts of the environment viewed from different vantage points *correspond* to the same physical space – the correspondence problem, and to distinguish parts that look the same when they are in fact different – the perceptual aliasing problem. This is regarded as one of the hard localisation problems in cognitive and robot mapping. In robot mapping it is known as the discrete component of the Simultaneous Localisation and Mapping (SLAM) problem, and is often termed cycle or loop closing. The robot traverses a cycle in it environment and must recognise that it has returned to a place it has already visited (i.e. it has closed a loop).

C. Freksa et al. (Eds.): Spatial Cognition IV, LNAI 3343, pp. 434–454, 2005.
© Springer-Verlag Berlin Heidelberg 2005

The problem is encountered in both topological and absolute metric maps. For absolute metric maps, current localisation methods provide consistent enough local maps but residual error accumulates over large distances. By the time a large cycle is encountered the map will contain significant inconsistencies. Current approaches use some form of probability evaluation to estimate the most likely pose (the robot's x-y location and its heading direction) of the robot given its current observations and the current state of its map [7-10]. Detecting the cycle allows the map to be aligned correctly but also means that the error has to be corrected backwards through the map.

Most topological approaches to robot spatial mapping partition the environment in some way and link these partitions as they are experienced to form a topological map [1, 11-13]. The advantage of this approach is that global consistency is not an issue because the error cannot grow unbounded, as in absolute metric maps. Consistency is not a problem within the partitions as they are usually around the size of a local environment. State of the art localisation methods are good enough for local environments.

Recently hybrid topological/metric approaches have emerged [12-14]. Hybrid approaches are popular in the cognitive mapping community [1, 11, 15]. However, the metric and topological maps do not have equal status. The topological map is the dominant representation in their models. Cognitive maps are often regarded as being like a "map in the head" that an agent (human, animal or robot) has for its experience of its spatial environment. In absolute metric maps the need to match the local map associated with a particular pose and the need to propagate error corrections backwards through the map has seen the introduction of topologically linked local metric maps for sequences of poses [7-9]. However, these are ultimately intended as a means to achieve more consistent absolute metric maps.

Cognitive mapping researchers have been interested in the correspondence problem for some time but it was not clear from their computer simulations that their algorithms would handle all the uncertainties that a robot faces in the real world [1, 16]. Recently cognitive mapping researchers have begun to adapt their theories and algorithms for the real world problem robots encounter [17-20]. Both Kuipers [17, 18, 21] and Jefferies [19, 20, 22] adhere to the dominant cognitive mapping tenet, that the prime representation is the topological map (see [1, 11] for a discussion on why this is so). However, they differ in their view of the global metric map. Modazil and Kuipers [21] use a topological map with local metric consistency to build a global metric map with minimal uncertainty. Jefferies et al. [22] use the global metric map to improve the consistency of the topological map although as in Modazil's global map it is built from the local space representations which make up the topological map.

Our approach to mapping the robot's environment extends the hybrid model of [1]. Yeap and Jefferies' [1] topological map of metric local space descriptions has been implemented on a mobile robot with minor adaptations to handle input from a laser range sensor. The local space is the space which appears to enclose the robot and is termed the Absolute Space Representation (ASR) to reflect each representation having its own independent frame of reference. The use of "absolute" in this sense is confined to an ASR.

In this paper we demonstrate how topological matching can be used to solve the correspondence problem and at the same time reduce the false positives which are due to perceptual aliasing. We discuss two approaches implemented in our cognitive mapping system: matching 2D landmarks computed from laser scan data, and matching 3D visual landmarks constructed from successive camera views. We present them as separate systems but the intention is to eventually combine evidence from both systems for more reliable predictions about the robot's location.

The nodes in the topological map are the individual local spaces that the robot visits, which are connected as they are experienced. For the 2D approach the landmarks are distinctive features on the boundary of the local space. Correspondences in the topological map are detected using feature matching. However, we can not match every feature in a local space because when it is approached from different view points, some parts of the local space may be occluded. Therefore, a standard backpropagation neural network is trained to learn a signature for each local space. The signature is composed of the subset of features that are viewable from whichever direction the ASR is approached. New local spaces are classified according to these signatures. If the classification process indicates a match then the neural network should be retrained to account for the different views the robot will have of the same space when it is approached from different routes. The key to solving the perceptual aliasing problem is to recognise that the nodes in the topological map do not exist on their own. They are organised according to their topological connections, and the neighbourhood provides a specific context for any node within the map. When neural network classification indicates a correspondence, subsequent local spaces that the robot visits should also match nodes in the topological map where appropriate.

For the 3D approach the landmarks are the distinctive "faces" of objects in the scene located in 3D space. These landmarks need not be on the boundary of the ASR as in the 2D case. Some of them will be above or below the line of sight of the 2D laser range scanner which provides the data from which the local space description is computed (see below for a description of the mapping algorithm). Furthermore, the landmarks need not be inside the local space, merely visible from within it.

The 3D landmarks, then, are computed from camera views of the local space. Using multiple 2D views, identified landmarks are projected, with their correct location and orientation into 3D world space by scene reconstruction. As the robot moves around the local space, extracted landmarks are integrated into the ASR's scene representation which comprises the 3D landmarks, ie. the faces of objects. The landmarks for an ASR scene are compared against the landmark scenes for previously constructed ASRs to determine when the robot is revisiting a place it has been to before.

2 The Basic Mapping Approach

The topological map comprises a representation for each local space visited with connections to others which have been experienced as neighbours. The local space is defined as the space which "appears" to enclose the robot. This local space representation is referred to as an Absolute Space Representation (ASR), a term emphasising the separateness and independence of each individual local space. Each ASR in the topological map has its own local coordinate frame. Note that these are *local* absolute spaces in contrast to the *global* absolute metric representations referred to in section 1.

Support for the notion of a local space comes from studies that demonstrate how humans [5, 6, 23] and rats [3] manipulate spatial information on the basis of a larger

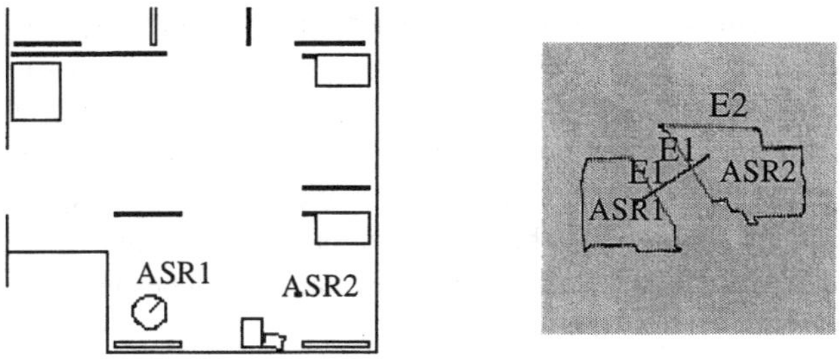

Fig. 1. (a) A section of the robot's environment. (b) The ASRs constructed correspond to the labelled sections of the environment in (a). E1 and E2 are exits, E1 links ASR1 and

spatial context, albeit the space the human or rat occupies. Cheng's [3] experiments demonstrated, that to locate a goal, a rat relies mostly on the geometric relations between the goal and the shape of the environment. In Huttlenlocher and Presson's [23] rotation of the spatial array versus rotation of the viewer tasks, children found it very difficult to construct an internal representation of the spatial array when, in the initial presentation, the room or some crucial element in it moved in relation to the viewer. Thus one's spatial ability to construct a reliable representation for what one sees depends on the spatial array having a fixed position in its surroundings. This suggests that it is represented in relation to the viewer's surroundings which in turn suggests a spatial framework that is based on these surroundings.

The basic algorithm described in [1] was modified to handle input from a laser range sensor and to handle accumulating odometric and sensor errors. However the fundamentals of the algorithm remain. Yeap and Jefferies [1] argued that the exits should be constructed first because they are the gaps in the boundary which tell the robot how it can leave the current space. An exit will occur where there is an occlusion and is formed by creating the shortest edge which covers the occlusion. Once the exits are formed it is a straightforward process to connect the surfaces, which lie between them, to form the boundary of the ASR. At the same time surfaces which are viewed through the exits, and are thus outside the ASR, are eliminated as they are not part of its boundary. Fig. 1(b) shows a sequence of two ASRs so computed. See [1] for an in-depth description of the basic algorithm and [19, 24] for the details of how it is implemented on an autonomous mobile robot using laser range sensing.

Rofer's [25] histogram correlation localisation method is used to provide consistency within ASRs. New ASRs are computed whenever the robot crosses an exit into an unexplored region and ASRs are linked, as they are experienced, via the exits which connect them to their neighbours in the topological map. The ASRs are the nodes of the topological map and the exits are its edges. Fig. 1(b) shows an example of a topological map constructed in this way.

3 Landmarks in Cognitive Mapping

The dominant theory regarding the way spatial knowledge of an environment progresses is that of Siegel and White [26]. According to this theory the developmental progression of a cognitive map is from landmark map to route map (also known as a topological map) to survey map (also known as an absolute metric map). That is, landmarks are first remembered and early on they are the key to localising in an environment. Landmarks are followed by an initial topological network, and then a much more expanded one, and finally metric information about direction and distances becomes available. This view has been popular in computational cognitive mapping with Siegel and White's theory underpinning several approaches [11, 15].

An alternative view is that metric knowledge can be learned very quickly at least alongside landmark and route knowledge [27, 28]. Montello [28] argued that pure landmark or route knowledge always coexists alongside metric knowledge about distance and direction. We argue further that landmarks have a location and extension in space and we know that space, ie. the local space and its manifestation in the ASR, before we compute the landmarks. Thus at least for the local space, metric knowledge precedes landmark knowledge. The progression of spatial knowledge, therefore, is from metric local space to landmark to topological map. We in fact compute a global metric map alongside the topological map but this is outside the focus of this paper (a discussion on this aspect of our work can be found in [22]).

4 Using 2D Landmark Signatures to Recognise ASRs in a Topological Map

In this section we present a feature based matching approach to closing cycles in a topological map. As the robot enters a local space and constructs an ASR for it, the set of features for the ASR are classified by the neural network. The neural network returns its prediction, in the form of a score for each ASR in the topological map, indicating its degree of similarity with the ASR that the robot currently occupies. If all the values are below a chosen threshold then the current ASR is treated as a new ASR. The neural network is then trained with the new ASR's feature set to find a signature that will be used to recognise it when it is revisited. The subset of features the neural network selects to characterise the ASR can be considered as landmarks and hence the term *landmark signatures*.

4.1 Feature Selection

The feature set needs to accommodate sensing errors and be able to handle partial matches resulting from occlusions. We divide the ASR into segments, where each segment is a region of the ASR boundary which has a consistent gradient. The segments are divided into minor (short) segments and major (long) segments. Minor segments often result from spurious effects therefore, they are not included in the feature set. The remaining segments are used to form the initial set of features given to the neural network. In addition to the segment, a feature comprises the angles corresponding to the change in gradient between adjacent segments, traversing the ASR in a clockwise direction. Fig. 2(c) shows the segments extracted for the ASR depicted in Fig. 2(a).

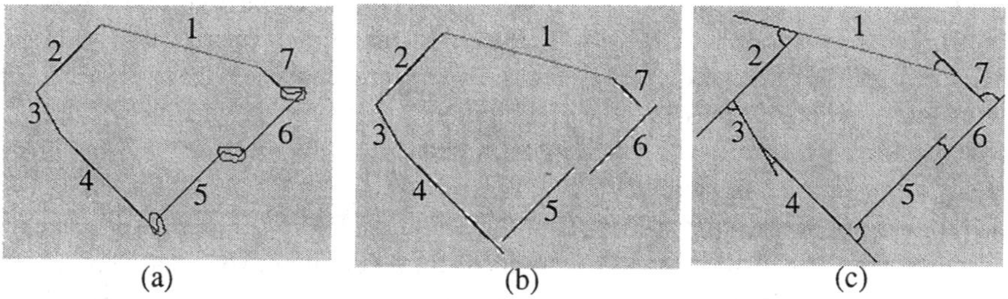

Fig. 2. The Features extracted from an ASR. (a) The ASR with minor segments encircled and major segments labelled 1-7. (b) Minor segments are removed. (c) The segments and angles which comprise the initial feature set

There are seven major segments labelled 1-7, and three minor segments. Segment 1 denotes an exit. Segment 3 represents a gap in the boundary but is turned into a surface because it is too small for the robot to pass through. The features extracted are listed in Table 1.

Table 1. The initial features extracted for the ASR in Fig. 2

Segment No	Length (mm)	Angle 1 (degrees)	Angle 2 (degrees)
1	1800	32.86	58.81
2	1008.9	58.81	-281.17
3	522.3	-281.17	10.6
4	1506.9	10.6	90.38
5	1014.9	90.38	0.07
6	991.3	0.03	88.44
7	392.8	88.44	32.86

4.2 Signature Learning and ASR Classification

The requirements of the learning algorithm were as follows. The learning algorithm needed to be incremental and be able to add in new classes (ASR signatures) online as the new ASRs are encountered. There could be no restriction on either the number of boundary segments or the number of distinct ASRs in the environment. The algorithm needed to be able to weaken the effect of features common to many ASRs while strengthening the effect of those that distinguish ASRs as each new ASR was encountered. Thus because the features of interest are those which *distinguish* ASRs, whenever a new ASR was encountered the signatures for all the ASRs needed to be relearned in the context of the feature set for the new ASR. This is what neural networks do well. While the learning process could run in the background a fast prediction process was essential if it was to run in real time. Back-propagation neural networks provide fast prediction times and were thus chosen to learn the ASR signatures and predict matches of newly computed ASRs with previously visited ASRs. Nguyen-Widrow initialisation, momentum and batch updating of weights are used along with a bipolar sigmoid activation function.

The ranges of the input values (10m for length, and 360° for angles) are discretised into intervals. This is a practical requirement for a neural network but also accommodates sensor error. In the current implementation, a length interval of 200mm and angle interval of 45° are used. Each input neuron represents a particular <length, angle, angle> combination. When classifying an ASR, the output neuron associated with each ASR outputs a value between -1 and 1 indicating the similarity of the new ASR with the visited ASR.

An example of a cycle is shown in Fig. 3. In this example, the environment is a replica of the laboratory our Pioneer 2DX traverses, from our simulator using laser range sensing. The robot has traversed the environment depicted in Fig. 3(a) constructing ASRs in the topological map (Fig. 3(b)) in the order that they are numbered. The robot now re-enters ASR2 via ASR7. The newly computed ASR2* is shown in Fig. 3(c). The similarity predictions for ASRs 1 - 11 are shown in Table 2. Five values stand out, .78, .94, .89, .71, and .72 for ASRs 1, 2, 3, 4, and 6 respectively. If the threshold value were set at 0.7, say, then these would all be candidate matches. One cannot simply choose the best match because in many environments the ASRs for different local spaces will look similar (the perceptual aliasing problem). More evidence is needed to choose between them if indeed any of them should be chosen. In this case it is appropriate to choose the largest value. However this is not always so as can be seen in the next example. While we are gathering empirical evidence as to what is a good threshold value, currently we take a conservative approach and reject similarity values below 0.9.

Table 2. The similarity values for ASR 2* in Fig. 3

ASR	1	2	3	4	5	6	7	8	9	10	11
pred	.78	.94	.89	.71	-.11	.72	.18	.51	.34	.36	.04

In the example in Fig. 4. the robot re-enters ASR3 via ASR10. The similarity values for ASRs 1-10 are shown in Table 3. Four values stand out, .97, .91, .88, and .77 for ASRs 2, 3, 8, and 10 respectively. With a threshold value of 0.9 we need to choose between 0.97 for ASR2 and 0.91 for ASR3. The highest value, for ASR2, is an example of a false positive.

Clearly in this case, the new ASR, ASR3*, overlaps both ASR3 and ASR2, the ASRs with the highest predictions. If the new ASR does match a previously visited ASR then one would expect that its neighbours would match neighbours of the matched

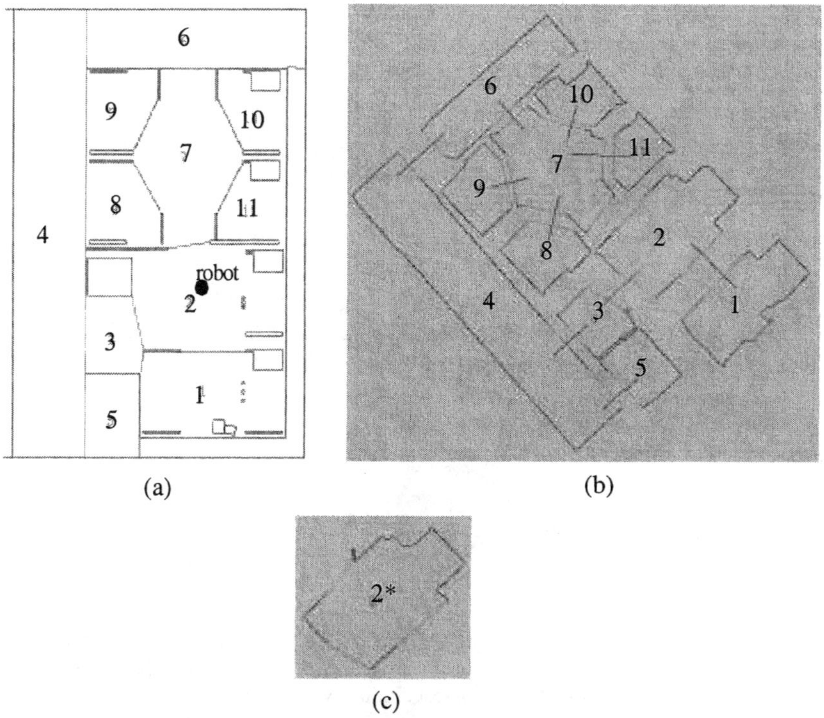

Fig. 3. A positive match. (a) The environment. (b) the topological map constructed in the order the ASRs are numbered. (c) The robot has re-entered ASR2 via ASR7. ASR2* depicts the newly computed ASR to be matched

ASR. We currently gather evidence in this way for sequences of three ASRs, combining their predictions (see Section 4.3).

Table 3. The similarity values for ASR3* in Fig. 4

ASR	1	2	3	4	5	6	7	8	9	10
pred	.46	.97	.91	.48	.64	.26	.57	.88	.15	.77

In the example in Fig. 4 there is evidence to suggest that the new ASR is a combination of both ASR3 and ASR2. This evidence comes in the form of the high predictions for ASR2 and ASR3 which are linked in the topological map and the overlap which occurs in the global metric map. However we need to do further testing to determine if there is any gain in matching under these circumstances. It may be that taking the conservative approach of rejecting the match would be less problematic. Note that missing a match in topological mapping is not catastrophic – an opportunity for a shortcut is missed but reliable (not necessarily optimal) navigation is still possible.

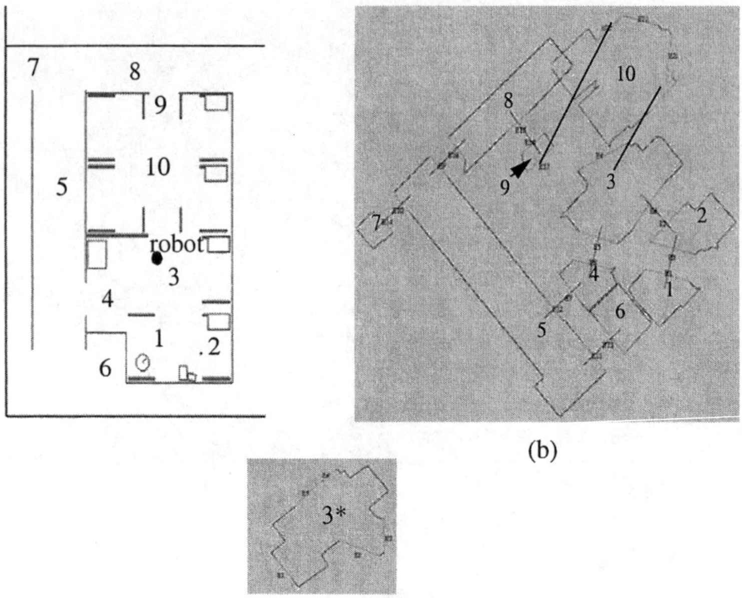

Fig. 4. An example of a false positive prediction (a) the environment (b) the topological map (c) the robot re-enters ASR3 and computes the ASR, ASR3*, as depicted. It covers both ASR3 and ASR4 and extends into ASR2. The highest similarity value is for ASR2

4.3 Topological Matching

The idea behind topological matching is to delay committing to a match in the topological map until it can be determined if a sequence of ASRs containing a new ASR matches a sequence of the previously computed ASRs. In other words we focus on matching 2 or more consecutive ASRs instead of single ASRs. The approach is motivated by the cost involved in unravelling the topological map, once a link has been made, if at some later stage it is discovered to be in error. An alternative approach would be to make the link at the outset and monitor the next few ASRs to see if they are consistent with what should follow in the cycle. This is the approach taken in [13]. In practice one would need to on the alert for mismatches at least until one's certainty with regards the match reaches an acceptable level. This would depend on the environment and how different the individual ASRs were. For the moment we take an ad hoc approach having found that sequences of order 3 give good results in the environments our robot navigates. A simple simulated environment is used in Fig. 5 to demonstrate the process. The robot traverses the environment computing ASRs which are numbered in the order that they are encountered. In Fig. 5(b) the robot has re-entered ASR1 via ASR5. A new ASR is constructed and labelled ASR6, but it is not clear whether or not this is a new ASR or a previously visited one. The similarities of ASR6 to the others are listed in Table 4. The robot delays committing to a match and continues to explore, visiting ASRs 6, 7 and 8 in Fig. 5 (c). The sequence of ASRs 1-3 is the only

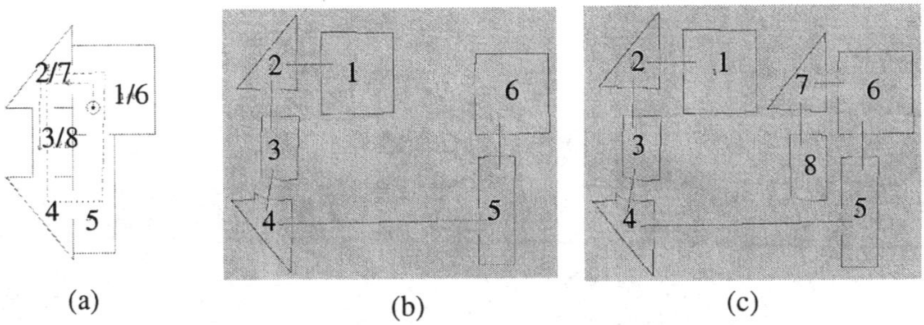

Fig. 5. Topological matching. (a) A simple environment from our simulator (b) The topological map after the robot has re-entered ASR 1 via ASR5. A new ASR5 for the same space, ASR5 is linked to ASR5. (c) The sequence ASR6, ASR7, ASR8 match the sequence ASR1, ASR2, ASR3 confirming the match of ASRs 1 and 6

sequence of order 3 containing ASR1. Classifying ASRs 7 and 8 give the predictions 0.92 and 0.93 respectively, that they match ASRs 3 and 2. All three predictions are above the 0.9 threshold, indicating a positive match of ASR1 and ASR6, and the topological map can be adjusted to reflect this.

Table 4. The similarity values for ASR6 in Fig. 5

ASR	1	2	3	4	5
pred	.9	.73	.74	.58	.55

In this example there was only one prediction to be validated. In more complex environments multiple hypotheses would be carried. We are currently investigating how best to converge to a winning hypothesis particularly in environments with a high similarity. Sequences of higher order may be needed in these environments.

5 Using 3D Visual Landmark Configurations to Recognise ASRs in a Topological Map

There are three main components to the visual landmark recognition. First the coordinate projections required to transform the 2D views to full 3D data are computed. These are determined by recognising strong matching corner features in a pair of images taken at different view points, usually with the robot having moved forward into the scene between the two images. Stereographic reconstruction is used to project the corners into 3D world space. Second, distinctive landmarks of uniform colour and texture are located in the views and using the known projections are projected into their correct location and orientation in 3D world space. Lastly, recorded distinctive landmarks in previously visited ASRs are tested against those detected in the current ASR for matches.

For the moment the projection of the 2D camera views into 3D world space has been kept separate from the segmentation and projection of distinctive landmarks in the

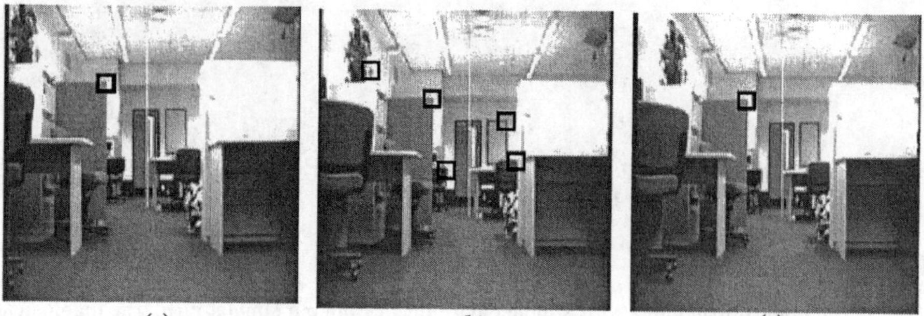

(a) (b) (c)

Fig. 6. Finding a corresponding corner in consecutive views (a) the corner in one image (b) candidate corners in the other image (c) The corresponding corner in the other image

camera view. This reductionist approach has been taken, at this stage, to provide a clear systematic route to implementation of the visual landmark recognition system. Ultimately one would envision developing an approach that calculates the scene reconstruction projection as part of the segmentation of distinctive landmarks, thus providing for better efficiency. We will now discuss each of the three main components of the system in detail.

5.1 Projection from 2D Camera Views to 3D Coordinate Space

To calculate the scene reconstruction projections required for projecting 2D camera views into 3D world space, matching points in two camera views are identified. The corners of intensity disparities which lie on landmark boundaries in camera views provide the necessary matching points. The robot's odometry provides the relative position of the two camera views. Fig. 6 shows the corresponding corners in two views from a camera mounted on our Pioneer 2DX robot. Pollefeys [29] method with one modification is used for corner detection and projection. The Harris corner detector [30] is employed to identify and extract corners. It proceeds by applying the Prewitt edge detector in the horizontal and vertical directions, and constructs the smoothed squared image derivatives:

$$l_x = G(dx^2)$$

$$l_y = G(dy^2)$$

$$l_{xy} = G(dx \cdot dy)$$

where *dx* and *dy* are the results of the Prewitt operator and G is a smoothing operator that consists of a convolution by a 5x5 pixel kernel of a circularly symmetric Gaussian of radial extension $\sigma = 1$.

The corner intensity measure [31]

$$c = \frac{l_x l_y - l_{xy}^2}{l_x + l_y}$$

is calculated. The local maxima in c are identified; these correspond to significant corners in the camera view.

The corner detector is applied to two consecutive views between which the robot has moved a small distance. To identify matching corners a small neighbourhood is extracted about each detected corner from the camera view and a similarity measure is evaluated between corners of the two consecutive views. We take u to be the pixel values of the neighbourhood of the first corner and v to be the pixel values of the neighbourhood of the second corner. The means ($\bar{u}$ and $\bar{v}$) are calculated for each corner, as is the Pearson cross-correlation χ between the two corners. Corners in the first view are compared to those of the second view by calculating the similarity value.

$$s = \chi(1 - |\bar{u} - \bar{v}|)$$

This similarity value is a modification and improvement to that specified by [29]. High values of s indicate potentially matching corners. By using the translation of the robot determined from odometry, a prediction of the likely position in the second view of a corner detected in the first view is made and this is used to reject false matches with a high value of s. In this manner unique corner matches between the two images are identified.

A perspective projection camera model is assumed [32]. With a corner matched between two views, and knowledge of the relative position of the camera of the two views, one can uniquely determine the location of the corner in 3D world space (see Fig. 6(c). Each matching corner is projected into 3D world space giving a set of corresponding points in the two camera views that are fully located in 3D world space. This constitutes the determination of the 3D scene reconstruction projection.

5.2 Landmark Construction

The second part of the problem is to identify distinctive landmarks in camera views that can be used for recognising previously visited ASRs. We treat distinctive landmarks as being contiguous regions of relatively constant colour and texture in camera views. A straight-forward method of segmentation, satisfactory for the structured indoor environments the robot is currently being tested in, has been devised to identify the landmarks

First, the magnitude of the gradient is calculated via the Sobel operator over each of the three colour components (red, green and blue) of a camera view. An edge image, constructed by taking the pixelwise maximum of the three Sobel gradient images, is then thresholded to indicate the boundaries of similar-colour regions. The boundary image is inverted (so that edge is background and connected regions are foreground) and each region is identified and uniquely labelled [33].

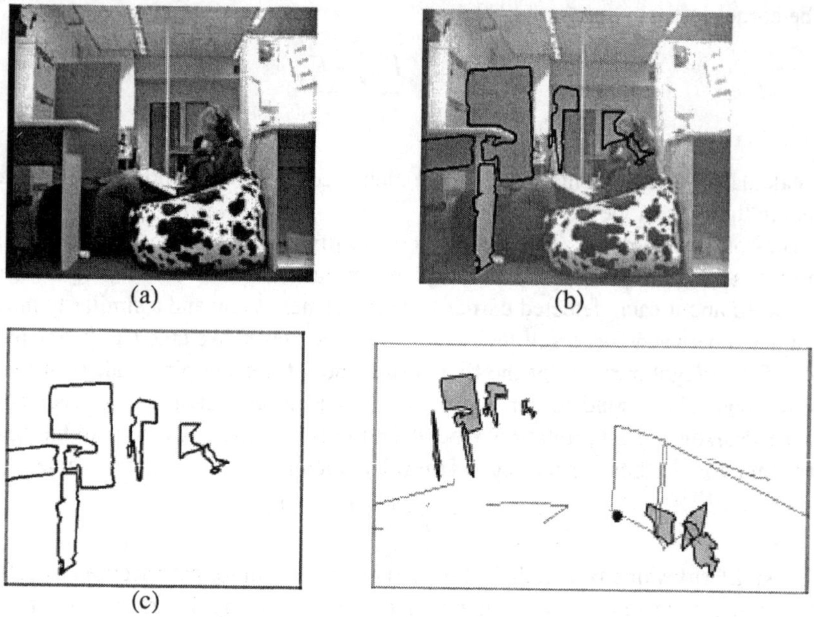

Fig. 7. The landmarks in an ASR scene (a) a view of the room (b) some landmarks overlaid the view (c) the landmarks for the view (d) the landmarks projected into the 3D scene of the ASR

Two methods are used to project the identified regions into 3D world space. If the region is in a position of the camera view that intersects the laser ranger data (a single horizontal range scan at a fixed height in the world) then the range data is used to locate the region in 3D world space. Should the region not intersect the available range data then the corners of the region are compared to the corners detected as part of the Harris corner detector and information from the previously detected corners are used to project distinctive regions into 3D world coordinates. Fig. 7 shows the landmarks computed for the room depicted.

5.3 Matching Landmark Configurations

Having detected and localised distinctive landmarks in world coordinates it remains to compare the landmarks of the current ASR with those of previous ASRs for a match. The colour and the shape of the landmark are used for matching. The colour of landmarks are compared by calculating the Euclidean distance in RGB colour space between the two landmarks. The shape of the landmarks are compared by the histogram correlation method described below.

A landmark from a camera view is projected into 3D space then reprojected on to a 2D plane parallel to the plane of the landmark, thus giving the front on view down the normal to the landmark. The boundary of the reprojected landmark is decomposed into straight line segments and the length and angle of each segment is calculated and inserted into a histogram in which the lines are sorted by orientation along the x-axis of

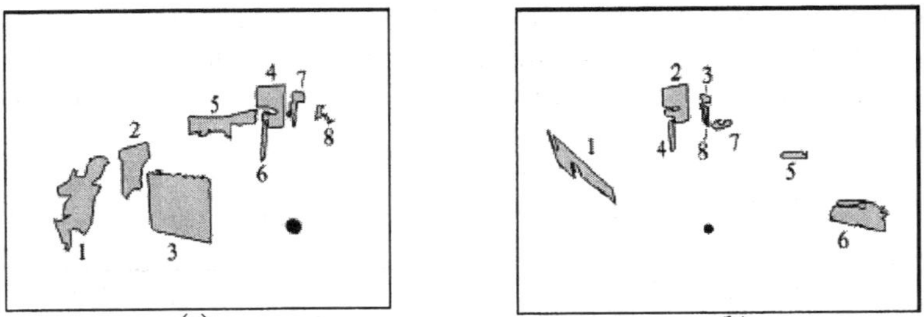

(a) (b)

Fig. 8. The landmark scene representation for two different encounters with the same environment. In (a) and (b) the landmarks are numbered as they are encountered in each separate encounter

the histogram and the lengths of the lines of the same binned orientation are summed to give the frequency axis of the histogram. For shape comparison the correlation of the histograms of reprojected landmarks involves aligning pairs of landmark and then computing the root mean square (RMS) error. The RMS is normalised by the energy of the two histograms. This constitutes the shape disparity value. The disparity values for the colour of landmark pairs are computed from the Euclidean distance between the means of the pairs in RGB colour space. Fig. 8 shows the 3D landmark scene representation for two different visits to the room depicted in Fig. 7. The comparison of the landmark pairings is shown in Tables 5 and 6. The most distinguishing feature is that the match of landmark 4 of Fig. 8(a) and landmark 2 of Fig. 8(b). has the smallest

Table 5. Comparison of colour disparities

landmark	1.a	2.a	3.a	4.a	5.a	6.a	7.a	8.a
1.b	**0.019**	0.045	0.307	0.188	**0.020**	0.356	0.202	0.130
2.b	0.161	0.222	0.133	**0.012**	0.197	0.183	0.052	0.052
3.b	0.252	0.314	0.048	0.096	0.289	0.088	0.068	0.141
4.b	0.313	0.375	**0.025**	0.146	0.350	**0.027**	0.131	0.200
5.b	0.350	0.412	0.063	0.184	0.387	**0.019**	0.168	0.237
6.b	0.076	**0.014**	0.365	0.247	0.039	0.414	0.260	0.189
7.b	0.073	**0.018**	0.362	0.244	0.038	0.410	0.256	0.186
8.b	0.249	0.311	0.041	0.085	0.286	0.089	0.068	0.137

shape disparity value by far and the smallest colour disparity. Looking at Fig. 8(a) and (b), it is obvious that these two landmarks have the most distinctive shape and are really the only two objects that should match. The fact that there are a number of low colour disparity values in the table is not of great concern; it is the combined evidence of both the colour and shape disparities that give confidence of this match. It is also to be noted that only one good matching pair landmarks is needed. This provides the transformation from one view's coordinate system to the other, hence a combined matching of all landmarks could now be performed in 3D space. This is an area we are currently investigating.

Table 6. Comparison of shape disparity

landmark	1.a	2.a	3.a	4.a	5.a	6.a	7.a	8.a
1.b	0.241	0.286	0.226	0.112	0.052	0.133	0.155	0.254
2.b	0.127	0.137	0.097	**0.017**	0.089	0.050	0.067	0.120
3.b	0.100	0.100	0.065	0.054	0.138	0.056	0.061	0.087
4.b	0.135	0.135	0.089	0.048	0.123	0.054	0.065	0.125
5.b	0.065	0.076	0.054	0.115	0.210	0.097	0.078	0.073
6.b	0.140	0.148	0.112	0.059	0.083	0.068	0.074	0.134
7.b	0.056	0.072	0.064	0.093	0.170	0.081	0.070	0.055
8.b	0.119	0.101	0.063	0.100	0.194	0.085	0.076	0.099

6 Related Work

We will first consider approaches which match 2D representations in a topological map. Just as there are many different ideas as to what constitutes a node in a topological map there are various approaches to *landmark* matching. Bosse et al.'s [13]mapping system (ATLAS) constructs a topological map which comprises interconnected local maps each of the same fixed size, each with its own local coordinate frame. Restricting the local maps to a certain size has the advantage that their complexity is limited and known. However, partitioning the environment in this arbitrary way, rather than exploiting the natural structure inherent in the environment to identify each local space, adds complexity to the transitions from one local map to another. In our system, exits determine the boundary of the local space, and are then the transition points between adjacent local maps. These exits carry an expectation that crossing a particular exit will take the robot into a particular neighbouring ASR. ATLAS constructs a signature for its local maps, as we do in [20, 24] but in ATLAS's local map these comprise non repetitive features from within the local frame. Cycles are detected by matching the local map signatures. The idea of using a subset of distinctive features within the local map to recognise places the robot is revisiting is similar to the topological matching approach that we employ in [20, 24]. However our features are chosen on the basis of how well they distinguish an ASR from other ASRs. While removing repetitive features from the signature reduces the complexity of the local map matching process, the remaining non-repetitive features could easily be the features which are common to many local maps, giving a higher likelihood of false positive matches. The repetitive structure which has been removed might contain the features which would make the local map recognisable, thus increasing the likelihood of false negative matches. The map-matching process is a search for a coordinate transformation that brings overlapping frames into alignment.

The nodes in Kuipers and Beeson's [17] topological map, on the other hand, are distinctive states. The edges connecting them are a description of the actions required to travel between adjacent distinctive states. A k-means clustering algorithm clusters different images of the same distinctive state in the same cluster, thus reducing image variability due to noise. However if the image variability problem is to be addressed it assumes that sufficient images have been captured at each distinctive state.

Kouzoubov and Austin's [34] mapping approach is similar to Bosse's [13] but FastSLAM [35] is used to build the local maps in their topological map. Cycle closing is achieved by matching local maps. Local maps are converted to fully connected graphs whose nodes are landmarks and the edges are the distances between connected landmark pairs. The graphs for two local maps are compared by finding the likelihood that the edges correspond. Thus it is the way the landmarks are related rather than the landmarks themselves that seems important. It is feasible that the landmarks could merely be locations in space - landmarks are often defined in this way in robot mapping. However, it is not clear how the landmarks are selected so that corresponding structures are identified in local maps that represent the same physical space.

Tomatis et al.'s [12] model comprises local metric maps and a separate topological layer. Nodes in the topological map are topological locations, and comprise edges and landmarks such as openings and corners, which are useful for distinguishing different locations in the map. Edges connect topological places and are the transitions which switch from topological navigation to navigation within the metric local map. Cycles are closes in the topological map. Multiple hypothesis tracking is used to determine if two similar nodes in the map belong to the same or different locations.

Most approaches which use vision to recognise locations in the robot's environment have used artificial landmarks [36] or easily detected features such as can be found in ceilings [37]. Recently some approaches have appeared which compute naturally occurring landmarks or features in the robot's surroundings. While not specifically engineered for robot localisation Lowe's [38] approach to matching different views of an object or a scene could well be applied in the robot mapping domain. Lowe matches Scale Invariant Feature Transform (SIFT) features, an approach which transforms image data into scale-invariant coordinates relative to local features. A database of these features is compiled from a set of reference images. Matching of new views is achieved by comparing each feature in the new image against the database of features and finding the best candidate match. Lamon et al. [39] also store a database of features, but in this case they are stored as groupings called fingerprints which characterise a location in the robot's environment. The features are ordered in the fingerprint as they appear in the robot's immediate surroundings. A new fingerprint is computed for each new view and matched against existing fingerprints in the database. Kosecka and Li [40] represent individual locations in the environment by a set of characteristic views and the SIFT features which are extracted from these views. Hidden Markov Localization is applied to the characteristic views and the SIFT features to determine the robot's location.

7 Discussion

To summarise, we have described two methods by a which a robot arriving at a new local space can either recognise that the local space is already visited, or realise that it is a new local space that must be added to its cognitive map. The first method is 2D-based, using laser scan data that is processed into surfaces and exits and then matched geometrically using a neural network, and the second method is 3D-based, in which images from a camera mounted on the robot are captured and features in the consecutive images of the place, and their 3D position, are determined and matched.

In the future, we plan to improve both methods and combine their predictions as to whether an ASR is previously visited or new to come up with a single overall prediction. Our plans for improvements to both the 2D and 3D approaches, and our methods for combining the predictions, are now discussed.

7.1 2D Landmark Matching

Currently a sequence of ASRs of length 3 are matched. The rationale for this was to reduce the likelihood of false positive matches when only single ASRs are compared. For example, suppose that the robot visits a sequence of ASRs 1-9, then ASRs 1* and 2*. The robot now enters a new ASR 3*. If ASR 3* has a strong match to a previously visited ASR, say 9, but none of the ASRs connecting to ASR 9 are similar to ASRs 1* and 2*, then the robot's confidence that ASRs 3* and 9 are the same can be reduced. In other words, the robot looks for a path of length 3 such there are good matches to all of the ASRs 1*, 2* and 3* that it has just traversed. It may turn out that the best matching path in fact comprises ASRs 1, 2, and 3, so ASR 3* matches ASR 3and not ASR 9.

This approach can be generalised so that paths of arbitrary length are maintained, so, in turn, the accuracy of the matching process will be improved. Such an approach may turn out to be helpful when the ASRs are geometrically highly similar with few distinguishing landmarks - this may occur, for example, if the sensors have a coarse resolution and miss many of the fine details, or if the environment is homogoneous, such as an office building, and comprises many identical rectangular offices.

The basic idea is that if a robot traverses N ASRs, then it incrementally maintains an array of dimension N, adding a dimension every time it enters a new or previously visited ASR. Every cell in the array represents a different unique path amongst N ASRs of length N. Each path can also be assigned a probability. For example, suppose the robot traverses ASRs 1-2-3-4 in that order. It is possible that one or more of the ASRs that are previously visited (for example, ASRs 1 and 3) might be the same. Table 7 below depicts what the resulting array would look like. Although there should be 44 possible entries in the array depicted in Table 7, most of them will have zero probability because of that fact that a robot cannot exit an ASR and arrive at the same ASR. In other words, adjacent ASRs in the path must be different. In Table 7, the five possible paths with non-zero probability are shown. The first path, determined by looking at the first row in Table 7, gives the probability of the robot having visited four different ASRs. The second row gives the probability of the robot visiting three different ASRs, and then returning to the original ASR. The fourth row gives the probability of the robot visiting only two ASRs, but each ASR is visited twice (this could happen if, for example, two ASRs are connected by three different exits). The probabilities can be calculated from similarity values produced by the neural network in a straightforward manner.

Table 7. An array of possible paths traversed by the robot

ASR1	ASR2	ASR3	ASR4	prob
1	2	3	4	0.3
1	2	3	1	0.35
1	2	3	2	0.1
1	2	1	2	0.2
1	2	1	3	0.05

This approach means that the matching process is not limited by a fixed value of N. Of course, if N becomes too large, then the array will inevitably also become large: in this case, a consolidation can take place in which the most probable path is selected and N is reduced to the number of ASRs in that path. For example, in Table 7, the most probable path comprises 3 ASRs, so N can be reduced to 3 at some point and the array recalculated. This corresponds to the robot making a decision that ASRs 1 and 4 are the same, but the decision is delayed until a convenient time, or a time when the decision is absolutely necessary. This is a natural extension to the currently implemented 2D matching process, but with an arbitrary delay.

7.2 3D Visual Landmark Matching

An important issue in computing 3D visual objects which need to be matched is that of representation and how to obtain invariance with respect to pose, illumination, and scale and how to deal with occlusion.

Because we project the landmarks into 3D world space we can handle variations in pose and scale. We currently avoid illumination problems by using an environment which has consistent illumination at all times. Using some colour space that separates chroma from intensity will be a first step to dealing with shading and illumination variances. Then we will consider how we could reason with the residual variances as for example, in some of the SIFT approaches.

We have not yet implemented a mechanism for reasoning with configurations of landmarks. One good match provides the coordinate transformation that will allow the process to begin. Given the difficulties in computing reliable shape information across different viewpoints of an object, colour and relative location of landmarks should provide sufficient information for matching configurations. The issue then is the underlying uncertainty which results from occluded and partially occluded landmarks.

7.3 Combining Predictions from 2D and 3D Matching

The next problem is how to combine the predictions of the 2D and 3D matching procedures. If both procedures predict that the robot has entered a new ASR, or agree that the robot has entered the same previously visited ASR, then we can accept that the ASRs match with a high degree of certainty. A problem arises if the predictions are mismatched.

We suggest a simple solution to this problem, which involves converting the predictions from each module into probabilities, and then taking the single prediction with the highest joint probability. For example, suppose the robot has visited three ASRs and then enters a fourth ASR. The probabilities of a match of the new ASR to each of the three previously visited ASRs from the 2D module are, respectively, 0.55, 0.4 and 0.05. However, from the 3D module, the probabilities are 0.25, 0.5 and 0.25. Clearly, the 2D module is picking ASR 1 as the best match to the new ASR, while the 3D module is picking ASR 2. Combining the probabilities by multiplication and renormalising yields 0.4, 0.57, and 0.03, so that clearly ASR 2 is the overall best match to the new ASR. This match can then be accepted (i.e. the robot decides that ASR 2 is the same as ASR 4), or it can be rejected because the probability of 0.57 is below the threshold, in which case the robot assumes that ASR 4 is a new ASR.

8 Conclusion

In this paper we have shown how topological matching of 2D landmarks can be used to solve the correspondence problem and at the same time reduce the effect of false positives which are due to perceptual aliasing. ASRs in a topological map can be recognised from a characteristic subset of their features (the landmarks). Context plays an important role in eliminating false positive matches. The context of a matched node (its neighbourhood) is used to verify that it is in fact a true positive match. We are currently investigating how the robot can bale out of a committed match at some later time when it discovers a mismatch. It should be able to return to an alternative high prediction and test its validity against the accumulated data.

We have shown how 3D landmarks, the faces of objects in a scene, are computed from camera views of the local space. Using multiple 2D views, identified landmarks are projected, with their correct location and orientation into 3D world space by scene reconstruction. The landmarks for an ASR scene are compared against the landmark scenes for previously constructed ASRs to determine when the robot is revisiting a place it has been to before. A successful match is dependent on at least one pair of corresponding landmarks having been reliably extracted. Removing some of the variance in illumination and a less rigorous shape matching approach should provide more matches than we have obtained here.

References

[1] Yeap, W.K. and Jefferies, M.E. Computing a representation of the local environment. *Artificial Intelligence*, 107: 265-301, 1999

[2] Yeap, W.K. Towards a Computational Theory of Cognitive Maps. *Artificial Intelligence*, 34: 297-360, 1988

[3] Cheng, K. A purely geometric module in the rat's spatial representation. *Cognition*, 23: 149-178, 1986

[4] Gallistel, C.R. and Cramer, A.E. Computations on metric maps in mammals: getting oriented and choosing a multi-destination route. *The Journal of Experimental Biology*, 199: 211-217, 1996

[5] Huttenlocher, J., Newcombe, N., and Sandberg, E.H. The coding of spatial location in young children. *Cognitive Psychology*, 27: 115-147, 1994

[6] Wang, R.F. and Spelke, E.S. Updating egocentric representations in human navigation. *Cognition*, 77: 215-250, 2000

[7] Hähnel, D., Burgard, W., Fox, D., and Thrun, S. An efficient fastSLAM algorithm for generating maps of large-scale cyclic environments from raw laser range measurements. *in Proceedings Intelligent Robots and Systems*, 2003.

[8] Thrun, S., Hahnel, D., Ferguson, D., Montemerlo, M., Triebel, R., Burgard, W., Baker, C., Omohundro, Z., Thayer, S., and Whittaker, W. A system for volumetric robotic mapping of abandoned mines. *in Proceedings International Conference on Robotics and Automation*, 2003.

[9] Hähnel, D., Thrun, S., Wegbreit, B., and Burgard, W. Towards lazy data association in SLAM. *in Proceedings 10th International Symposium of Robotics Research*, 2003.

[10] Gutmann, J.-S. and Konolige, K. Incremental mapping of large cyclic environments. *in Proceedings International Symposium on Computational Intelligence in Robotics and Automation*, 1999.

[11] Kuipers, B. The spatial semantic hierarchy. *Artificial Intelligence*, 119: 191-233, 2000

[12] Tomatis, N., Nourbakhsh, I., and Siegwart, R. Hybrid simultaneous localization and map building: Closing the loop with multi-hypotheses tracking. *in Proceedings International Conference on Robotics and Automation*, 2002.

[13] Bosse, M., Newman, P., Leonard, J., Soika, M., Feiten, W., and Teller, S. An Atlas framework for scalable mapping. *in Proceedings International Conference on Robotics and Automation*, 2003.

[14] Thrun, S. Learning metric-topological maps for indoor mobile robot navigation. *Artificial Intelligence*, 99(1): 21-71, 1998

[15] Chown, E., Kaplan, S., and Kortenkamp, D. Prototypes, Location, and Associative Networks (PLAN): Towards a Unified Theory of Cognitive Mapping. *Cognitive Science*, 19: 1-51, 1995

[16] Kuipers, B.J. and Byun, Y.-T. A Robust, Qualitative method for Robot Spatial learning. *in Proceedings of the National Conference on Artificial Intelligence (AAAI-88)*, 1988.

[17] Kuipers, B. and Beeson, P. Bootstrap learning for place recognition. *in Proceedings 18th International Conference on Artificial Intelligence*, 2002.

[18] Kuipers, B., Modayil, J., Beeson, P., and MacMahon, M. Local metrical and global topological maps in the hybrid Spatial Semantic Hierarchy. *in Proceedings IEEE International Conference on Robotics and Automation (ICRA-04)*. 2004.

[19] Jefferies, M.E., Baker, J., and Weng, W. Robot cognitive mapping: A role for a global metric map in a cognitive mapping process. *in Proceedings Workshop on Robot and Cognitive Approaches to Spatial Mapping*, 2003.

[20] Jefferies, M.E., Weng, W., Baker, J.T., and Mayo, M. Using context to solve the correspondence problem in simultaneous localisation and mapping. *in Proceedings 2004 Pacific Rim Conference on Artificial Intelligence*, 2004.

[21] Modazil, J., Beeson, P., and Kuipers, B. Using the topological skeleton for global metric map-building. *in Proceedings IEEE/RSJ International Conference on Intelligent Robots and Systems (IROS-04)*, 2004.

[22] Jefferies, M.E., Cosgrove, M.C., Baker, J.T., and Yeap, W.K. The correspondence problem in topological metric mapping - Using absolute metric maps to close cycles. *in Proceedings 2004 International Conference on Knowledge-Based Intelligent Information and Engineering Systems*, 2004.

[23] Huttenlocher, J. and Presson, C.C. The Coding and Transformation of Spatial Information. *Cognitive Psychology*, 11: 375-394, 1979

[24] Jefferies, M.E., Weng, W., Baker, J.T., Cosgrove, M.C., and Mayo, M. A hybrid approach to finding cycles in hybrid maps. *in Proceedings Australian Conference on Robotics and Automation*, 2003.

[25] Rofer, T. Using histogram correlation to create consistent laser scan maps. *in Proceedings IEEE International Conference on Intelligent Robotics Systems*, 2002.

[26] Siegel, A.W. and White, S.H. The Development of Spatial Representations of Large-Scale Environments. *Advances in Child Development & Behavior by H.W. Reese*, 10, 1975

[27] Moar, I. and Carleton, L.R., Memory for routes. *Quarterly Journal of Experimental Psychology*, 34A: 381-394, 1982

[28] Montello, D.R. A new framework for understanding the acquisition of spatial knowledge in large-scale environments, in *Spatial and Temporal Reasoning in Geographic Information Systems*, Egenhofer, M.J. and Golledge, R.G., Editors. Oxford University Press: New York. p. 143-154, 1998

[29] Pollefeys, M. 3D modeling from inages. *in Proceedings Tutorial Notes 2000 European Conference on Computer Vision*, 2000.

[30] Harris, C. and Stephens, M. A combined corner and edge detector. *in Proceedings 4th ALVETY Vision Conference*, 1998.

[31] Noble, J., Descriptions of Image Surfaces. PhD thesis *Robotics Research Group, Department of Engineering Science*. 1989, Oxford University.

[32] Klette, R., Schlüns, K., and Koschan, A. *Computer Vision: Three-Dimensional Data from Images*. Singapore: Springer Verlag, 1998

[33] Gupta, G.S., Win, T.A., Messom, C., Demidenko, S., and Mukhopadhyay, S. Defect analysis of grit-blasted or spray-painted surface using vision sensing techniques. *in Proceedings Image and Vision Computing New Zealand*, 2003.

[34] Kouzoubov, K. and Austin, D. Hybrid Topological/Metric Approach to SLAM. *in Proceedings 2004 IEEE international Conference on Robotics and Automation*, 2004.

[35] Montemerlo, M., Thrun, S., Koller, D., and Wegbreit, B. Fastslam: A factored solution to the simultaneous localization and mapping problem. *in Proceedings AAAI-02*, 2002.

[36] Taylor, C.J. and Kriegman, D., Vision-based motion planning and exploration algorithms for mobile robots. *IEEE Transactions on Robotics and Automation*, 14(3): 417-427, 1998

[37] Thrun, S. Robot mapping: A survey, in *Exploring Artificial Intelligence in the New millennium*, Lakemeyer, G. and Nebel, B., Editors. Morgan Kaufmann, 2002

[38] Lowe, D. Distinctive image features from scale-invariant keypoints. *International Journal of Computer Vision*, 60(2): 91-110, 2004

[39] Lamon, P., Tapus, A., Glauser, E., Tomatis, N., and Siegwart, R. Environmental Modeling with Fingerprint sequences for topological global localization. *in Proceedings IEEE/RSJ International Conference on Intelligent Robots and Systems*, 2003.

[40] Kosecka, J. and Li, F. Vision based topological markov localization. *in Proceedings 2004 International Conference on Robotics and Automation*, 2004.

Treemap: An $O(\log n)$ Algorithm for Simultaneous Localization and Mapping

Udo Frese

Fachbereich 3 Mathematik – Informatik, Universität Bremen,
Bibliothekstr. 1, 28359 Bremen, Germany
ufrese@informatik.uni-bremen.de

Abstract. This paper presents a very efficient SLAM algorithm that works by hierarchically dividing the map into local regions and subregions. At each level of the hierarchy each region stores a matrix representing some of the landmarks contained in this region. For keeping the matrices small only those landmarks are represented being observable from outside the region. A measurement is integrated into a local subregion using $O(k^2)$ computation time for k landmarks in a subregion. When the robot moves to a different subregion a global update is necessary requiring only $O(k^3 \log n)$ computation time for n overall landmarks. The algorithm is evaluated for map quality, storage space and computation time using simulated and real experiments in an office environment.

1 Introduction

The problem of making a map from local observations is a very old one basically as old as maps themselves. While geodesy, the science of surveying in general dates back to 8000 B.C., it was the achievement of C.F. Gauss to first formalize the problem from the perspective of statistical estimation in his article *"Theoria combinationis observationum erroribus minimis obnoxiae"* [1](1821).

1.1 Simultaneous Localization and Mapping

In the much younger realm of robotics the corresponding problem is that of simultaneous localization and mapping (SLAM). It requires the robot to continuously build a map from sensor data while traveling through the environment. It has been under research since the mid 80ies gaining enormous popularity in recent years. A majority of approaches adhere to the Gaussian formalization. They estimate a vector of n features, e.g. landmarks or laserscan reference frames by minimizing a quadratic error function, i.e. by solving implicitly a linear equation system. With this well established methodology the main question is how to compute or approximate the estimate efficiently. To make this more explicit, there are three important requirements an ideal SLAM algorithm should fulfill that were first proposed by the author in [2] and further discussed in [3]:

C. Freksa et al. (Eds.): Spatial Cognition IV, LNAI 3343, pp. 455–477, 2005.
© Springer-Verlag Berlin Heidelberg 2005

(R1) Bounded Uncertainty. *The uncertainty of any aspect of the map should not be much larger than the minimal uncertainty that could be theoretically derived from the measurements.*

(R2) Linear Storage Space. *The storage space of a map covering a large area should be linear in the number of landmarks ($O(n)$).*

(R3) Linear Update Cost. *Incorporating a measurement into a map covering a large area should have a computational cost at most linear in the number of landmarks ($O(n)$).*

(R1) states that the map shall represent nearly all information contained in the measurements, thus binding the map to reality and limiting approximations. The other postulates (R2) and (R3) regard efficiency, requiring linear space and time consumption. The contribution[1] of this paper is a hierarchical SLAM algorithm that meets the above mentioned requirements. It works by dividing the map into regions and subregions. When integrating a measurement it needs $O(k^2)$ computation time for updating the estimate for a region with k landmarks, $O(k^3 \log n)$ when the robot moves to a different region and $O(kn)$ to compute an estimate for the whole map. There is an extension to the algorithm not covered in this paper that applies "nonlinear rotations" to individual regions to greatly reduce the linearization error caused by error in the robot orientation [4]. The algorithm is landmark (feature) based, requires known data association and assumes a "topologically suitable building" (§6).

1.2 Spatial Cognition

As formalized above the task can be described as computing global coordinates from local measurements. This corresponds to the distinction between egocentric and allocentric spatial memories reported in cognitive psychology [5]. It is a way of integrating spatial observations that is very suited for mobile robots, because the result is a single estimate that incorporates all information available, i.e. one concrete map that is (statistically) consistent with all observations. Together with the corresponding uncertainty information, generally provided as a covariance matrix, such an estimate is very useful. Any derived spatial quantity, like distances and angles can be directly computed from the estimate together with its uncertainty without any complex inference process. Especially most existing algorithms that use a map, like path planning, navigation and localization are designed with global coordinates. Even when the task involves human robot communication, for instance matching natural language descriptions ("after passing the entrance hall turn right") it appears to be promising to directly match qualitative predicates resulting from natural language processing ("right") with an estimated metrical map using empirical definitions of the predicates [6, 7].

[1] This article is based on research conducted during the author's Ph.D. studies at the German Aerospace Center (DLR) in Oberpfaffenhofen.

The paper is organized as follows. After a brief review of related work (§2) the algorithm is presented (§3...§8). It follows an investigation of map quality and computation time based on simulations (§9) and experiments on a real robot in an 60m × 45m office building (§10).

2 State of the Art

After the fundamental paper by Smith et al. [8] in 1988 most work on SLAM was based on the Extended Kalman Filter (EKF) that allows to treat SLAM theoretically thorough as an estimation problem. However, the problem of large computation time remained. The most time consuming part is to update the EKF's covariance matrix after each measurement, taking $O(n^2)$ time for n landmarks. This limited the use to small environments ($n \lessapprox 100$ landmarks).

Recently, interest in SLAM has increased drastically and several, more efficient algorithms have been developed. Many approaches exploit, that observations are *local* in the sense that from a single robot pose only few k landmarks are visible. In the following the more recent contributions will be briefly reviewed. A general overview is given by Thrun [9] and a discussion of the inherent structure of SLAM by Frese [3].

To the authors knowledge the first SLAM algorithm achieving computation time below $O(n^2)$ per measurement while maintaining a consistent estimate for the whole map was the relaxation algorithm by Duckett et al. [10, 11]. They employed an iterative equation solver called *relaxation* to the linear equation system appearing in maximum likelihood estimation. One iteration is applied after each measurement with computation time $O(kn)$ and $O(kn)$ storage space. After closing a loop, more iterations are necessary leading to $O(kn^2)$ computation time in the worst case. This was later improved by the Multilevel Relaxation (MLR) algorithm [12]. It optimizes the map at different levels of resolution similar to multigrid methods used for numerical solution of partial differential equations leading to $O(kn)$ computation time even when closing loops.

Montemerlo et al. [13] derived an algorithm called *FastSLAM* from the observation that the landmark estimates are conditionally independent given the robot pose. Basically, the algorithm is a particle filter (M particles) in which every particle represents a sampled robot trajectory plus a set of n Kalman filters estimating the position for each landmark. The number of particles M is a difficult tradeoff between computation time and quality, especially since it is not clear how M scales with the complexity of the environment. However, the algorithm can handle uncertain landmark identification, which is a unique advantage over the other algorithms discussed in this section.

Guivant and Nebot [14] developed a modification of the EKF called *Compressed EKF* (*CEKF*) that allows the accumulation of measurements in a local region with k landmarks at cost $O(k^2)$ independent from the overall map size n. When the robot leaves this region, the accumulated result must be propagated to the full EKF (*global update*) at cost $O(kn^2)$. An approximate global update can be performed more efficiently in $O(kn^{3/2})$ with $O(n^{3/2})$ storage space needed.

Thrun et al. [15] presented a "constant time" algorithm called the *Sparse Extended Information Filter (SEIF)*, which uses an information matrix instead of a covariance matrix to represent uncertainty. The algorithm exploits the observation that the information matrix is approximately sparse[2] requiring $O(kn)$ storage space. The information matrix representation allows integration of a new measurement in $O(k^2)$ computation time, but to produce a map estimate a system of n linear equations must be solved. Thrun et al. use relaxation but updating only $O(k)$ landmarks after each measurement (using so-called amortization). In general this can negatively affect map quality, since in the numerical literature, relaxation is reputed to need $O(n^2)$ time for reducing the equation error by a constant factor [16].

Bosse et al. [17] avoid the computational problem of updating an estimate for n landmarks in their *Atlas* framework by dividing the map into submaps. There is no global coordinate system, rather each submap performs estimation in its own local frame.

Paskin [18] views the estimation problem as a Gaussian graphical model. He proposed the *Thin Junction Tree Filter (TJTF)* based on the observation that if a set of node separates the graph into two parts, then these parts are conditionally independent given estimates for the separating nodes. The algorithm maintains a junction tree ($O(k^2n)$ space), where every edge corresponds to such a separation. Estimation is performed in $O(k^3n)$ time by passing marginalized distributions along the edges of the junction tree. This algorithm is closely related to the treemap algorithm proposed in this paper although both have been independently developed from completely different perspectives. The correspondence is basically that both use a tree and pass marginalized distributions (TJTF) equivalent to Schur complements (treemap) along edges.

In the next section the treemap algorithm proposed in this paper will be introduced. It can be used in the same way as CEKF providing an estimate for k landmarks of a local region but with only $O(k^3 \log n)$ computation time when changing the region instead of $O(kn^{3/2})$ for CEKF. Alternatively the algorithm can also compute a global estimate for all n landmarks with computation time $O(kn)$. As reported in the experiments, the prefactor in the $O(kn)$ computation is so small, that this can be done for almost "arbitrarily" large maps (12.37ms for $n = 11300$) being the main contribution from a practical perspective.

3 Basic Idea of the Algorithm

The basic idea of the treemap algorithm is to organize the map hierarchically by decomposing the information into small parts called *information blocks* (IBs) and distributing these IBs along the hierarchy. Then each update involves only a small part of the information. For verification, consider figure 1a with a building that is divided into two parts A and B. Now consider the following question:

If the robot is in part A, what is the information needed about B?

[2] This property has later been proven by the author [4, 3].

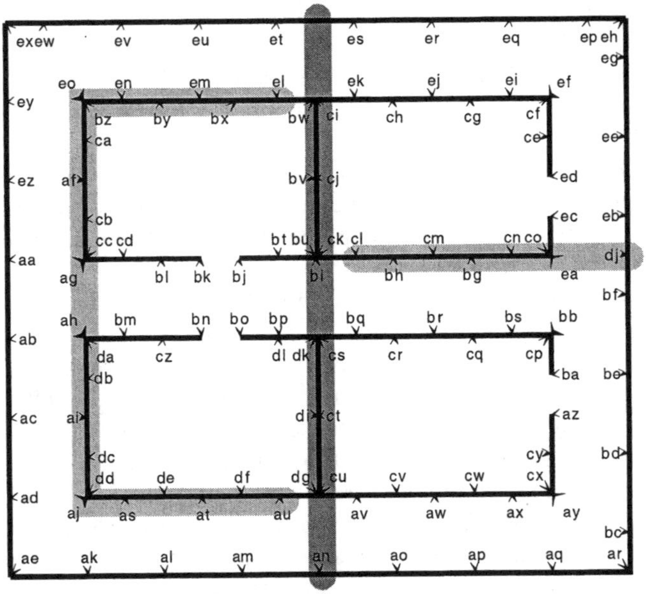

(a)

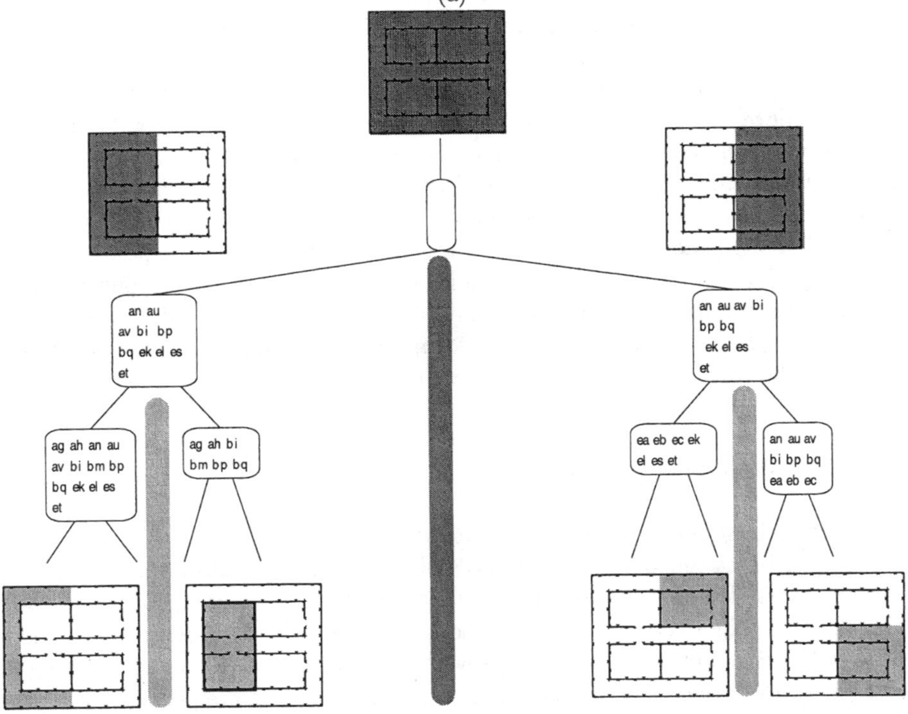

(b)

Fig. 1. First two levels of a hierarchically decomposed building (a) and respective tree representation (b). The first level is indicated by bold dark-gray lines, the second level by bold light-gray lines. The region corresponding to a node is shown next to the node

Some landmarks of B are observable from A and thus may be involved in measurements while the robot is in A. For integrating these measurements, the algorithm must have all information about these landmarks explicitly available. It is important that this information comprises more than just the measurements that directly involve those landmarks. Rather all measurements in B can indirectly yield information about the landmarks observable from A. So the information needed about B is the whole integrated information of all measurements made while the robot is in B on landmarks observable from A. In the following this information is said to be *condensed*, since it comprises everything from the measurements made in B that is needed outside of B.

The idea can be applied recursively by dividing the building into a hierarchy of regions (Fig. 1a). The recursion stops when the size of a region is comparable to the robot's field of view. The condensed information for the different regions can be computed by recursion. For a specific region condensed information for the two subregions is integrated. After that, all landmarks not being observable from outside the region are removed from representation. This process is called *elimination* of landmarks. This is how the information is decomposed into two parts: Part 1 contains information about eliminated landmarks and is stored at the region and not considered further. Part 2 contains information about the landmarks observable from outside and is passed to the next region above. This part contains every information about the region that is necessary when the robot is outside of the region.

At each moment the robot position corresponds to a particular region on the lowest level of hierarchy called the *actual* region into which new landmark observations can be integrated. When the robot is moving the actual region changes from time to time and a global update has to be performed. The key advantage of the hierarchical decomposition is that therefor only the condensed information of the actual region and all regions above need to be updated.

In a similar way an estimate for the local landmarks can be computed. The final integrated information about a landmark is stored in the region where the landmark has been eliminated. So the information about landmarks of the actual region can be collected by traversing the hierarchy down to the actual region.

4 Treemap Data Structure

This section introduces the *treemap* data structure used by the algorithm. At first, it will be assumed that the robot's observations are landmark–landmark measurements. Under this assumption the algorithm is exact up to linearization. In §7 the algorithm will be extended to integrate also landmark–robot and robot–robot (odometry) measurements with a small approximation when changing regions. Both linearization and odometry approximation are performed when storing a measurement in the treemap. The actual computation of the least square estimate from the stored information is performed exactly without further approximations. Thereby the algorithm computes a consistent estimate that is statistically compatible with all measurements following requirement (R1).

4.1 Data Structure

The hierarchy is realized by a binary tree. Each node corresponds to a region and stores information about the landmarks of this region in so called information blocks (IBs). These IBs are quadratic error functions that describe the negative log-likelihood for a vector of landmark positions given the information represented by the IB. Internally they are represented by a small matrix (the information matrix) and a vector. It is said that an information block, a matrix or a vector respectively *represents* a landmark, if it contains information about it. This means that a row / column of the matrix or an entry of the vector corresponds to the landmark.

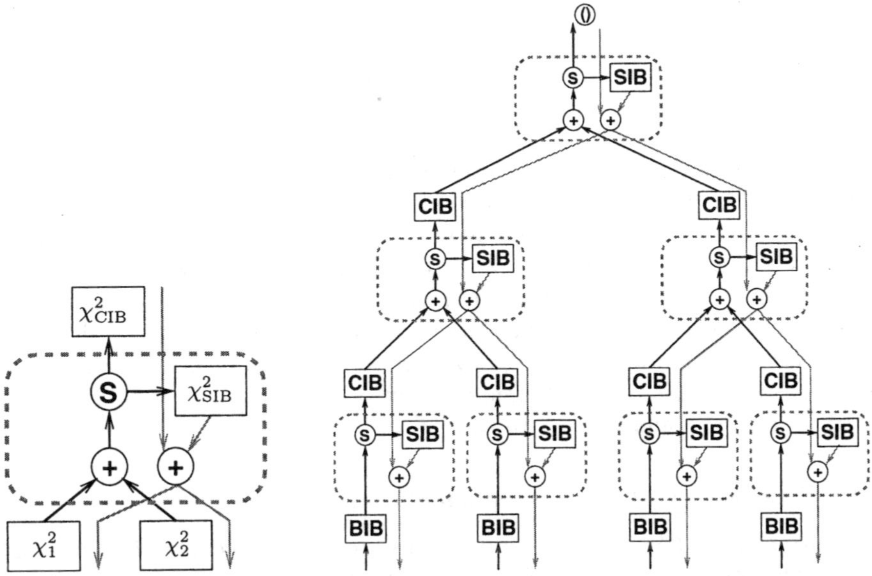

Fig. 2. Integration and decomposition of information in a single node (oval) and a three level tree: Two IBs χ^2_1 and χ^2_2 from the nodes children are integrated $(+)$ and then decomposed $(\mathbf{S})$ into a CIB χ^2_{CIB} and a SIB χ^2_{SIB}. The CIB is passed to the parent and the SIB stored at the node (black arrows). Later, an estimate for the landmarks represented at the node is combined $(+)$ with the SIB, resulting in an estimate for the landmarks represented at the children nodes (gray arrows)

The regions corresponding to nodes are not defined geometrically, but rather as a set of landmarks being close to each other. At each moment there is one leaf called the *actual leaf* that corresponds to the region where the robot is currently located. All leaves hold a *Basic Information Block* (BIB). New measurements are integrated into the BIB of the actual leaf called the *actual BIB*. Thus, integration of all BIBs constitutes the complete information contained in the treemap. The information is recursively integrated and decomposed along the tree as described in the previous section: Each node holds a *Condensed Information Block* (CIB) for the information about landmarks observable from outside the region. The node is said to *represent* these landmarks, since the nodes CIB contains all information about this region needed from outside the region. Furthermore, each node holds a *Substitution Information Block*[3] (SIB) containing the information about eliminated landmarks, needed when the robot is inside the region.

Definition 1 (Node). *A node represents those landmarks that are represented both in some BIB inside and in some BIB outside the subtree below this node. It stores a* Condensed Information Block *(CIB) containing the integrated information of all BIBs below this node on the landmarks represented at this node. It further stores a* Substitution Information Block *(SIB) that contains the information from the childrens' CIBs (leaf's BIB resp.) that is not contained in the nodes CIB.*

According to this definition, a landmark is represented from each leaf where the BIB represents the landmark up to the least common ancestor of all those leaves. The least common ancestor is called *elimination node* of the landmark, since it is that node the landmark is eliminated from the CIB and finally stored into a SIB. The different elimination nodes are maintained in an array.

Figure 2 shows the role of the different IBs (BIB, CIB, SIB) and how a nodes CIB and SIB are computed recursively from the children's BIB resp. CIB. For the moment, the symbols **(+)** and **(S)** can be viewed as black box operations integrating and decomposing information. A detailed explanation will follow. Altogether the intention of this approach is to eliminate landmarks as early as possible, so all CIBs and SIBs represent only few landmarks and all involved matrices are small and efficient to handle.

4.2 Integration of a Measurement

It is currently assumed that all observations are measurements of relative landmark positions (see §7). As long as the observed landmarks are represented in the actual BIB, the measurement can be integrated there and the local estimate can be updated by EKF equations. Such an update does not use the treemap at all and its computation time $O(k^2)$ is independent from the size of the map.

When a landmark is observed that is not represented in the actual BIB, a new BIB must be made the actual one and a global update is required. Since the actual BIB has changed all CIBs and SIBs of ancestor nodes are invalid and must be updated. However most CIBs and SIBs remain unaffected, so computation is highly efficient. After that an estimate for the landmarks represented in the new actual BIB has to be computed. This is done proceeding from the root down to the actual BIB. At each node an estimate for landmarks represented at the childrens' nodes is computed by combining an estimate for the node's landmarks with the nodes' SIB. In order to compute an estimate for all landmarks the tree is traversed recursively.

[3] The name is explained in §5.

4.3 Representation of IBs

The purpose of the algorithm is to compute a maximum likelihood estimate for the map. This is equivalent to finding the minimum of the *negative* log-*likelihood* given the statistical information known from the measurements. Since Gaussian noise is assumed, this is a quadratic error function $\chi^2_{\text{all}}(x)$. Each information block also represents a quadratic error function $\chi^2_{\text{IB}}(x)$ referring to the conditional likelihood of landmark position vector x given the information represented by the IB. $\chi^2_{\text{IB}}(x)$ is the negative logarithm of this likelihood and stored using a constant γ, a vector b and a so called information matrix A being symmetric positive semidefinite (SPSD), as

$$\chi^2_{\text{IB}}(x) := x^T A x + x^T b + \gamma = \sum_{i,j} A_{ij}\, x_i x_j + \sum_i b_i\, x_i + \gamma. \tag{1}$$

This is the usual representation of a quadratic function. Each row / column of A and each entry of b corresponds to a landmark's x- or y-coordinate or the robot's x-, y-coordinate or orientation ϕ.

5 Elimination of Landmarks by Schur Complement

This section presents how to use a mathematical technique called Schur complement to compute a node's CIB and SIB from the CIB of both children. The first step is to integrate the CIB from both children by simply adding ((+) in figure 2). The second step is to eliminate some landmarks by decomposing the result into two parts (**S**). The first part does not depend on eliminated landmarks any more (CIB). The second part is a maximum likelihood substitution of eliminated landmarks by the remaining ones with a known uncertainty (SIB). The structure of the SIB as a substitution with uncertainty is the reason for the second part of the decomposition being called substitution information block.[4]

This operation is a redistribution of information, since the integrated information of both input CIBs is equal to the integrated information of the resulting CIB and SIB. Figure 2 illustrates the underlying data flow. In the following, the formulas for the integration and decomposition are given ([4] for a derivation).

Lemma 1. *Let $\chi^2_1(x)$ and $\chi^2_2(x)$ be two stochastically independent information blocks. Then the integrated information is*

$$\chi^2(x) = \chi^2_1(x) + \chi^2_2(x). \tag{2}$$

If both IBs represent different sets of landmarks, the matrices and vectors have to be permuted and extended, so the same columns / rows correspond to

[4] From an abstract statistical perspective, this is just decomposing $P(x) = P\left(\begin{smallmatrix} y \\ z \end{smallmatrix}\right)$ as $P(z)P(y|z)$, i.e. as the product of a marginalized distribution of z and a conditional distribution of y with parameter z.

the same landmark. For ease of notation it is assumed that A is decomposed into 2×2 blocks such that block row / column 1 corresponds to landmarks to be eliminated and stored in the SIB:

$$\chi^2(x) = x^T A x + x^T b + \gamma \tag{3}$$

$$= \chi^2 \begin{pmatrix} y \\ z \end{pmatrix} = \begin{pmatrix} y \\ z \end{pmatrix}^T \begin{pmatrix} P & R^T \\ R & S \end{pmatrix} \begin{pmatrix} y \\ z \end{pmatrix} + \begin{pmatrix} y \\ z \end{pmatrix}^T \begin{pmatrix} c \\ d \end{pmatrix} + \gamma. \tag{4}$$

The following lemma provides the formulas for decomposing χ^2 into χ^2_{CIB} and χ^2_{SIB} performing the elimination.

Lemma 2 (Schur Complement). *Let $\chi^2\left(\begin{smallmatrix} y \\ z \end{smallmatrix}\right)$ be an information block as in (4) with P being symmetric positive definite (SPD). Then $\chi^2(x)$ can be uniquely decomposed into an information block $\chi^2_{CIB}(z)$ on z and an information block $\chi^2_{SIB}(Hz + h - y)$ on $Hz + h - y$, with $\chi^2_{SIB}(0) = 0$:*

$$\chi^2_{SIB}(w) = w^T P w, \quad H = -P^{-1} R^T, \quad h = -P^{-1} c/2. \tag{5}$$

An estimate can be easily computed from the SIBs: Since the root node represents no landmark, start with an empty estimate $\hat{x} = ()$, with covariance $C = ()$. Proceed down and use the estimate for a node's landmarks and the SIB stored there to derive an estimate for the landmarks represented at the node's children applying lemma 3:

Lemma 3. *Let $\chi^2(x)$ be decomposed as in lemma 2 and let $\hat{z}$ be an estimate with covariance C. Then the optimal estimate for y is*

$$\hat{y} = H\hat{z} + h, \text{ with } \operatorname{cov} \begin{pmatrix} y \\ z \end{pmatrix} = \begin{pmatrix} HCH^T + P^{-1} & HC \\ CH^T & C \end{pmatrix}. \tag{6}$$

With lemma 1, 2 and 3 the necessary tools for using a treemap are available (Fig. 2). Lemma 1 and 2 are used from the leaves up to the root (black arrows) and lemma 3 from the root down to the leaves (gray arrows). When a global update is performed only the way from the old actual BIB up to the root and down again to the new actual BIB has to be computed. Even when an estimate for all landmarks is desired, computation is extremely efficient, since lemma 3 (without covariance) requires just a small matrix-vector multiplication.

6 Assumptions on Topologically Suitable Buildings

The time needed for the computation discussed above depends on the size of the matrices involved, which is determined by the number of landmarks represented at the node's children. So for the algorithm to be efficient it is crucial that each node represents only a few landmarks. Thus, the tree must hierarchically divide the building in a way that each node, i.e. each region, contains only a few landmarks observable from outside the region. Achieving this goal requires some sophisticated optimization of the tree, since it is not a simple bookkeeping task. As experiments and the following considerations confirm, this is possible for typical buildings, which will be called *"topologically suitable"*.

Typical buildings allow such a hierarchical partitioning because they are hierarchical themselves, consisting of floors, corridors and rooms. Different floors are only connected through a few staircases, different corridors through a few crossings and different rooms most often only through a single door and the adjacent parts of the corridor. Thus, on the different levels of hierarchy natural regions are: rooms, part of a corridor including adjacent rooms, one or several adjacent corridors and one or several consecutive floors (Fig. 3).

To allow a thorough theoretical analysis of the algorithm it is formally assumed that the building is topologically suitable:

Definition 2 (Topologically Suitable Building). *Let the building be decomposed into a hierarchy of regions according to definition 1. Let k ("number of local landmarks") be the maximum number of landmarks represented in a BIB. Then the building is said to be topologically suitable if the following holds:*

1. *For each node only $O(k)$ landmarks exist that are represented both in BIBs inside and in BIBs outside the subtree of this node.*
2. *Each BIB shares landmarks only with $O(1)$ other BIBs.*

The parameter k is small, since the robot can only observe a few landmarks simultaneously because its field of view is limited both by walls and sensor range. In particular, k does not increase when the map gets larger ($n \to \infty$). Although by this argument $k = O(1)$, the asymptotical expressions in this paper explicitly show the influence of k and do not formally assume k to be constant.

A counter-example for a not topologically suitable building is a large open storeroom with many boxes, where the robot can navigate arbitrarily not confined to designated paths. A region corresponding to one half of the hall will have a whole border line with the region corresponding to the other half and thus violate condition 1. For cross-country navigation, the same problem appears, when the robot builds an area-wide map covering every detail. However, in most cases the goal is to explore a large area rather than mapping a small area in detail. Thus, the robot will use passable paths once it has found them. So again, each region will be connected to the remaining map only with a few of these paths and definition 2 is fulfilled.

Condition 1 is powerful. The fact that buildings have such a loosely connected topology is a key property distinguishing SLAM from other estimation problems.

6.1 Computational Efficiency

By condition 2 there are $O(\frac{n}{k})$ nodes in the tree each storing matrices of dimension $O(k \times k)$ (condition 1). Thus, the storage requirement of the treemap is $O(k^2 \cdot \frac{n}{k}) = O(nk)$ meeting requirement (R2).

Computation time depends: When a measurement involves only landmarks represented in the actual BIB it can be integrated into this BIB and the estimate

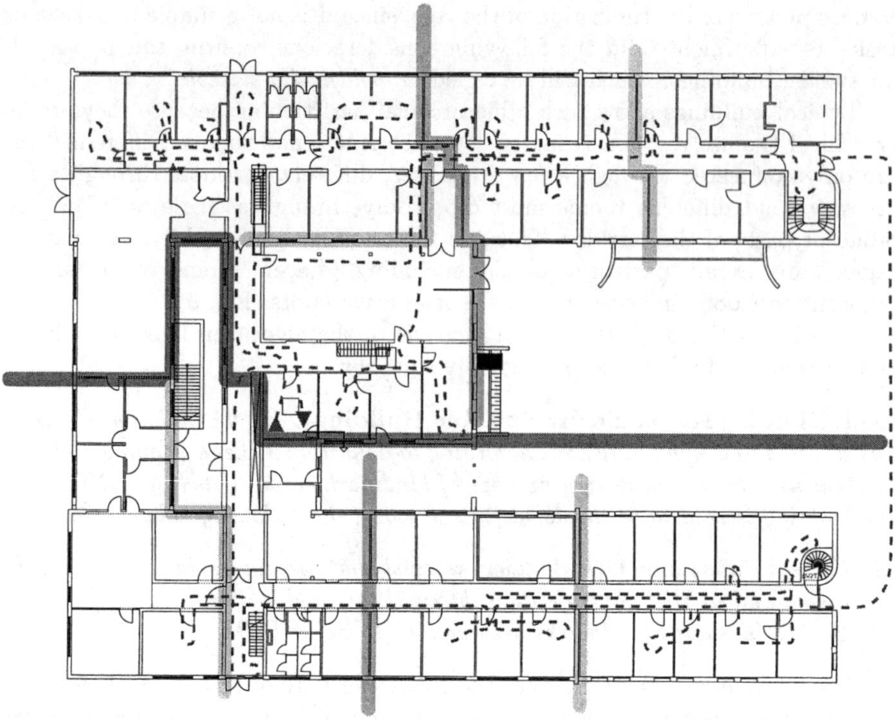

Fig. 3. DLR Institute of Robotics and Mechatronics – A typical topologically suitable building with the first three level of a suitable hierarchical partitioning. The building has been mapped in the experiments reported in §10, with the dashed line sketching the robots trajectory. Start and finish are indicated by small triangles

can be updated using EKF equations. Similar to CEKF this needs $O(k^2)$ computation time, independent from n. Otherwise, a different BIB is made actual one and a global update has to be performed. The update basically requires recomputing the CIB and SIB from the old actual BIB up to the root and compiling an estimate from the root down to the new actual BIB ($O(k^3)$ per node). There are $O(\log n)$ nodes to be updated, so the overall time is $O(k^3 \log n)$. Under some circumstances, more nodes are involved and additional computation is necessary for bookkeeping but still with the same asymptotical complexity.

In order to compute an estimate not only for local but for all landmarks, lemma 3 must be applied recursively from the root down to all BIBs taking $O(kn)$. It will turn out in the experiments in §9 that the prefactor involved is extremely small. So while from a theoretical perspective the possibility to perform updates in sublinear time is most appealing, practically the algorithm allows computing an estimate for all landmarks in extremely large maps.

7 Integration of Odometry Measurements

Up to now the observations have been assumed to consist of landmark–landmark measurements, i.e. information about the relative locations of a group of landmarks. Lu & Milios [19] established a well known approach utilizing this kind of information, where laserscan reference frames are treated as "landmarks" and the relative pose of two scans is "measured" by scan matching. Indeed the treemap algorithm could be readily used to solve the linear equation system derived by Lu & Milios reducing computation time from $O(n^3)$ to $O(kn)$ or to $O(k^3 \log n)$ for an incremental local estimate.

Most often measurements are landmark–robot measurements, i.e. information about the relative location of a landmark with respect to the robot. Another source of information is odometry, i.e. robot–robot measurements providing information of the current robot pose relative to a previous robot pose. These information could be processed as well by the algorithm described so far if all robot poses were explicitly represented as random variables to be estimated, like Lu & Milios did. This, however, violates (R2) because it leads to map size growing even if moving through an area already mapped.

In the following it will be discussed how to avoid represantation of old robot poses by using an EKF as preprocessing stage.

7.1 Landmark–Robot Measurements

First assume that odometry can be neglected, i.e. the robot's motion is evident from the landmark observations alone: Each robot pose is considered as a separate random variable that can be eliminated since it will not appear in any further measurement: The landmark measurements made at a certain robot pose are integrated into an IB representing the robot pose and all involved landmarks as random variables. Then the robot pose is eliminated from the IB using Schur complement (lemma 2). The resulting IB does not represent the robot pose any more and can be integrated into the actual BIB just the same way like pure landmark–landmark measurements.

The precondition of this approach is that at least two common landmarks are being observed from successive robot poses. If this condition is met, odometry can often be neglected [20]. Theoretically, this is even appealing, since the assumption of stastistical independence between successive odometry measurements is hardly true in reality. Although this is not a theoretically optimal approach, it will presumably be a good choice in practice and considerably simpler than the more general approach described below.

7.2 Robot–Robot Measurements (Odometry)

When odometric measurements have to be integrated, it is necessary to represent the robot pose as a random variable. Thus old robot poses have to be eliminated later to prevent the map size from growing. This leads to new couplings intro-

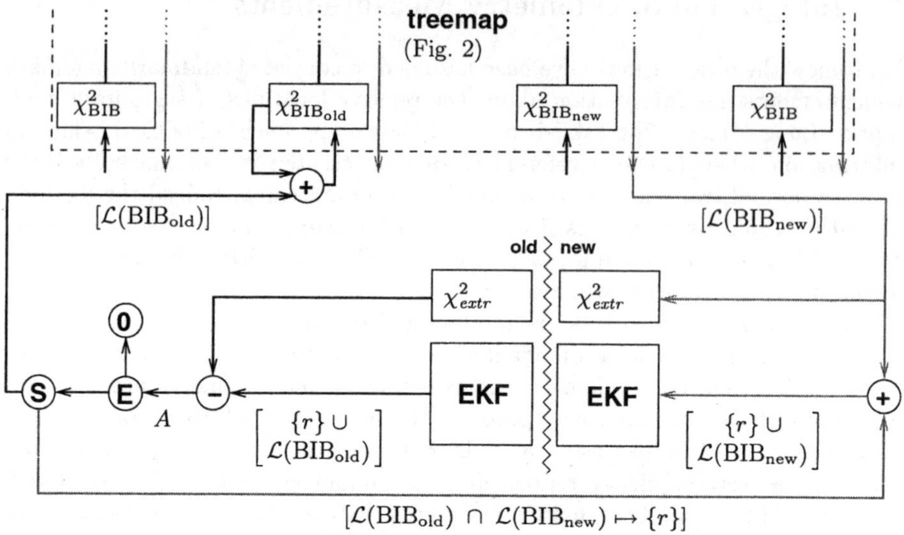

Fig. 4. Data flow between EKF and treemap when changing the actual BIB from $\mathrm{BIB}_{\mathrm{old}}$ to $\mathrm{BIB}_{\mathrm{new}}$. Represented landmarks are shown in brackets (Tab. 1). Black arrows depict information matrices, gray arrows covariance matrices

duced between all landmarks observed from an eliminated robot pose and in the end between all pairs of landmarks.

To avoid this dilemma a conservative approximation is performed. All coupling coefficients are eliminated except those with landmarks represented in the actual BIB. This means to deliberately discard the information contained in the eliminated coupling coefficients to make the representation less complex. It has been proven [3], that the occurring information matrices are approximately sparse. This theorem ensures, that couplings decay exponentially with distance traveled and not too much information is discarded by the elimination.

The measurements are integrated by an EKF as a preprocessing stage (Fig. 4). It represents the robot pose and all landmarks of the actual BIB and can directly integrate odometry and landmark observations. The information about the robot pose is exclusively contained in the EKF and not transferred into the treemap. When a global update becomes necessary all coupling coefficients between the robot pose and landmarks not represented in the new actual BIB are eliminated:

First, the EKF state is converted into an information block χ^2. Then, the information χ^2_{extr} is subtracted **(-)**. This is the information obtained from the tree map the last time the EKF was initialized and must not be integrated a second time. The resulting difference A is the information gained from measurements since then. Next, the couplings between robot pose and landmarks not represented in $\mathrm{BIB}_{\mathrm{new}}$ are eliminated **(E)** by subtracting a SPSD matrix cancelling the necessary coefficients from A. This means, a part of the information is deliberately discarded **(0)** to give the remaining information a simpler structure. After this the robot pose is eliminated from the IB by Schur complement

Table 1. Random variables corresponding to different block rows / columns of A. $A_{21} = A_{12}^T$ is to be eliminated. $\mathcal{L}(\text{BIB})$ denotes the landmarks represented in BIB

Blockrow	Notation	Random variables
1	$[\{r\}]$	Robot pose
2	$[\mathcal{L}(\text{BIB}_{\text{old}}) - \mathcal{L}(\text{BIB}_{\text{new}})]$	Landmarks represented in the old but not in the new actual BIB
3	$[\mathcal{L}(\text{BIB}_{\text{old}}) \cap \mathcal{L}(\text{BIB}_{\text{new}})]$	Landmarks represented in both the old and new actual BIB

(S) and the resulting CIB is added to BIB_{old} (+) replacing it in the treemap. The corresponding SIB defines the robot pose as a function of landmarks which are both in BIB_{old} and BIB_{new} (due to **(E)**). After the estimate for BIB_{new} has been generated by updating the treemap, the SIB can be integrated (+). The result is the estimate for the landmarks of the new actual BIB and the robot pose. Together with the corresponding covariance matrix the estimate is used as a new EKF state.

7.3 Stepwise Optimal Elimination of Off-Diagonal Entries

In this section the procedure **(E)** is derived. It eliminates some coupling entries in an information matrix A by subtracting a so called elimination matrix B. The key idea is to make B small so as little information as possible is discarded. This task is similar to the *sparsification* procedure used by Thrun et al. [15] in their Sparse Extended Information Filter (SEIF) algorithm. Their approach optimally approximates the original distribution in the sense of Kullback-Leibler (KL) divergence.

The problem is reduced from a k-D problem to k 1-D problems by eliminating different coupling entries columnwise, where the following theorem gives an optimal solution for eliminating a single column.

Theorem 1 (Elimination Matrix). *Let A be a 3×3 block SPD matrix being decomposed as* $A = \begin{pmatrix} \psi & r^T & w^T \\ r & S & W^T \\ w & W & X \end{pmatrix}$ *with 1-dimensional first block row / column. Then the best elimination matrix for A_{21} is xx^T with x defined as*

$$x = A \begin{pmatrix} \gamma \\ \delta S^{-1} r \\ 0 \end{pmatrix}, \text{ with} \tag{7}$$

$$\alpha = r^T S^{-1} r, \quad \beta = (\psi - \alpha)^{-1},$$
$$\lambda = \sqrt[4]{\psi (r^T S^{-1} r)}, \; \gamma = \beta(\lambda - \lambda^{-1}\alpha), \; \delta = \beta(-\lambda + \psi\lambda^{-1}). \tag{8}$$

The result is optimal with respect to (R1) since it minimizes the *worst factor* by which the covariance of *any* aspect of the map is increased. For lack of space the reader is referred to [4] for a mathematical discussion.

Each measurement is affected by the elimination operation only once, namely the next time when the actual BIB changed. So the elimination procedure preserves topological information, i.e. when measurements report two landmarks to be close to each other this information will be included in the BIB although less precisely. Since propagation of information through the tree is extact, a loop will be closed in the estimate immediately after integrating the corresponding measurement. This indicates although does not proof that the algorithm complies with (R1) and will be further investigated with simulation experiments in §9 reporting the actual increase of error encountered.

8 Maintenance of the Hierarchy

Up to now the linear algebra part of the algorithm has been described. It provides the subalgorithms for manipulating IBs and in the end for computing an estimate from the measurements. The bookkeeping part of the algorithm takes care to update CIBs and SIBs as necessary using the subalgorithms decribed before. It further optimizes the tree, so that it is balanced and hierarchically partitions the set of BIBs in a way that at any level of hierarchy a partition shares only a few landmarks with BIBs not belonging to the partition. Thus the node corresponding to the partition represents only a few landmarks, and computation at this node is efficient. This is problem is in theory NP-complete, with many established heuristic approaches existing[21]. The algorithm incrementally optimizes the tree by moving a single subtree to a different location whenever a global update is performed[4].

There exists a nonlinear extension to the algorithm that corrects the linearization error resulting from large error in the robot orientation by applying "Nonlinear Rotations" to individual IBs before integrating them. The extension is omitted here due to lack of space refering the reader to [4] for an extensive discussion and experimental results handling up to 140° orientation error.

9 Simulation Experiments

This section presents the simulation experiments conducted to verify the algorithm with respect to the requirements (R1)-(R3). For this purpose, a simulation approach is advantageous because ground truth is available and it allows to repeat the same experiment with identical measurements but new independent measurement noise.

All experiments have been conducted on an Intel Xeon, 2.67 GHz with $2.5\%, 2^\circ$ noise for the landmark sensor, $0.01\sqrt{\mathrm{m}}$ noise for the odometry sensor (proportionally to square root of distance traveled) and a robot radius of 0.3m. The algorithm's parameters are $optHTPSteps = 5$ steps of tree optimization per global update and $maxDistance = 5\mathrm{m}$ as maximum diameter of a region.

Clearly space (R2) and time (R3) consumption are straightforward to measure but how should one assess map quality with respect to requirement (R1)?

9.1 Assessment of Map Quality

With known ground truth the estimation error can readily be computed. But while it is a good measure for the overall system performance, it doesn't tell anything about the algorithm. An error, for example of 1m, could either be caused by large sensor noise despite an optimal algorithm or it could be caused by crude approximations in the algorithm despite precise sensor measurements. To assess the performance of the algorithm with respect to requirement (R1) the error must be compared to the "*minimal uncertainty that could be theoretically derived from the measurements*" as evident from the optimal nonlinear Maximum Likelihood estimate. So if, for instance the ML estimate has an error of 0.5m it can be concluded, that the algorithm has increased the error by 100%. This number i.e. the relative error indicates the prize to pay for using the algorithm instead of ML estimation and characterizes the algorithm's map quality with respect to (R1). Another point to consider when interpreting absolute error specifications is that the absolute error is accumulating and thus depends on the map size.

To summarize: When the focus is on the core estimation algorithm not on the overall system, relative not absolute error is the quantity to be considered.

It is well known [3] that relative aspects of a map e.g. the distance between two landmarks have much less uncertainty than absolute landmark positions. Since the uncertainty of absolute landmark positions is often several meters navigation would be impossible otherwise. Thus it is essential, not only to look at the relative error of different landmarks but at the relative error of *any aspect* of the map as required by (R1). It has been derived [3] that this can be done by computing a generalized eigenvalue spectrum

$$Cv = \lambda \, C_{\mathrm{ML}} v \tag{9}$$

of the covariance of the algorithm's estimate C relative to the covariance of the maximum likelihood estimate C_{ML}. The generalized eigenvalue λ corresponding to an eigenvector v gives the squared relative error in the two estimates for the aspect corresponding to the eigenvector v. These eigenvalues characterizes the relative error encountered in different aspects of the map just the same way as ordinary eigenvalues characterize the absolute error in different aspects.

9.2 Small Map Experiment

The small map simulation experiment allows statistical evaluation of the estimation error and comparison with EKF and ML (Fig. 5). At first sight all three basically appear of same quality (except for the left upper room in the treemap estimate) and perfectly usable for navigation. Quantitative inspection however will still show a notable difference:

Figure 5d compares the relative error in the three estimates in *all aspects* of the map. The error covariances C for treemap, C_{EKF} for EKF and C_{ML} for ML are approximatively determined by Monte Carlo simulation with 1000 runs. To limit the number of runs necessary only eight selected landmarks are evaluated.

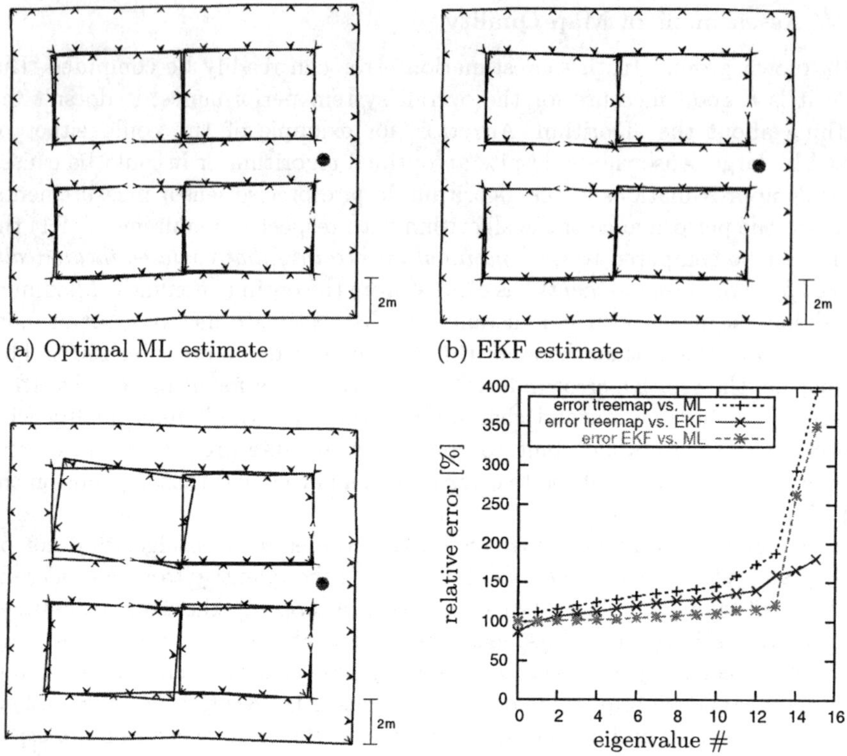

(a) Optimal ML estimate (b) EKF estimate

(c) Treemap estimate (d) Relative error spectrum

Fig. 5. Small map simulation experiment results

The square root of the smallest eigenvalue is 110% (87% vs. EKF) and the largest 395% (181% vs. EKF). This means that the map estimate computed by treemap has an error 10% larger in the best aspect and 295% larger in the worst aspect than the ML estimate. The typical (median) relative error is 137% compared to ML with two outliers of 395% and 293% and typically (median) 125% compared to EKF. The outliers are also apparent in the plot comparing EKF to ML, so they are probably caused by linearization errors occurring in EKF and treemap. This is surprising since at visual inspection the EKF map is so good one would hardly suspect linearization problems.

9.3 Large Scale Map Experiment

The second experiment uses an extremely large map consisting of 10×10 copies of the building used before (not shown for its size). The experiment encompasses $n = 11300$ landmarks, $m = 312020$ measurements and $p = 63974$ robot poses. The EKF experiment was aborted earlier due to large computation time.

In Fig. 6a storage space consumption is clearly shown to be linear for treemap ($O(kn)$) and super-linear ($O(n^2)$) for EKF. Overall computation time was 31.34s

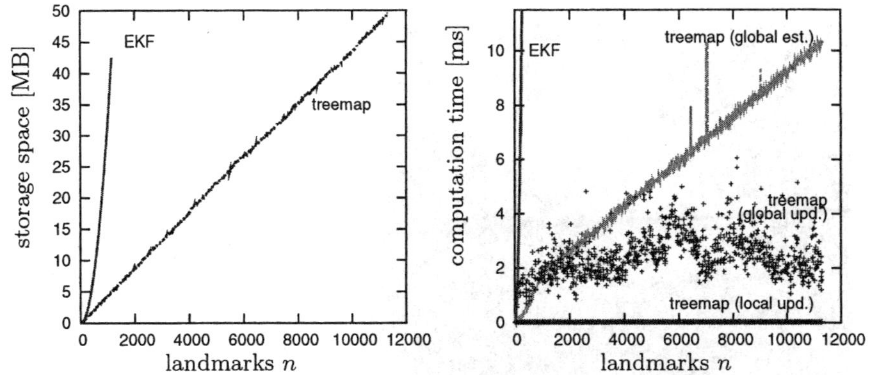

r p p n p r r n

Fig. 6. Large scale simulation experiment: Storage space and computation time over number of landmarks n. Observe different computation time for a local update, a global update and for computing a global estimate

for treemap and 18.89 days (extrapolated $\sim n^3$) for EKF. Computation time per measurement is shown in figure 6b. Time for three different computations is given: Local updates (dots below < 0.5ms), global updates computing a local map (scattered dots above 0.5ms) and the additional cost for computing a global map are plotted w.r.t. n. The algorithm is extremely efficient updating an $n = 11300$ landmark map in 12.37ms. Average time is $1.21\mu\text{s} \cdot k^2$ for local update, $0.38\mu\text{s} \cdot k^3 \log n$ for global update and $0.15\mu\text{s} \cdot kn$ for a global map respectively. The latter is surely the most impressive result from a practical perspective.

10 Real World Experiments

The real world experiments reported in this section are used to demonstrate how to apply the treemap algorithm in practice by mapping the DLR Institute of Robotics and Mechatronics' building (Fig. 3). It is used as an example for a typical office building and indeed turns out to be "topologically suitable" as defined in §6. The algorithm is generating a balanced and well partitioned tree representation online and closes three large loops during mapping.

In the experiments a wheeled mobile robot was moved manually through the building. The robot is equipped with a camera system (field of view: $\pm 45°$) at a height of 1.55m. As maximum diameter of a region *maxDistance* = 7m was used. For the purpose of conducting this experiments circular artificial landmarks were set throughout the floor of the building (Fig. 7a) and visually detected by a combination of Hough-transform and a gray-level variance criterion.

Since the landmarks are identical, identification is based on their relative position employing two different strategies in parallel: Local identification is performed by simultaneously matching all observations from a single robot pose to the map taking into account both error in each landmark observation and

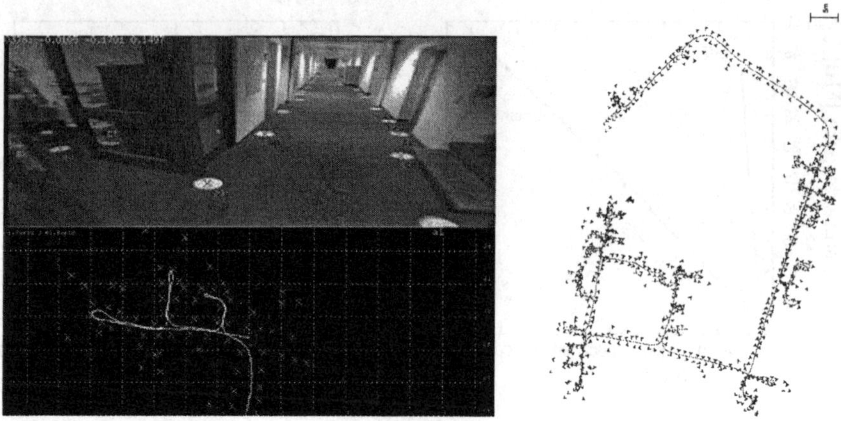

(a) Screen shot with live image and map estimate. (b) Before closing the loop.

Fig. 7. Real world experiments: Implementation mapping the DLR building

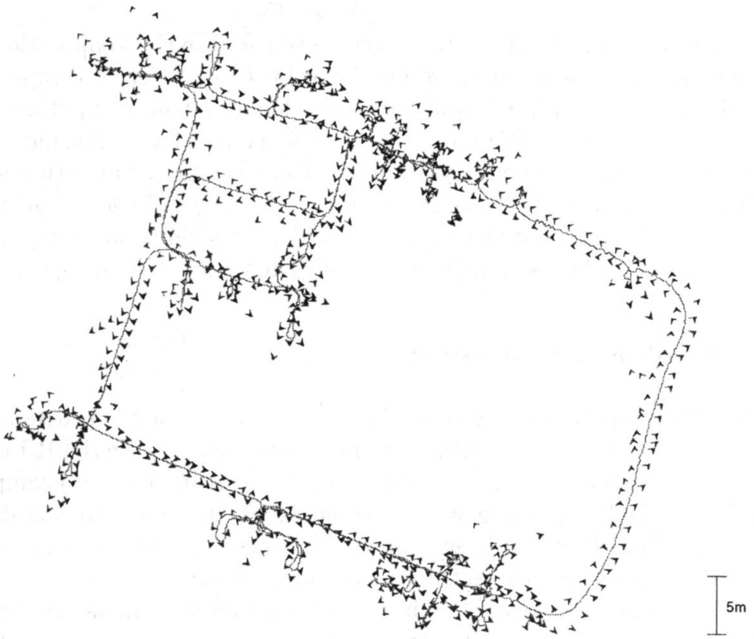

Fig. 8. Real world experiments: Final map estimate

error in the robot pose. For global identification considerable difficulties were encountered in detecting closure of a loop: Before closing the largest loop the accumulated robot pose error was 16.18m (Fig. 7b, 8) and the average distance between adjacent landmarks was $\approx$ 1m. With indistinguishable landmarks matching observations from a single image was not reliable enough.

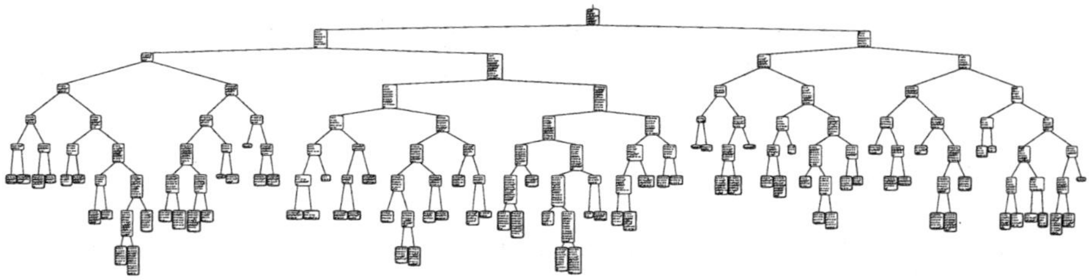

Fig. 9. Tree representation of the map. Size of the node ovals is proportional to number of represented landmarks

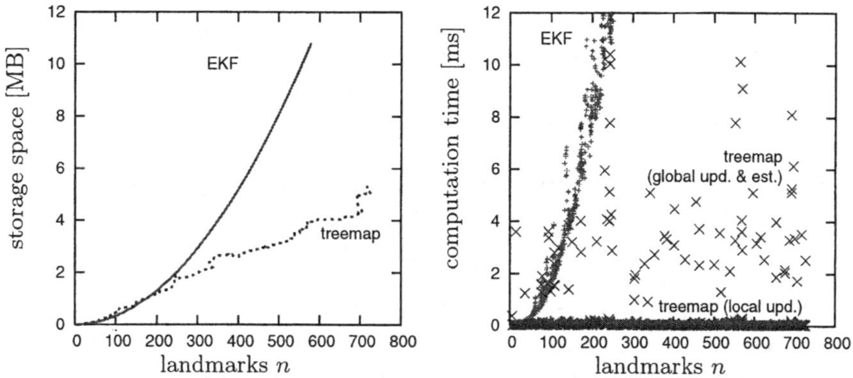

r p p n p r r n

O r p ≈ 0.07ms

Fig. 10. Real experiment performance

Instead, the algorithm has been designed to match a map patch of radius 5m around the robot. When the map patch is recognized somewhere else in the map, the identity of all landmarks in the patch is changed accordingly and the loop is closed. It is a particular advantage of the treemap algorithm to be able to change the identity of landmarks already integrated into the map (referred to as *lazy data association* by Haehnel et al. [22]). Technical details of computer vision and landmark identification can be found in [4]. The final map contains 725 landmarks, 29142 measurements and 3297 robot poses (Fig.7b, 8). The results highlight the advantage of using SLAM because after closing the loop the map is much better and at visual inspection impressively good for such a large building. Figure 9 shows the internal tree representation used by the algorithm. On the average there are $k \approx 16.39$ landmarks represented in each BIB. The tree is balanced and well partitioned, i.e. no node represents too many landmarks. It can be concluded that the building is indeed topologically suitable in the sense discussed in §6. Computation time is extremely low (0.07ms per measurement) if, only a local update is performed as is the case most often. The average time is $0.77\mu s \cdot k^2$ for local update, $0.02\mu s \cdot k^3 \log n$ for global update and $0.04\mu s \cdot kn$

for a global map respectively (Fig. 10). Accumulated computation time is 2.95s for treemap and 601s (extrapolated $\sim n^3$) for EKF.

11 Conclusion

The treemap SLAM algorithm proposed in this paper works by dividing the map into a hierarchy of regions represented as a binary tree. With this data structure, the computations necessary for integrating a measurement are limited essentially to updating a leaf of the tree and all its ancestors up to the root. From a theoretical perspective the main advantage is that a local map can be computed in $O(k^3 \log n)$ time. Practically, it is equally important that a global map can be computed in $O(kn)$ additional time allowing computation of a map with $n = 11300$ landmarks in 12.37ms on an Intel Xeon, 2.67 GHz.

With respect to the three proposed criteria the algorithm was verified theoretically, by simulation experiments, and by experiments with a real robot. A precondition is a typical, topologically suitable building as explained in §6.

From the author's perspective a drawback is the algorithms complexity necessary for performing bookkeeping in $O(k^3 \log n)$. Consequently a promising idea currently investigated is to simplify the algorithm for computing a global map in $O(kn)$ rather than a local in $O(k^3 \log n)$.

Apart from computation time, the most important challenge is landmark identification. Multi-Hypothesis tracking is generally seen as a promising idea to tackle situations where identification is difficult. With such an approach, efficiency of the core algorithm becomes even more crucial as it has to handle all hypotheses simultaneously, multiplying the computation time needed.

References

1. Gauss, C.: Theoria combinationis observationum erroribus minimis obnoxiae. Commentationes societatis regiae scientiarum Gottingensis recentiores **5** (1821) 6–93
2. Frese, U., Hirzinger, G.: Simultaneous localization and mapping - a discussion. In: Proceedings of the IJCAI Workshop on Reasoning with Uncertainty in Robotics, Seattle. (2001) 17 – 26
3. Frese, U.: A discussion of simultaneous localization and mapping. Autonomous Robots (2004) (submitted).
4. Frese, U.: An $O(\log n)$ Algorithm for Simulateneous Localization and Mapping of Mobile Robots in Indoor Environments. PhD thesis, University of Erlangen-Nürnberg (2004)
5. McNamara, T.: How are the locations of objects in the environment represented in memory? In Freksa, C., Brauer, W., Wender, C.H.K., eds.: Spatial cognition III: Routes and navigation, human memory and learning, spatial representation and spatial reasoning, Berlin: Springer-Verlag (2003) 174–191
6. Moratz, R., Tenbrink, T., Bateman, J., Fischer, K.: Spatial knowledge representation for human-robot interaction. In Freksa, C., Brauer, W., Habel, C., Wender, K., eds.: Spatial Cognition III. Springer (2003) 263 – 286

7. Skubic, M., Perzanowski, D., Blisard, S., Schultz, A., Adams, W., Bugajska, M., Brock, D.: Spatial language for human-robot dialogs. IEEE Transactions on System Man Cybernetics Part C **34** (2004) 154 – 167
8. Smith, R., Self, M., Cheeseman, P.: Estimating uncertain spatial relationships in robotics. In Cox, I., Wilfong, G., eds.: Autonomous Robot Vehicles. Springer Verlag, New York (1988) 167 – 193
9. Thrun, S.: Robotics mapping: A survey. Technical report, School of Computer Science, Carnegie Mellon University (2002)
10. Duckett, T., Marsland, S., Shapiro, J.: Learning globally consistent maps by relaxation. In: Proceedings of the IEEE International Conference on Robotics and Automation, San Francisco. (2000) 3841–3846
11. Duckett, T., Marsland, S., Shapiro, J.: Fast, on-line learning of globally consistent maps. Autonomous Robots **12** (2002) 287 – 300
12. Frese, U., Larsson, P., Duckett, T.: A multigrid algorithm for simultaneous localization and mapping. IEEE Transactions on Robotics (2004) (to appear).
13. Montemerlo, M., Thrun, S., Koller, D., Wegbreit, B.: FastSLAM: A factored solution to the simultaneous localization and mapping problem. In: Proceedings of the AAAI National Conference on Artificial Intelligence, Edmonton. (2002) 593–598
14. Guivant, J., Nebot, E.: Solving computational and memory requirements of feature based simultaneous localization and map building algorithms. Technical report, Australian Centre for Field Robotics, University of Sydney, Sydney (2002)
15. Thrun, S., Koller, D., Ghahramani, Z., Durrant-Whyte, H., A.Y., N.: Simultaneous mapping and localization with sparse extended information filters: Theory and initial results. In: Proceedings of the Fifth International Workshop on Algorithmic Foundations of Robotics, Nice. (2002)
16. Press, W., Teukolsky, S., Vetterling, W., Flannery, B.: Numerical Recipes, Second Edition. Cambridge University Press, Cambridge (1992)
17. Bosse, M., Newman, P., Leonard, J., Teller, S.: SLAM in large-scale cyclic environments using the Atlas framework. International Journal on Robotics Research (2003) (to appear).
18. Paskin, M.: Thin junction tree filters for simultaneous localization and mapping. In Gottlob, G., Walsh, T., eds.: Proceedings of the 18th International Joint Conference on Artificial Intelligence (IJCAI-03), San Francisco, CA (2003) 1157–1164
19. Lu, F., Milios, E.: Globally consistent range scan alignment for environment mapping. Autonomous Robots **4** (1997) 333 – 349
20. Gutmann, J., Konolige, K.: Incremental mapping of large cyclic environments. In: Proceedings of the IEEE International Symposium on Computational Intelligence in Robotics and Automation, Monterey. (1999)
21. Fiduccia, C., Mattheyses, R.: A linear-time heuristic for improving network partitions. In: Proceedings of the 19th ACM/IEEE Design Automation Conference, Las Vegas. (1982) 175–181
22. Hähnel, D., Burgard, W., Wegbreit, B., Thrun, S.: Towards lazy data association in SLAM. In: In Proceedings of the 10th International Symposium of Robotics Research (ISRR'03). (2003)

Towards Dialogue Based Shared Control of Navigating Robots*

Robert J. Ross, Hui Shi, Tillman Vierhuff, Bernd Krieg-Brückner, and John Bateman[†]

FB3 Mathematik und Informatik, Universität Bremen
{robertr, shi, tv, bkb}@informatik.uni-bremen.de
[†]FB10 Sprach und Literaturwissenschaften, Universität Bremen
bateman@uni-bremen.de
http://www.sfbtr8.uni-bremen.de/project/I3/

Abstract. Establishing a clean relationship between a robot's spatial model and natural language components is a non-trivial task, but is key to designing verbally controlled, navigating service robots. In this paper we examine the issues involved in the development of dialogue controlled navigating robots. In particular, we treat our robots as so-called *Shared Control Systems*, where robot and user cooperate to achieve a shared goal. We begin by characterising four categories of Shared Control Problems that affect verbally controlled navigating robots. Producing solutions to these problems requires a clear methodology in the linking of 'common-sense' representations of space used by the robots, and the language interface. To this end, we present the *SharC Cognitive Control Architecture* as a general purpose, agent-based dialogue control system that provides a suitable framework for relating spatial information to natural language communication. To illustrate our approach, we focus in particular on natural language understanding, and show how natural language utterances may be mapped to formally modelled spatial concepts, thus helping to overcome problems in shared control.

1 Introduction

With increased applicability in the domestic and office domains, service robots are becoming more and more interesting for both industrial and academic research. A characteristic of service robots, as distinct from heavy industrial or exploratory robots, is that they will often operate in partially known and dynamic environments, moving between locations while performing their assigned duties. Thus, these embodied, situated robots require a working understanding of their spatial environment.

* We gratefully acknowledge the support of the Deutsche Forschungsgemeinschaft (DFG) through the Collaborative Research Centre SFB/TR 8 Spatial Cognition, Project I3-[SharC].

C. Freksa et al. (Eds.): Spatial Cognition IV, LNAI 3343, pp. 478–499, 2005.
© Springer-Verlag Berlin Heidelberg 2005

Service robots are examples of *shared-control systems* where a human operator and an automated technical system are interdependently in charge of control. For this to be effective, user and robot must be able to share spatial knowledge and goal information. Since users may be technically naive, and manual interfacing may not be feasible, information should be exchanged through natural modalities. Natural language dialogues have long been acknowledged as a potentially fruitful modality in human-machine interfaces. However, practical connection of natural language and spatial information is a non-trivial task. Representations of space are often created with little concern for language oriented control, and the mapping between language, space and action is a formidable problem.

The study of cognitive control systems has led to a number of proposed architectures. On one hand, the past decade has seen the development of hybrid Robot Control Architectures [1, 2] that provide an autonomous system with multiple layers of intelligence in order to cope with both deliberative and reactive requirements. On the other hand, research based in the discourse community has led to the development of intelligent conversational systems such as Allen's TRIPS [3], and Lemon's WITAS [4], providing complex discourse models to mediate control of an autonomous system between user and automaton.

In this paper we give a detailed description of the *SharC Cognitive Control Architecture*, and show its relevance in addressing so-called *Shared Control Problems* in the navigating robots domain. We begin in Section 2 by reviewing the state of the art in dialogue-capable navigating robots. This is followed by Section 3 which addresses *Shared Control Systems*, discussing four classes of problems related to verbal interaction with spatially aware systems. Section 4 then introduces the SharC architecture, as a general framework for bridging language systems and spatial representations. We discuss the architecture both from a conceptual and an implementation perspective, giving outlines of relevant components. Since the matching of a natural language utterance to an interpertation is of utmost importance to user-friendly robotics, Section 5 examines our natural language understanding model in more detail – showing how a user utterance can be mapped to a natural language independent action and spatial representation. This is followed by summary and a discussion of future work.

2 Related Work

Work on the linguistic control of robots dates back to the pioneering research of Shakey [5] and SHRDLU [6]. More recent work has either focused on the intentration of natural language components with sophisticated service robot design [7, 8, 9, 1], or have developed models of dialogue management and control from a more theoretic perspective [10, 11, 4].

One of the more recent examples of complete language enabled service robots is Jijo-2 from Matsui et. al. [7]. Jijo-2 was an office assistant robot designed for the Japanese market. As such, Jijo-2's development was more concerned with the integration of a complete system than with the improvement of any one piece of language technology. Notable characteristics of its language systems include: inference of under-specified referents and zero pronouns using the attentional states; context-sensitive construction of semantic frames from fragmented utterances; a modular speech recogniser that swapped recognition vocabularies at runtime. A state based *Dialogue Manager* was used to control both dialogue progress and overall robot action. Dialogue decisions were partially decided upon by the output of a frame like language analyser. The language analyser and speech recognition systems were based around a restrictive grammar; while this improved speech recognition accuracy, it limited overall robustness to partial and ungrammatical utterances which are common in natural spoken language. Also, there was no attempt to recover or process ambiguous utterances through dialogues or other processes.

The relationship between verbal movement commands and action abstractions in human-robot interaction has recently been studied by Bugmann et. al. [10, 12]. The Instruction Based Learning for Mobile Robots (IBL) project has used a miniature remote-brained robot in a model town to build a model of corpus-based instructions for mobile robots. Research has focused on the production of a number of primitive movement actions that map directly to phrases used by humans to direct robots in spatial environments. While an important part of IBL's research has been on the detection and analysis of errors in the analysis process, treatment of how to deal with such errors is only lightly considered, with a mention of the probable importance of confirmation dialogues being noted without any concise treatment of how such a dialogue confirmation would be implemented.

3 Shared Control Problems in Spatially Aware Assistants

In today's shared-control systems, such as intelligent service robots, human operators are no longer continuously in control of the technical system. Instead, they monitor the behaviour of the automation, making command level decisions, and sometimes taking over control of the system in unforseen critical situations. In our navigating robots domain, natural language communication is an important modality in maintaining shared control (e.g., [13, 14]). The shared control of navigating robots will present a distinct range of problems related to the exchange of spatial knowledge with intelligent robots. In this section we discuss a number of shared control problem types that can occur in dialogue based spatially aware systems. The categories presented are not intended to be a complete taxonomy of shared control issues, but serve instead to illustrate the types of problems faced in this domain.

Our discussions here, and approaches presented later in the paper, are made with reference to our chief experimental scenario: the shared control of Rolland, the Bremen Autonomous Wheelchair [15]. Rolland, depicted in Figure 1, makes use of laser range finding sensors to construct spatial representations that can later be used in shared control interactions between user and wheelchair. While

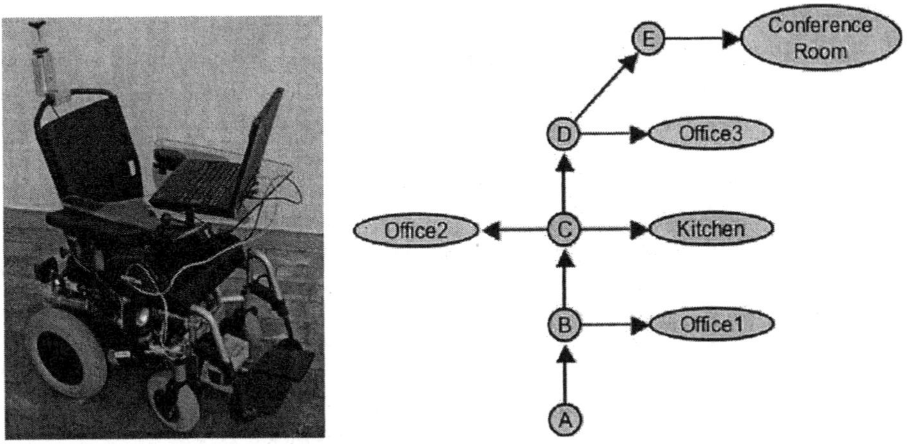

Fig. 1. (a) Rolland – the autonomous wheelchair (b) A user-level route graph – internal space representation used by Rolland (nodes and edges denote decision points and route segments respectively)

such capabilities are modest, they are sufficient for the exploration of shared control problems such as those presented in the rest of this section.

3.1 Linguistic Ambiguity

One of the well-known problems in human language technology is so-called *linguistic ambiguity*: some natural language inputs cannot be allocated a unique semantic representation. In the context of human-robot communication for navigation, such problems must be solved before later stages of processing can reasonably be addressed. Linguistic ambiguities may occur at any level of natural language processing, the most prominent examples being attachment of prepositional phrases and quantifier scopes. The latter can be modelled using minimal recursion semantics [16], however other ambiguities are more difficult to handle. As an example, consider the following short dialogue of an ambiguity at the lexical level between the user and Rolland when approaching a crossroad:

Rolland: Drive left?
User: Right!

The user's answer is ambiguous, it could either be interpreted to acknowledge or to correct the robot's suggestion; hence, there are two conflicting interpretations which cannot be resolved by linguistic knowledge only. Consequently, for this kind of ambiguity, a clarification dialogue must be raised in order to negotiate the user's intention.

3.2 Mode Confusion

For several years, research in the human factors and aviation psychology communities has focused on the issue of *mode awareness* and *mode confusion* (see for example [17, 18, 19]). A mode represents a set of system behaviours defined by transitions between system states. Mode confusions occur if the human operator loses track of the mode transitions performed by the automation. Consider a scenario in which Rolland is driving down a corridor when a person suddenly steps into its path. Upon seeing the colleague, the user may decide to stop and talk for a moment; hence uttering "please halt". However, unbeknownst to the user, Rolland did not actually respond to the user's utterance, but decided to come to a stop of its own accord – having viewed the colleague as an obstacle. Thus, when the colleague moves on, the user will be surprised that the wheelchair continues on its path, despite the user not having instructed it to continue.

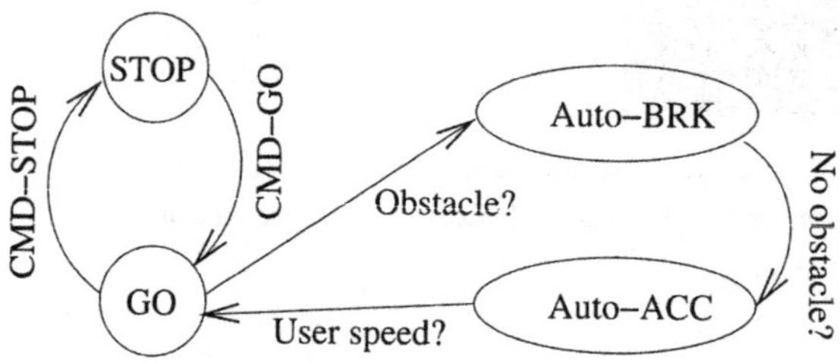

Fig. 2. A simple model of wheelchair behaviour

Figure 2 presents an abstracted model of Rolland automation. In normal circumstances, a stop command, but not a halt command, will cause the robot to move to state STOP that can only be left through a direct command CMD-GO to proceed. Conversely, the detection on an obstacle during the movement state will cause the automation to move to the brake state (Auto-BRK), which will automatically be exited and Rolland will accelerated in state Auto-ACC when the obstacle is no longer present.

In recent approaches [20, 21], formal methods are used to systematically detect and avert these conflict situations in which the human operator assumes the technical system to be in a different situation than it actually is. After detecting the general categories of situations where such 'mismatches' may occur, audio or visual feedback from the system can be used to notify the user when state initiated behaviours have occurred, and indeed to confirm when a command from the user has been recognised.

Formalisation of machine models has made extensive progress in the last 10 years. There are a number of successful examples of applying formal methods for handling complex industrial systems (e.g., [22]). The challenge here is the construction of the user's mental models, discussion of which goes beyond the scope of the present paper.

3.3 Spatial Representation Disparity

Routes are a concept commonly encountered when dealing with human spatial navigation. When planning a trip from one point to another, a number of intermediate points can usually be identified, thus allowing the total trip to be broken down into a number of different segments. The routes we take are often physically defined in our environment, e.g. by determining the sequence of cities and exits when driving in a particular direction. The set of navigational strategies found in artificial moving agents, such as service robots, mirrors the complexity of those employed by human beings. The Route Graph ([23, 24]) is a simple model describing key elements for route based navigation as part of an agents' general knowledge, in which a *route* is a concatenation of *route segments* from one place to another; and a *place* is a tactical decision point. Figure 1 (b) gives a part of a route graph for an office building. Such a representation can be used as the internal representation for a navigating robot such as an autonomous wheelchair. A to E are positions in a corridor.

While the route graph shown above can be viewed as the robot's internal representation of its environment, users will often have their own internal representation, or mental model, of their surroundings. If the user is mistaken in the understanding of the office space layout, or, indeed, if the robot's representation is not accurate, then a command issued by the user may conflict with what the robot 'believes to be true'. To illustrate this, suppose that, at position (D) in Figure 1 B, the user orders Rolland to "Follow the corridor, and turn left at the end", then clearly this is not actually possible based on what Rolland believes. Further dialogue is necessary to resolve such conflicts in order to coordinate the user's and the robot's representation of the environment.

3.4 Spatial Concept Disparity

The ontological modelling of space is considered to be particularly necessary for facilitating qualitative spatial reasoning in general, and for ontologically grounding the spatial expressions found in natural language. For example, in the so-called 'perspectivalist' approach of Smith and colleagues (e.g., [25]), objects, events and locations such as 'rooms', 'corridors', 'robot movements' and 'at the end of the corridor' are considered as real as quantum flows. Cognitively, there is little doubt that the kinds of everyday objects, events and places found in commonsense views of the world play an important role in all aspects of cognitive behaviour. From the robotics point of view, well formed ontologies of conceptual knowledge provide the robot with a common-sense viewpoint that can be related to other artificial or human agents. Inevitably, there are mismatches between the conceptual information held by the robotic agent, and information held by the user. Identification of such 'mismatches' is undoubtedly vital to developing a robot which does what the user expects it to do.

To illustrate this, we will consider an example involving the perceived categorisation of real world objects. Such an example is a very simplistic view of the role of ontology in cognitive robotics, but serves to illustrate the issues involved. Figure 3 (a) shows the common-sense taxonomy that might be used directly or indirectly by a robot in its representation of and reasoning about the outside world. Figure 3 (b) depicts a typical taxonomy of rooms for a German speaking user. On examination we see that there is a fundamental difference between the

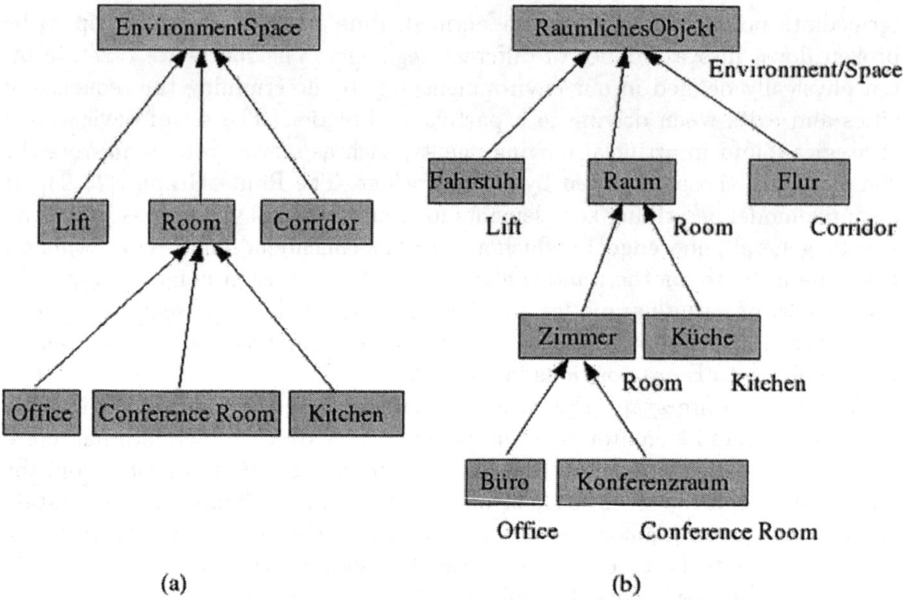

Fig. 3. Two contrasting views of commonsense reality

classification of spatial areas as viewed by a German user, typical in a domain ontology.

To see the effect of such a conceptual difference, consider the simple example where a user at position (A) of the route graph (see Figure 1 (b)) instructs Rolland with "Ich möchte zum dritten Zimmer rechts" (meaning "I'd like to go to the third room on the right"). If the user's mental model is that of Figure 3 (b) then it is likely that she/he intended to go to the conference room. However, if Rolland has a flat conceptualisation such as that shown in 3 (a), then Rolland should decide to drive the user to the kitchen.

As another example we once again examine the route graph of Figure 1 (b) and consider what it means if an English speaking user were to say "go right" at point (D). At D the corridor both veers to the right and has a clear right turn. Thus, the user's utterance can be seen as either underspecification, or that the user has a very clear understanding that "go right" means that the robot should veer to the right. In either case, pragmatics dictate that the analysis of the user's command should at least identify that there are two possible interpretations to this utterance (at least as far as the robot is concerned), and that further dialogue, task oriented reasoning, or dialogue history is required to explicitly determine the underlying intent of the user. To that extent, this situation can actually be viewed as an advanced form of linguistic ambiguity where the utterance, although specifying a clear meaning (turning right) cannot be mapped to an action, since multiple interpretations are possible at that point. Once again, a dialogue can be initiated to clear up such a misunderstanding.

4 The SharC Cognitive Control Architecture

In the above examples, we saw that clarification dialogues and contextual knowledge can be used to circumvent many shared control problems. It is therefore important that robot control systems be developed that provide a well defined cognitive interface to behaviours and sensory capabilities, thus facilitating interaction with users in a natural manner. In this section we are going to introduce the *SharC Cognitive Control Architecture* to bridge these human-robot interaction issues, with the more traditional deliberative and reactive requirements. The approach taken to the architecture's construction has been discussed elsewhere [26]. Here we will detail the components used within the architecture, showing their relevance to implemented shared control tasks such as those described above.

The SharC architecture, shown in Figure 4, splits system control amongst a number of deliberative agents. Each agent encapsulates a central component – often inherited from legacy applications) – with a Belief-Desire-Intention (BDI) abstraction. These abstractions are made in the AgentFactory Agent Programming Language (AF-APL) [27], and allow each agent to perform high-level introspective reasoning in a similar manner to hybrid architecture design. But, by distributing control amongst a number of agents, the architecture also provides robustness and scalability gains. The SharC Cognitive Control Architecture, presented here, is based on a more abstract MultiAgent Architecture for Robot Control (MARC) [28].

The SharC architecture is a high-level control architecture for Rolland, as opposed to Rolland's lower level automation architecture, which has previously been described in [15]. The automation control architecture, which addresses low-level automation and safety issues, is encapsulated in one SharC agent. Although primarily developed for the Rolland platform, our agent oriented approach will allow us to easily migrate SharC to other platforms as needed.

Figure 4 presents the SharC architecture for Rolland. Rounded blocks represent complete control agents that encapsulate a system component. Arrows between the agents show primary information flow. All information exchange is via messages rather than more tightly coupled method calls. This provides a loosely coupled distributed system which can be implemented across a number of different machines. Where possible, we have based the agents on off-the-shelf components. This code re-use approach was essential in precuring the tools for speech synthesis and recognition. However with integrating legacy components, there is always a risk that some components may not behave as expected. In such cases it is important that the overall architecture should be robust to fault. SharC's AF-APL based agent design is ideally suited to such occurrences.

The architecture is being developed for both German and English use. This bilingual requirement is facilitated with linguistic components that perform mapping from either German or English to internal representations and vice versa. Key to this mapping is the use of formally specified Linguistic and Domain Ontologies [29, 30]. These two bodies of knowledge provide the agents with a common ontological viewpoint, based on which they can also reason about

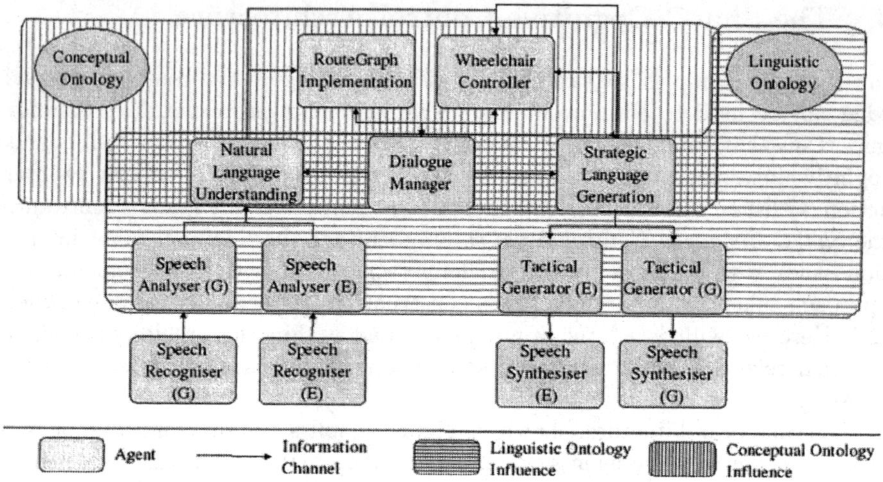

Fig. 4. The SharC Architecture for Rolland

the environment and internal states. Two hatched regions in Figure 4 show where these Spatial and Linguistic Ontologies are principally used. The vertically hatched region depicts the conceptual ontology that provides SharC agents with a common-sense style of spatial and action knowledge. It is used in the definition of Rolland's internal map representation, the RouteGraph. The horizontally hatched region shows the influence of the Linguistic Ontology over the SharC architecture. Concepts from the Linguistic Ontology, including the Generalized Upper Model [31], form the cornerstone of SharC's handling of natural language. As can be seen from the ontological overlaps, the SharC architecture can be split between a natural language independent, internal representation, and a language dependent section. The task of natural language generation and understanding is to mediate between these different viewpoints. In the following we will briefly outline each of the SharC agents, as well as the relationship between them. This is followed in Section 5 with a more detailed description of our language understanding approach and its use of ontology.

4.1 Natural Language Synthesis and Recognition

Automatic Speech Recognition (ASR) is the process of extracting lexical representations of a users' utterances from acoustic input. Despite considerable improvements in the last decade, ASR systems still encounter many problems including: interference by ambient or electronic noise; speaker-dependent pronunciation characteristics such as dialects; phonetic ambiguities; determination of word boundaries; unknown words; and multi-user interference (so-called Cocktail Party Effect). Despite these problems, commercially available ASR systems can provide a reasonable degree of reliability, as long as they are provided with a description of which words and combinations of words are likely to occur. For our wheelchair application, we wish to have both German and English communication available. Based on the recognition rates of available recognisers, we decided to employ the Nuance [1] speech recognition system for both English and German recognition.

Conversely, the *Speech Synthesis* process generates acoustic signals (audible speech) from a lexical input representation (string of text). As with ASR, there are a number of commercially available systems both for German and English. While *Concatenative Speech Systems* are often employed in telephone exchanges and public address systems, they are dependent on a vocabulary of input utterances having been created and joined together. While this approach is practical for systems based on numeric output (such as those mentioned), they are not practical for systems that require a large number of dynamic sentence constructions, particularly in the experimental domain where additions to the vocabulary are often required. For these reasons, we employ the MARY speech synthesis system [32] for both German and English synthesis.

4.2 Natural Language Analysis

While keyword-spotting can provide control levels acceptable to some limited voice controlled applications, the communication of task and spatial information to mobile robots requires a more sophisticated analysis approach. As mentioned before, in the highly dynamic scenario of robot navigation, humans are likely to produce ungrammatical sentences or even sentence fragments. Furthermore, while ASR systems have grown more reliable, recognition mistakes are always possible. Hence, language analysis must be robust enough to produce at least some meaningful representation of the user's intention, from which contextual resolution and confirmation dialogues can be used to further interpret meaning.

An important requirement related to analysis robustness is incrementality – allowing for immediate processing of partial input, and, thus, providing interpretations as early as possible. Incremental handling of ambiguities and inconsistencies requires that syntactic and semantic analysis be done simultaneously, or at least alternately. This approach is supported by the cognitive observation that humans integrate all available information as soon as possible (*rule-to-rule hypothesis*, [33]). Our analysis approach achieves this by integrating both syntactic and semantic information into one formalism, utilising the well-known notion of *unification* [34]. Furthermore, we are currently augmenting the formalism with a probabilistic approach inspired by Graded Unification [35], thus allowing the language analyser to achieve robustness against minor mismatches at either level. In the SharC context, natural language processing will be done in English and German, hence the parsing algorithm must be flexible enough to deal with different grammars.

Taking into account the requirements of incrementality and lexicalisation, we opted for a variant of *Combinatory Categorial Grammar* (CCG) [36]. The key concept of CCG is to regard linguistic entities as functions, and to provide a formal model of how these functions can be applied and composed with each

[1] www.nuance.com

other, thereby determining the syntactic roles of phrases. The same formalism can be easily extended to cover semantic roles in the same way, thus enabling the desired incremental integration of different information. Concerning expressivity, CCG falls into the class of mildly context-sensitive grammars which seem to be sufficient to capture any natural language construct while still being tractable [37]. CCG defines a set of abstract reduction rules together with a lexicon of possibly complex categories for each word. In our grammar, the set of rules used for processing has been slightly altered in order to include some phenomena of German word order [38].[2]

At the semantics level, the implications of all sorts of ambiguities must be dealt with correctly, which calls for information about how syntactic and semantic categories behave and combine with each other. This leads to the notion of a linguistic ontology that defines the interactions of linguistic entities. The semantic side of processing is aligned to the concepts defined in a linguistic ontology. Because of the alternative use of German and English, a language-independent approach is needed, that allows interfacing between natural language components and domain ontologies. We adopt the Generalized Upper Model [39] that meets these requirements. During analysis, a semantic structure is constructed according to the *Sentence Planning Language* (SPL) [40] which was originally designed for language generation using the Generalized Upper Model, but is equally suitable as a representation language for communication between natural language components in general. Furthermore, to account for complex quantifications, we are currently considering moving this representation the way of minimal recursion semantics [16]. Ambiguities that persist after the analysis of a sentence is complete are delegated to the *Natural Language Understanding* agent in order to consider contextualisation for resolution, or to have a clarification dialogue raised.

While natural language analysis provides a semantic representation of a user's input, it cannot by itself decide on actions to be performed; nor can it decide whether sufficient or appropriate information has been provided to initiate a particular action. Within the SharC architecture, the *Natural Language Understanding* component processes the output of language analysis to build complete queries for particular domain components. The Semantic Interpreter makes use of a slot-filling strategy to build requests to domain components including the *RouteGraph* and *Wheelchair Controller*. The Natural Language Understanding, making use of the *Dialogue Manager* for reference management and dialogue history, can initiate clarification dialogues when necessary. In Section 5, the interaction between the language analysis, understanding and domain components is discussed in greater detail with an example from the wheelchair control domain.

4.3 Natural Language Generation

In order to produce dialogue contributions, including spatial descriptions and announcements to a user of potential problems in interpretation, and so on, we have

[2] This work is partially based on results contributed by the Collaborative Reasearch Centre "Situated Artifical Communicators" at the University of Bielefeld.

decided also to employ general purpose natural language generation technology. The functionality of these components involves mapping between a semantic specification and annotated text strings suitable as input to the adopted speech synthesis component. The functionality aimed at within our system demands flexible solutions to the generation task for two principal reasons.

First, generated dialogue utterances must be appropriate for their particular contexts of use within the unfolding dialogue between user and robotic system. This cannot be achieved with canned text or restricted template generation, since an essential property of naturally produced dialogue-contributions is that they show in their design just how an interlocutor's statement is being interpreted. This is one of the main methods by which smooth dialogic interaction is achieved: possible misinterpretations can be signalled very early and the corresponding dialogue partner can provide further information or corrected information in order to keep the interaction on track. Building in such implicit feedback signals of interpretations can only be done with a fully flexible generation component that has extensive grammatical competence in the languages being produced.

Second, we are also exploring empirical methods by which particular selections of lexical items and grammatical constructions can influence the linguistic behaviour of a user so as to channel that behaviour along predictable lines. This channelling involves both purely linguistic properties, such as keeping the forms of language used within the interpretative capabilities of the system, and the selection of spatial perspectives. The latter is again potentially of significant value for improving the perceived robustness and utility of the complete system. If the user is implicitly directed towards spatial description strategies that align well with the functionalities supported by the perceptive systems of the robot, then the result is an increase in successful interactions. Previous work [41] has shown that mismatches in the expectations of the user concerning spatial strategies and the actual spatial mechanisms employed by a robot can cause severe communication problems and task failures. The flexible generation of interactionally appropriate utterances tailored specifically to avoiding such misperceptions promises to improve on this substantially.

To meet these requirements we are employing a general purpose generation system that already has reached a robust and mature stage of development, the KPML multilingual generation system developed over the past 10 years and currently maintained at Bremen [42]. This system also has the advantage that its semantic input has strong semantic typing and these types are drawn exclusively from the linguistic ontology that we are employing within the SharC architecture overall, the Generalized Upper Model. The general purpose orientation of the system and the grammars that it supports also allows extensive control of the precise phrasing adopted for any particular semantic content. This is essential for ensuring the precise interactional appropriateness of the dialogue contributions produced.

The basis of the generation process within KPML is a deterministic traversal of grammatical resources written within the linguistic framework of systemic-functional linguistics [43], a linguistic theory with a long tradition both of computational instantiation and of attending to interactional and other non-propositional aspects of meaning [44]. Substantial computational grammars for several languages are available within this framework, making it a natural choice for an initially bilingual system such as is envisaged for SharC. Systemic-functional grammars are used extensively in natural language generation because of their orientation to organising linguistic resources around communicative function and intentions rather than autonomous structure. The deterministic implementation makes the use even of very large grammars unproblematic. Even relatively long (for dialogue) utterances, such as running descriptions of scenes or way finding explanations, are readily produced in real-time, once the semantic specifications are available. The output of the generation process is a sentence structure that can also include functionally-motivated annotations for guiding intonational choice by the speech synthesis component–which is another prerequisite for achieving natural dialogic interaction [45].

4.4 Dialogue Management

Mode Confusions, such as those introduced in Section 3.2, are a considerable issue in the deployment of service robotics. The dialogue manager described below enables us to construct a formal model of a robotic or complex service system at a high level. This model can then be compared and contrasted against the user's perceived model of the system's operation. Within the context of natural language interaction, the Dialogue Manager is responsible for controlling the flow of dialogue between user and robot – deciding at a high level what questions should be asked of the user, what information is required, and which information should be passed onto domain specific components such as the *RouteGraph* or *Wheelchair Controller*. The Dialogue Manager is also responsible for maintaining a history of user and robot dialogue acts, thus providing an essential resource for the evaluation of otherwise ambiguous references.

Since Shared Control systems are often embedded in safety-critical devices, such as aircraft, power plants or service robots, it is crucial that safety-critical requirements be accounted for in Dialogue Manager design. Experience with safety-critical systems shows that the quality of such systems can be significantly improved through the application of formal methods. Since dialogue management plays a central role in the shared-control of the whole system, we have chosen to apply the well developed method *Communicating Sequential Processes* (CSP) to model the component. In fact, in [21] CSP is used to model and detect the mode confusion problems discussed in Section 3.2. Once detected, a mode confusion situation can be avoided through generating some proper dialogues by the dialogue manager.

CSP has been designed to describe systems of cooperating processes such as reactive systems. In general, processes proceed by engaging into events, where synchronisation of such events is required. This, rather than assignments to shared state variables, is the fundamental means of interaction between agents. CSP can be seen as a very abstract, highly readable and easily maintainable language to specify finite state automata. It is not a very strong specification

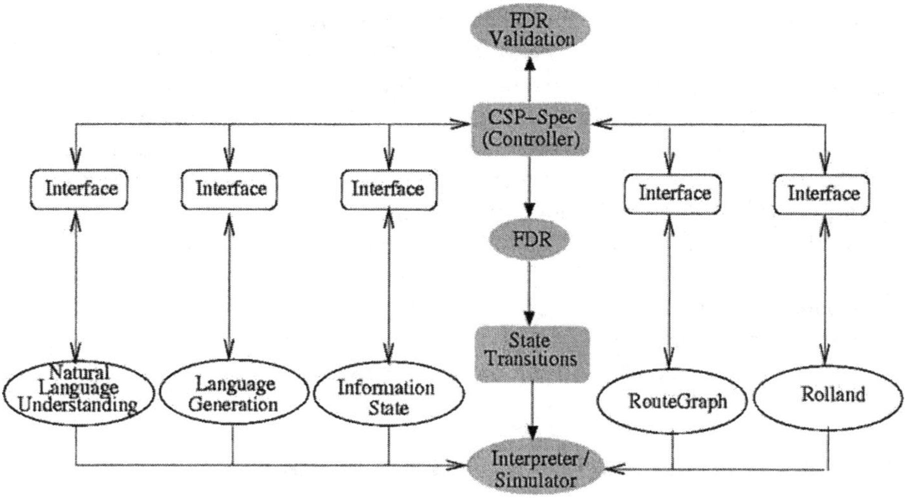

Fig. 5. A formal framework for dialogue management

language, indeed it lacks the ability of a more abstract temporal logic to specify liveness properties, but it is executable and comes with good support, there is extensive experience with it (e.g., [21, 46, 47, 22]). Nevertheless, our approach is not restricted to CSP. Other formal methods such as SPIN [48], Kronos [49, 50], SMV [51] and so forth, could also be applied.

A formal method based framework for implementing theories of information state is shown schematically in Figure 5. This framework splits control over a number of different elements and interfaces, including: the Dialogue Management control module that incorporates a CSP specification and a validation tool to perform verification using the model-checker FDR; a set of interfaces for integrating and communicating with information states, natural language input and output, and other domain specific components, e.g. RouteGraph [23] and wheelchair control for Rolland; and, finally, a state machine module including a interpreter for state transitions with development tools including a simulator using the graphical editor *daVinci* [52] to view state transition graphs (generated by FDR) dynamically.

4.5 Domain Components

The components outlined above constitute the linguistic elements of the SharC architecture. These, along with the Generalized Upper Model provide a general purpose dialogue framework for natural language based shared control systems. In principle, this general design can be applied to a number of different application domains. Here we will discuss the domain components used for the Rolland navigation demonstrator. This is important in illustrating where spatial concepts are relevant to the general architecture implementation.

As indicated earlier, the SharC architecture is being primarily developed for the autonomous wheelchair Rolland (depicted in Figure 1), but is intended for use beyond this single experimental platform. Rolland's low-level control is encapsulated within an *Wheelchair Controller* agent. The automation control regulates low-level control issues including basic trajectory management, movement and obstacle-avoidance behaviours, and low-level safety-critical issues. The automation control's interface is defined via a set of primitive movement actions.

The *Wheelchair Controller* agent is not responsible for maintaining a representation of the robot's behaviour. Instead, such representations are maintained within the RouteGraph component. This component, based broadly on the Route Graph described in Section 3.3 and [24], provides a spatial representation along with query and information update algorithms. The component is a dynamic data structure which may be updated and queried either by the *Wheelchair Controller* or the user through the natural language interface.

5 Interpreting Navigational Instructions

The control architecture presented above provides a general framework for the mapping of linguistic representations to domain specific concepts. In this section we further detail the relationship between linguistic and domain knowledge in the architecture, and show how this yields a generalised model of natural language understanding and action invocation. While our approach is illusrated via the Rolland use scenario, i.e, spatially aware service robots, the approach can be generalised to other application domains.

Rather than applying ad-hoc representations of spatial concepts, or, alternatively, spatial representations influenced solely by linguistic concerns, our approach is to apply well defined conceptual ontologies in the creation and description of spatial representations. It is our hope that by adopting this rigorous approach, we can simplify the process of mapping from natural language to internal representation, and vice versa.

In recent years a number of so-called "upper" ontologies have been developed, most notably the OpenCyc upper ontology [53], the Suggested Upper Merged Ontology (SUMO) [54], and the Descriptive Ontology for Linguistic and Cognitive Engineering (DOLCE) [55]. Upper ontologies are an attempt to formally specify and constrain knowledge at the *ontological level* of knowledge-based systems [56]. Here principles are set out for the kinds of formal properties that we demand of 'concepts' and the kinds of formal properties that we demand of 'roles' – this is then intended to lead to more consistent modelling decisions being taken for domains as a whole and, as a consequence, more reliable and re-usable representations. This level of representation is no longer arbitrary: ontology is very much constrained. Using these upper ontologies as a formal basis, domain ontologies can then be developed for multiple software systems, e.g., autonomous agents, to take advantage of shared knowledge to negotiate in some context.

Within the context of navigating robots, domain knowledge is most often required to model representation of space, motion, and action. As outlined in [57], the cognitive robotics literature is now becoming more endowed with formally modeled representations of spatial knowledge. The basic calculi of regional

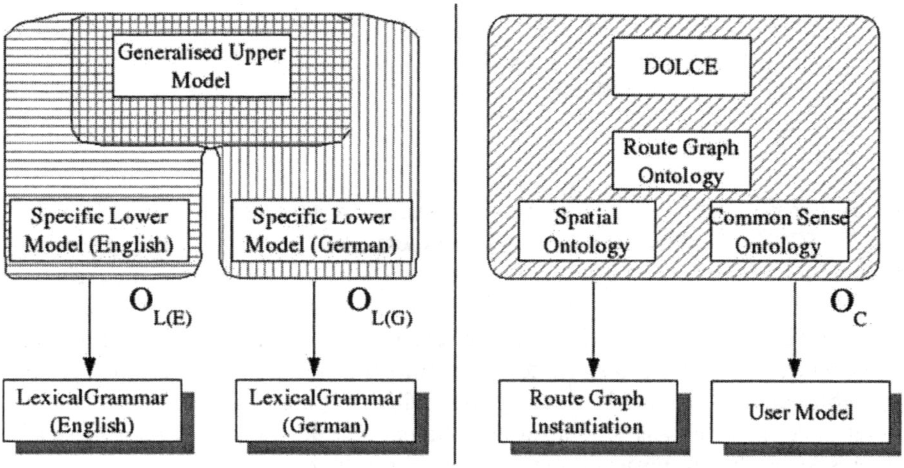

Fig. 6. Relationships between conceptual information

representation [58] have recently been augmented with more focused models of route type information [24] and spatial relationships [59]. While such formal representations are currently heterogeneous in nature, there is a drive towards a unified approach using upper spatial ontologies, below which more specified domain ontologies can be developed for individual mobile robots.

Such a well defined representation of spatial concepts should then make it possible to implement more concrete and extensible systems that relate language to spatial knowledge. To illustrate the application of this approach within the SharC architecture, Figure 6 schematically depicts the composition of ontologies within the SharC architecture, along with a number of components and knowledge sources that are constrained by those ontologies. In general, we see the linguistic ontology can actually be partitioned into two overlapping linguistic ontologies for English and German respectively; in addition, we see that the conceptual ontology can be viewed as a construction from three distinct ontologies, all constrained by a formal upper ontology. The details of the relationships between these ontologies is discussed further in [24, 57]; here, we describe the general approach and explicit flow of information. Consider the processing of the following request from a user while the wheelchair is situated at point (D) in Figure 1 (b) : `go right`. Assuming clean language recognition, an analyser, making use of a suitable linguistic ontology, can produce a shallow semantic representation such as the following[3]:

```
[syn: [type: sentence, mood: imperative],
 sem: [type: motion-process,
```

[3] For simplicity, we illustrated the example with a relatively neutral structure, rather than using the actual, more complex, analysis output – the details of which are not relevant here.

```
        actor: hearer,
        spatial-direction: right,
        speechact: command
      ]
]
```

where syn and sem denote syntactic and semantic information about the utterance respectively. Here, the linguistic ontology defines the semantic types and terms, including: *motion-process*, and *command*and *right*. Although analysis provides an initial semantic model of user input, it does nothing to determine the effect of the user input on the complete system. To do this, the semantic input must be examined with respect to constraints and requirements of domain specific components such as the route graph. This mapping and examination of user input is vital to the task of relating to coherent spatial concepts, and is generally the responsibility of natural language understanding.

Table 1. A sample of the conceptual interface used by dialogue management and natural language understanding to interact with the route graph component. All parameters are formalised as first order sentences

Action	**Parameters**	**Description**
isKnownLocation	place	Determine if place is known
addPlanInformation	place description	Add annotation information about place
findRoute	place place	Determine the Route from one place to another
isValidRoute	route	Determine whether route is valid with respect to current knowledge

The natural language understander, along with the dialogue manager, interact with domain components through abstract conceptual interfaces. These interfaces, adhering to the types, relations, and predicates defined in the conceptual ontology provide a consistent and natural language independent abstraction of low level functionality [27]. Table 1 presents an excerpt from the Route Graphs interface. The interface defines a number of different query types along with valid parameters for these queries; parameters, along with implicit return values, are statements of first order knowledge. It should be noted that while such abstract protocols define an conceptual interface into a component, they do not by necessity constrain the component implementation to be based on the conceptual ontology.

The dialogue manager, in conjunction with the natural language understander, performs contextual resolution, and also implements a proceedural knowledge based frame filling system. Queries to the RouteGraph or Robot Controller, composed under the constaints of the conceptual ontology, can then be passed to the domain components as appropriate.

In our example, the resolution of 'hearer' to Rolland is relatively straightforward, whereas the results of a `go right` request are entirely dependent on the spatial reasoning ontologies adopted for the route graph. For example, one approach would be to decide that all corridors leading from 15° to 75° relative the forward axis of the wheelchair's motion correspond to "right". In such a case, dialogue based resolution would most likely be required to confirm the user's intentions. If other domain information – or user modelling – cannot be used to determine which of the two possibilities is most accurate, then the *Dialogue Manager* must initiate a clarification dialogue with the user.

Our approach is intended to produce a degree of natural language independent human-robot interaction, thus augmenting a future localisation process. Such an approach is in contrast to either a *Canned Speech* approach, which is most common in the localisation industry, or indeed the development of internal robotic representations which are directly linked to a specific natural language; this so called *Corpus Based Robotics* approach of Bugmann et al [10] may however prove fruitful in the provision of a corpus of instruction types that are often used – to be incorporated in the production of a natural language independent internal representation of action.

6 Summary and Future Work

In this paper, we reported on our initial investigations into the problems of verbally controlled navigational robots. In particular, we presented a clarification of the types of Shared Control Problems that can be encountered in our application domain. To address these problems we have presented a partially implemented framework for testing dialogue based control methods. Although initially designed for a semi-autonomous wheelchair, the ontology-centric, agent-oriented nature of the architecture should make application to other hardware and domain examples feasible. Internally, this architecture has made extensive use of state-of-the-art language technology components to implement a clean separation between the robot's language interface, and the representations which underpin its internal reasoning systems.

In future work, we will be applying the architectural framework to investigate many shared control problems, looking in particular at how formal ontology definitions can be used to identify and solve ambiguities on-line, and then using dialogues with the user to solve issues that could not be handled otherwise. In immediate work, we will be investigating the direct translation of route instructions to a suitable representation for the robot's own spatial representations.

References

1. Konolige, K., Myers, K.L., Ruspini, E.H., Saffiotti, A.: The Saphira Architecture: A Design for Autonomy. Journal of experimental & theoretical artificial intelligence: JETAI **9** (1997) 215–235
2. Peters II, R.A., Wilkes, D., Gaines, D., Kawamura, K.: A software agent based control system for human-robot interaction. In: Proc Second International Symposium on Humanoid Robotics, Tokyo, Japan (1999)
3. Allen, J., Ferguson, G., Stent, A.: An architecture for more realistic conversational systems. In: Proceedings of Intelligent User Interfaces 2001 (IUI-01), Sante Fe, New Mexico, Association for Computing Machinery (2001) 1–8
4. Lemon, O., Gruenstein, A., Peters, S.: Collaborative activities and multi-tasking in dialogue systems. Traitement Automatique des Langues (TAL) **43** (2002) 131–154 Special issue on dialogue.
5. Nilsson, N.J.: A mobile automaton: An application of artificial intelligence techniques. In Walker, D.E., Norton, L.M., eds.: Proceedings of the 1st International Joint Conference on Artificial Intelligence, Washington, D. C., William Kaufmann (1969) 509–520
6. Winograd, T.: Understanding Natural Languages. 2 edn. Academic Press, New York (1976)
7. Matsui, T., Asoh, H., Fry, J., Motomura, Y., Asano, F., Kurita, T., Hara, I., Otsu, N.: Integrated Natural Spoken Dialogue System of Jijo-2 Mobile Robot for Office Services. In: AAAI/IAAI. (1999) 621–627
8. Bischoff, R., Jain, T.: Natural Communication and Interaction with Humanoid Robots. In: Proceedings of The Second International Symposium on Humanoid Robots, Tokyo, Japan (1999)
9. Roy, N., Pineau, J., Thrun, S.: Spoken Dialog Management for Robots. In: Proceedings of the ACL 2000. (2000)
10. Bugmann, G., Klein, E., Lauria, S., Kyriacou, T.: Corpus-Based Robotics: A Route Instruction Example. In: In Proceedings of IAS-8. (2004)
11. Porzel, R., Gurevych, I.: Towards context adaptive utterance interpretation. In: Proceedings of the 3rd SIGDial Workshop on Discourse and Dialogue. Association for Computational Linguistics, Philadelphia. (2002)
12. Bugmann, G.: Challenges in Verbal Instruction of Domestic Robots. In: Proceeding of the ASER'03, the 1st International Workshop on Advances in Service Robotics. (2003)
13. Stopp, E., Blocher, A.: Spatial information in instructions and questions to an autonomous system. Technical report, Universität des Saarlandes (1997)
14. Fong, T., Thorpe, C., Baur, C.: Collaboration, dialogue, and human-robot interaction. In: 10th Int'l Symp. of Robotics Research. (2001)
15. Lankenau, A., Röfer, T.: A versatile and safe mobility assistant. IEEE Robotics and Automation Magazine **7** (2001) 29 – 37
16. Copestake, A., Flickinger, D.P., Sag, I.A.: Minimal recursion semantics: An introduction. Manuscript, Stanford University: CSLI (1999)
17. Starter, N., Woods, D.: How in the world did we ever get into that mode? mode error and awareness in supervisory control. **37(1)** (1995) 5–19
18. Leveson, N., et al.: Analyzing software specifications for mode confusion potential. In Johnson, C.W., ed.: Workshop on Human Error and System Development. Volume 97-2 of GAAG-TR., Glasgow Accident Analysis Group (1997) 132–146

19. Degani, A., Heymann, M.: Pilot-autopilot interaction: A formal perspective. In Abbott, K., S.J.J., Boy, G., eds.: International Cnference on Human-Computer Interaction in Aeronautics: HCI-Aero 2000. (2000) 157–168
20. Rushby, J.: Using model checking to help discover mode confusions and other automation surprises. **75(2)** (2002) 167–177
21. Bredereke, J., Lankenau, A.: A rigouous view of mode confusion. In: SAFECOMP 2002, 21st Int. Conf. on Computer Safety, Reliability and Security. Volume 2434 of LNCS., Springer Verlag (2002)
22. Shi, H., Peleska, J., Kouvaras, M.: Combining methods for the analysis of a fault-tolerant system. In: Pacific Rim International Symposium on Dependable Computing (PRDC), IEEE Computer Society (1999) 135–142
23. Werner, S., Krieg-Brückner, B., Herrmann, T.: Modelling navigational knowledge by route graphs. In Freksa, C., H.C., Wender, K., eds.: Spatial Cognition II. Volume 1849 of Lecture Notes in Artificial Intelligence., Springer Verlag (2000) 295–317
24. Krieg-Brückner, B., Frese, U., Lüttich, K., Mandel, C., Mossakowski, T., Ross, R.J.: Specification of route graphs via an ontology. (2004) In Proceedings of Spatial Cognition 2004, Chiemsee, Germany.
25. Smith, B., Grenon, P.: The cornucopia of formal-ontological relations. (Technical report)
26. Krieg-Bruckner, B., Shi, H., Ross, R.: A safe and robust approach to shared-control via dialogue. (Submitted to the Chinese Journal of Software, Beijing, China)
27. Ross, R., Collier, R., O'Hare, G.: [af-apl]: Bridging princples & practices in agent oriented languages. In: Proc. The Second International Workshop on Programming Multiagent Systems Languages and Tools (PROMAS 2004). Held at AAMAS'04, New York, USA (2004)
28. Ross, R.: MARC - applying multiagent systems to service robot control. Master's thesis, University College Dublin (2004)
29. Bateman, J., Farrar, S.: Spatial ontology baseline. SFB/TR8 internal report I1-[OntoSpace]: D2, Collaborative Research Center for Spatial Cognition, Universität Bremen, Germany (2004)
30. Mossakowski, T., Lttich, K., Kried-Brückner, B.: Ontologies for the semantic web in casl. Submitted to WADT 04 (2004)
31. Bateman, J.A., Kasper, R.T., Moore, J.D., Whitney, R.A.: A general organization of knowledge for natural language processing: the PENMAN upper model. Technical report, USC/Information Sciences Institute, Marina del Rey, California (1990)
32. Schröder, M., Trouvain, J.: The German text-to-speech synthesis system MARY: A tool for research, development and teaching. In: Proceedings of the 4th ISCA Tutorial and Research Workshop on Speech Synthesis, August 29 - September 1, Perthshire, Scotland (2001) 131–136
33. Just, M., Carpenter, P.: The psychology of reading and language comprehension. Allyn and Bacon (1987)
34. Shieber, S.M.: An Introduction to Unification Based Grammar. Center for the Study of Linguistics and Information (1986)
35. Kim, A.: Graded unification: A framework for interactive processing. (1994)
36. Steedman, M.: The Syntactic Process. MIT Press (2000)
37. Joshi, A., Vijay-Shanker, Weir, D.: The convergence of mildly context-sensitive grammar formalisms. In Sells, P., Shieber, S., Wasow, T., eds.: Foundational Issues in Natural Language Processing, Cambridge, MA (1991)

38. Vierhuff, T., Hildebrandt, B., Eikmeyer, H.J.: Effiziente Verarbeitung deutscher Konstituentenstellung mit der Combinatorial Categorial Grammar. Linguistische Berichte **194** (2003) 213–237
39. Bateman, J.A., Henschel, R., Rinaldi, F.: Generalized upper model 2.0: documentation. Technical report, GMD/Institut für Integrierte Publikations- und Informationssysteme, Darmstadt, Germany (1995)
40. Kasper, R., Whitney, R.: SPL: A sentence plan language for text generation. Technical Report forthcoming, Information Sciences Institute (1989) 4676 Admiralty Way, Marina del Rey, California 90292-6695.
41. Moratz, R., Fischer, K., Tenbrink, T.: Cognitive modelling of spatial reference for human-robot interaction. International Journal On Artificial Intelligence Tools **10** (2001)
42. Bateman, J.A.: Enabling technology for multilingual natural language generation: the KPML development environment. Journal of Natural Language Engineering **3** (1997) 15–55
43. Halliday, M.A.K.: An Introduction to Functional Grammar. Edward Arnold, London (1985) (2nd. edition 1994).
44. Matthiessen, C.M.I.M., Bateman, J.A.: Text generation and systemic-functional linguistics: experiences from English and Japanese. Frances Pinter Publishers and St. Martin's Press, London and New York (1991)
45. Teich, E., Hagen, E., Grote, B., Bateman, J.A.: From communicative context to speech: integrating dialogue processing, speech production and natural language generation. Speech Communication **21** (1997) 73–99
46. Buth, B., Kouvaras, M., Peleska, J., Shi, H.: Deadlock analysis for a fault-tolerant system. In Johnson, M., ed.: Algebraic Methodology and Software Technology (AMAST). Volume 1349 of Lecture Notes in Computer Science., Springer Verlag (1997) 60–75
47. Buth, B., Peleska, J., Shi, H.: Livelock analysis for a fault-tolerant system. In Haeberer, A.M., ed.: Algebraic Methodology and Software Technology (AMAST). Volume 1548 of Lecture Notes in Computer Science., Springer Verlag (1998) 124–135
48. Holzmann, G.: Design and Validation of Computer Protocols. Prentice-Hall (1991)
49. Daws, C., Yovine, S.: Two examples of verification of multirate timed automated with kronos. In: 1995 IEEE Real-Time Systems Symposium., IEEE Computer Society (1995) 66–77
50. Henzinger, T., Nicollin, X., Sifakis, J., Yovine, S.: Symbolic model checking for real-time system. **111** (1994) 193–244
51. McMillan, K.: Symbolic Model Checking: An Approach to the State Explosion Problem. Kluwer Academic Publishers (1993)
52. daVinci: (http://www.tzi.de/ davinci)
53. Cycorp: Opencyc 0.7.0. Technical report (2004)
54. Niles, I., Pease, A.: Toward a standard upper ontology. In Welty, C., Smith, B., eds.: Proceedings of the 2nd International Conference on Formal Ontology in Information Systems (FOIS-2001), Ogunquit, Maine, Association for Computing Machinery (2001)
55. Masolo, C., Borgo, S., Gangemi, A., Guarino, N., Oltramari, A., Schneider, L.: The WonderWeb library of foundational ontologies: preliminary report. WonderWeb Deliverable D17, ISTC-CNR, Padova, Italy (2002)
56. Guarino, N.: The ontological level. In Casati, R., Smith, B., White, G., eds.: Philosophy and the Cognitive Sciences. Hölder-Pichler-Tempsky, Vienna (1994)

57. Bateman, J., Farrar, S.: Modelling models of robot navigation using formal spatial ontology. (2004) In Proceedings of Spatial Cognition 2004, Chiemsee, Germany.
58. Randell, D., Cui, Z., Cohn, A.: A spatial logic based on regions and connection. In: Proceedings of the 3rd. International Conference on Knowledge Representation and Reasoning, San Mateo, Morgan Kaufmann (1992) 165–176
59. Moratz, R., Wallgrün, J.O.: Spatial reasoning about relative orientation and distance for robot exploration. In Kuhn, W., Worboys, M., Timpf, S., eds.: Spatial Information Theory: Foundations of Geographic Information Science. Conference on Spatial Information Theory (COSIT). Lecture Notes in Computer Science, Springer-Verlag; D-69121 Heidelberg, Germany; http://www.springer.de (2003) 61–74

Perception and Tracking of Dynamic Objects for Optimization of Avoidance Strategies in Autonomous Piloting of Vehicles

Lía García-Pérez[1], María C. García-Alegre[1], Ángela Ribeiro[1], Domingo Guinea[1], and Jose María Cañas[2]

[1] Industrial Automation Institute, Consejo Superior de Investigaciones Científicas, 28500, Madrid, Spain
{lia, maria, angela, domingo}@iai.csic.es
[2] Universidad Rey Juan Carlos, Madrid, Spain
jmplaza@gsyc.escet.urjc.es

Abstract. In the autonomous piloting of vehicles, the characterization of nearby dynamic object motion by perception and tracking techniques aids in the optimization of avoidance strategies. Knowledge of the features of object motion in goal-driven navigation allows for accurate deviation strategies to be implemented with appropriate anticipation. This perceptual competence is a fundamental issue in the design of unmanned commercial outdoor vehicles with an often reduced capability for maneuvering. To this aim, a grid map representation of the local panorama is proposed such that laser rangefinder images are converted into grid cells that are segmented and assigned to objects, allowing classification and monitoring. The motion properties of objects are thus used to establish avoidance behavior to smartly control the vehicle steering, such that a safe and optimal detour maneuver is carried out while driving to a target. The developed perceptual ability is demonstrated here in several tests performed in a relatively clutter-free area to detect and track walking pedestrians. Some results are also shown to highlight the modulation of moving object properties on trajectories described by a robot when a fuzzy avoidance strategy is used to control the steering angle.

Keywords: Detection and tracking, moving objects, obstacle avoidance, human spatial cognition, dynamic environments, laser rangefinder, fuzzy control, outdoor vehicles.

1 Introduction

The detection of moving objects has been extensively analyzed in Ethology [1], and has inspired the development of artificial perception strategies [2], [3]. Attention and pre-attention visual mechanisms have been studied by psychophysicists, with some authors arguing the existence of motion filters in the brain as a sign of evolutionary utility: a moving target symbolizes either food or danger, so motion detection is a cue for animal survival. An additional, well-known and investigated natural phenomenon involves the maintenance of the moving object in the visual field, i.e., an object tracking strategy [4], [5]. The visual

C. Freksa et al. (Eds.): Spatial Cognition IV, LNAI 3343, pp. 500–517, 2005.
© Springer-Verlag Berlin Heidelberg 2005

perception of object motion is fundamental to our capacity to understand and interact with our surrounding environment.

In the past, perception algorithms, supported by instantaneous sensor readings, have been implemented in holonomic indoor robots for collision avoidance independently of their motion state, relying both on fast feedback reactive control loops and on a high maneuverability; however these algorithms are unable to generate optimal deviation trajectories [6], [7], [8], [9], [10]. Other approaches on indoor vehicles require that the working area be equipped with a well-distributed set of scanners [11] or team of robots [12], to derive kinematic information to identify moving objects.

Several attempts have been made to develop complex movement strategies based on models of object dynamics [13]; however these models were hampered by restrictions imposed by real systems (sensors, vehicle, environment). In recent years, research has focused on the detection and tracking of moving objects based on the data provided by real sensors. To this aim, a stochastic map building method was proposed to model quasi-static environments, in real time, using a 2D laser [14]. The detection and tracking of walking persons in a cluttered railway station was addressed in [15], but without any reference to their effect on the piloting actions to be performed . In [16] detection, tracking and avoidance of persons in an office-like environment is addressed using a probabilistic model for person locomotion. In a similar manner, the detection and tracking of moving objects, is formulated only as a monitoring system to warn and assist bus and car drivers in advance, from a high temporal resolution laser rangefinder integrated on an urban vehicle [17].

Up to now, research dealing with moving objects and non-holonomic, car-like, vehicles has departed from the formulation of object motion models and has not engaged in the connection between perception and action to locally optimize navigation strategies. However, the optimization of avoidance strategies for car-like vehicles in dynamic environments still remains a challenge in mobile robotics. In this paper, moving objects present in a scene are characterized by a set of features, as a first step to optimize a collision avoidance strategy in goal-driven navigation. Techniques dealing with the visual detection and tracking of moving objects, which have been extensively developed in the field of artificial vision, are out of the scope of the current work, although some visual image processing approaches are used here and adapted to planar laser scanner images [18], [19], [20], [21]. In the following section, algorithms developed to detect, track and characterize moving objects are depicted. In Section 3, an avoidance strategy, formulated by incorporating knowledge on moving object features into a fuzzy controller in order to derive a more appropriate steering action, is described. Examples of the results obtained with the proposed approach are presented and commented upon in Section 4, while conclusions are presented in Section 5.

2 Detection, Tracking and Characterization of Moving Objects

The detection and tracking of moving objects, in a dynamic environment, is a key issue to be addressed in order to achieve the collision-free navigation required for non-holonomic commercial vehicles with limited maneuverability. To this aim, a local grid map representation [22] has been proposed to detect, track and characterize moving objects that could potentially obstruct the vehicle's intended trajectory, accounting for characteristics of both the vehicle and its immediate environment [23]. This grid map (2D cartesian coordinates) is a gross grain representation that acts as a short term memory accumulating the occupancy evidence. Its resolution (cell dimension) can be changed according with the expected objects size. A flow diagram of the processes developed to derive moving object features is displayed in Figure 1.

The dynamic environment is perceived with a laser rangefinder (SICK-LMS 291) that delivers a 2D polar coordinates representation of close objects, at a rate of 5 Hz (RS232 protocol), with 0.5 degrees resolution over a 180 degrees visual field and a maximum measurement range of 30 m. A grid map is then generated by mapping the obtained polar coordinates to a Cartesian coordinates representation of 30×30 meters , composed of cells of 20×20 cm, both selected as a trade-off between the estimated speed of the most likely moving objects and the computational cost. One of three states is then assigned to the grid map cells: occupied, free or unknown. The grid occupancy latency is fixed to 400 ms, corresponding to two complete laser scans at maximum frequency rate. The position of the robot in the grid is updated by means of a Differential Global Position System (DGPS) integrated in the vehicle.

An image segmentation algorithm then groups occupied cells in ensembles using an adjacency criterion. Thus, the grid map is scanned by searching for adjacent occupied cells, recording simultaneously all visited cells, and exploring only those cells marked as occupied. From the segmentation process a collection of clusters is obtained, namely objects, which are ascribed to the $Object\_List(t) = \{O_1(t), O_2(t), ..., O_n(t)\}$, with n being the number of detected objects at time t. Each object of the Object_List is described by the set of adjacent occupied cells and its motion is represented by its centroid motion (1).

$$O_i(t) = \{ID_i(t), cells_i(j,k)(t), \boldsymbol{r}_{cent_i}(t), \boldsymbol{v}_{cent_i}(t)\} \tag{1}$$

$ID_i(t)$ is the identification of object $O_i(t)$, and $cells_i(j,k)(t)$, the locations of the cells that represent object i. The two last descriptors, position and speed of the object centroid (2), are defined by the following expressions.

$$\begin{aligned} \boldsymbol{r}_{cent_i}(t) &= \sum \boldsymbol{r}_{cell_i}(j,k)(t)/M \\ \boldsymbol{v}_{cent_i} &= (\boldsymbol{r}_{cent_i}(t) - \boldsymbol{r}_{cent_i}(t-1))/\Delta t \end{aligned} \tag{2}$$

Correspondence between objects in Object_List (t) and in Object_List (t-1) is completed according to a Nearest-Neighbor criterion, widely used for its sim-

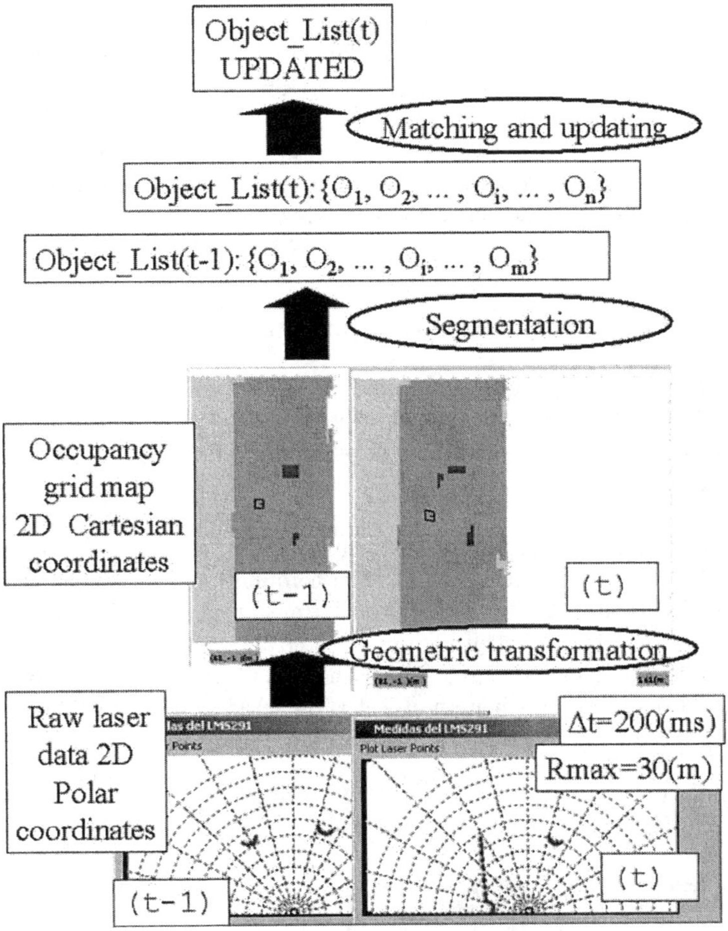

Fig. 1. Flow diagram of representations and processes developed for object detection, tracking and characterization

plicity and good performance in real-time applications, in contrast to more complex approaches such as the Hungarian algorithm [24]. The matching similarity function is the Cartesian metric between object centroid current and predicted position. The Object_List (t-1) is scanned by searching for a correct correspondence among objects at (t-1) and objects at (t), extracting only those that are within a circle of radius R centred at the predicted position at (t) of objects at (t-1). The distance threshold R has been set to one meter according to the maximum speed that could be detected. The predicted position at (t) of objects belonging to the Object_List (t-1), is calculated as follows (3):

$$r_{PREDICTED_{cent_i}}(t) = r_{cent_i}(t-1) + v_{cent_i} \cdot \Delta t \tag{3}$$

An object having a maximum number of overlapping cells, among all possible candidates, is selected, and object features are updated. The first selection of candidates greatly reduces the search space, as only the closest obstacles are investigated in the matching process. The proposed overlapping criterion for candidate selection helps to prevent incorrect obstacle associations being made. The updating of the object features in the Object_List is performed at a rate of $\Delta t = 200$ ms, although the algorithm computational cost is around 5 ms.

All described processes are embedded in a perceptual agent, namely MOVING_OBSTACLES, which entails the moving objects characterization competence.

3 Avoidance Strategy Optimization

The avoidance strategy designed is devoted to perform smart deviations when objects, either static or dynamic, unexpectedly appear within a bounded ring region in the laser scanner viewing angle while the vehicle is en route to a defined target. The avoidance strategy proposed here, AVOID, has been encapsulated as an agent of the multiagent architecture already implemented in an unmanned lawnmower for use in outdoor environments [25]. Three concurrent basic motion agents, ADVANCE, AVOID and STOP, in conjunction with a perceptual agent MOVING_OBSTACLES, determine the piloting competence. The processes contained in the ADVANCE strategy are directed to drive the robot to a specific location through free-space areas, with the STOP agent acting only in emergency situations, that is, when an object is detected at a distance less than 2m.

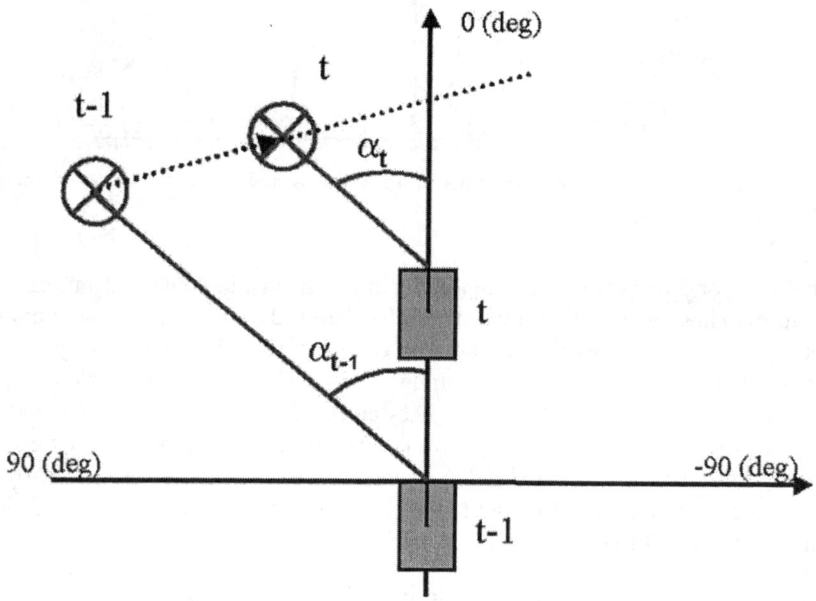

Fig. 2. Collision_angle α, in t and $t-1$, measured with respect to the vehicle heading. Constant values of the *collision_angle* variable in a time period indicates collision risk

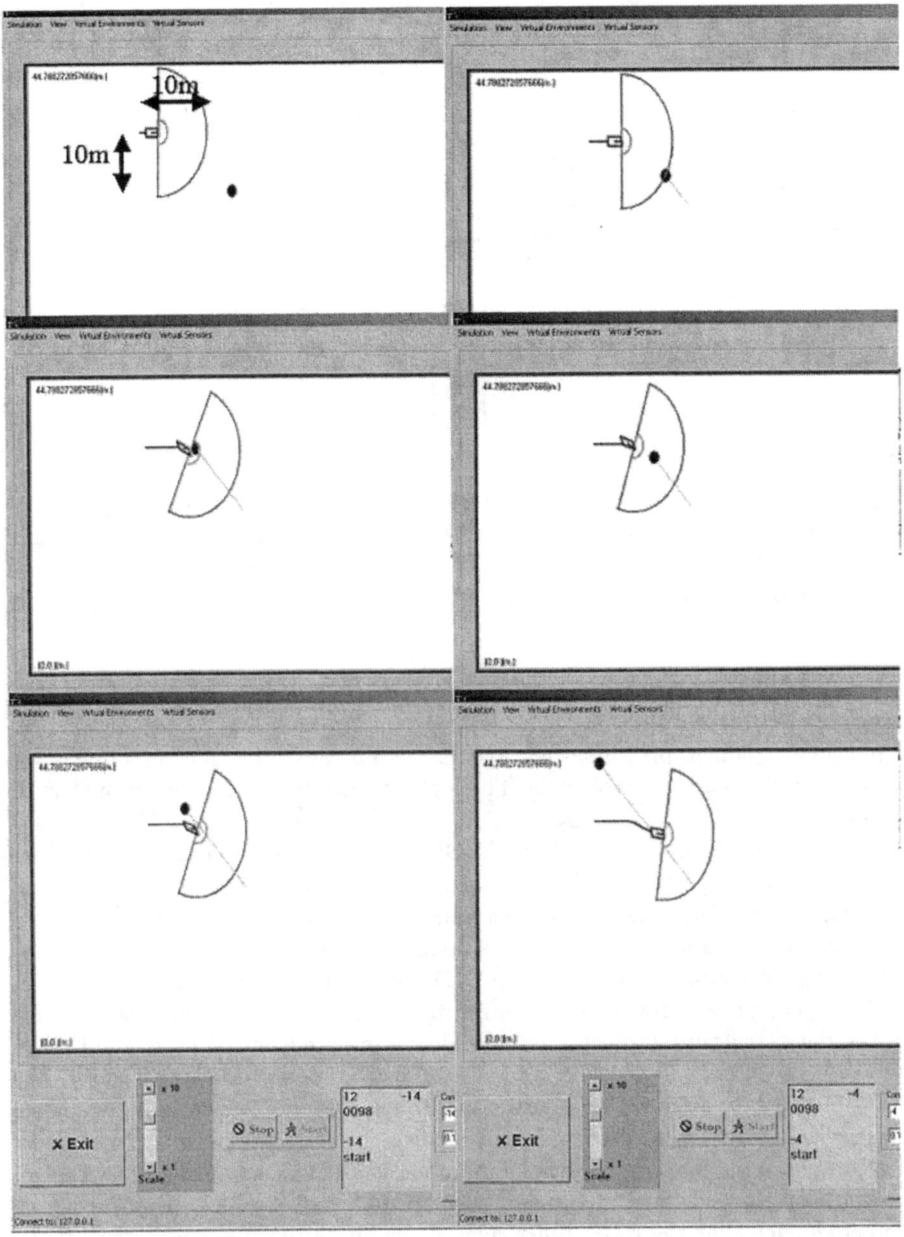

Fig. 3. Robot (black) and object (grey straight line) trajectories. The goal-driven robot trajectory is slightly deviated from its target through the activation of the AVOID strategy as collision risk is detected. ($v_{robot} = 0.3$ m/s, $v_{object} = 0.5$ m/s)

The AVOID agent detours the vehicle in order to maintain collision-free piloting when an object enters the ringed region of radius 2-10m from the laser-centred

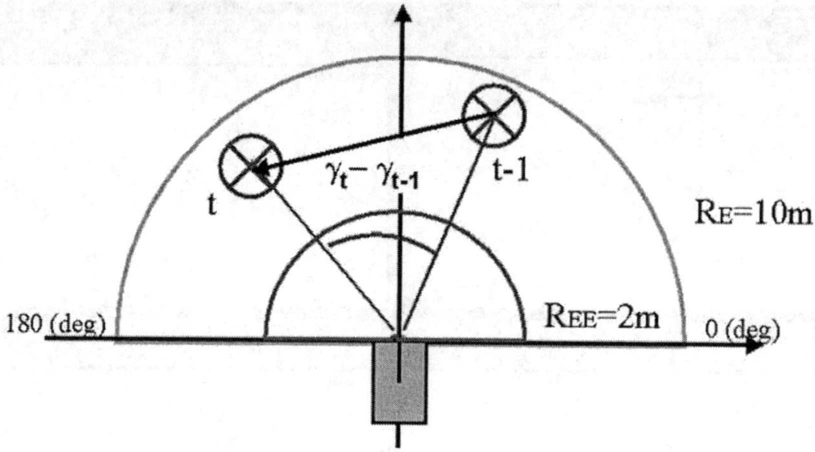

Fig. 4. Inputs to the AVOID strategy :$\gamma(t)$ moving object angular distance in t, relative to a laser-centred reference system, and $[\gamma(t-1) - \gamma(t)]$ angular distance variation of the moving object location in two successive steps. R_{EE} is the radius of the extreme emergency zone, while R_E corresponds to the emergency area

reference system. At the piloting level, motor agents are designed so that they are activated by mutually exclusive perceptions. In addition, a redundant coordination mechanism prioritises the agents, thereby ensuring that only one agent is active within each control cycle. The highest priority is assigned to the STOP agent.

To disallow the activation of the AVOID behaviour in those situations where there is no risk of collision, in spite of the moving object is within the emergency bounded ring area, the *collision_angle* α, Figure 2, is calculated in two consecutive instants elapsed 500 miliseconds. Consequently, only when the *collision_angle* remains constant, the AVOID behaviour is activated, Figure 3.

The perceptual information required by the three motor agents is available in the data structure, Object_List(t), which is updated by the perceptual agent MOVING_OBJECTS, and is recorded in a shared memory among agents. The AVOID agent is modeled as a fuzzy controller that mimics human avoidance strategies in response to detected moving obstacles. Basic driving control strategies proposed in the fuzzy heuristic approach is a deviation to the left when an object is on the right and vice-versa. This strategy varies when object motion is detected, such that the robot must turn in the opposite direction to that of the moving object, independently of its angular location.

Only two input variables are required: the angular object position γ, *obstacle_position*, and the object motion direction calculated as $[\gamma(t) - \gamma(t-1)]$, namely *movement_direction*, displayed in Figure 4. The output of the fuzzy controller is the robot wheel angle β, *wheel_angle*. The linguistic labels associated to the former variables and their definitions by trapezoidal membership functions are displayed in Figure 5. A “decision” on the *wheel_angle* is made each 500 ms.

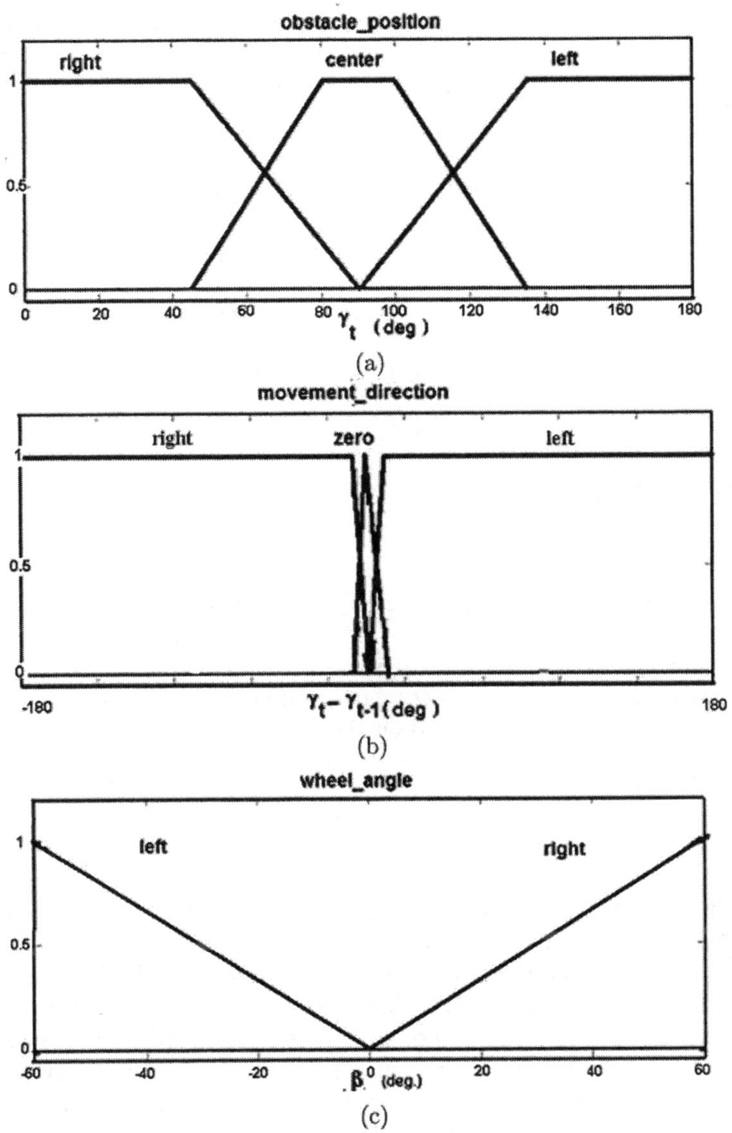

Fig. 5. Membership function definitions for the linguistic labels assigned to the AVOID fuzzy controller inputs (*obstacle_position*, *movement_direction*) and output variables (*wheel_angle*)

The unique output control variable of the detour strategy is the *wheel_angle*, which is related to the steering system, as the lawnmower mechanical design inhibits speed changes.

The knowledge base of the AVOID fuzzy control system is composed of the following nine rules:

```
R1: IF obstacle_position  IS left AND movement_direction IS right
    THEN wheel_angle IS left
R2: IF obstacle_position  IS left AND movement_direction IS zero
    THEN wheel_angle IS right
R3: IF obstacle_position  IS left AND movement_direction IS left
    THEN wheel_angle IS right
R4: IF obstacle_position  IS center AND movement_direction IS right
    THEN wheel_angle IS left
R5: IF obstacle_position  IS center AND movement_direction IS zero
    THEN wheel_angle IS right
R6: IF obstacle_position  IS center AND movement_direction IS left
    THEN wheel_angle IS right
R7: IF obstacle_position  IS right AND movement_direction IS right
    THEN wheel_angle IS left
R8: IF obstacle_position  IS right AND movement_direction IS zero
    THEN wheel_angle IS left
R9: IF obstacle_position  IS right AND movement_direction IS left
    THEN wheel_angle IS right
```

In those scenarios where unexpected obstacles are static only three rules of the proposed knowledge base are fired, R2, R5 and R8, corresponding to a zero value for the variable *movement_direction*. The *movement_direction* variable takes into account the moving object motion direction, so as to overcome undesirable avoidance trajectories, based on the dynamic objects features previously calculated.

4 Results

Some results of different trials are presented here to show the performance of the perceptual agent MOVING_OBJECTS in the detection and tracking of dynamic objects in an outdoor scenario. Experiments were performed with a commercial lawnmower in a garden-like area of the IAI-CSIC Institution Campus, located between office buildings and warehouses, where pedestrians cross at speeds ranging from 0.25 to 1.50 m/s. Experiments were aimed at demonstrating the performance and limitations of the proposed approach under different conditions. The laser rangefinder was mounted on the front of the lawnmower in a fixed position at a height of 50 cm above the ground. Two representative examples have been selected and are presented here, highlighting situations where two moving objects are approaching the vehicle and where the intended trajectory of the objects is occluded in front of the vehicle.

In the first experiment, two pedestrians describe parallel trajectories towards the vehicle, Figure 6a. In the second one, a straight and curved trajectories, intersecting in front of the vehicle Figure 6b, are displayed. In both trials the moving objects were accurately detected and tracked during the complete run, even during the occlusion phase. The results of first experiment reflect the extended time that the pedestrians remained in their initial position before beginning to move.

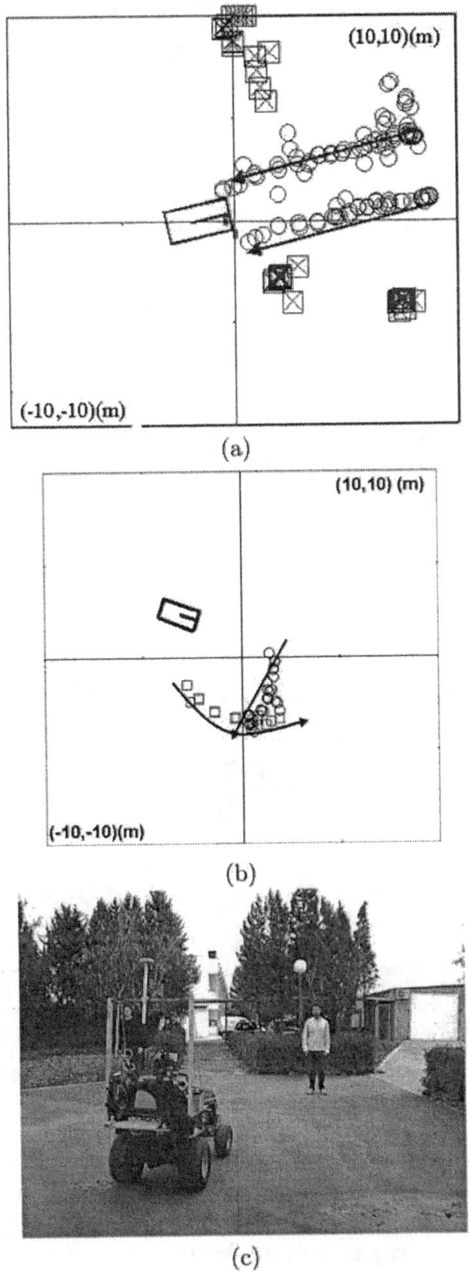

Fig. 6. Real time detection and tracking of two pedestrians: (a) walking in parallel toward the lawnmower, and (b) walking in trajectories that intersect in front of the lawnmower. Pedestrians are represented by circles, and static objects, such as bushes or walls, by a cross within a square, in a 2D Cartesian representation of the detected objects in a time period. A visual snapshot of test (a) departure positions is displayed in image (c)

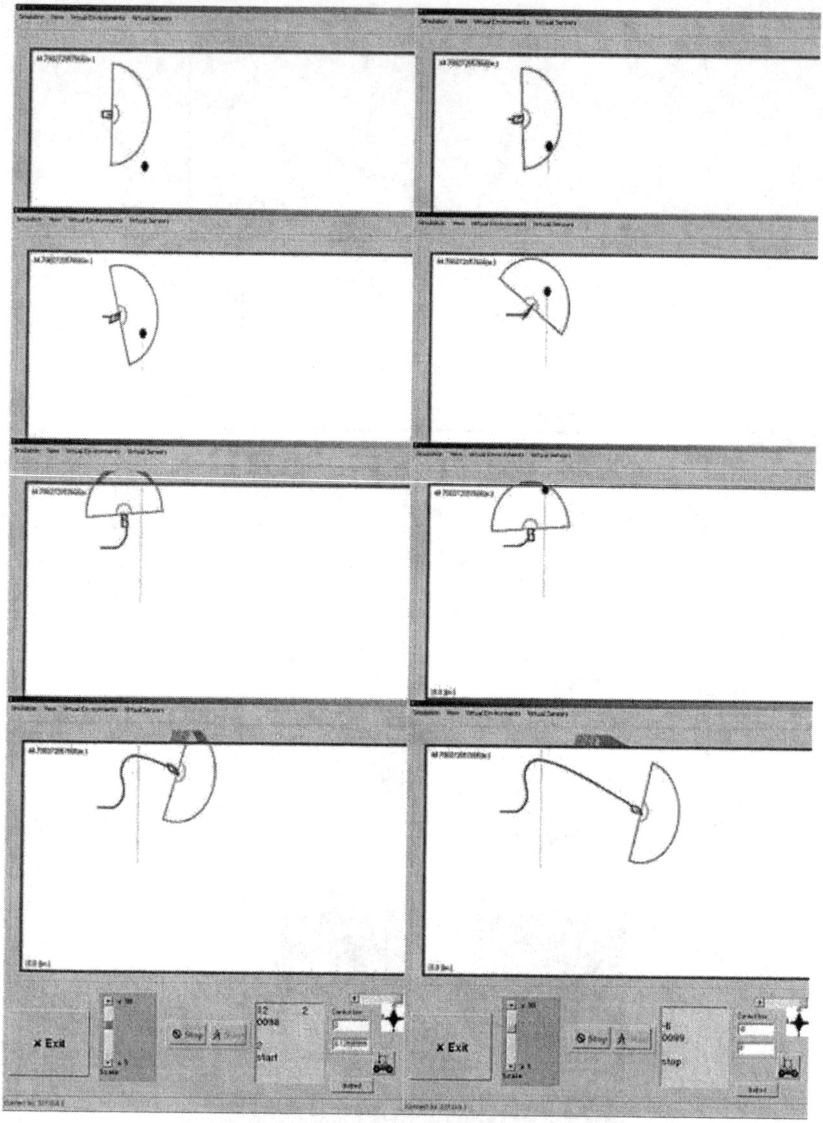

Fig. 7. Robot trajectory (black) described by the activation of only three rules from the AVOID strategy, due to the lack of knowledge on the moving object. Vehicle and obstacle speed are 0.3 and 0.5 m/s, respectively. Object trajectory is represented by a grey straight line, and semicircles limit the emergency areas

That is, a higher density of circles are evident at each pedestrian's departure location than in the remainder of their trajectory, Figure 6a. In second experiment, both pedestrians are correctly detected and tracked by the proposed algorithms, with higher density of circles at the end of the trajectory corresponding to pedestrian stop. Correct classification of each object just after the obstruction relies

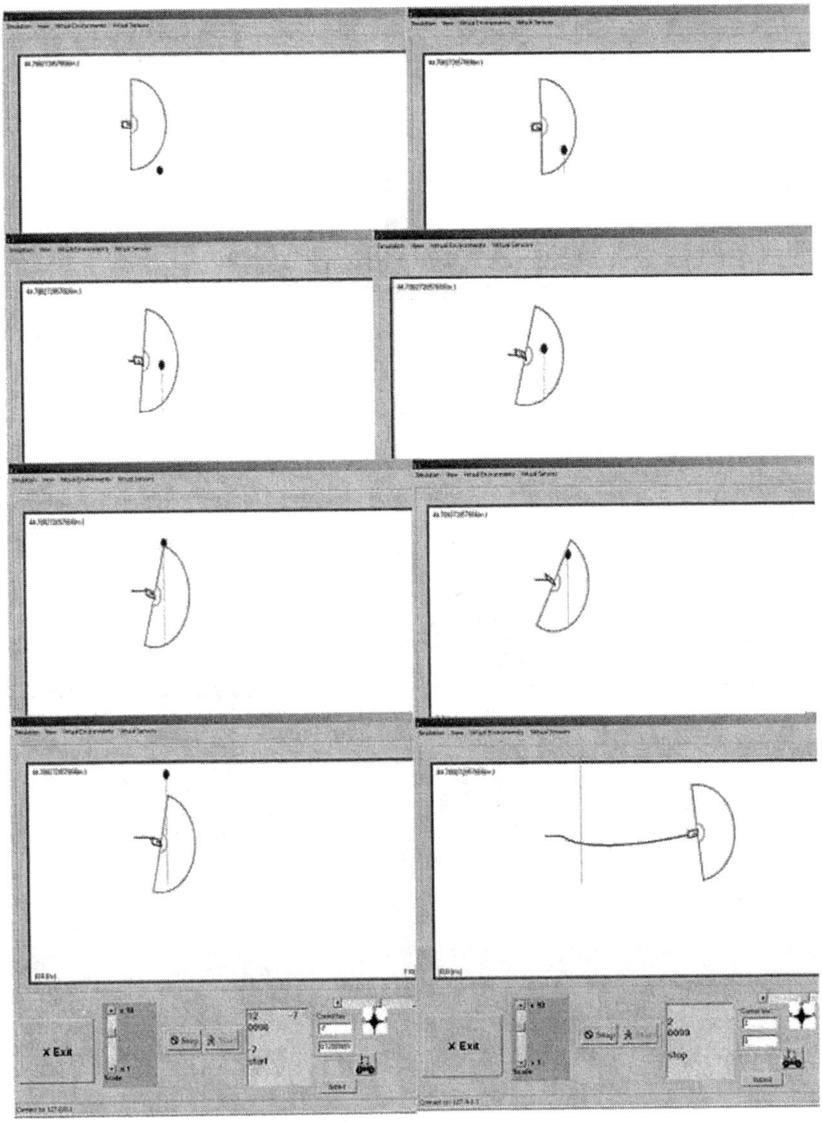

Fig. 8. Robot (black) and object (grey) trajectories. Deviations from the robot goal-driven tentative trajectory are optimised due to the existing knowledge on moving object descriptors. Vehicle and obstacle speed are 0.3 and 0.5 m/s, respectively

on previous knowledge on motion features, that is, it is expected that the second pedestrian keeps going along the curved path.

The limit to low speed detection is related to the grid map cell size, meaning that objects moving at a speed below 10 cm/s are considered to be static. High speed object detection was limited in current work by the maximum sampling rate of the sensing system (laser RS-232 serial connection), 5 Hz, which corres-

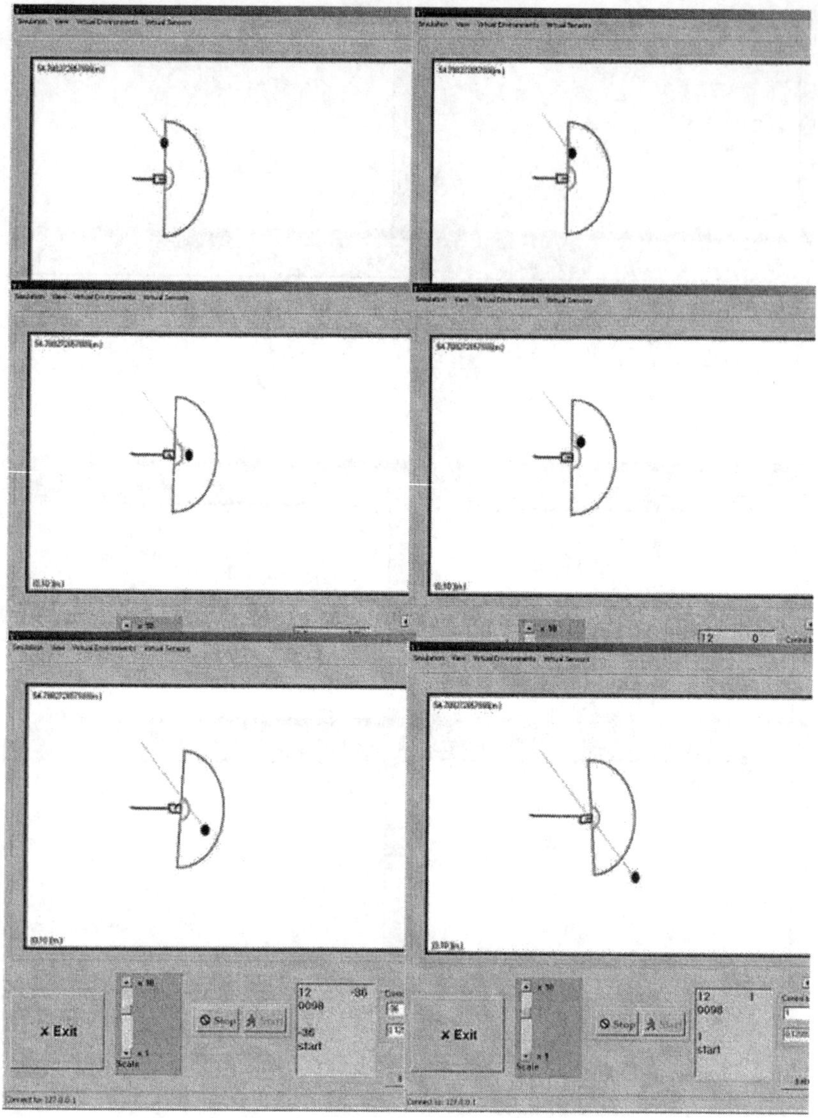

Fig. 9. Robot (black) and object (grey) straight line trajectories. The AVOID agent is not activated as collision risk is not detected, in spite of a speedy object is within the emergency area. (v_{robot} = 0.3 m/s, v_{object} = 0.5 m/s)

ponds to an object speed of 5 m/s. The correspondence between moving objects in two consecutive laser images for objects moving faster than 5 m/s would obviously fail due to the maximum laser sampling rate. Nevertheless, this is not a major drawback as in most outdoor applications, with the exception perhaps of sporting competitions, objects could be expected to move at speeds far below 5 m/s (18 km/h). A high speed harvester, for example, moves at 6 km/h. The

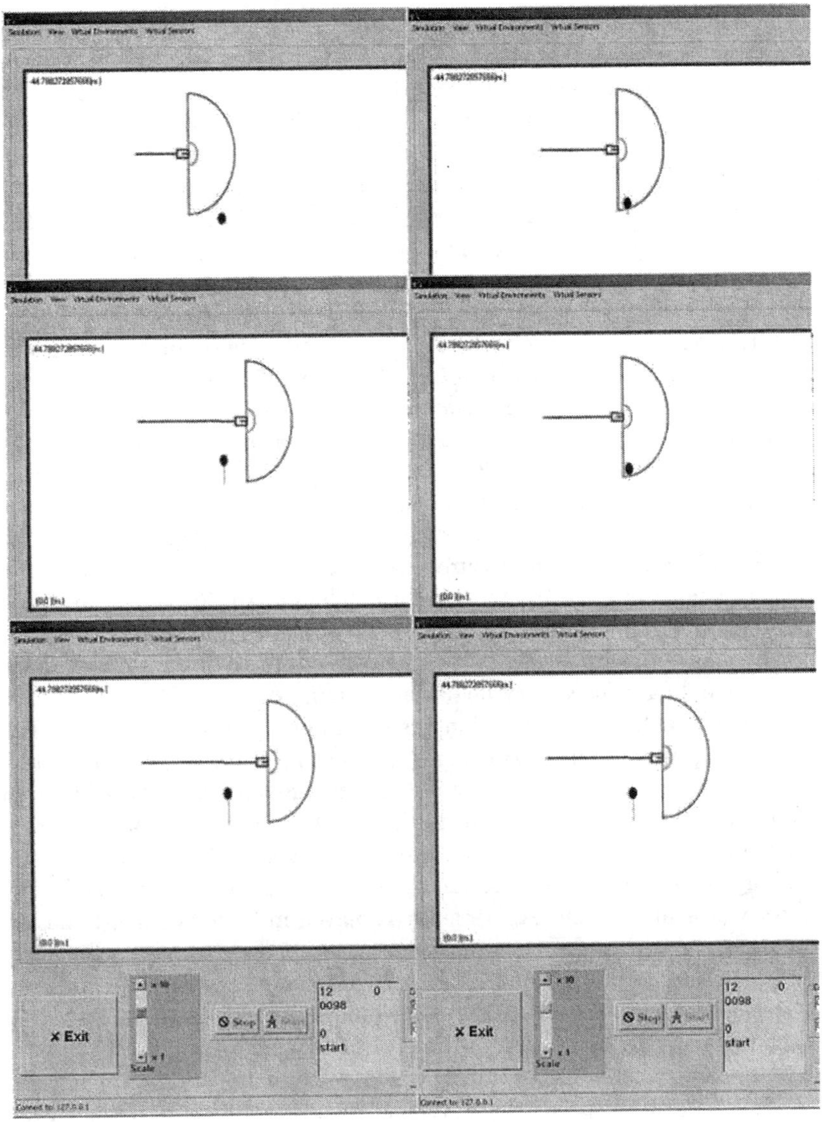

Fig. 10. Robot (black) and object (grey bottom-up) straight line trajectories. Optimized robot trajectory in the presence of an object in the emergency area. No detour is performed as there is no collision risk ($v_{robot} = 0.3$ m/s, $v_{object} = 0.05$ m/s)

robot-centred grid map representation employed, moves jointly with the vehicle and allows for ease of use of global positioning (DGPS) in outdoor scenarios for both the vehicle and objects. A further challenge concerns the relatively improbable situations wherein objects move at speeds less than 5 m/s but change suddenly of direction in a time period of less than 200ms.

Experimental work with outdoor robots to correctly tune each navigation behaviour is hard and bothersome, due to mechanical and energy failures in addition to adverse weather conditions. To speed up the tunning process a simulator of sensors and worlds, namely AGRO-SIM, has been developed to test and debug the different navigation strategies. One of the relevant features of current simulator is the use in the robot of the same code generated in simulation, by defining a client/server communication architecture. Both, robot and simulator are server applications allowing the connection of client programs to each one indistinctly.

Some preliminary results of the modulation of moving object descriptors on the AVOID agent deviation behaviour are displayed. The simulation environment [26], AGRO-SIM, has been used to test the different models that linguistically describe the avoidance strategies learned from human experience. The AVOID knowledge base rules are defined from the object motion descriptors calculated in real-time by the MOVING_OBJECTS agent. The *movement_direction* variable takes into account the moving object motion direction, so as to overcome undesirable "persecution" trajectories such as the one displayed in Figure 7, when knowledge on object motion features lacks and therefore unexpected objects appearing in the scene are treated as static. The integration of the fuzzy *movement_direction* variable permits to reason on object motion features to make a correct decision on next *wheel-angle* turn direction, to optimise the resulting trajectory avoiding unnecessary deviations, Figure 8.

The AVOID behaviour activation pre-conditions greatly reduce the changes in the steering angle to perform a safe an optimum navigation goal-driven in dynamic environments, under the general model embedded in the AVOID fuzzy knowledge base. Spatio-temporal evolution of the vehicle and object for configurations of either high or low speed object, relative to the vehicle motion, are successfully solved. Two cases, of unnecessary activation of the AVOID agent, uncorrectly activated in the experiments displayed in Figures 7 and 8, despite of the object being within the emergency areas 2m. $\leq d <$ 10 m., are displayed in Figure 9 for a speedy object and in Figure 10 for a very slow object relative to the robot speed. In these situations the *collision_angle* shows a temporal variation and thus prevents the activation of the AVOID agent.

In those cases where objects move parallel to the robot within the emergency area and at the same speed of the vehicle, risk of collision is detected as the collision angle remains constant and consequently the AVOID behaviour is activated to slightly deviate the lawnmower. However if the obstacle is not within the emergency area the detour strategy is not fired.

5 Conclusions

Identification of the dynamic state of moving objects has a crucial influence on the implementation of safe and optimized avoidance strategies in robot navigation. This point is a major issue for non-holonomic vehicles operating in outdoor scenarios, such as in agriculture, horticulture or gardening.

Likely outdoor obstacles could be people, animals or commercial vehicles. Pedestrian and animal motion is difficult to characterize as its characteristics may abruptly change in a short time period. In such scenarios, neither static nor quasi-static probabilistic techniques are able to cope with dynamic and unpredictable situations, meaning that the matching of objects in successive grid maps on a spatio-temporal representation would better serve as an appropriate and flexible model. To this aim, a detection, tracking and characterization algorithm for moving objects, based on laser rangefinder readings and grid maps, has been proposed and demonstrated here in relatively uncluttered outdoor scenarios.

Current approach relies on the calculated moving-object features to search for the best match among objects in different time steps. The moving objects matching algorithm proposed here has been selected to deal with real-time applications, being a simple and efficient process with low computational cost. The updating of object features in the Object_List is performed at a rate of 200 ms, but detection and tracking algorithm computing cost is about 5 ms, which is well below the maximum scan frequency of the laser rangefinder.

Preliminary tests show the adequateness of the calculated moving object features to optimise avoidance maneuvers in goal-driven navigation, heuristically defined by means of a set of fuzzy rules. Strategies are tested and tuned with the aid of the AGRO-SIM application that simulates real systems and behaviours. The piloting tests clearly show the improvement of the vehicle detour trajectory, when the AVOID strategy integrates moving objects descriptors. More simulated and real experiments are now being performed to derive a general framework to optimise detour trajectories, particularly in those cases where more than one object is present in the working scenario.

Other fuzzy control agents such as FOLLOW_WALL or APPROACH_OBJECT can be easily implemented, using the same sensor systems and multi-agent proposed approach.

Acknowledgements

This research has been fully funded by the Spanish Science and Technology Commission through grants CICYT-TAP98-0781: Multiagent architecture: complex behavior generation for an outdoor robot (AMARA II), and MCYT-AGL-2002-04468-C03-01: Artificial vision and spatio-temporal reasoning for specific site operations with an autonomous tractor (GEA II). Lía García Pérez has been sponsored by a predoctoral grant from the Spanish Ministry of Science and Technology.

References

1. MacFarland D. and Bosser T. *Intelligent behavior in animals and robots* MIT Press, 1993.
2. Arbib M. A. and Cobas A. Schemas for prey-catching in frog and toad. In *From Animals to Animats: Proceedings of the first International Conference on Simulation of Adaptive Behavior* , 142–151, 1990.

3. Arkin, R.C., Fujita M., Takagi T., Hasegawa R. En ethological and emotional basis for human-robot interaction. *Robotics and Autonomous Systems*, 42: 191-201, 2003.
4. Nordlund P. and Uhlin T. Closing the loop: Detection and pursuit of a moving object by a moving observer. *Image and Vision Computing*, 14:265–275.
5. Guinea D. and Recio F. and Vicente J. Framing autonomy: A visual tracking architecture. *Proceedings of the First International Conference on Autonomous Agents*, Marina del Rey, CA, 1997.
6. Borenstein J. and Koren Y. The vector field histogram fast obstacle avoidance for mobile robots. *IEEE Journal of Robotics and Automation*, 7(3):278–288, 1991.
7. García-Alegre M.C. and Ribeiro A. and Gasós J. and Salido J. Optimization of fuzzy behavior-based robots navigation in partially known industrial environments. In *Proceedings of the IEEE Inter.Conf.on Industrial Fuzzy Control and Intell.Syst.*, Houston,TX, 50–54, 1993.
8. Simmons R. The curvature-velocity method for local obstacle avoidance. In *Proceedings of the 1996 International Conference on Robotics and Automation*, Minneapolis, USA, 1996.
9. Fox D. and Burgard W. and Thrun S. The dynamic window approach to collision avoidance. *IEEE Robotics and Automation Magazine*, 23–33, 1997.
10. Benayas J.A. and Fernández J.L. and Sanz R. and Diéguez A.R. The beam-curvature method: a new approach for improving local real time obstacle avoidance. In *Proceedings of the 15th Triennial World Congress of the International Federation of Automatic Control IFAC*, Barcelona, Spain, 2002.
11. Fod A. and Howard A. and Mataric M. J. A laser-based people tracking. In *Proceedings of the IEEE International Conference on Robotics and Automation (ICRA)*, 2002.
12. Montesano L., Montano L. Identification of moving objects by a team of robots based on kinematic information. In *Proceedings of IROS2003*, 2003.
13. Cruz A. and Muñoz V. and García-Cerezo A. and OlleroA. Moving obstacles avoidance algorithm for mobile robots under speed restrictions. In *Proceedings of the IFAC, Intelligent Components for Vehicles, ICV'98*, Madrid, Spain, 233–238, 1998.
14. Kwon Y. D. and Lee J. S. A stochastic map building method for mobile robot using 2-D laser range finder. *Autonomous Robots*, 7:187–200, 1999.
15. Prassler E. and Scholz J. and Elfes A. Tracking multiple moving objects for real-time robot navigation. *Autonomous Robots*, 8:105–116, 2000.
16. Bennewitz, M. and Burgard W., Thrun, S. Adapting Navigation Strategies Using Motions Patterns of People In *Proc. of the IEEE International Conference on Robotics and Automation (ICRA), vol. 2 2000-2005, May 2003*,
17. Wang C.C., Thorpe C., Thrun S. Online Simultaneous localization and mapping with detection and tracking of moving objects: Theory and results from a ground vehicle in crowded urban areas. In *Proc. IEEE Intern Conf on Robotics and Automation, May 2003.*
18. Papanikolopoulos N. K. 1993. Visual tracking of a moving target by a camera mounted on a robot: A combination of control and vision. *IEEE Transactions on Robotics and Automation*, 9(1):14–35, 1996.
19. Guinea D. and Sánchez G. and Bustos P. and García-Alegre M.C. A distributed architecture for active perception in autonomous robots. *Proceedings IEEE International Conference on Systems, Man and Cybernetics*, Vancouver, Canada, 1995.
20. Nair D. and Aggarwal J. K. Moving obstacle detection from a navigating robot. *IEEE Transactions on Robotics and Automation*, 14(3):404–416, 1998.

21. Ning H., Tan T., Wang L., Hu W. People tracking based on motion model and motion constraints with automatic initialisation. *Pattern Recognition*, 37: 1423-1440, 2004.
22. Elfes, A. Occupancy grids: a stochastic representation for active robot perception In *Proc. 6th Intern Conf on Uncertainty in A.I.*, 1990.
23. Cañas J. M. and García-Alegre M.C. Modulated agents for autonomous robot piloting. In *Proceedings of 8 Conf. Española para la Inteligencia Artificial (CAEPIA'99)* Murcia, Spain, 98–106, 1999.
24. J.R.Munkres. Algorithms for the assignment and transportation problems. *Journal of. SIAM* 5:32–38, 1957.
25. García-Alegre M.C., Ribeiro A., García-Pérez L., Martínez R., Guinea D., Pozo-Ruz A.. Autonomous Robot in Agriculture Tasks. In*Proccedings of the 3ECPA European Conf. On Precision Agriculture*, Montpellier. June, 2001.
26. García-Pérez L. and García-Alegre M.C. A Simulation Environment to Test Fuzzy Navigation Strategies. In *Proceedings of the 10th IEEE International Conference on Fuzzy Systems.*, Melbourne, Australia, 2001.

Author Index

Lecture Notes in Artificial Intelligence (LNAI)

Vol. 3416: M. Böhlen, J. Gamper, W. Polasek, M.A. Wimmer (Eds.), E-Government: Towards Electronic Democracy. XIII, 311 pages. 2005.

Vol. 3403: B. Ganter, R. Godin (Eds.), Formal Concept Analysis. XI, 419 pages. 2005.

Vol. 3398: D.-K. Baik (Ed.), Systems Modeling and Simulation: Theory and Applications. XIV, 733 pages. 2005.

Vol. 3397: T.G. Kim (Ed.), Artificial Intelligence and Simulation. XV, 711 pages. 2005.

Vol. 3396: R.M. van Eijk, M.-P. Huget, F. Dignum (Eds.), Advances in Agent Communication. X, 261 pages. 2005.

Vol. 3374: D. Weyns, H.V.D. Parunak, F. Michel (Eds.), Environments for Multi-Agent Systems. X, 279 pages. 2005.

Vol. 3369: V.R. Benjamins, P. Casanovas, J. Breuker, A. Gangemi (Eds.), Law and the Semantic Web. XII, 249 pages. 2005.

Vol. 3366: I. Rahwan, P. Moraitis, C. Reed (Eds.), Argumentation in Multi-Agent Systems. XII, 263 pages. 2005.

Vol. 3359: G. Grieser, Y. Tanaka (Eds.), Intuitive Human Interfaces for Organizing and Accessing Intellectual Assets. XIV, 257 pages. 2005.

Vol. 3346: R.H. Bordini, M. Dastani, J. Dix, A.E.F. Seghrouchni (Eds.), Programming Multi-Agent Systems. XIV, 249 pages. 2005.

Vol. 3345: Y. Cai (Ed.), Ambient Intelligence for Scientific Discovery. XII, 311 pages. 2005.

Vol. 3343: C. Freksa, M. Knauff, B. Krieg-Brückner, B. Nebel, T. Barkowsky (Eds.), Spatial Cognition IV. XIII, 519 pages. 2005.

Vol. 3339: G.I. Webb, X. Yu (Eds.), AI 2004: Advances in Artificial Intelligence. XXII, 1272 pages. 2004.

Vol. 3336: D. Karagiannis, U. Reimer (Eds.), Practical Aspects of Knowledge Management. X, 523 pages. 2004.

Vol. 3327: Y. Shi, W. Xu, Z. Chen (Eds.), Data Mining and Knowledge Management. XIII, 263 pages. 2005.

Vol. 3315: C. Lemaître, C.A. Reyes, J.A. González (Eds.), Advances in Artificial Intelligence – IBERAMIA 2004. XX, 987 pages. 2004.

Vol. 3303: J.A. López, E. Benfenati, W. Dubitzky (Eds.), Knowledge Exploration in Life Science Informatics. X, 249 pages. 2004.

Vol. 3275: P. Perner (Ed.), Advances in Data Mining. VIII, 173 pages. 2004.

Vol. 3265: R.E. Frederking, K.B. Taylor (Eds.), Machine Translation: From Real Users to Research. XI, 392 pages. 2004.

Vol. 3264: G. Paliouras, Y. Sakakibara (Eds.), Grammatical Inference: Algorithms and Applications. XI, 291 pages. 2004.

Vol. 3259: J. Dix, J. Leite (Eds.), Computational Logic in Multi-Agent Systems. XII, 251 pages. 2004.

Vol. 3257: E. Motta, N.R. Shadbolt, A. Stutt, N. Gibbins (Eds.), Engineering Knowledge in the Age of the Semantic Web. XVII, 517 pages. 2004.

Vol. 3249: B. Buchberger, J.A. Campbell (Eds.), Artificial Intelligence and Symbolic Computation. X, 285 pages. 2004.

Vol. 3248: K.-Y. Su, J. Tsujii, J.-H. Lee, O.Y. Kwong (Eds.), Natural Language Processing – IJCNLP 2004. XVIII, 817 pages. 2005.

Vol. 3245: E. Suzuki, S. Arikawa (Eds.), Discovery Science. XIV, 430 pages. 2004.

Vol. 3244: S. Ben-David, J. Case, A. Maruoka (Eds.), Algorithmic Learning Theory. XIV, 505 pages. 2004.

Vol. 3238: S. Biundo, T. Frühwirth, G. Palm (Eds.), KI 2004: Advances in Artificial Intelligence. XI, 467 pages. 2004.

Vol. 3230: J.L. Vicedo, P. Martínez-Barco, R. Muñoz, M. Saiz Noeda (Eds.), Advances in Natural Language Processing. XII, 488 pages. 2004.

Vol. 3229: J.J. Alferes, J. Leite (Eds.), Logics in Artificial Intelligence. XIV, 744 pages. 2004.

Vol. 3228: M.G. Hinchey, J.L. Rash, W.F. Truszkowski, C.A. Rouff (Eds.), Formal Approaches to Agent-Based Systems. VIII, 290 pages. 2004.

Vol. 3215: M.G.. Negoita, R.J. Howlett, L.C. Jain (Eds.), Knowledge-Based Intelligent Information and Engineering Systems, Part III. LVII, 906 pages. 2004.

Vol. 3214: M.G.. Negoita, R.J. Howlett, L.C. Jain (Eds.), Knowledge-Based Intelligent Information and Engineering Systems, Part II. LVIII, 1302 pages. 2004.

Vol. 3213: M.G.. Negoita, R.J. Howlett, L.C. Jain (Eds.), Knowledge-Based Intelligent Information and Engineering Systems, Part I. LVIII, 1280 pages. 2004.

Vol. 3209: B. Berendt, A. Hotho, D. Mladenic, M. van Someren, M. Spiliopoulou, G. Stumme (Eds.), Web Mining: From Web to Semantic Web. IX, 201 pages. 2004.

Vol. 3206: P. Sojka, I. Kopecek, K. Pala (Eds.), Text, Speech and Dialogue. XIII, 667 pages. 2004.

Vol. 3202: J.-F. Boulicaut, F. Esposito, F. Giannotti, D. Pedreschi (Eds.), Knowledge Discovery in Databases: PKDD 2004. XIX, 560 pages. 2004.

Vol. 3201: J.-F. Boulicaut, F. Esposito, F. Giannotti, D. Pedreschi (Eds.), Machine Learning: ECML 2004. XVIII, 580 pages. 2004.

Vol. 3194: R. Camacho, R. King, A. Srinivasan (Eds.), Inductive Logic Programming. XI, 361 pages. 2004.

Vol. 3192: C. Bussler, D. Fensel (Eds.), Artificial Intelligence: Methodology, Systems, and Applications. XIII, 522 pages. 2004.

Vol. 3191: M. Klusch, S. Ossowski, V. Kashyap, R. Unland (Eds.), Cooperative Information Agents VIII. XI, 303 pages. 2004.

Vol. 3187: G. Lindemann, J. Denzinger, I.J. Timm, R. Unland (Eds.), Multiagent System Technologies. XIII, 341 pages. 2004.

Vol. 3176: O. Bousquet, U. von Luxburg, G. Rätsch (Eds.), Advanced Lectures on Machine Learning. IX, 241 pages. 2004.

Vol. 3171: A.L.C. Bazzan, S. Labidi (Eds.), Advances in Artificial Intelligence – SBIA 2004. XVII, 548 pages. 2004.

Vol. 3159: U. Visser, Intelligent Information Integration for the Semantic Web. XIV, 150 pages. 2004.

Vol. 3157: C. Zhang, H. W. Guesgen, W.K. Yeap (Eds.), PRICAI 2004: Trends in Artificial Intelligence. XX, 1023 pages. 2004.

Vol. 3155: P. Funk, P.A. González Calero (Eds.), Advances in Case-Based Reasoning. XIII, 822 pages. 2004.

Vol. 3139: F. Iida, R. Pfeifer, L. Steels, Y. Kuniyoshi (Eds.), Embodied Artificial Intelligence. IX, 331 pages. 2004.

Vol. 3131: V. Torra, Y. Narukawa (Eds.), Modeling Decisions for Artificial Intelligence. XI, 327 pages. 2004.

Vol. 3127: K.E. Wolff, H.D. Pfeiffer, H.S. Delugach (Eds.), Conceptual Structures at Work. XI, 403 pages. 2004.

Vol. 3123: A. Belz, R. Evans, P. Piwek (Eds.), Natural Language Generation. X, 219 pages. 2004.

Vol. 3120: J. Shawe-Taylor, Y. Singer (Eds.), Learning Theory. X, 648 pages. 2004.

Vol. 3097: D. Basin, M. Rusinowitch (Eds.), Automated Reasoning. XII, 493 pages. 2004.

Vol. 3071: A. Omicini, P. Petta, J. Pitt (Eds.), Engineering Societies in the Agents World. XIII, 409 pages. 2004.

Vol. 3070: L. Rutkowski, J. Siekmann, R. Tadeusiewicz, L.A. Zadeh (Eds.), Artificial Intelligence and Soft Computing - ICAISC 2004. XXV, 1208 pages. 2004.

Vol. 3068: E. André, L. Dybkjær, W. Minker, P. Heisterkamp (Eds.), Affective Dialogue Systems. XII, 324 pages. 2004.

Vol. 3067: M. Dastani, J. Dix, A. El Fallah-Seghrouchni (Eds.), Programming Multi-Agent Systems. X, 221 pages. 2004.

Vol. 3066: S. Tsumoto, R. Słowiński, J. Komorowski, J.W. Grzymała-Busse (Eds.), Rough Sets and Current Trends in Computing. XX, 853 pages. 2004.

Vol. 3065: A. Lomuscio, D. Nute (Eds.), Deontic Logic in Computer Science. X, 275 pages. 2004.

Vol. 3060: A.Y. Tawfik, S.D. Goodwin (Eds.), Advances in Artificial Intelligence. XIII, 582 pages. 2004.

Vol. 3056: H. Dai, R. Srikant, C. Zhang (Eds.), Advances in Knowledge Discovery and Data Mining. XIX, 713 pages. 2004.

Vol. 3055: H. Christiansen, M.-S. Hacid, T. Andreasen, H.L. Larsen (Eds.), Flexible Query Answering Systems. X, 500 pages. 2004.

Vol. 3048: P. Faratin, D.C. Parkes, J.A. Rodríguez-Aguilar, W.E. Walsh (Eds.), Agent-Mediated Electronic Commerce V. XI, 155 pages. 2004.

Vol. 3040: R. Conejo, M. Urretavizcaya, J.-L. Pérez-de-la-Cruz (Eds.), Current Topics in Artificial Intelligence. XIV, 689 pages. 2004.

Vol. 3035: M.A. Wimmer (Ed.), Knowledge Management in Electronic Government. XII, 326 pages. 2004.

Vol. 3034: J. Favela, E. Menasalvas, E. Chávez (Eds.), Advances in Web Intelligence. XIII, 227 pages. 2004.

Vol. 3030: P. Giorgini, B. Henderson-Sellers, M. Winikoff (Eds.), Agent-Oriented Information Systems. XIV, 207 pages. 2004.

Vol. 3029: B. Orchard, C. Yang, M. Ali (Eds.), Innovations in Applied Artificial Intelligence. XXI, 1272 pages. 2004.

Vol. 3025: G.A. Vouros, T. Panayiotopoulos (Eds.), Methods and Applications of Artificial Intelligence. XV, 546 pages. 2004.

Vol. 3020: D. Polani, B. Browning, A. Bonarini, K. Yoshida (Eds.), RoboCup 2003: Robot Soccer World Cup VII. XVI, 767 pages. 2004.

Vol. 3012: K. Kurumatani, S.-H. Chen, A. Ohuchi (Eds.), Multi-Agents for Mass User Support. X, 217 pages. 2004.

Vol. 3010: K.R. Apt, F. Fages, F. Rossi, P. Szeredi, J. Váncza (Eds.), Recent Advances in Constraints. VIII, 285 pages. 2004.

Vol. 2990: J. Leite, A. Omicini, L. Sterling, P. Torroni (Eds.), Declarative Agent Languages and Technologies. XII, 281 pages. 2004.

Vol. 2980: A. Blackwell, K. Marriott, A. Shimojima (Eds.), Diagrammatic Representation and Inference. XV, 448 pages. 2004.

Vol. 2977: G. Di Marzo Serugendo, A. Karageorgos, O.F. Rana, F. Zambonelli (Eds.), Engineering Self-Organising Systems. X, 299 pages. 2004.

Vol. 2972: R. Monroy, G. Arroyo-Figueroa, L.E. Sucar, H. Sossa (Eds.), MICAI 2004: Advances in Artificial Intelligence. XVII, 923 pages. 2004.

Vol. 2969: M. Nickles, M. Rovatsos, G. Weiss (Eds.), Agents and Computational Autonomy. X, 275 pages. 2004.

Vol. 2961: P. Eklund (Ed.), Concept Lattices. IX, 411 pages. 2004.

Vol. 2953: K. Konrad, Model Generation for Natural Language Interpretation and Analysis. XIII, 166 pages. 2004.

Vol. 2934: G. Lindemann, D. Moldt, M. Paolucci (Eds.), Regulated Agent-Based Social Systems. X, 301 pages. 2004.

Vol. 2930: F. Winkler (Ed.), Automated Deduction in Geometry. VII, 231 pages. 2004.

Vol. 2926: L. van Elst, V. Dignum, A. Abecker (Eds.), Agent-Mediated Knowledge Management. XI, 428 pages. 2004.

Vol. 2923: V. Lifschitz, I. Niemelä (Eds.), Logic Programming and Nonmonotonic Reasoning. IX, 365 pages. 2003.